Freeform Optics for LED Packages and Applications

Freeform Optics for LED Packages and Applications

Kai Wang
Southern University of Science and Technology
Guangdong, China

Sheng Liu
Wuhan University
Hubei, China

Xiaobing Luo
Huazhong University of Science and Technology
Hubei, China

Dan Wu
Nanyang Technological University
Singapore

Chemical Industry Press

This edition first published 2017 by John Wiley & Sons Singapore Pte. Ltd under exclusive licence granted by CIP for all media and languages (excluding simplified and traditional Chinese) throughout the world (excluding Mainland China), and with non-exclusive license for electronic versions in Mainland China.

Registered Offices
John Wiley & Sons, Inc., 111 River Street, Hoboken, NJ 07030, USA
John Wiley & Sons Singapore Pte. Ltd, 1 Fusionopolis Walk, #07-01 Solaris South Tower, Singapore 138628

Editorial Office
The Atrium, Southern Gate, Chichester, West Sussex, PO19 8SQ, UK

For details of our global editorial offices, customer services, and more information about Wiley products visit us at www.wiley.com.

Wiley also publishes its books in a variety of electronic formats and by print-on-demand. Some content that appears in standard print versions of this book may not be available in other formats.

Library of Congress Cataloging-in-Publication Data Applied for

ISBN: 9781118749715

A catalogue record for this book is available from the British Library.

Cover image: © The-Tor/Gettyimages
Cover design: Wiley

Set in 10/12pt WarnockPro by SPi Global, Chennai, India
Printed in Singapore by C.O.S. Printers Pte Ltd

10 9 8 7 6 5 4 3 2 1

Contents

Preface

Light-emitting diodes (LEDs), having superior characteristics such as high luminous efficiency, low power consumption, compact size, long lifetime, high reliability, and being environmentally friendly, have been widely accepted by the industry as the new-generation light source in the 21st century as well as the future developing trend in lighting technology, possessing great market potential. With increasing luminous efficiency (e.g., more than 300 lm/W in lab and 150 lm/W in market in 2014) and cost performance in recent years, LEDs have more and more applications in general and special lighting, such as LED indoor lighting, LED road lighting, LED backlighting for liquid crystal displays (LCDs), and LED automotive headlamps. The LED market is growing rapidly. Optical design, thermal management, and power supply are three key issues for LED lighting. With the development of LED lighting, high-quality LED lighting has attracted more and more attention from customers, which means that higher optical efficiency, more controllable light patterns, and higher spatial color uniformity will be needed for the optical design of LED packages and applications. Freeform optics is an emerging optical technology in LED lighting, with the advantages of having high design freedom and precise light irradiation control, and it provides a more promising way to realize high-quality LED lighting. In the next 5 years, with the development of LED chip and packaging technologies, the luminous efficiency of high-power white LEDs will reach a higher level, which will broaden the application markets of LEDs furthermore and also will change the lighting concepts in our lives. Moreover, people also will pay more attention to the quality of LED lighting. Therefore, more new algorithms of freeform optics and advanced optical design methods will be needed to meet the requirements of LED lighting in the future. Thus, a book discussing freeform optics for LED lighting is expected and needed by LED researchers and engineers at present.

This book will introduce the freeform optics for LED packages and applications, from new algorithms of freeform optics to detailed design methods and then to advanced optical designs and case studies. A series of basic freeform optics algorithms specialized for LED packages and applications will be introduced in detail in this book, such as circular symmetry, noncircular symmetry, point sources, extended sources, array illumination, and so on. Most algorithms have been validated by the industry. Many algorithms and designs will be proposed systematically for the first time in this book, and there is no similar book in the market yet. Moreover, besides these core algorithms, detailed and practical freeform lens design methods derived from these algorithms also will be introduced. Novel and advanced optical designs for various

LED packages and applications will be introduced in detail, too, such as noncircularly symmetrical freeform lenses for LED road lighting, application-specific LED packaging (ASLP) integrated with petal-shaped freeform lenses for direct-lit large-scale LED backlighting, Fresnel freeform lenses for low-beam automotive LEDs, total internal reflection (TIR) freeform lenses for an LED pico-projector, and a freeform lens for high spatial color uniformity. Moreover, codes of fundamental freeform optics algorithms will be included as appendices.

Chapter 1 gives an introductory basic background in the area of LED packages and applications, especially the three key issues of optical design of LED lighting. Chapter 2 reviews existing major algorithms of freeform optics. Chapter 3 demonstrates a series of basic freeform optics algorithms and design methods specialized for LED packages and applications. Following chapters will apply these algorithms in specific LED packaging and applications, including ASLP in Chapter 4, LED indoor lighting in Chapter 5, LED road lighting in Chapter 6, LED backlighting for large-scale LCD displays in Chapter 7, LED automotive headlamps in Chapter 8, emerging LED applications in Chapter 9, and high spatial color uniformity of LED modules in Chapter 10.

From this book, readers can gain an overall understanding of the application of freeform optics, which is regarded as one of the most powerful design methods in nonimaging optics, for different LED packages and applications. Readers also can obtain detailed algorithms and design methods of freeform optics systematically, and they will be able to develop advanced optical designs for LED lighting by studying this book. This book will be very helpful for enhancing the abilities of optical designs of LED researchers and engineers, and it will accelerate the research and development of LED packages and applications. Many algorithms and design methods of freeform optics for LED lighting proposed in this book have been validated by the industry, which gives this book important value not only academically but also in practical instruction. More importantly, with the open source codes of algorithms and case studies, readers will be able to learn and master the design skills of freeform optics much easier and more conveniently.

This book is to be used as a textbook for undergraduate senior and postgraduate courses. Also, researchers and engineers with interest in LED lighting will find this book particularly useful. This book includes content from fundamentals of optics to advanced algorithms of freeform optics and novel freeform optical designs for LED packages and applications. Readers just need a basic understanding of LED and optics, and then they can read the book and learn freeform optics for LED lighting step by step. It is our hope that information in this book may assist in removing some barriers, helping researchers avoid unnecessary false starts, and accelerating the applications of these techniques.

Development and preparation of *Freeform Optics for LED Packages and Applications* were facilitated by a number of dedicated people at John Wiley & Sons, Chemical Industry Press, Southern University of Science and Technology (SUSTech), Wuhan University (WHU), Huazhong University of Science and Technology (HUST), and Nanyang Technological University (NTU). We would like to thank all of them, with special mentions of Gang Wu of Chemical Industry Press and Mustaq Ahamed and Victoria Taylor of John Wiley & Sons. Without them, our dream of this book could not have come true, as they have solved many problems during this book's preparation. It has been a great pleasure and fruitful experience to work with them in transferring our manuscript into a very attractive printed book.

The material in this book clearly has been derived from many sources, including individuals, companies, and organizations, and we have attempted to acknowledge the help we have received. We would like to thank several professional societies, in whose publications we previously published some of the materials in this book, for permission to reproduce them in this book. They are the Optical Society (OSA), the International Society for Optics and Photonics (SPIE), the Institute of Electrical and Electronic Engineers (IEEE), and the Society for Information Display (SID). We also appreciate the support and help from China Solid State Lighting Alliance (CSA), Guangdong Solid State Lighting Industry Innovation Center (GSC), China CESI (Guangzhou) Laboratory, Guangdong Real Faith Enterprises Group, and Tianjin Zhonghuan Electronic and Information Group.

We also would like to thank those colleagues who reviewed some chapters and gave suggestions and comments on the manuscript. They are Prof. Chingping Wong of the Chinese University of Hong Kong, Prof. Xiaowei Sun of SUSTech, Prof. Xinhai Zhang of SUSTech, Prof. Mingxiang Chen of HUST, Prof. Zhiying Gan of HUST, Prof. Jiangen Pan of Everfine Optoelectronics, Prof. Muqing Liu of Fudan University, and Dr. Hugo Cornelissen of Philips. We also would like to thank our colleagues Dr. Fei Chen, Dr. Zong Qin, Dr. Shang Wang, Dr. Shuang Zhao, Dr. Shuiming Li, Dr. Zongyuan Liu, Dr. Run Hu, Mr. Bing Xie, Mr. Chuangang Ji, Dr. Zhili Zhao, and Mr. Yanbing Yang. Their depth of knowledge and their dedication have been demonstrated throughout the process of reviewing this book.

We also would like to acknowledge the support of many funding agencies over the past several years, such as the National Natural Science Foundation of China, the Ministry of Science and Technology of China, Hubei Department of Science and Technology, Guangdong Department of Science and Technology, Shenzhen Commission on Innovation and Technology, Nanshan Bureau of Science and Technology, Foshan Bureau of Science and Technology, and Nanhai Bureau of Science and Technology.

We wish to thank most warmly all our friends who have contributed to this book. We also wish to express our appreciation to our family members for continuous support and encouragement. It would be quite impossible for us to express our appreciation to everyone concerned for their collaboration in producing this book, but we would like to extend our gratitude.

Kai Wang, PhD, Assistant Professor
Department of Electrical and Electronic Engineering
Southern University of Science and Technology
Shenzhen, Guangdong, China

Sheng Liu, PhD, IEEE/ASME Fellow
ChangJiang Scholar Professor
School of Power and Mechanical Engineering
Wuhan University
Wuhan, Hubei, China

Xiaobing Luo, PhD, Professor
School of Energy and Power Engineering
Huazhong University of Science and Technology
Wuhan, Hubei, China

Dan Wu, PhD
School of Electrical and Electronic Engineering
Nanyang Technological University
Singapore

1

Introduction

1.1 Overview of LED Lighting

Solid-state lighting is a new type of lighting technology based on the high power and brightness of LEDs (light-emitting diodes), and it is commonly known as *LED lighting*. LED is a semiconductor luminescence device based on the P–N junction electroluminescence principle. Compared with conventional light sources, LED has advantages such as low power consumption, high luminous efficiency, long lifetime, compact size, low weight, high reliability, and being environmentally friendly; it is regarded by the industry as the twenty-first-century green light source as well as a future developing trend in lighting technology, possessing great market potential.[1–5]

For the past two centuries, natural and energy resource consumption by human beings has increased, and the corresponding environmental pollution and ecological damage have drawn much attention. Reducing energy consumption and adopting energy conservation and environmental protection products are becoming common beliefs of international society. Lighting alone consumes around one-fifth of the total electricity generated globally. In the current energy shortage situation, adoption of energy conservation and environmentally friendly LED-lighting products is meaningful for conservation of energy and reduction of petroleum imports.

To cope with development trends of the semiconductor lighting industry, the United States, the European Union, Japan, and China have given high priority to LED as a primary and strategic industry. They have unveiled their national solid-state lighting plans and devoted huge labor and material resources to research and development (R&D). Take the United States as an example: in 2000, they promoted the National Solid State Lighting Research Program, which invested $500 million for the development of the solid-state lighting industry from 2000 to 2010. In July 2000, the European Union constructed the Rainbow Project, which grants commissions by six large companies and two universities to promote the application of LED. In 2003, China launched the National Solid State Lighting Project. China also announced officially in 2006 that the Solid State Lighting Project will be included in the Tenth Five-Year Plan, and China expected to invest 350 million RMB for semiconductor lighting technology development and promotion. In 2010, solid-state lighting was listed as one of the most important strategic emerging industries in China; there are nearly 30 research institutes and more than

Freeform Optics for LED Packages and Applications, First Edition. Kai Wang, Sheng Liu, Xiaobing Luo and Dan Wu.

2000 companies nationwide that will focus their research and development on LED lighting related to epitaxial growth, chip fabrication, packaging, and application. In the meantime, the world's three lighting giants, General Electric (GE) in the United States, Philips in the Netherland, and OSRAM in Germany, not only established respective solid-state lighting divisions or subsidiaries, but also carry out a wide range of global cooperative efforts and strive to seize the high ground of the future market.

Under the government's strong support and guidance, as well as large investments in R&D from both institutes of companies and universities, solid-state lighting technologies (including epitaxy growth, chip manufacturing, and packaging technologies) are developing very fast, and the performance of LEDs is increasing dramatically. The luminous efficiency of an LED module reached more than 300 lm/W in the lab[6] and 150 lm/W in the market in 2014, much higher than the luminous efficiencies of most traditional light sources (e.g., 15~25 lm/W for incandescent, 70~100 lm/W for fluorescent, and 100~120 lm/W for high-pressure sodium) and creating significant energy saving of more than 50% for lighting power consumption. Moreover, with rapid improvements in the cost-effectiveness of LED packaging modules and application products, the global solid-state lighting market is growing rapidly. Take China's LED-lighting market as an example. In 2014, the gross output of the LED industry of China reached as high as 344.5 billion RMB (about US$55.1 billion), an increase of 30.6% compared with the gross output in 2013 of 263.8 billion RMB (about US$42.2 billion).[7]

In Figure 1.1, high-power LEDs have been vigorously developed in general lighting and specialty lighting applications. For example, through the 10 City 10,000 Lamps project, LED road lighting and LED tunnel lighting are currently being promoted nationwide in China. LED-backlighted cell phone screens, computer monitors, and large-size TVs are gradually becoming familiar to a huge number of families. Emerging LED automotive headlamps and LED-based pico-projectors are under rapid development. A variety of LED lamps for indoor lighting (e.g., LED bulbs, spotlights, downward lights, and panel lights) are applied in commercial lighting, factory lighting, office lighting, museum lighting, house lighting, and other lighting occasions to get more and more applications. Also, LED lamps also are widely used in urban lighting, military and police special lighting, and other occasions for a large number of applications.

The solid-state lighting industry chain includes upstream LED substrate material, epitaxy growth, chip design, and manufacturing; midstream includes LED packaging and testing; and downstream is reserved for specific applications, including LED general lighting and special lighting as shown in Figure 1.2. The ultimate LED lighting product performance is closely correlated with every aspect of upstream, midstream, and downstream: from epitaxy growth, chip manufacturing, packaging, and testing to application product design and manufacturing, where each process step will exert important impacts on the system's lighting efficiency, the lighting quality, and the reliability of LED lighting products. In the meantime, for a high-power LED manufacturing process, each process step is interrelated that will have a coupling impact on LED output efficiency, reliability, and so on.

Therefore, solid-state lighting technology is a combination of the semiconductor, physics, optical, thermal, mechanical, materials, electronics, and other interdisciplinary research fields. For LED packaging and applications, optical design, thermal design, drive control, and reliability are the four key factors. Among them, the lighting performance is the ultimate design purpose of LED lighting, and good heat dissipation

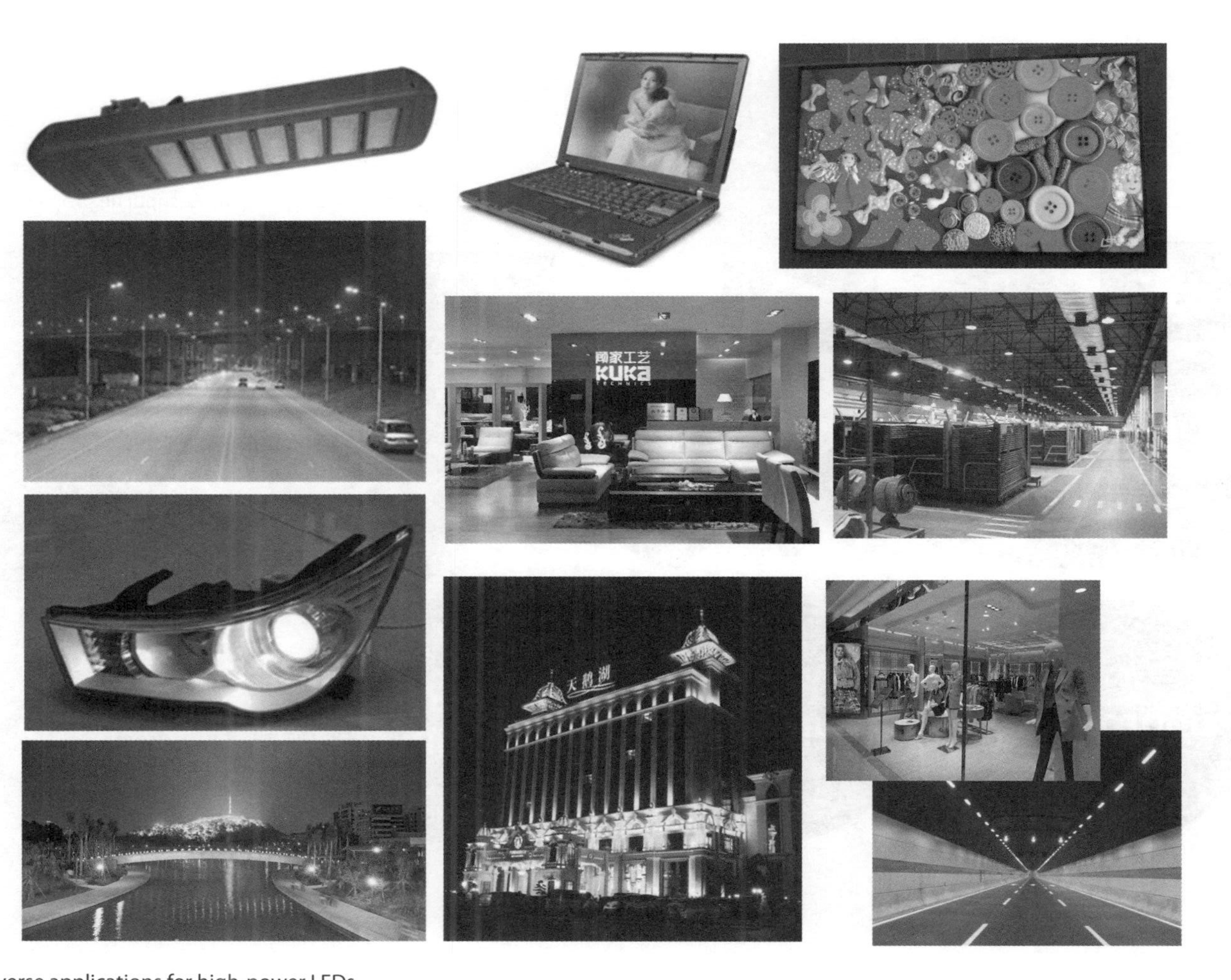

Figure 1.1 Diverse applications for high-power LEDs.

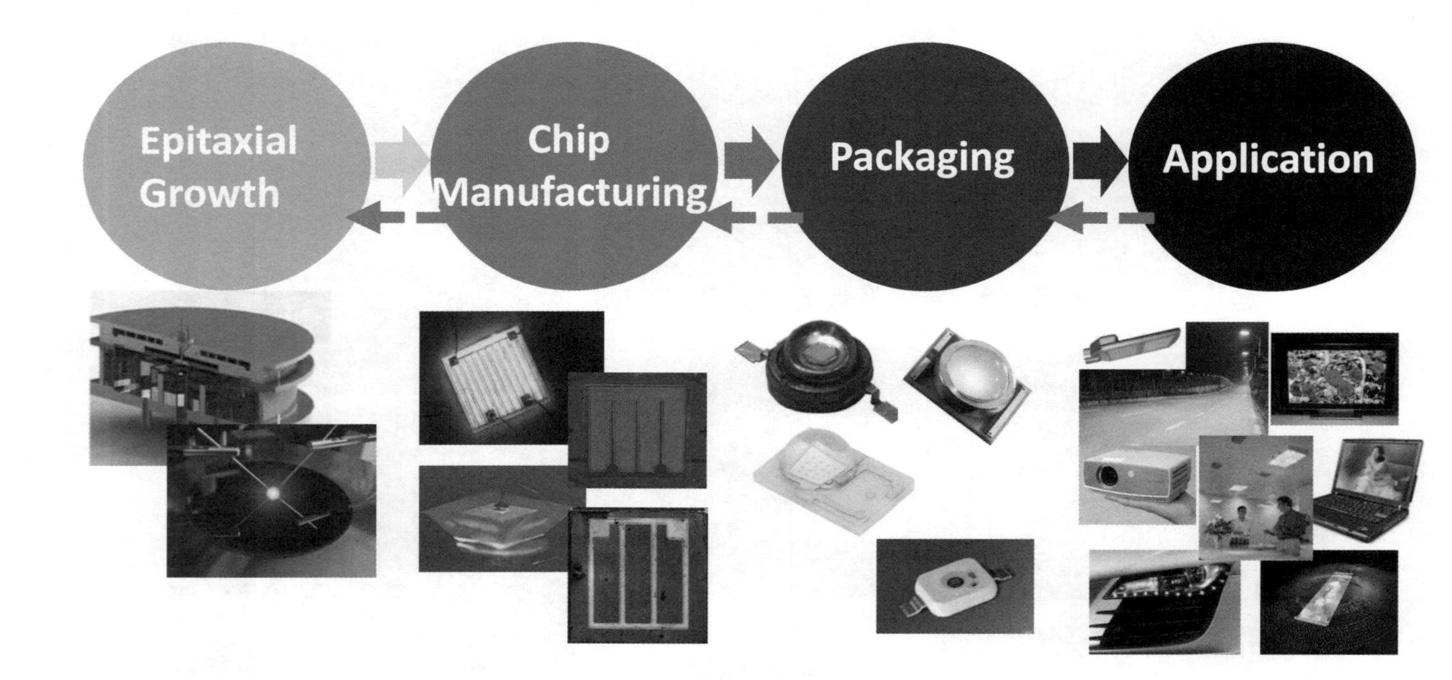

Figure 1.2 Schematic of an industry chain for high-power LED manufacturing.

and high reliability design offer a guarantee for long-life LED lighting products. Moreover, in the LED industry chain, packaging is the key connecting link between the upstream and downstream, and application is the major driving force for the whole LED industry. According to GG-LED, the gross output of LED packaging and applications accounts for more than 90% of the total LED industry's gross output.[7] This book will focus on the emerging freeform optics technologies for LED packaging and various applications.

1.2 Development Trends of LED Packaging and Applications

To gain a better understanding of the design concept of freeform optics, which will be discussed in later chapters in this book on LED lighting, we will first have a brief introduction to the development trends of LED packaging and applications in this section.

As shown in Figure 1.3, brighter, smaller, smarter, and cheaper are always the development trends of LED packaging. With the rapid development in the last decade of LED chip manufacturing, such as patterned sapphire substrate (PSS) and thin film flip chip (TFFC); LED packaging materials, such as direct plated copper (DPC) ceramic, white epoxy molding compound (EMC), and high-stability silicone; and LED packaging processes, such as wafer-level packaging (WLP) and chip-scale packaging (CSP), the performance of LED packaging modules, whatever their luminous efficiency, thermal stability, or reliability, has improved significantly. For example, the largest size of a 1 W high-power LED packaging module based on a metal lead frame was over 10 mm in 2006. However, with the emerging flip chip and CSP technologies, one LED packaging module measuring 1.5 × 1.5 × 1.4 mm in size was able to emit more than 220 lm light when driving by typically 700 mA, and it could reach 300 lm by a maximum of 1000 mA in 2014.[8] Therefore, freeform optics should be integrated with LED packaging modules to achieve a new kind of application-specific LED packaging (ASLP) module that can be adopted directly in various LED applications to meet illumination requirements without large-sized secondary optics. Moreover, with the size decrease of LED packaging modules, more compact freeform lenses should be

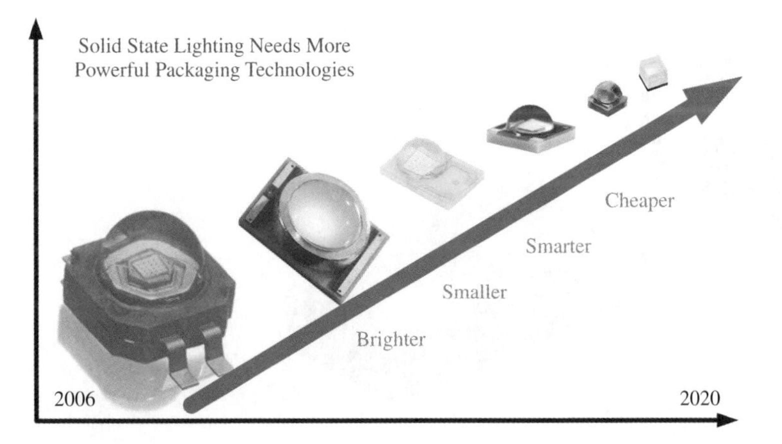

Figure 1.3 Development trends in LED packaging.[1]

designed for new-generation LED packaging, which requires improved freeform optics algorithms to deal with the extended source problem to achieve good illumination performance. In this book, these emerging and key issues of freeform optical design for new LED packaging will be presented in detail.

For LED applications, the standardization of an LED module is the key issue to promote the wide application of LED lighting worldwide. The standard LED light module, also called the *LED light engine,* is integrated with functions of a high-luminous-efficiency LED light source, precise optical design, efficient heat dissipation, power driving, and intelligent control; it is widely recognized as the trend of LED applications and has developed rapidly in recent years.[1] Figure 1.4 shows the design concept of the LED light engine and some LED modules released by CREE.[9] Freeform optics design methods mentioned in this book will follow this development trend to meet the demands of the fast-growing LED industry. Moreover, specific LED light engines integrated with freeform optics for applications of LED road lighting, LED automotive headlamps, and LED backlighting of liquid crystal displays (LCDs) also will be demonstrated in this book.

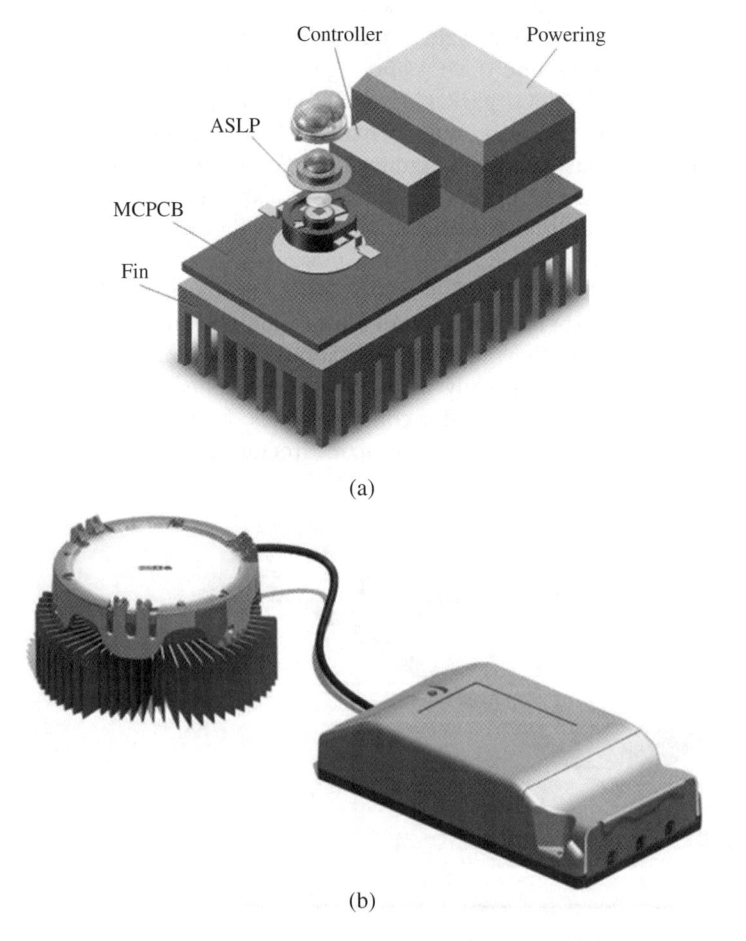

Figure 1.4 (a) Design concept; and (b) CREE LED light engine.[1,9]

1.3 Three Key Issues of Optical Design of LED Lighting

For LED lighting, an LED lamp's illumination performance is the final objective that people see and feel, so optical design is very important for LED lighting. There are three key issues in optical design of LED lighting: high system luminous efficiency, a controllable light pattern, and uniform spatial color distribution. High luminous efficiency is the most fundamental issue of LED sources and lamps to achieve great energy saving. A controllable light pattern is the most basic as well as important issue to meet the requirements of various LED lighting applications with different specifications. In addition, besides efficiency, recently people have paid more attention to the comfort level of LED lighting sources. High spatial color uniformity (SCU) will enhance the comfort and the lighting experience of LED lighting significantly.

1.3.1 System Luminous Efficiency

In order to realize optimized LED illumination performance, effective and efficient optical design is indispensable. The system luminous efficiency (lm/W) of an LED illumination luminaire is a very critical factor in optical design. Due to the fact that the cost per lumen for LED sources is still higher than that of some conventional sources, improvements in the system luminous efficiency of LED lighting devices can reduce the number of LEDs in use, lower system costs, and improve the competitiveness of products. The system luminous efficiency (lm/W) of an LED illumination luminaire is the product of the LED package module's luminous efficiency (lm/W), the system optical efficiency of the lighting device, and the efficiency of driving power. Among them, the luminous efficiency of an LED package module is influenced by LED chip design, phosphor material selection, the package lens, the lead frame, board surface reflectivity, and so on.[1,10] System optical efficiency is affected by the optical efficiencies of optical components (e.g., lens and reflector) within the illuminating system.[1]

In this book, we first introduce the concepts of primary optics and secondary optics in LED lighting. Primary optics is the optical design that is closely correlated with LED packaging modules, including the LED chip and package material coupling light output, the lighting property of phosphor, the packaging lens, and so on. At the same time, it also includes LED chip microstructures on the top surface with collaborative design for the packaging process and integrated design for the packaging lens and secondary optics. The quality of primary optics design directly affects the LED packaging module's luminous efficiency. Secondary optics, in contrast, includes the optical design except for the LED package module–related optics. Typical examples include reflectors, lenses, and diffusers in an illumination system. Note that the system efficiency of an illuminating device is determined by secondary optics, not only the light output efficiency of the secondary optics element itself but also accurate light pattern control to reduce light waste in lighting applications. The discussion range of this book mainly focuses on both the packaging primary optics lens design and secondary optics design in LED illumination applications.

1.3.2 Controllable Light Pattern

For LED illumination luminaire design, precise control over light pattern is another critical factor. Generally, the spatial light intensity distribution of an LED packaging module

Figure 1.5 Illumination performances for LED road lighting with a (left) circular light pattern and (right) rectangular light pattern.

is very close to Lambertian, which will generate a circle-shaped light pattern. This light pattern has a nonuniform illuminance distribution that decreases with $cos^4\theta$ from the center to the edge. Uniform illumination is required in many lighting applications. Due to the nonuniform nature of LED light intensity distribution, LED packaging modules usually should be combined with a secondary optics design. Therefore, whether the light pattern of an LED illuminating device can be controlled directly results in whether we can achieve functional LED lighting such as a uniform circular light pattern, uniform rectangular light pattern, automotive low-beam with a special light pattern, and the like. Figure 1.5 shows a comparison of LED road lighting performance. On the left-hand side, it shows the LED road lighting illumination performance with a circular light pattern, which has spots of light and dark pavement staggered with lower uniformity. On the right-hand side, it shows the LED road lighting illumination with a rectangular pattern, and the pavement has bright uniform lighting effects that are much better than the performance shown on the left.

1.3.3 Spatial Color Uniformity

Currently, white-light high-power LEDs are mainly realized through high-power blue light emitted from GaN LEDs stimulating the yellow phosphor. With the efficiency improvement of LED chips and phosphor, the luminous efficiency of commercialized LED packaging modules has reached 150 lm/W and 300 lm/W in the laboratory, which is much higher than that of conventional lighting, and the energy-saving advantage is obvious. However, with the promotion of LED lighting, especially for outdoor lighting applications such as road lighting and tunnel lighting and indoor lighting applications such as bulb light and down light, customers have set high standards on the quality of LED lighting performance in both brightness and spatial color uniformity. The drawback of the spatial color nonuniformity of LED packaging modules is revealing, and it is one of the bottleneck problems hindering the further promotion of LED applications. Figure 1.6 shows the light pattern of a high-power LED in the market

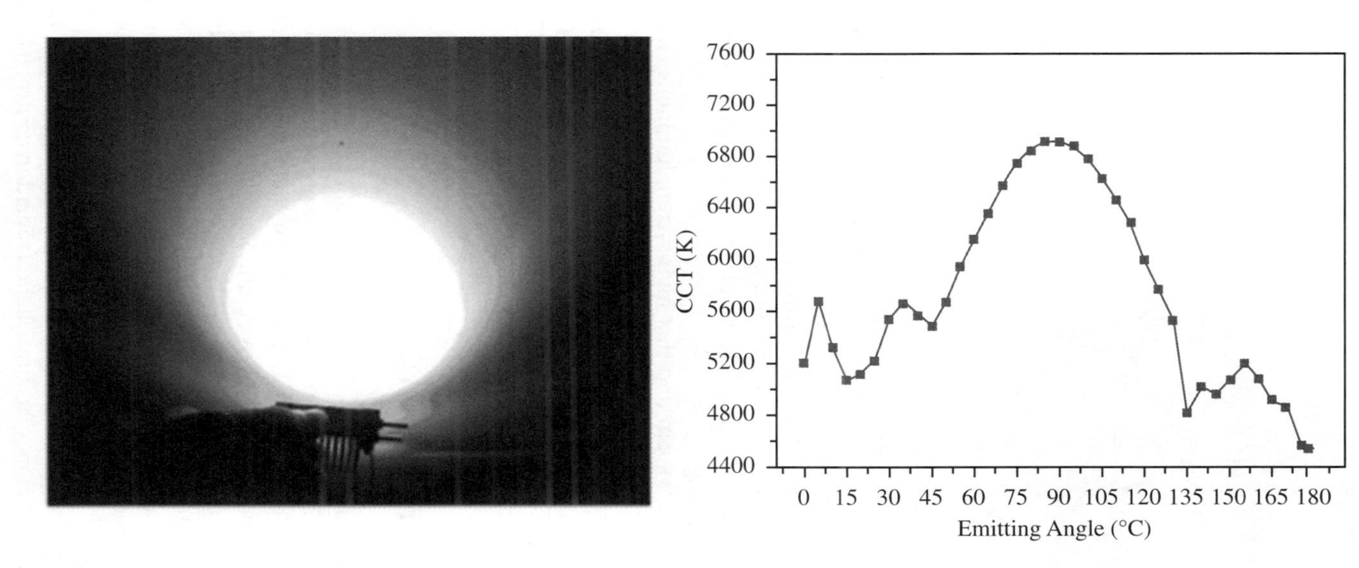

Figure 1.6 Nonuniform light pattern of a high-power white LED and its correlated color temperature distribution.

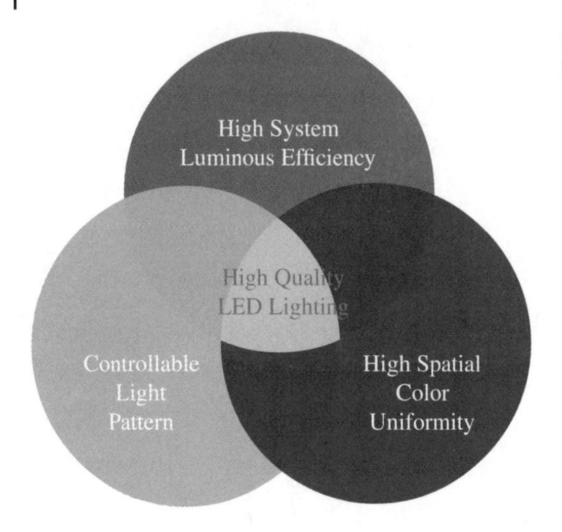

Figure 1.7 Three key optical issues for LED lighting.

manufactured by a famous LED company. We can find that both the light pattern and the color are not uniform: there are obvious yellow rings at the edge of the light pattern, while in the center the color is bluish, which shows the low spatial color uniformity of LEDs emitting light. The uniformity of correlated color temperature barely reaches 67%. Low quality of illumination and color uniformity is acceptable for outdoor application, but for indoor lighting application or backlighting application, this will generate visual error and discomfort and will not meet the standard requirements, which greatly limits the application of LED lighting.

From the above analyses, we can find that high system luminous efficiency, lighting pattern controllability, and high spatial color uniformity are the three key issues to realize high-quality LED lighting, as shown in Figure 1.7. LED optical design research is carried out in these areas, and this will help to improve LED lighting performance to satisfy a number of functional lighting application demands and to promote the popularization of LED lighting and wide application of great significance.

1.4 Introduction of Freeform Optics

There are several ways to address the three key issues of optical design of LED lighting mentioned in this chapter. Most important is choosing a proper and powerful way to create optical designs for various kinds of LED packaging and applications. For example, a rectangular light pattern can be achieved by combining multiple reflectors with different inclined angles. However, is it the most efficient, compact, and convenient way?

Freeform surfaces are those that cannot be expressed analytically and are not suitable for description in uniform equations, as shown in Figure 1.8. Freeform optics, including reflectors and lenses, based on freeform surfaces was first developed for concentrated solar power in the 1960s.[11] Freeform optics is an important means to realize nonimaging optics. Freeform optics is able to redistribute lighting energy, and it has advantages such as high light output efficiency, accurate light irradiation control (shape of light pattern, uniformity, etc.), high design freedom, compact size, convenient assembly, and so on.[11–23] Readers will become aware of these characteristics gradually after reading the

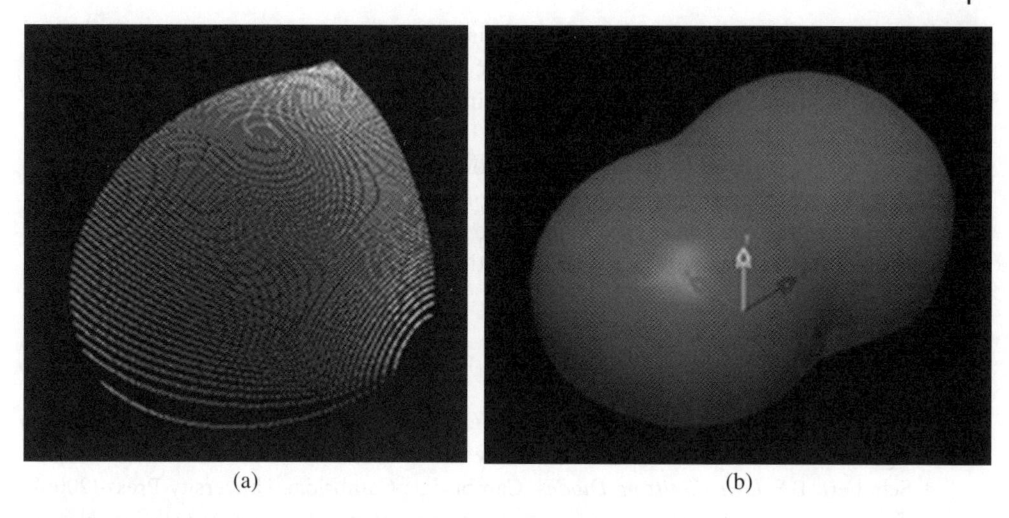

Figure 1.8 Schematic of freeform optics.

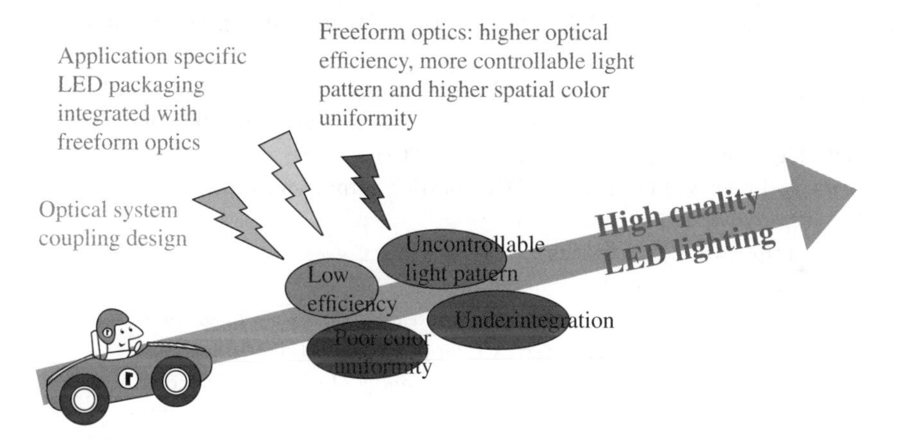

Figure 1.9 Freeform optics as an emerging optical design method for LED lighting.

chapters in this book about freeform optics algorithms and design methods. With these advantages, as shown in Figure 1.9, freeform optics becomes a powerful and advanced solution to overcome problems existing in LED lighting, such as low system efficiency, uncontrollable light pattern, poor spatial color uniformity, underintegration, and so on, and to achieve high-performance and high-quality LED lighting.

With the rapid development of LED lighting in the last decade, freeform optics has been widely recognized as an emerging optical design method for LED lighting, and it has increasing applications in most fields of LED lighting, such as ASLP, LED indoor lighting, LED road lighting, LED backlighting for LCD displays, LED automotive headlamps, LED pico-projectors, high-SCU LED lighting, and so on. Therefore, freeform optics has become the main trend of nonimaging optics design for LED lighting.

In this book, after reviewing existing major freeform optics algorithms in Chapter 2, a series of basic freeform optics algorithms and design methods specialized for LED packaging and applications will be introduced in detail in Chapter 3, such as circularly symmetrical, noncircularly symmetrical, point source, extended source, and array illumination. Following chapters will apply these algorithms in specific LED packaging and applications, including ASLP in Chapter 4, indoor lighting in Chapter 5, road lighting in Chapter 6, backlighting for large-scale LCD displays in Chapter 7, automotive headlamps in Chapter 8, emerging LED applications in Chapter 9, and high SCU in Chapter 10. Not only detailed freeform optical design methods but also many practical design cases will be presented in these chapters, so that readers can have a complete understanding of the freeform optics for LED lighting.

References

1 Liu, S., and Luo, X.B. *LED Packaging for Lighting Applications: Design, Manufacturing and Testing*. Hoboken, NJ: John Wiley & Sons (2010).

2 Schubert, E.F. *Light-Emitting Diodes*. Cambridge: Cambridge University Press (2006).

3 Zukauskas, A., Shur, M.S., and Gaska, R. *Introduction to Solid-State Lighting*. New York: John Wiley & Sons (2002).

4 Schubert, E.F., and Kim, J.K. *Solid-State Light Source Getting Smart. Science* **308**, 1274–1278 (2005).

5 Krames, M.R., Shchekin, O.B., Mueller-Mach, R., Mueller, G.O., Zhou, L., Harbers, G., and Craford, M.G. Status and Future of High-Power Light-Emitting Diodes for Solid-State Lighting. *J. Disp. Tech.* **3**, 160–175 (2007).

6 CREE. First to Break 300 Lumens-Per-Watt Barrier, http://www.cree.com

7 GG-LED. http://www.gg-led.com

8 UNISTARS. http://www.unistars.com.tw

9 CREE. http://www.cree.com

10 Liu, Z., Liu, S., Wang, K., and Luo, X. Status and Prospects for Phosphor-Based White LED Packaging. *Front. Optoelect. China* **2**, 119–140 (2009).

11 Winston, R., Miñano, J.C., and Benítez, P. *Nonimaging Optics*. San Diego, CA: Elsevier (2005).

12 Chaves, J. *Introduction to Nonimaging Optics*. Boca Raton, FL: CRC Press (2008).

13 Koshel, R.J. *Illumination Engineering Design with Nonimaging Optics*. Hoboken, NJ: IEEE Press and John Wiley & Sons (2013).

14 Benítez, P., Miñano, J.C., Blen, J., Mohedano, R., Chaves, J., Dross, O., Hernández, M., and Falicoff, W. Simultaneous Multiple Surface Optical Design Method in Three Dimensions. *Opt. Eng.* **43**, 1489–1502 (2004).

15 Benítez, P., Miñano, J.C., Blen, J., Mohedano, R., Chaves, J., Dross, O., Hernández, M., Alvarez, J.L., and Falicoff, W. SMS Design Method in 3D Geometry: Examples and Applications. Proc. SPIE 5185 (2004).

16 Dross, O., Mohedano, R., Benítez, P., Miñano, J.C., Chaves, J., Blen, J., Hernández, M., and Muñoz, F. Review of SMS Design Methods and Real World Applications. Proc. SPIE 5529 (2004).

17 Ding, Y., Liu, X., Zheng, Z., and Gu, P. Freeform LED *Lens for Uniform Illumination. Opt. Expr.* **16**, 12958–12966 (2008).

18 Wang, L., Qian, K., and Luo, Y. Discontinuous Free-Form Lens Design for Prescribed Irradiance. *Appl. Opt.* **46**, 3716–3723 (2007).

19 Wang, K., Liu, S., Chen, F., Qin, Z., Liu, Z., and Luo, X. Freeform LED Lens for Rectangularly Prescribed Illumination. *J. Opt. A: Pure Appl. Opt.* **11**, 105501 (2009).
20 Aslanov, E., Doskolovich, L.L., and Moiseev, M.A. Thin LED Collimator with Free-Form Lens Array for Illumination Applications. *Appl. Opt.* **51**, 7200–7205 (2012).
21 Wang, K., Chen, F., Liu, Z., Luo, X., and Liu, S. Design of Compact Freeform Lens for Application Specific Light-Emitting Diode Packaging. *Opt. Expr.* **18**, 413–425 (2010).
22 Chen, F., Wang, K., Qin, Z., Wu, D., Luo, X., and Liu, S. Design Method of High-Efficient LED Headlamp Lens. *Opt Expr.* **18**, 20926–20938 (2010).
23 Wang, K., Wu, D., Qin, Z., Chen, F., Luo, X., and Liu, S. New Reversing Design Method for LED Uniform Illumination. *Opt Expr.* **19**(Suppl. 4), A830–A840 (2011).

25 Wang K, Liu S, Chen L, Zhao J, Dong Z, [illegible] [illegible] Photocatalytic [illegible]. *Opt. [illegible]*. [illegible]

26 Alimov S, [illegible] *[illegible]* [illegible] (2013).

27 Wang K, Chen R, [illegible] [illegible] [illegible] [illegible] (2013).

28 Li, [illegible] Determination of [illegible] *Opt. Express* [illegible] (2010).

29 Wang K, [illegible] [illegible] [illegible]

2

Review of Main Algorithms of Freeform Optics for LED Lighting

2.1 Introduction

Nonimaging optics, as an important category of optical design methods, has very wide applications in concentrated solar energy, display, and illumination. Many algorithms of nonimaging optics have been fundamentally and specifically described, and many related design cases are well presented, in many excellent books about nonimaging optics.[1–4] *Nonimaging Optics* written by Winston *et al.*[1] has systematacially introduced design concepts and detailed methods of nonimaging optics, including basic ideas in geometrical optics, the edge-ray principle, compound parabolic concentrators (CPCs), the flow-line method, the simultaneous multiple surface (SMS) design method, and global optimization of high-performance concentrators, which are mainly designed for concentrated solar energy. In particular, the famous SMS method is described in extensive detail, such as in XR concentrators, RX concentrators, XX concentrators, RXI concentrators, and so on, which provide rich effective ways to realize solar concentrators with high concentration ratios. Furthermore, *Introduction to Nonimaging Optics* written by Chaves[2] introduces algorithms as well as design cases of nonimaging optics, such as fundamental concepts of nonimaging optics, design of two-dimensional concentrators, the Winston–Welford design method, SMS design methods, and so on. *Illumination Engineering Design with Nonimaging Optics*, edited by Koshel,[3] introduces basic concepts and standards of illumination, such as photometry, Etendue, Etendue squeezing, and so on, as well as detailed design methods of illumination and solar energy, such as SMS 3D design methods, solar concentrators, lightpipe and backlight design, and so on. Moreover, sampling, optimization, and tolerancing of nonimaging optics also have been introduced. *Nonimaging Optics in Solar Energy* by O'Gallagher[4] mainly introduces algorithms of CPCs and detailed design methods for solar thermal and solar photovoltaic applications.

As mentioned in Chapter 1, light-emitting diode (LED) lighting has been developing at a high speed in the last decade and brings us new concepts and experiences of lighting, energy saving, and illumination quality in both general and specific lighting (e.g., LED backlighting and automotive headlamps). Freeform optics, as an important category of nonimaging optics, has the ability to address the three key issues of optical design of LED lighting, and it has developed very quickly in recent years. In this book, we will focus

Freeform Optics for LED Packages and Applications, First Edition. Kai Wang, Sheng Liu, Xiaobing Luo and Dan Wu.

on the technology of freeform optics for emerging LED lighting, including both LED packaging and applications. Before systematically introducing algorithms and design methods of freeform optics for various LED applications, we will briefly introduce the main design methods of freeform optics for LED lighting in this chapter, to give readers an overall understanding of freeform optics for LED applications.

2.2 Tailored Design Method

In 2001, Ries *et al.* proposed the tailored design method for freeform optics in three-dimensional space, to solve the archetypal problem of illumination design: redistribute the radiation of a given small light source onto a given reference surface, thus achieving a desired irradiance distribution on that surface.[5–8] The basic design concept is to create a set of partial nonlinear differential equations describing the optical surface based on the illuminance distribution on the target plane and the characteristics of the light source. Then, the shape of the optical surface is obtained by solving the partial nonlinear differential equations. This method has introduced the concept of wavefront. Wavefront is a series of equiphase surfaces, and each face is perpendicular to the light ray propagation direction. When incident wavefront reaches the optical interface, there will be a distortion and it will no longer be an ideal spherical shape. Along the direction of light transmit, the energy density is proportional to the absolute value of wavefront Gaussian curvature. The schematic of wavefronts passing through one optical surface is shown in Figure 2.1. Figure 2.2 shows a freeform lens designed according to the tailored method and its illumination performance, which is the logo of OEC Company. Advantages of the tailored method include following aspects. Firstly, this method changes the focus from finding the solution of the optics surface to solving a set of partial nonlinear differential equations, and this can obtain a complex freeform surface. Secondly, adoption of a Gaussian curvature continuous method will guarantee the smoothness of local as well as overall surfaces. Thirdly, within a small angle range, light rays can be precisely controlled, and with specific light patterns even a picture can be obtained. Software for freeform optics design based on this tailored method has also been released recently.[9]

The freeform lens design method based on the tailored method proposed by Hao *et al.* has generated circular-symmetrical freeform optical lenses with a small

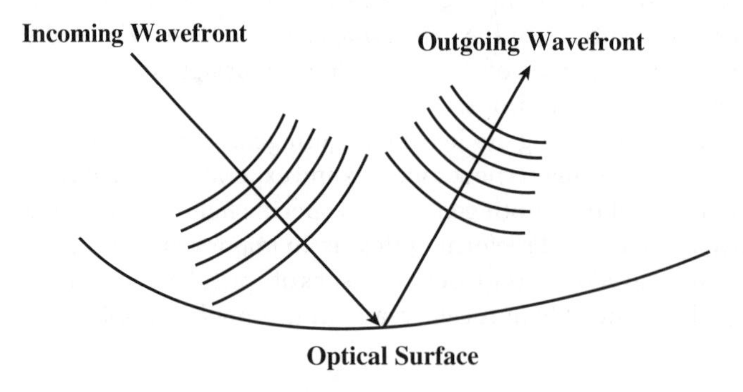

Figure 2.1 Schematic of wavefront reflections from an optical interface.[6]

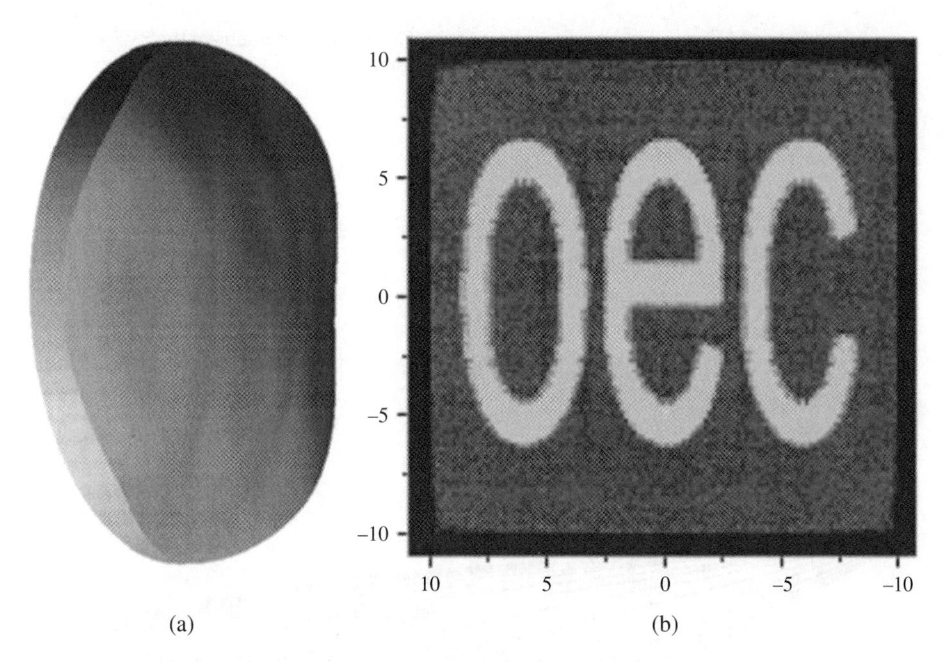

Figure 2.2 Freeform lens designed according to the tailored method, and its illumination performance.[6]

light-emitting angle, and a uniform circular light pattern is realized.[10,11] In this design, the freeform surface is obtained through solving the partial differential equations. The input, the output light rays, and the normal vector on the surface are expressed in a derivative way in this method. These vectors then are substituted into a vector Snell equation, and partial differential equations are obtained. The discrete points on the freeform surface that are the solutions to these equations are obtained by an appropriate numerical solution method. Through curve fitting and rotation of the obtained curve, a circular-symmetrical freeform lens design is achieved. By this method, one realizes the uniform illumination at the target plane with a small light-emitting angle when the light intensity distribution curve (LIDC) of the light source is $I(\theta) = I_0 \cos^m\theta$.

2.3 SMS Design Method

For a noncircular-symmetrical light pattern (e.g., rectangular light pattern), besides the tailored method, there are other freeform optics design methods. SMS, which is one of the most famous design methods and is described in great detail in Ref. [13], was proposed by Benítez *et al.* in 2004.[12–15] The basic design concept is establishing the relationship between a pair of input wavefronts W_{i1} and W_{i2} and output wavefronts W_{o1} and W_{o2}, which is W_{i1}–W_{o1} and W_{i2}–W_{o2}. Then, two freeform surfaces of an optical system are built up simultaneously, the function of which is to make the input wavefronts W_{i1} and W_{i2} match up with the output wavefronts W_{o1} and W_{o2} through reflection and refraction, as shown in Figure 2.3. The SMS method provides

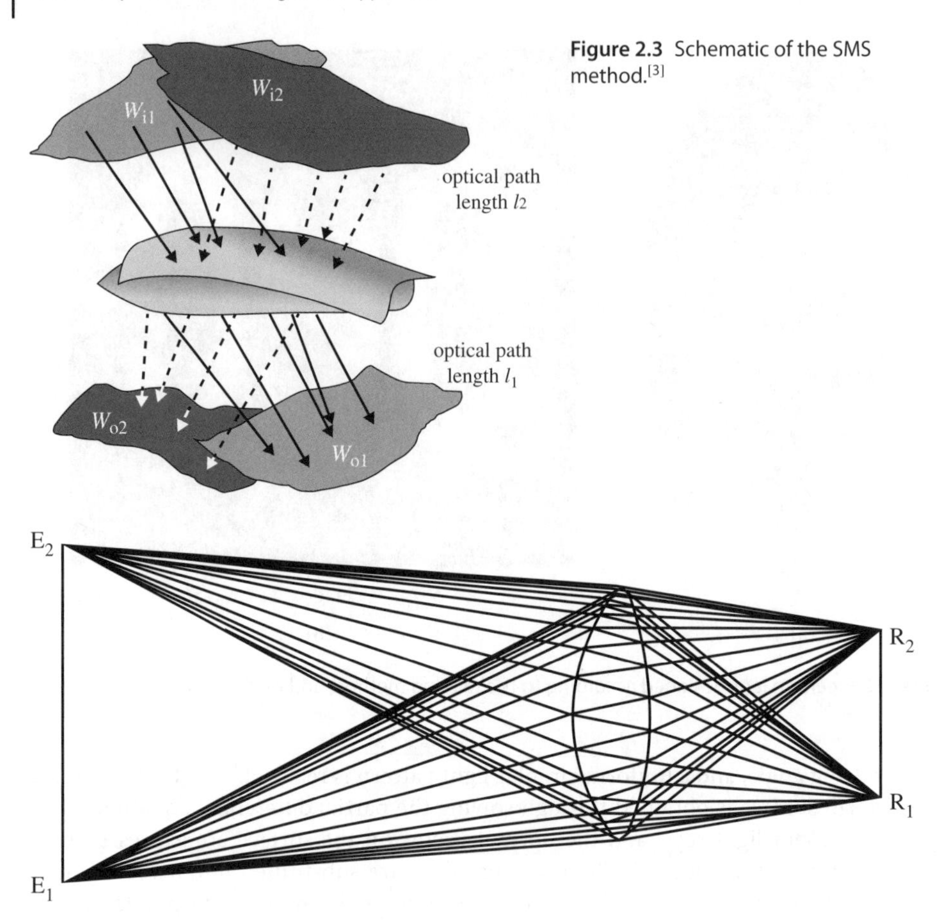

Figure 2.3 Schematic of the SMS method.[3]

Figure 2.4 Matching relationship between an extended source (E_1E_2) in the SMS method and the target plane (R_1R_2).[1]

a solution that can design multiple freeform surfaces simultaneously. This method can establish a relationship between the incident light rays with the expected output light rays, and vice versa. Advantages of the SMS method include simultaneous multiple-surfaces design and control over the edge light rays of the extended source. At the same time, as shown in Figure 2.4, since freeform surfaces are calculated by the mapping of two edge rays of an extended source and two points on the target plane.

2.4 Light Energy Mapping Design Method

Both the tailored method and SMS method are based on the coupling relationship between input wavefronts and the output wavefronts, and the calculation is relatively complicated. For freeform optics lenses, another practical method is to build a light energy mapping relationship between the light source and target plane.[16–18] The design idea is: in the first place, an assumption that all the light energy emitted through the light source will reach the target plane without any loss, that is, the light source optical energy is the same as the optical energy on the target plane. Secondly, the light energy distribution of the light source and the target plane is divided into multiple grids. According to the edge ray principle,[19] we can obtain the solution of differential equations of light energy conservation, and the relationship between light energy in each source grid and area in each area grid on the target plane is acquired. In the last step, according to Snell's law and certain surface construction methods, we can calculate the point coordinates on the freeform surface, and freeform surface is acquired that will be input to simulation tools for verification.

Based on the design concept of involving solutions of differential equations of light energy conservation, Wang *et al.* proposed a discontinuous freeform optics design method in 2007.[20] The construction of freeform surfaces has intrinsic deviations, and the accumulation of these deviations will lead to deterioration of the illumination performance. Involvement of deviation control during construction of the freeform optical lens can restrict the deviation within a certain range and achieve better illumination performance. Specifically, when the deviation of the lens is larger than a certain value, the initial line is reset to begin a new run of lens construction. Figure 2.5 shows an E-shaped light pattern of this discontinuous freeform optics lens based on this algorithm. This algorithm has the advantage of precise light pattern control.

Also based on a similar algorithm, in 2008 Ding *et al.* proposed a discontinuous freeform lens design method based on multiple tiny surfaces.[21] They discovered that there is poor illumination performance when the freeform is smooth and continuous, and therefore a multiple-surfaces construction method is adopted and a discontinuous freeform surface is obtained to realize precisely controllable illumination. Figure 2.6 shows the multiple-surfaces discontinuous freeform lens and the illumination performance. This freeform surface is composed of 450 pieces of multiple tiny surfaces. The advantage of this design method is reduction of construction error to realize uniform illumination.

2.5 Generalized Functional Design Method

In order to overcome the extended source problem, researchers have done a lot of work. In 2006, Bortz *et al.* proposed a *generalized function method*.[22,23] This method considers the extended source as the assembly of point sources. For each point source, a light ray is emitting. Through multiple reflection and deflection surfaces, these light rays are optimized, and control over the deterioration problem caused by the extended source and uniform illumination is realized. This method can effectively solve the extended source light pattern deterioration problem, but the optical system is complicated. Usually, it will comprise at least two lenses, and since for each optical interface there is Fresnel loss, the light efficiency of the system is low. In 2008, Fournier *et al.* proposed an optimization reflector cup design method based on an extended source.[24] On the basis of the point source design, through optimizing input parameters such as the light output position of the light source, the lighting distribution on the target plane for an extended source is optimized and ultimately realizes the expected design requirement.

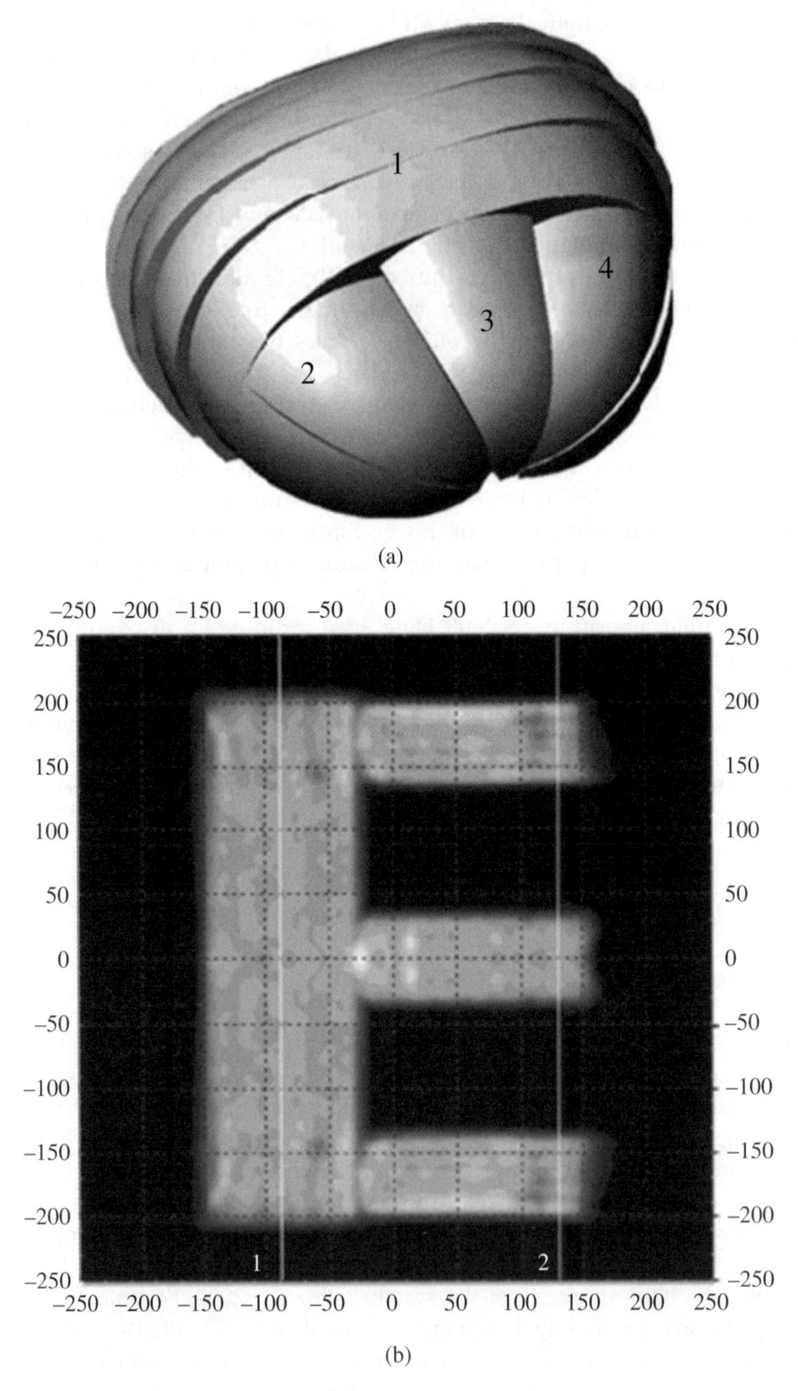

Figure 2.5 Discontinuous freeform lens and its E-shaped light pattern.[20]

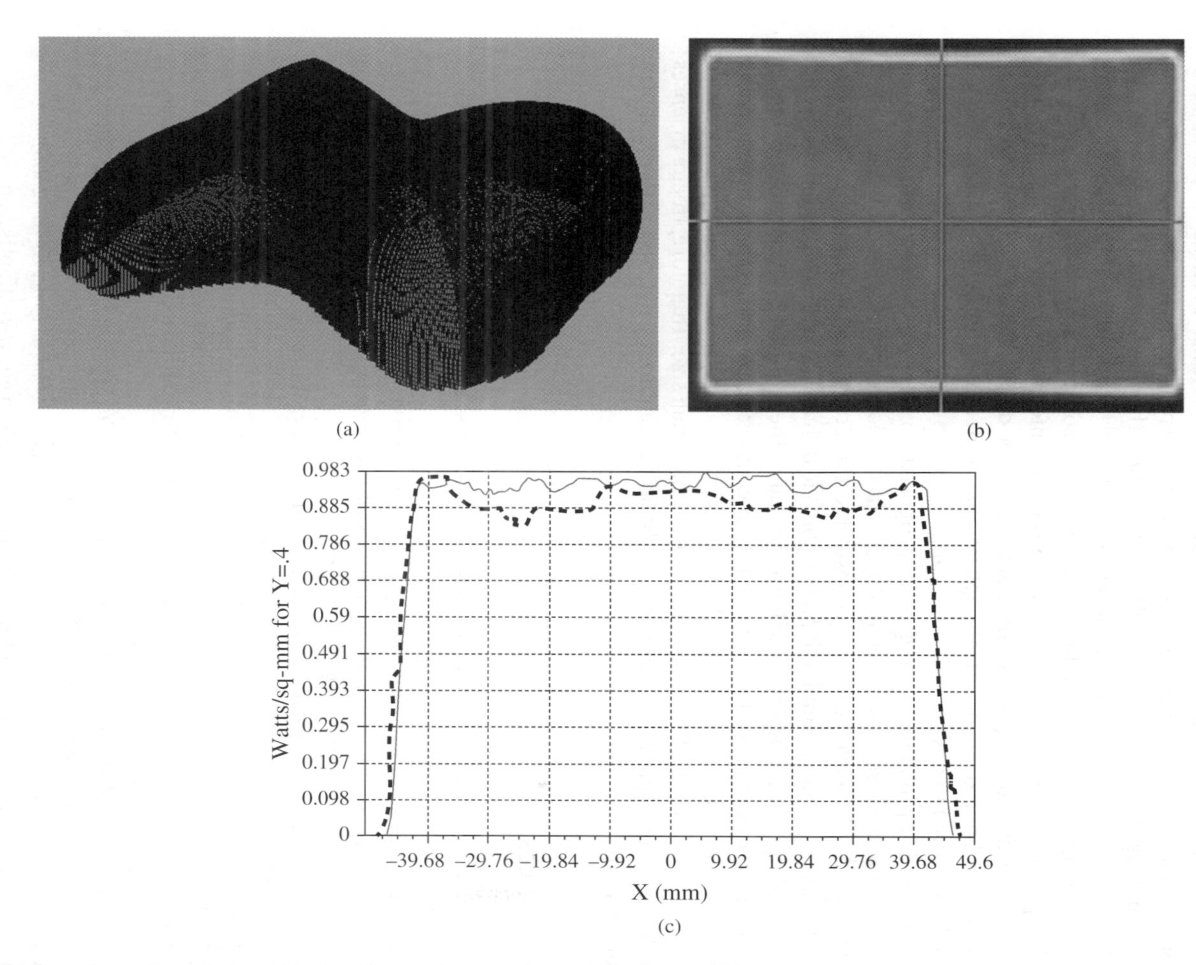

Figure 2.6 Multiple-surfaces discontinuous freeform lens and its rectangular light pattern.[21]

2.6 Design Method for Uniform Illumination with Multiple Sources

Within LED applications, uniform illumination of the LED array is crucial, and it is required by most applications. A lot of research work has been done in this area. Moreno *et al.* proposed a method to achieve uniform illumination by optimizing the distance–height ratio (DHR) of LED arrays when the irradiance distribution $E(r, \theta)$ of LED has an expression of $E(r, \theta)=E_0(r)cos^m\theta$.[25,26] Zong *et al.* improved Moreno's method, making it also suitable when the light source has a large view angle.[27] Although these methods are effective to obtain uniform illumination, the LIDC of a light source should be given as an input parameter during design, which makes their applications limited. In practical applications, sometimes the DHR is given and the LIDC is required to design and to optimize.

To solve this problem, Whang *et al.* proposed a reversing design method for uniform illumination systems by integrating surface-tailored lenses with configurations of LED arrays.[28] However, they only considered the situation when $E(r, \theta) = E_0(r)cos^m\theta$, which makes the maximum DHR $[4/(m+2)]^{1/2}$ less than 1.41, resulting in limited applications. Moreover, an algorithm of lenses to realize the required LIDC is another important issue that is needed during design. In recent years, freeform lenses have become popular for optical design of LED lighting, with advantages of high efficiency, precise light irradiation control, high design freedom, and small size. Therefore, freeform lenses are more likely to provide a new way to obtain the required LIDC to achieve uniform illumination.

References

1 Winston, R., Miñano, J.C., and Benítez, P. *Nonimaging Optics*. Amsterdam: Elsevier (2005).
2 Chaves, J. *Introduction to Nonimaging Optics*. Boca Raton, FL: CRC Press (2008).
3 Koshel, R.J. *Illumination Engineering Design with Nonimaging Optics*. Hoboken, NJ: IEEE Press and John Wiley & Sons (2013).
4 O'Gallagher, J.J. *Nonimaging Optics in Solar Energy*. Williston, VT: Morgan & Claypool (2008).
5 Ries, H., and Muschaweck, J. *Tailoring Freeform Lenses for Illumination. Proc.* SPIE (2001).
6 Ries, H., and Muschaweck, J. Tailored Freeform Optical Surfaces. *J. Optical Society Amer. A* **19**(3), 590–595 (2002).
7 Timinger, A., Muschaweck, J., and Ries, H. *Designing Tailored Free-From Surfaces for General Illumination. Proc.* SPIE, 128–132 (2003).
8 Jetter, R., and Ries, H. *Optimized Tailoring for Lens Design. Proc.* SPIE (2005).
9 ffTOP Freeform Tailored Optics Package. www.ffTop.eu, www.ffoptik.com
10 Hao, X. *Study of LED Illumination System Based on Freeform Optics*. Master's thesis, Zhejiang University, 2008.
11 Wang, W. *Study of Uniform Illumination Design Method Based on Freeform Optics*. Master's thesis, Zhejiang University, 2008.

12 Benítez, P., Miñano, J.C., Blen, J., Mohedano, R., Chaves, J., Dross, O., Hernández, M., and Falicoff, W. Simultaneous Multiple Surface Optical Design Method in Three Dimensions. *Opt. Eng.* **43**, 1489–1502 (2004).

13 Muñoz, F., Benítez, P., Dross, O., Miñano, J.C., and Parkyn, B. Simultaneous Multiple Surface Design of Compact Air-Gap Collimators for Light-Emitting Diodes. *Opt. Eng.* **43**(7), 1522–1530 (2004).

14 Dross, O., Mohedano, R., Benítez, P., Miñano, J.C., Chaves, J., Blen, J., Hernández, M., and Muñoz, F. *Review of SMS Design Methods and Real World Applications. Proc.* SPIE (2004).

15 Benítez, P., Miñano, J.C., Blen, J., Mohedano, R., Chaves, J., Dross, O., Hernández, M., Alvarez, J.L., and Falicoff, W. *SMS Design Method in 3D Geometry: Examples and Applications. Proc.* SPIE (2004).

16 Parkyn, W.A. *The Design of Illumination Lenses via Extrinsic Differential Geometry. Proc.* SPIE (1998).

17 Parkyn, W.A. *Illumination Lenses Designed by Extrinsic Differential Geometry. Proc.* SPIE (1998).

18 Parkyn, W.A. *Segmented Illumination Lenses for Step Lighting and Wall-Washing. Proc.* SPIE (1999).

19 Ries, H., and Rabl, A. Edge-Ray Principle of Nonimaging Optics. *J. Opt. Soc. Amer. A* **11**(10), 2627–2632 (1994).

20 Wang, L., Qian, K.Y., and Luo, Y. Discontinuous Free-Form Lens Design for Prescribed Irradiance. *Appl. Opt.* **46**(18), 3716–3723 (2007).

21 Ding, Y., Liu, X., Zheng, Z.R., and Gu, P.F. Freeform LED Lens for Uniform Illumination. *Opt. Expr.* **16**(17), 129–12966 (2008).

22 Bortz, J., and Shatz, N. Generalized Functional Method of Nonimaging Optical Design. *Proc. SPIE* **6338**, 633805 (2006).

23 Bortz, J., and Shatz, N. Iterative Generalized Functional Method of Nonimaging Optical Design. *Proc. SPIE* **6670**, 66700A (2007).

24 Fournier, F.R., Cassarly, W.J., and Rolland, J.P. Optimization of Single Reflectors for Extended Sources. *Proc. SPIE* **7103**, 710301 (2008).

25 Moreno, I. Configurations of LED Arrays for Uniform Illumination. *Proc. SPIE* **5622**, 713–718 (2004).

26 Moreno, I., Alejo, M.A., and Tzonchev, R.I. Design Light-Emitting Diode Arrays for Uniform Near-Field Irradiance. *Appl. Opt.* **45**, 2265–2272 (2006).

27 Qin, Z., Wang, K., Chen, F., Luo, X.B., and Liu, S. Analysis of Condition for Uniform Lighting Generated by Array of Light Emitting Diodes with Large View Angle. *Opt. Expr.* **18**, 17460–17476 (2010).

28 Whang, A.J.W., Chen, Y.Y., and Teng, Y.T. Designing Uniform Illumination Systems by Surface-Tailored Lens and Configurations of LED Arrays. *J. Disp. Technol.* **5**, 94–103 (2009).

3

Basic Algorithms of Freeform Optics for LED Lighting

3.1 Introduction

Many freeform optics algorithms, such as the simultaneous multiple surface (SMS) and tailored design methods, have been developed as introduced in Chapter 2, and they provide delicate optical design methods for illumination with precise control of illuminance distribution. In this chapter, we will introduce a series of basic freeform optics algorithms especially designed for light-emitting diode (LED) lighting based on light energy mapping relationship,[1–9] the edge ray principle,[10] and Snell's law.[11] Many specific application situations, such as circularly symmetrical illumination, noncircularly symmetrical illumination, point source, extended source, and so on, will be included in these algorithms, which cover most applications of LED lighting, such as LED spotlighting,[5,9] road lighting,[6–8] tunnel lighting,[6,7] backlighting for liquid crystal displays (LCDs),[12–14] automotive headlamps,[15] pico-projectors,[16] and so on. These algorithms will provide practical, effective, flexible, and concise ways for freeform optics designs of LED packages and applications with precise light energy distribution control.

In addition, as described in Chapter 1, reflectors converge light emitted from LEDs most of the time, which limits the applications of reflectors in LED lighting, especially for many LED applications with large emitting angles, such as LED road lighting, LED backlighting, and so on. Therefore, the algorithms to be discussed in this chapter will focus on the optical designs of freeform lenses. Moreover, both general single freeform lens design algorithms as well as the whole illumination performance with multiple freeform lens (e.g., LED backlighting) design algorithms will be introduced in this chapter. Specific design methods based on these fundamental algorithms for various kinds of LED packages and applications, such as LED road lighting, LED backlighting, LED automotive headlamps, and so on, will be introduced in details in later chapters.

3.2 Circularly Symmetrical Freeform Lens – Point Source

In LED lighting, the light spot shapes of most LED luminaires are circular, such as the LED MR16 spot light, LED down light, and LED bulb. At the same time, much of the illumination performance of LED lighting, such as LED backlighting, LED indoor lighting, and LED commercial lighting, is formed by overlapping multiple circular light

Freeform Optics for LED Packages and Applications, First Edition. Kai Wang, Sheng Liu, Xiaobing Luo and Dan Wu.

spots. Thus, circular light spots have wide applications in LED lighting. Moreover, uniform illumination is important in many different lighting applications. The illuminance uniformity is always to be considered as a critical parameter in various illumination standards. High illuminance uniformity will provide a comfortable and soft lighting space. On the contrary, nonuniform illuminance distribution can very possibly cause visual fatigue and discomfort. Therefore, it's important to realize uniform illumination of a single light spot as well as the whole lighting space by optical design. In this section, we will introduce circular freeform lens design methods for a single uniform light spot, including large emitting angles and small emitting angles.

3.2.1 Freeform Lens for Large Emitting Angles

The beam angle is usually defined as twice the angle between the emitting ray of half maximum light intensity and the center axis (maximum light intensity). The light intensity distribution curve (LIDC) reflects the light intensity or light energy spatial distribution of one luminaire. Therefore, it's easy to calculate the beam angle according to the LIDC. LIDCs of general LED luminaires are always the Lambert type or convergence type (i.e., its light intensity distribution is more convergent than the Lambert type), and the maximum light intensities of these kinds of LED luminaires appear at the directions of center axes. Therefore, beam angles could be adopted to roughly describe the light intensity spatial distribution of these LED luminaires. However, the freeform lenses discussed in this section are designed to achieve uniform illumination, and the maximum light intensities do not appear at the directions of center axes. Therefore, it is hard to adopt the traditional concept of beam angles to describe this situation. A new concept of emitting angles will be introduced in this section, which are defined as twice the angle between the edge ray of a light spot and the center axis. For example, if a light spot has an emitting angle of 120°, it means that the light emitting from a lens will be controlled within the area of ±60°. In addition, the algorithms of freeform lenses for large emitting angles (≥90°) and small emitting angles (<90°) are different, and we will discuss large emitting angles first.

The schematic of a design concept of a freeform lens for circularly symmetrical uniform illumination is shown in Figure 3.1. This illumination system consists of an LED light source, freeform lens, and target plane, within which freeform lenses could be secondary optics or primary optics integrated with an LED package. An LED light source adopts polar coordinates of (γ, θ, ρ), and the lens and target plane adopt Cartesian coordinates of (x, y, z). Ray $\boldsymbol{I}$ emitted from LED light source $S(\gamma, \theta, \rho)$ will incident at a point $P(x, y, z)$ at the outside surface of the freeform lens, and will become emergent ray $\boldsymbol{O}$ irradiating at a point $Q(x, y, z)$ on the circular target plane after refracting by the freeform lens. The direction of incident ray $\boldsymbol{I}$ is from point P to point S, and the direction of emergent ray $\boldsymbol{O}$ is from point Q to point P. Therefore, if the mapping relationship between incident ray $\boldsymbol{I}$ and emergent ray $\boldsymbol{O}$ could be established, we are able to figure out the coordinates and normal vector of point P according to Snell's law and the construction method of a freeform surface, thereby figuring out the whole freeform lens. In addition, considering the convenience of optical design and manufacturing of freeform lenses, the inside surface of a freeform lens is always designed as a sphere and the LED source is positioned at the center of the sphere. This guarantees that, in the point source situation, the propagation direction of a ray emitted from the LED source will be perpendicular

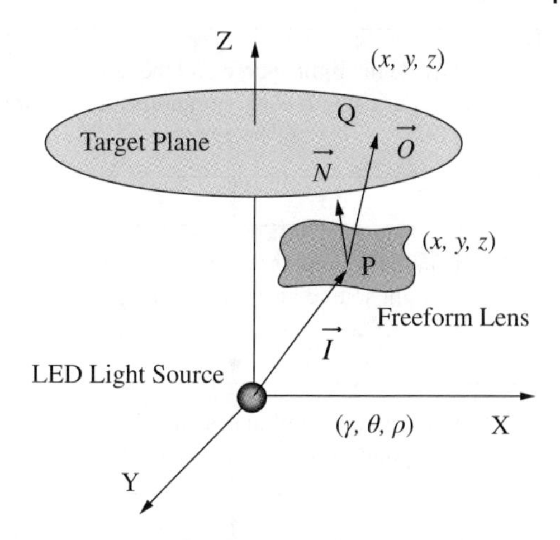

Figure 3.1 Schematic of a design concept of a freeform lens for circularly symmetrical uniform illumination.

to the inner surface of the lens, which will not change the propagation direction of the ray after being refracted by the interface of air and lens.

There are three main steps for the algorithm of freeform lenses to achieve circularly symmetrical uniform illumination. Firstly, divide the light energy distribution of the LED light source into N grids with equal luminous flux, and divide the target plane into N grids with equal area. Establish a mapping relationship between the light source grids and the target grids. Secondly, figure out the coordinates and normal vectors of points at the surface of the freeform lens, and construct the lens. Finally, validate the designed lens by the ray-tracing simulation method. If the design results cannot meet requirements, modify some parameters (e.g., division number N and position of the LED light source) to optimize the design until meeting its requirements, and then finish the design. The flowchart of this algorithm is shown in Figure 3.2.

3.2.1.1 Step 1. Establish a Light Energy Mapping Relationship between the Light Source and Target

First of all, divide the spatial distribution of light energy of the LED light source into N grids with equal luminous flux of Φ_0, and divide the target plane into N grids with equal area of S_0. As shown in Figure 3.3, if the mapping relationship between the light energy grid and the area grid could be established, then the luminous flux of each light energy grid is able to irradiate at the corresponding area grid, and each area grid has the same average illuminance of $E_0 = \eta\Phi_0/S_0$, wherein η is the efficiency of the freeform lens. Therefore, when the number of N is large, which means the ratio of the area of each area grid to the whole target plane becomes very small, it's able to realize illumination with uniform illuminance on the target plane.

Next, we will discuss the equal luminous flux division of an LED light source. As shown in Figure 3.4, the entire light energy spatial distribution Ω of an LED light source could be regarded as being composed of a number of unit light energy grids of Ω_0, which represents the luminous flux within the angular range with field angle $d\gamma$ in the latitudinal direction and $d\theta$ in the longitudinal direction. The calculation of luminous flux Φ_0 of the unit light energy grid of Ω_0 is:

$$\Phi_0 = \int I(\theta)d\omega \tag{3.1}$$

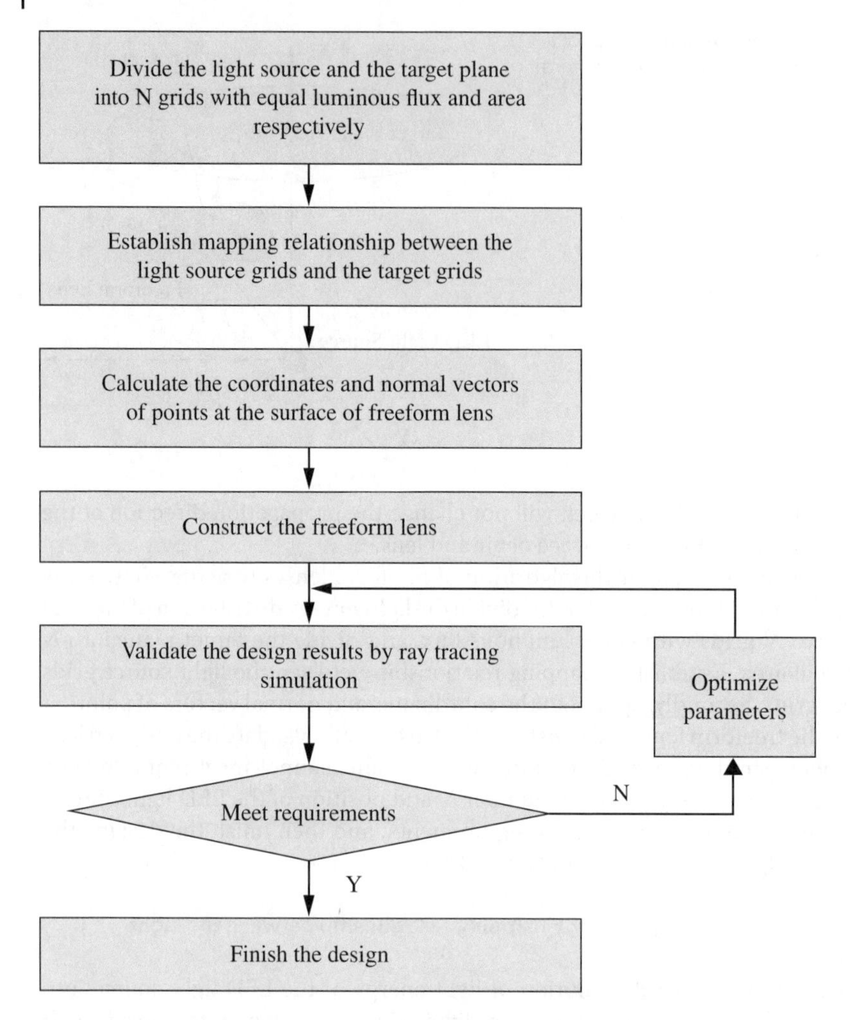

Figure 3.2 Flowchart of the algorithm of a freeform lens for circularly symmetrical uniform illumination.

wherein $I(\theta)$ is the light intensity distribution of the LED light source and $d\omega$ is the solid angle. As shown in Figure 3.4, the light source is set at the origin coordinates $\boldsymbol{O}$, and the solid angle $d\omega$ constructed by the small area dS on the light energy spatial distribution Ω to point O is as follows:

$$d\omega = \frac{dS}{\rho^2} \tag{3.2}$$

wherein ρ is the distance between small area dS and O. The position of small area dS is decided by the spatial polar coordinates of γ, θ, and ρ, and the area is decided by sides lengths of a and b. From Figure 3.4, we can find:

$$a = \rho \sin \theta d\gamma \tag{3.3}$$

$$b = \rho d\theta \tag{3.4}$$

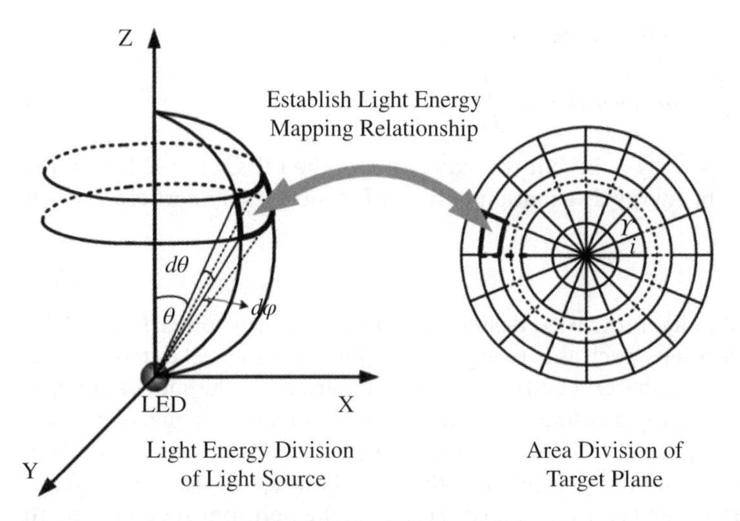

Figure 3.3 Schematic of a light energy mapping relationship between the LED light source and target plane for circularly symmetrical uniform illumination.

Figure 3.4 Schematic of light energy distribution of the LED light source with circularly symmetrical division.

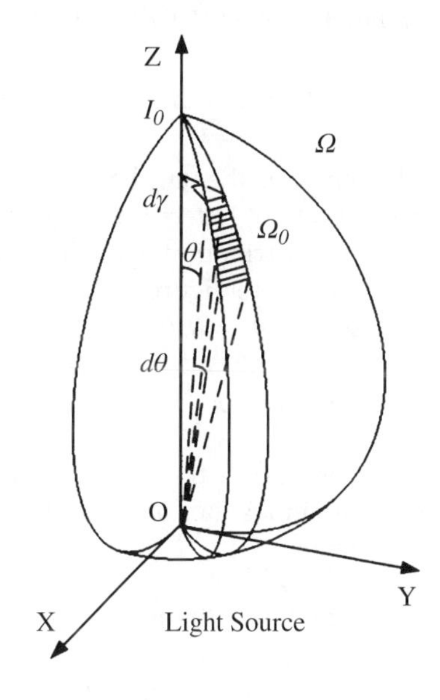

so $d\omega$ is:

$$d\omega = \frac{dS}{\rho^2} = \frac{ab}{\rho^2} = \frac{\rho^2 \sin\theta d\theta d\gamma}{\rho^2} = \sin\theta d\theta d\gamma \tag{3.5}$$

We substitute Eq. 3.5 into Eq. 3.1 and can obtain:

$$\Phi_0 = \int I(\theta) d\omega = \int_{\gamma_1}^{\gamma_2} d\gamma \int_{\theta_1}^{\theta_2} I(\theta) \sin\theta d\theta \tag{3.6}$$

As shown in Figure 3.3, since the light energy spatial distribution and the target plane are all circularly symmetrical, the unit light energy grid in this design could be regarded as a circled grid that is $\gamma_1 = 0$, $\gamma_2 = 2\pi$, so Φ_0 is:

$$\Phi_0 = \int_0^{2\pi} d\gamma \int_{\theta_1}^{\theta_2} I(\theta) \sin\theta d\theta = 2\pi \int_{\theta_1}^{\theta_2} I(\theta) \sin\theta d\theta \tag{3.7}$$

Here, we need to especially point out the expression of the LIDC of the LED source of $I(\theta)$. Usually, the light energy distribution of the LED source is Lambert type, which means:

$$I(\theta) = I_0 \cos\theta \tag{3.8}$$

wherein I_0 is the central light intensity. Besides Lambert type, many other LED LIDCs exist in practical applications, such as batwing and side-emitting, to meet different lighting requirements. Non-Lambert type LIDCs could be described by the sum of multiple Gaussian functions,[17] multiple cosine-power functions,[17] or a polynomial of θ with a primary function of $\{1,\theta,\theta^2,\ldots,\theta^m\}$.[12] The sum of multiple Gaussian functions is able to describe LIDCs with large emitting angles, and the sum of multiple cosine-power functions is able to describe small emitting angles. However, the polynomial of θ is able to describe various kinds of LIDCs with large or small emitting angles. Moreover, considering the requirement of reversing design and optimize LIDCs, we adopt the polynomial of θ to describe LIDCs in this chapter, as shown in Eq. 3.9:

$$I(\theta) = I_0' \sum_{i=0}^{m} a_i \theta^i \tag{3.9}$$

wherein I_0' is the unit light intensity and m is the order of the polynomial. This polynomial cannot reflect the value of light intensity directly but only to fit the spatial light intensity distribution of the light emitted from an LED.

According to Figure 3.4 and Eq. 3.7, when calculating the total luminous flux of an LED source, $\theta_1 = 0$, $\theta_2 = \pi/2$, so the total luminous flux Φ_{total} is:

$$\Phi_{total} = 2\pi \int_0^{\pi/2} I(\theta) \sin\theta d\theta \tag{3.10}$$

In this design, the luminous flux of the unit light energy grid is set as $1/N$ of the total luminous flux of the LED source. Therefore, according to Eqs. 3.7 and 3.10, we can obtain:

$$2\pi \int_{\theta_i}^{\theta_{i+1}} I(\theta) \sin\theta d\theta = \frac{2\pi}{N} \int_0^{\pi/2} I(\theta) \sin\theta d\theta \ (i = 0,1,\ldots,N-1,\ \theta_0 = 0). \tag{3.11}$$

wherein θ_i is the angle of the emitting ray that divides the luminous flux of the LED source equally. Since $\theta_0 = 0$, we can obtain each emitting angle θ_i and the included angle $\Delta\theta_{i+1}$ between two emitting angles, according to an iterative calculation as follows:

$$\Delta\theta_{i+1} = \theta_{i+1} - \theta_i \ (i = 0,1,\ldots,N-1) \tag{3.12}$$

So far, the spatial light energy distribution of the LED source has been divided into N disc-annulus parts with equal luminous flux, and each emitting angle θ_i also has been obtained.

Next, the circularly symmetrical target plane also will be divided into N parts of unit area grid S_0 with equal area. The emitting angle of the designed freeform lens is α, and the distance between the LED source and the target plane is d, so the radius R of the target plane is:

$$R = d \tan \alpha \tag{3.13}$$

As shown in Figure 3.3, the disc-annulus unit light energy grid has a mapping relationship with the disc-annulus unit area grid on the target plane. There are N disc-annulus unit area grids, and the radius is r_i $(i = 0, 1, \ldots, N-1)$ of each unit area grid, wherein r_0 represents the radius of the central point of the target plane and $r_0 = 0$. So, the area of each disc-annulus unit area grid S_0 is:

$$S_0 = \pi {r_{i+1}}^2 - \pi {r_i}^2 = \frac{\pi R^2}{N} \quad (i = 0,1,\ldots,N-1,\ r_0 = 0) \tag{3.14}$$

We can obtain the radius r_i of each disc-annulus as follows:

$$r_i = R\sqrt{\frac{i}{N}} \quad (i = 0,1,\ldots,N) \tag{3.15}$$

So far, the circularly symmetrical target plane has been divided into N disc-annulus parts with equal area of S_0, and the radius r_i of each disk-annulus also has been obtained.

Following, we will establish the light energy mapping relationship between the LED source and the target plane. Based on the divisions with equal light energy and equal target area mentioned in this chapter, emitting angles θ_i and θ_{i+1} of two edge rays (defining each unit light energy grid) and radiuses r_i and r_{i+1} (defining each unit area grid) have been obtained. According to the edge ray principle, the ray emitted from the edge of a light source will irradiate at the edge of a target plane after passing an optical system, and rays between two edge rays will irradiate at the target plane. Therefore, if we want the light energy of the unit light energy grid to irradiate at the corresponding unit area grid, we need to make sure the edge rays of one unit light energy grid could irradiate at the edge positions of the corresponding unit area grid. For this reason, rays emitted from a light source with emitting angles of θ_i and θ_{i+1} are required to irradiate at the positions of the radiuses of r_i and r_{i+1} on the target plane after having been refracted by the freeform lens. Through this method, we establish the light energy mapping relationship between the LED source and target plane.

3.2.1.2 Step 2. Construct a Freeform Lens

Since the freeform lens designed in this section is circularly symmetrical and the lens could be obtained by rotating a freeform curve around the central axis, in this section we will focus on the freeform curve design in the two-dimensional plane. As shown in Figure 3.5 and Figure 3.6, the design flow is as follows:

1. Set the start vertex P_0 of a freeform lens. Vertex P_0 $(0, 0, z_0)$ is the first point of freeform curve C_0, located in the central axis of the LED source, and the direction of its normal vector $\overrightarrow{N_0}$ is straight up. The height z_0 of the freeform lens is decided by the distance of P_0 and the LED source. Based on the body size of a regular LED packaging module, the height of the freeform lens usually is set between 6 and 10 mm.
2. Calculate the second point P_1 of freeform curve C_0. Point P_1 is the crossing point of a ray whose incident angle is θ_1 and a tangent plane of the start vertex P_0. As normal vector $\overrightarrow{N_0}$ is straight up, its tangent plane is perpendicular with $\overrightarrow{N_0}$, which is a horizontal plane; therefore, the altitude of P_1 and P_0 is the same, and the coordinates

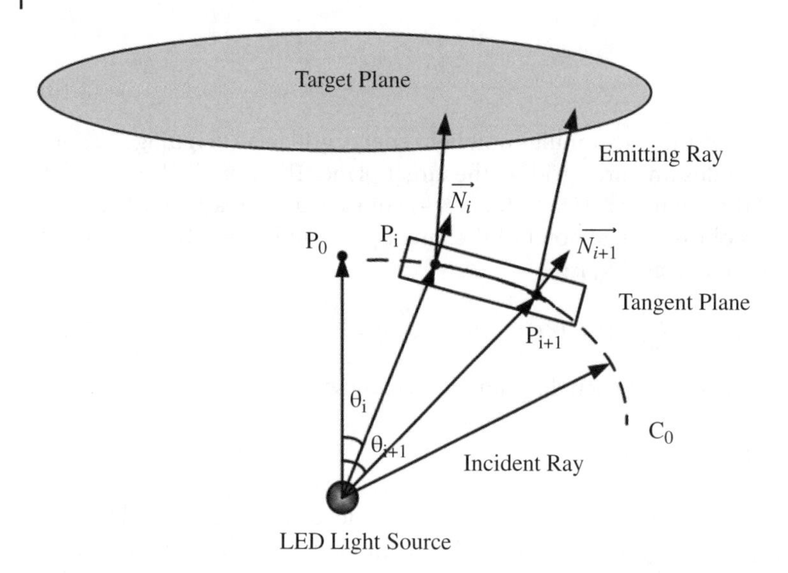

Figure 3.5 Schematic of the construction of a freeform curve.

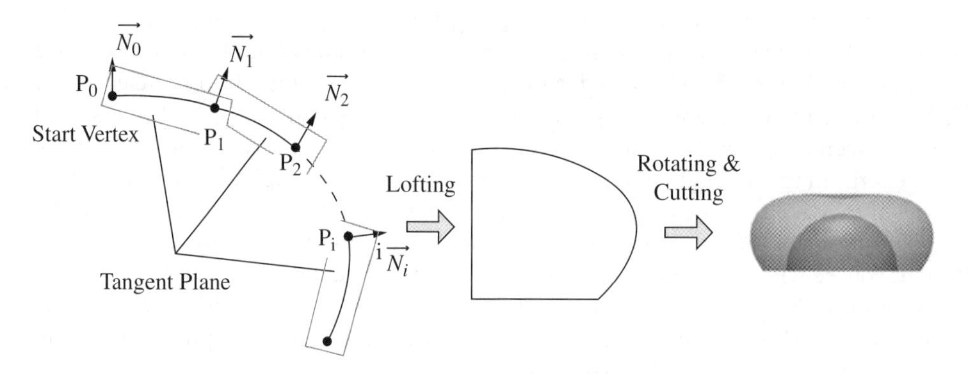

Figure 3.6 Schematic of the construction of a circularly symmetrical freeform lens.

of P_1 are $(x_1, 0, z_0)$. According to the energy mapping relationship mentioned in this chapter, an incident ray with incident angle θ_1 will be refracted at point P_1 and then reach the point with a radius of r_1 on the target plane, which is assumed as Q_1. Therefore, we can obtain that the direction of incident ray $\overrightarrow{I_1}$ is $\overrightarrow{P_1O}$, and the direction of exit ray $\overrightarrow{O_1}$ via point P_1 is $\overrightarrow{Q_1P_1}$. According to Snell's law (Eq. 3.16), we can obtain the normal vector $\overrightarrow{N_1}$ of point P_1:

$$[1 + n^2 - 2n(\overrightarrow{O}\cdot\overrightarrow{I})]^{1/2}\overrightarrow{N} = \overrightarrow{O} - n\overrightarrow{I} \tag{3.16}$$

wherein $\overrightarrow{I}$ and $\overrightarrow{O}$ are the unit vector of incident and refracted rays, respectively; $\overrightarrow{N}$ is the unit normal vector of the refracted point; and n is the ratio of the refractive index of the optical medium where the incident ray exists to the refractive index of the optical medium where the emergent ray exists. Since the refractive index of air is approximately 1, n is the refractive index of the lens in this section.

3. Calculate the subsequent points P_i of the freeform curve C_0. In the same way, we are able to get the coordinates of P_2 $(x_2, 0, z_0)$ by calculating the intersection of the incident ray with the incident angle θ_2 and the tangent plane of the point of P_1. Similarly, we are able to obtain the coordinates of P_i by calculating the intersection of the incident ray with the incident angle θ_i and the tangent plane of the point of P_{i-1}, and then get the normal vector of P_i according to the energy mapping relationship and Snell's law. Therefore, coordinates and normal vectors of all the points on the freeform curve C_0 can be obtained conveniently.
4. Construct the freeform curve C_0. As shown in Figure 3.6, we fit points P_i to form the freeform curve by three-dimensional modeling software.
5. Construct the freeform lens. We rotate the freeform curve C_0 around the central axis to obtain the circularly symmetrical freeform lens. Since the lens is integrated with the LED module, we construct a hemispherical cavity in the bottom of the lens, so that the rays emitted from the LED source will hardly change their directions after going through the inner surface of the freeform lens.

3.2.1.3 Step 3. Validation and Optimization

Before finalizing the freeform lens design, numerical simulation should be conducted. In this book, the Monte Carlo ray-tracing method is adopted to conduct the optical simulation. In this method, rays emitted from random locations of source surface, and then they exit with random directions. The initial emitting point and the exit direction are determined by a probability function that describes the optical properties of the source. In the transmitting procedure, rays will be reflected, refracted, or totally internally reflected at the interface of the optical medium. At the same time, in the simulation, each ray emitted from the light source carries certain light energy, which is determined by the light source property. The light energy of each ray also will be changed according to the optical loss (e.g., absorption, Fresnel loss, etc.) in the optical system during the transmitting procedure. Finally, statistically analyze all the rays at the set receive plane and obtain the light energy distribution of the receive plane. In the ray-tracing process, it always needs more than 1 million rays to guarantee the accuracy of simulation results. After several decades of development, the Monte Carlo ray-tracing method is already mature. In addition, this method is just one way to conduct optical validation; it doesn't involve a lens design algorithm. Therefore, research of ray tracing is not the key point of this book.

We use the designed freeform lens to conduct optical simulation and then analyze the results. If the results meet the design requirements, we finish the optical design. Otherwise, we should perform feedback optimization by adjusting some parameters (e.g., dividing number N, LED module location, etc.) until the final design meets the requirements, and then finish the design.

3.2.2 TIR-Freeform Lens for Small Emitting Angle

According to the freeform lens design method introduced in Section 3.2.1, we designed two poly(methyl methacrylate) (PMMA) freeform lens specific to the Philips Lumileds K2 LED module, and the emitting angles are 60° and 38°. These two angles are the most commonly used emitting angles of LED spotlights, and their lighting performance on the receive plane 1 m away are shown in Figure 3.7 and Figure 3.8, respectively. For the 60° lens, the illuminance distribution of the light spot center is uniform, while illuminance

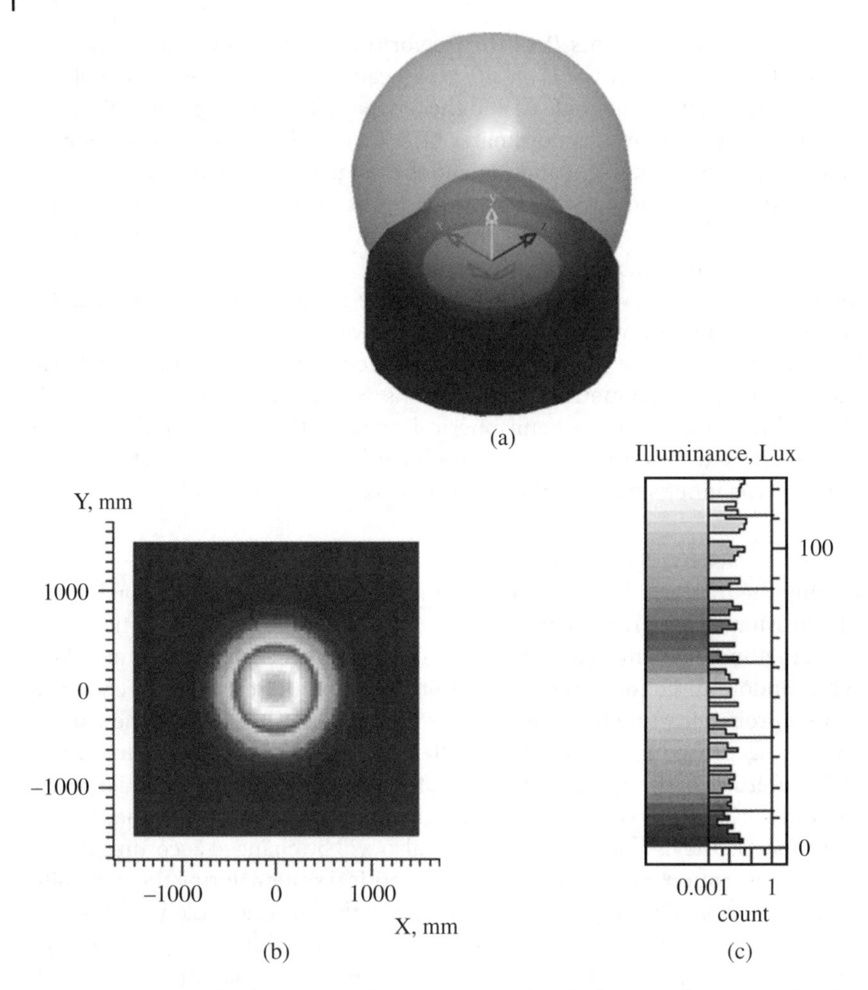

Figure 3.7 A PMMA freeform lens with an emitting angle of 60° and its illuminance distribution on the receive plane 1 m away.

uniformity declines significantly at the edge. The illuminance uniformity U_E of the whole expected lighting area is 0.417, which shows a great drop compared to the large-angle case. In addition, the 38° lens shows a similar illuminance distribution. The central area with uniform illuminance is smaller, and illuminance uniformity declines faster at the edge area. The light spot deteriorates dramatically, and the illuminance uniformity of the whole expected lighting area is only 0.196.

In small-emitting-angle lighting, the light deterioration is mainly caused by those with a large intersection angle between the incident ray and the corresponding exit ray. As shown in Figure 3.9, when the ray exited from the optically denser medium n_1 and refracted into the optically thinner medium n_2, the incident angle θ_1 is smaller than the refract angle θ_2; the intersection angle β between the incident ray and refract ray can be expressed as:

$$\beta = \theta_1 + 90^\circ + (90^\circ - \theta_2) = 180^\circ + \theta_1 - \theta_2 \tag{3.17}$$

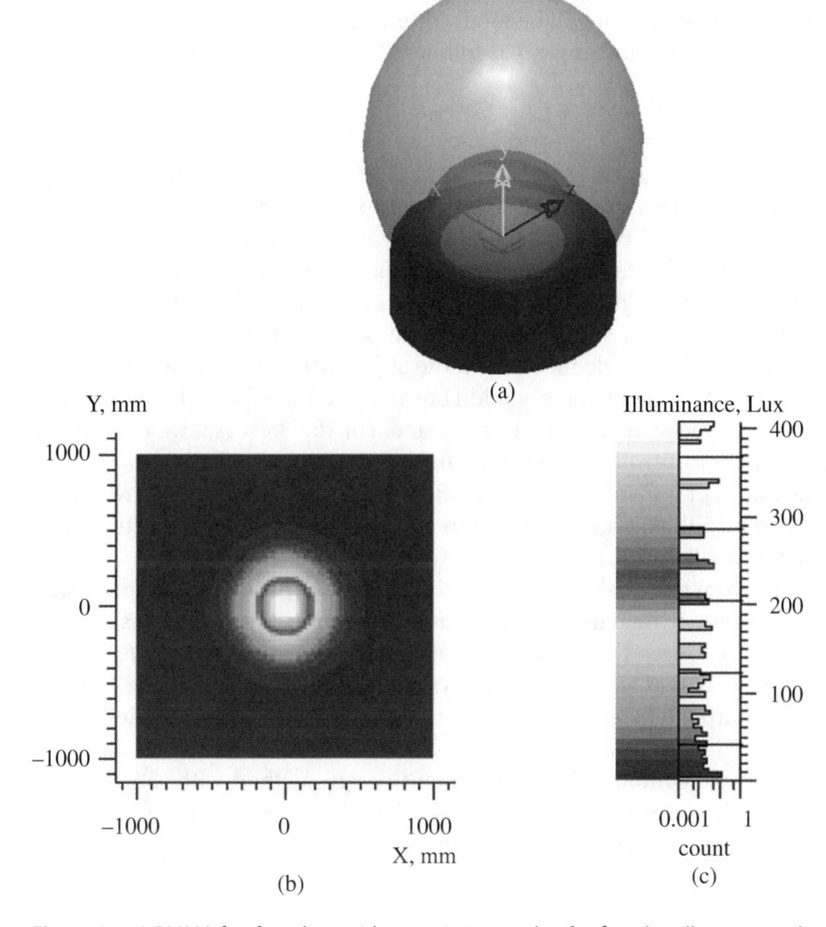

Figure 3.8 A PMMA freeform lens with an emitting angle of 38° and its illuminance distribution on the receive plane 1 m away.

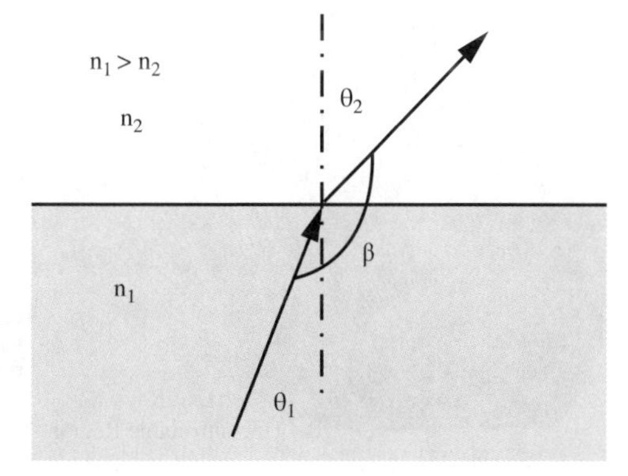

Figure 3.9 Ray exited from the optically denser medium n_1 and refracted into the optically thinner medium n_2.

According to the TIR law, when incident angle θ_1 equals critical angle θ_C, the incident ray will no longer be refracted, but will be totally internally reflected. Therefore, we can calculate the smallest intersection angle β as follows:

$$\beta_{\min} = 180^\circ + \theta_C - 90^\circ = 90^\circ + \theta_C = 90^\circ + \arcsin\frac{n_2}{n_1} \tag{3.18}$$

In this design, $n_1 = 1.49$, $n_2 = 1.0$ (air), so $\beta_{\min}$ is 132.16°. We can reasonably consider that when the calculated intersection angle is less than 132.16°, there will not be such an optical surface. The calculated coordinates and its normal vector are insignificant, and incident rays will not be refracted to the expected direction when they go through this surface. When the emitting angle is 60°, the intersection angle of the last incident ray and its refract ray is 90° + 30° = 120°, which is less than the minimum allowable intersection angle of 132.16°. We can deduce that in the 60° emitting-angle case, the part of incident rays with a large incident angle wouldn't be refracted as expected, thus causing the deterioration at the edge area of the target plane. For the 38° case, we can expect that more rays will be uncontrollable and cause further deterioration in lighting performance. The schematics of uncontrollable region in the 60° lens and 38° lens are shown in Figure 3.10. Therefore, the design method of freeform lens introduced in Section 3.2.1 will not apply to the small-emitting-angle case.

At the same time, from this analysis we can arrive at the conclusion that in the design of freeform lenses, in order to achieve a high output efficiency, we should ensure that the intersection angle between the incident ray and its exit ray is greater than $\beta_{\min}$, thus theoretically making sure that all the incident rays will not be dissipated by the TIR and improving the light output efficiency. These design principles also can be applied to the unsymmetrical freeform lens design.

In this section, we will combine the freeform lens design method with the TIR lens design method to realize small-emitting-angle lighting. As shown in Figure 3.11, this new freeform TIR lens consists of surface 1, surface 2, surface 3, and surface 4. Surface 1 is a flat surface, and surface 2, surface 3, and surface 4 all are freeform surfaces. Rays

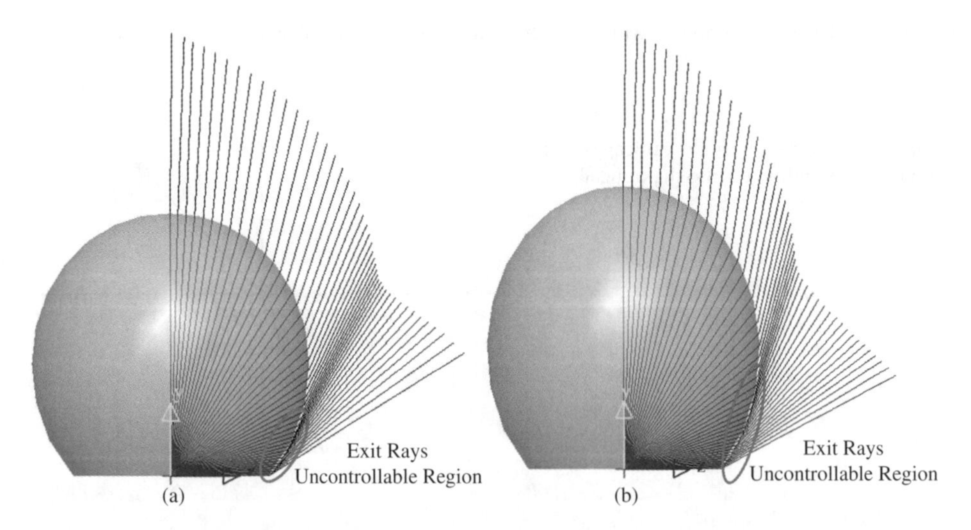

Figure 3.10 Schematic of uncontrollable exit rays of (a) 60° and (b) 38° freeform lens.

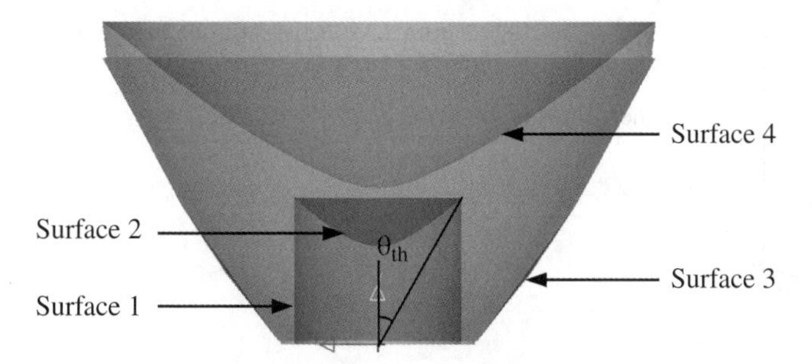

Figure 3.11 Schematic of a TIR freeform lens.

emitted from the source center can be divided into two parts. The first part ($0° \sim \theta_{th}$) is incident to surface 2 directly, then refracts into lens, finally exiting from surface 4. The second part ($\theta_{th} \sim 90°$) is incident to surface 1 and refracts into the lens, then is totally internally reflected by surface 3, and finally exits from surface 4. The surface of an ordinary TIR lens is a flat plane or array of convex lenses. In this design, surface 4 is designed to be a freeform surface, thus improving the accurate controlling ability and the design freedom and flexibility.

The design process of this freeform TIR lens also includes three procedures: establishing light energy mapping relationships, constructing a freeform lens, and validating the lens design, wherein establishing light energy mapping relationships is the key issue. In this design, two key design points are the exit directions of first-part rays after being refracted by surface 2 and the exit directions of second-part rays after being totally internally reflected by surface 3. To realize uniform lighting with a small emitting angle, we set the first and second parts' rays to incident perpendicularly to surface 4. According to $\beta_{min} = 132.16°$, after being refracted by surface 4, the largest emitting angle of those perpendicular incident rays is 47.84°. Therefore, by optimizing the design of freeform surface 4, we can realize uniform lighting within the emitting angle range of 0°~95.68°, thus realizing uniform lighting with a small emitting angle. Following are the design methods of the three freeform surfaces.

Firstly, the energy distribution of light source and the target plane are both divided into N grids equally, and corresponding dividing angles θ_i and dividing radius r_i are calculated. As for surface 2, we set a start vertex $P2_0$ as shown in Figure 3.12. Rays whose incident angle ranges from 0° to θ_{th} will incident to surface 2 directly, and every incident direction of every ray is obtained before. At the same time, we expect that the rays refracted by surface 2 will emit vertically. Therefore, according to Snell's law, we can calculate the normal vector of every point $P2_i$. According to the circularly symmetrical freeform lens three-step construction method discussed in Section 3.2.1, we can obtain surface 2.

For freeform surface 3, we also set the start vertex $P3_0$, as shown in Figure 3.13. Note that the start vertex $P3_0$ of surface 3 is located at the bottom of the surface and all the other points $P3_i$ are calculated accordingly. The rays emitting from a light source within the range of θ_{th}~90° will be refracted by vertical surface 1 and then incident on freeform surface 3, so the direction of each ray could be obtained. Rays incident on surface 3 will not be refracted but totally internally reflected. The rays that are totally internally reflected will emit vertically, and in this way the reflected rays' directions can

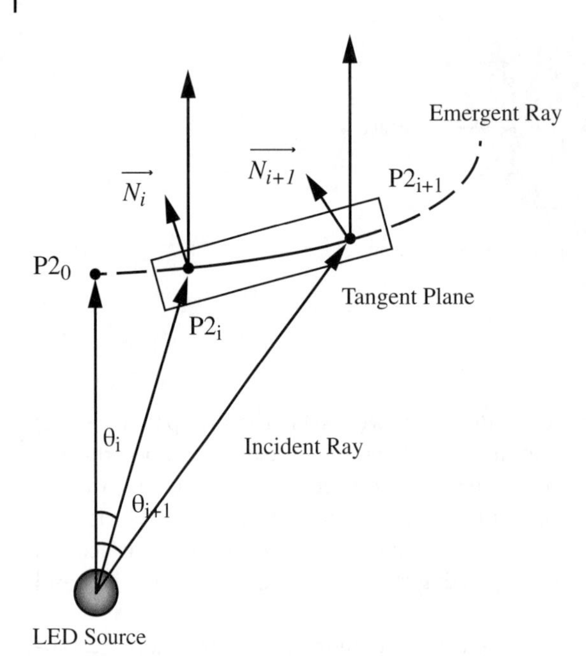

Figure 3.12 Schematic of a design of surface 2 of a TIR freeform lens.

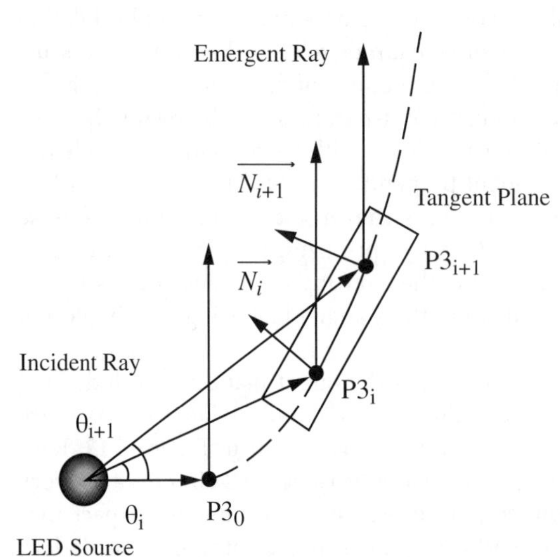

Figure 3.13 Schematic of a design of surface 3 of a TIR freeform lens.

be obtained. Set the n in Snell's law as 1, and Eq. 3.16 can be used for the TIR situation. Therefore, we can calculate the normal vector of each point $P3_i$ on surface 3. Based on the same surface construction method, freeform surface 3 could be calculated. In this design, the reason why we set the surface 1 as the vertical plane surface is to enable the rays emitting within $\theta_{th}{\sim}90°$ could be irradiated on surface 3 with small incident angles after being refracted by surface 1, so as to reduce the height of surface 3 as well as the height of the lens.

For freeform surface 4, the direction of each incident ray is vertical up and the position of r_i on the target plane of each corresponding exit ray is known. Thus, freeform surface

4 could be constructed according to the three steps mentioned in section 3.2.1. Through the analysis and design given in this chapter, the four main freeform surfaces of the TIR freeform lens have been constructed successfully, and the total lens could be obtained by combining and rotating these surfaces in 3D modeling software.

3.2.3 Circularly Symmetrical Double Surfaces Freeform Lens

We have proposed several circularly symmetrical freeform lens algorithms with a point source, within which the inner surface of the freeform lens is built as a hemispherical dome shape. This inner hemispherical dome shape is designed so that the secondary freeform lens can be integrated directly on the LED modules in the market that usually integrated with a hemispherical silicone lens. In some cases, we want to develop LED packages for specific applications, and the outer shape of the silicone lens is no longer restricted to a hemispherical dome shape. There is more design freedom, and we develop double-surface freeform lenses for LED uniform illumination with an extra condition.[18,19]

For an LED package, the light emanated from the LED light source is actually refracted twice when transmitting from the inner and the outer surfaces of the lens. As shown in Figure 3.14, the inner surface refracts the incident ray $\overrightarrow{OA}$ into the first output ray $\overrightarrow{AB}$, which is the incident ray for the outer surface. The outer surface refracts the second incident ray $\overrightarrow{AB}$ into the second output ray $\overrightarrow{BR}$, which irradiates at the corresponding point R on the target plane. According to Snell's law, the rays $\overrightarrow{OA}$, $\overrightarrow{AB}$, and $\overrightarrow{BR}$ must satisfy Eqs. 3.19 and 3.20:

$$n_1 \frac{\overrightarrow{OA}}{|\overrightarrow{OA}|} \times \overrightarrow{N_1} = n_2 \frac{\overrightarrow{AB}}{|\overrightarrow{AB}|} \times \overrightarrow{N_1} \tag{3.19}$$

$$n_2 \frac{\overrightarrow{AB}}{|\overrightarrow{AB}|} \times \overrightarrow{N_2} = n_3 \frac{\overrightarrow{BR}}{|\overrightarrow{BR}|} \times \overrightarrow{N_2} \tag{3.20}$$

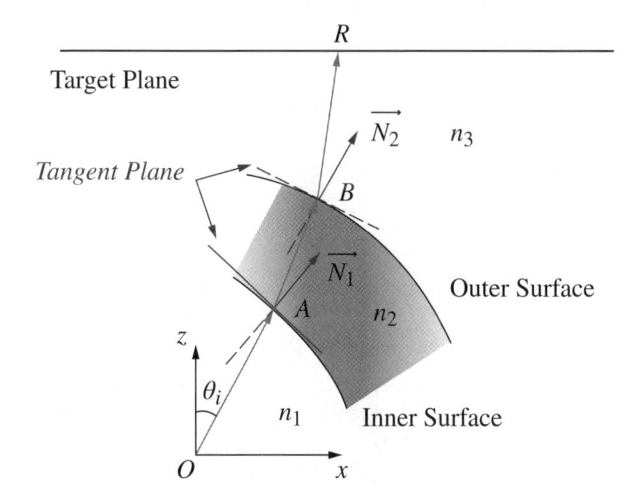

Figure 3.14 Schematic of a typical light path through a double freeform surface lens. θ_i is the edge angle of the incident ray.

where $\overrightarrow{N_1}$ and $\overrightarrow{N_2}$ are the normal vectors on the inner and outer surfaces, respectively. n_1, n_2, and n_3 are the refractive indices of each area, respectively. n_2 is the refractive index of the lens material, and $n_3 = 1.00$ because the corresponding area is air. When the lens is a primary lens, n_1 is the refractive index of the silicone or phosphor–silicone matrix; when the lens is a secondary lens, $n_1 = 1.00$ because the area is air.s

From Eqs. 3.19 and 3.20, we may discover that these two equations can't be solved because the number of unknowns is more than the number of equations. However, if we assume the inner surface is a hemispherical surface, $\overrightarrow{OA}$ and $\overrightarrow{AB}$ are parallel, and Eq. 3.19 can be neglected. This is the exact reason why most researchers simplify the problems by considering the inner surface as hemispherical. In this situation, we want to design the two surfaces simultaneously. We adopted the practical nonimaging optical design method to design the lens.[5–9] After the meshing of light source and target plane, we can obtain the vector of each incident ray (i.e., the edge angle θ_i) and the coordinates of corresponding points on the target plane. Then, we supposed that the inner and outer surfaces can be described by surface functions $f_1(x,y,z)$ and $f_2(x,y,z)$, respectively. The tangent plane at each point is the differential coefficient of the functions $f_1(x,y,z)$ and $f_2(x,y,z)$ at each point, and the unit normal vector is the orthogonal of the tangent plane at each point. To make the indefinite equations solvable, we can add appropriate extra conditions to balance the number of unknowns and equations. There may be a lot of extra conditions that can help solve the problem, like surface function relationship control and light deviation control. Different extra conditions correspond to different applications. Two kinds of extra conditions are illustrated and discussed here:

1. *Surface function relationship control*: As one kind of extra condition, we also can supply the relationship between the two surfaces. The relationship can be assigned based on appropriate assumptions. In the simplest case, we can give the function of one surface and solve the other surface of the freeform-surface lens. For example, we can give the function of the inner surface to calculate the function of the outer surface, and vice versa.

   As shown in Figure 3.15a, when the function of the outer surface is given (i.e., $f_2(x,y,z)$ is known), we firstly fix a point A_0 as the vertex of the inner surface of the

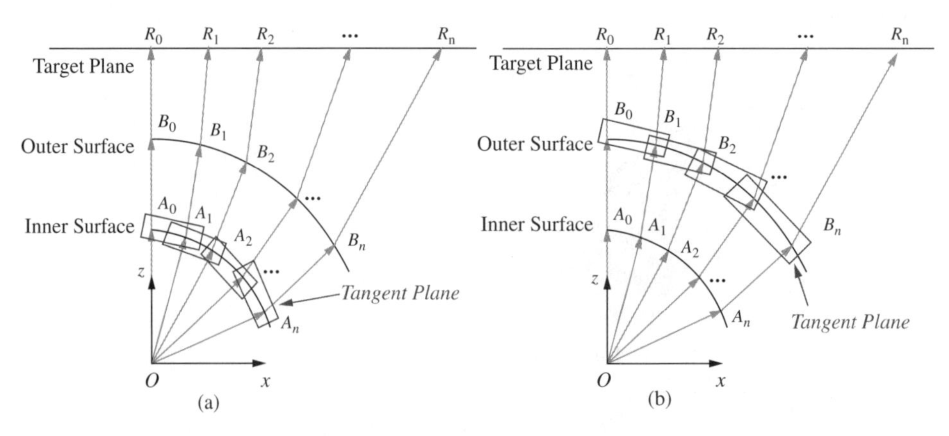

Figure 3.15 Schematic of (a) point generation on the inner surface when the outer surface is given and (b) point generation on the outer surface when the inner surface is given.

lens, and the normal vector at this point is vertical up. The second point A_1 can be calculated by the intersection of the incident ray $\overrightarrow{OA_1}$ and the tangent plane of the point A_0. Then, applying Eq. 3.20, we can calculate the point B_1 on the outer surface since the unit normal vector at point B_1 is determined by the function $f_2(x,y,z)$, and the corresponding point R_1 was determined previously. Then the unit normal vector at point A_1 is also calculated by applying Eq. 3.19, and the tangent plane at point A_1 is consequently determined. By repeating this process until the edge angle θ_i equals 90°, we can get all the points and their unit normal vectors on the inner surface. As shown in Figure 3.15b, when the function of the inner surface is given (i.e., $f_1(x,y,z)$ is known), we can calculate the outer surface with the same method. After obtaining all the coordinates on the inner or outer surface, we fit these points to form the contour line of the cross section of the lens.

2. *Light deviation control*: As another kind of extra condition, we can control the two light deviation angles on the two surfaces with a proper ratio. For each ray, the total light deviation angle from the incident ray $\overrightarrow{OA}$ to the output ray $\overrightarrow{BR}$ can be obtained. With a proper ratio, we can divide the total deviation angle as two components, corresponding to the two refractions on the inner and outer surfaces, respectively. As shown in Figure 3.16, we firstly fixed point A_0 and point B_0 as the vertex of the inner and outer surface of the lens, respectively. The normal vectors at these two points were vertical up. The second point A_1 on the inner surface could be calculated by the intersection of the incident ray $\overrightarrow{OA_1}$ and the tangent plane of the point A_0. With proper deviation angles, we could obtain the ray $\overrightarrow{A_1B_1}$ and ray $\overrightarrow{B_1R_1}$. The second point B_1 on the outer surface could be calculated by the intersection of the ray $\overrightarrow{A_1B_1}$ and the tangent plane of the point B_0. By applying Eqs. 3.19 and 3.20, we could calculate the unit normal vectors and the tangent planes at point A_1 and point B_1. Then we could obtain the third point A_2 on the inner surface by the intersection of the incident ray $\overrightarrow{OA_1}$ and the tangent plane of the point A_1. By repeating this process until the edge angle θ_i equaled 90°, we could get all the points and their unit normal vectors on the

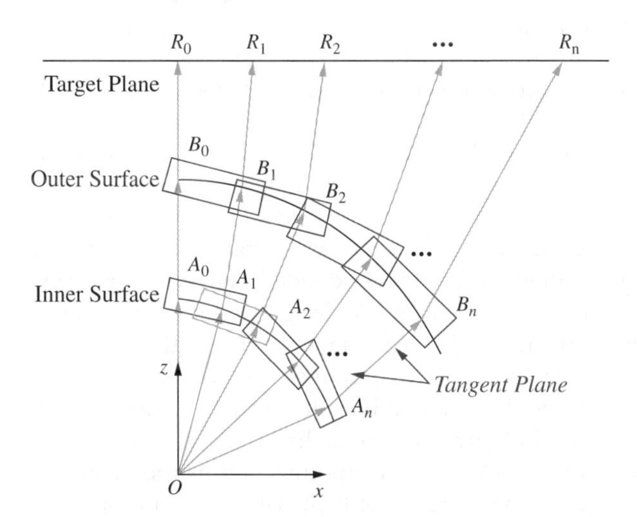

Figure 3.16 Schematic of point generations on the inner and outer surfaces simultaneously.

two surfaces. After obtaining all the coordinates on the inner and outer surfaces, we fit these points to form the contour line of the lens' cross section.

If the lens is designed as an axis-symmetrical lens, then we can get the freeform lens by rotating the contour line around the symmetry axis. If not, we can use a similar method to calculate the other contour lines. After obtaining all the contour lines, the lofting method can be used to form smooth surfaces of the lens between the contour lines.[16] What should be emphasized is that this general design method has much design freedom. The extra conditions could be varied according to the different requirements, and the resulting double freeform-surface lenses could have different applications. Even for the two aforementioned extra conditions, it also has much design freedom. As long as the optical performance, volume, and profile of the designed lens are acceptable, the inner surface could be given as spherical, cubic, cylinder, or even spheroidal and parabolic while the outer surface is freeform, and vice versa. The ratio of the two deviation angles also can be changed according to different applications. With other extra conditions, the method could be extended to design different double freeform-surface lenses.

3.3 Circularly Symmetrical Freeform Lens – Extended Source

The size of a high-power LED chip is 1×1 mm, and the distance between the LED source and the secondary optics is always as 7∼10 mm, so the distance from the LED source surface to the surface of the lens is five times larger than the size of the LED chip. Therefore, a single LED source is usually regarded as a point source in secondary freeform optics design. To meet the requirements of LED lighting with more luminous flux, a multichip LED package is one of the developing trends of LED packages. More and more LED modules are integrated with multichips such as 2×2 and 3×3. Compared with single-chip LED modules, multichip LED packages can emit more luminous flux from a single module and therefore are more suitable for high-power LED lighting applications such as LED down lamps, LED road lighting, and LED automotive headlamps. However, apart from the high luminous flux, the size of multichip LED packaging raises new challenges for optical design. The light-emitting area of multichip LED packages is much larger than that of single LED packages. For example, the light-emitting area of a 3×3 LED array package is nine times larger than that of a single LED package, and the ratio of distance between the light source and lens to the size of the light source will be less than 5. Therefore, during the design of secondary optics, we should consider the light source of a multichip LED package as an extended source rather than a point source. If we adopt the point source design method directly to an extended source, the light pattern will deteriorate severely. For example, a freeform lens designed, based on the point source freeform optics algorithm as mentioned in this chapter, with a height of 7 mm is applied to a single LED source and to a 3×3 LED array source, respectively. As shown in Figure 3.17, we can find that both the shape and uniformity of the radiation pattern deteriorated a lot when using the 3×3 LED array source as the extended source.

In this section, we will propose a feedback-reversing optimization method that can greatly improve extended source uniformity with little light pattern expansion. This method can be expanded into three specific algorithms: target plane grids division

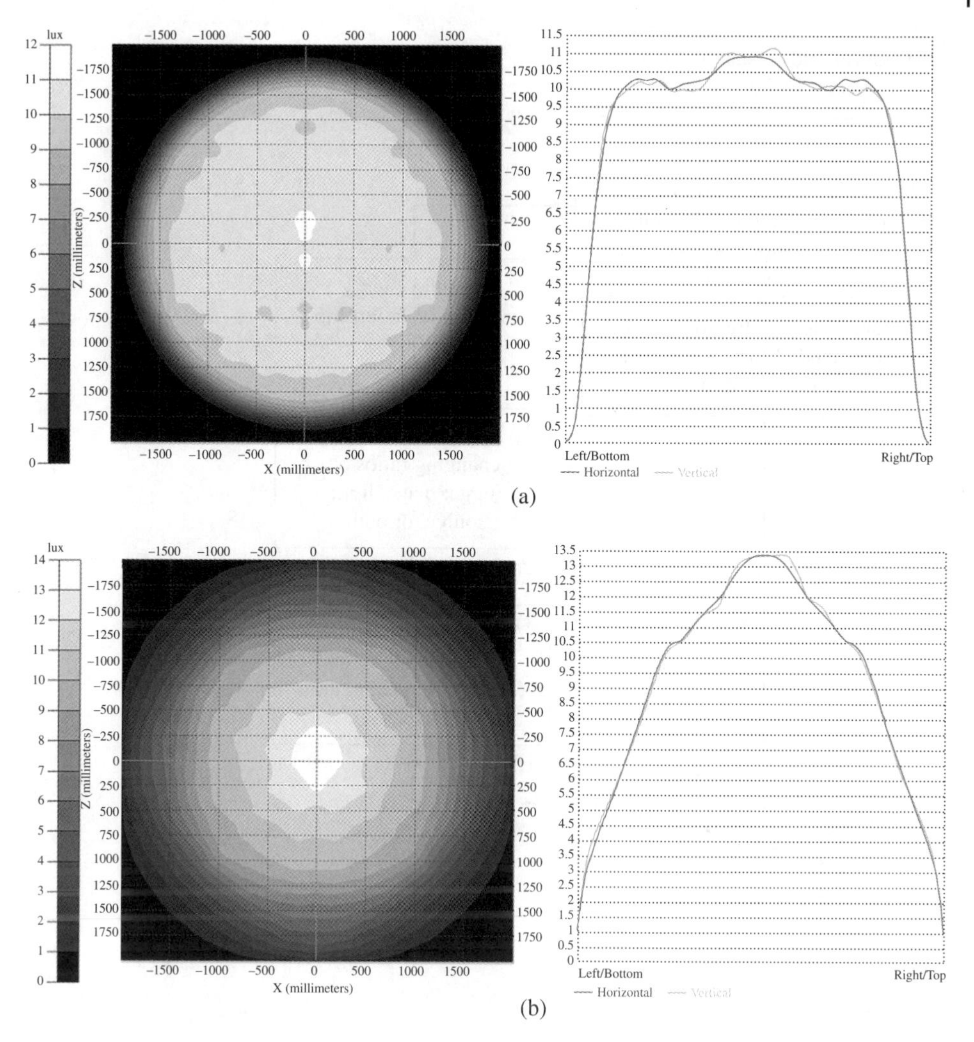

Figure 3.17 Lighting performance of a circularly symmetrical freeform lens with a (a) single LED source and (b) 3 × 3 LED array source.

optimization, light source energy grids division optimization, and both target plane and light source energy grids division optimization. It is proved that through any of these three algorithms, we can improve the lighting performance deterioration of a light pattern caused by size expansion of the light source, achieving higher uniformity and accurate control of the shape of the light pattern.

In this method, optimization of overall lighting performance is to be achieved by adjusting the target plane and/or light source grids division and reconstruction of the outside surface of the lens. Therefore, in order to give prominence to our key issue, the inner surface of the lens is designed as a spherical shape, which will little change the transmission directions of the incident light. Focus will be on the construction of the outside surface of the freeform lens. A feedback optimization design method of a circular-symmetrical freeform lens for an extended source is shown in Figure 3.18.

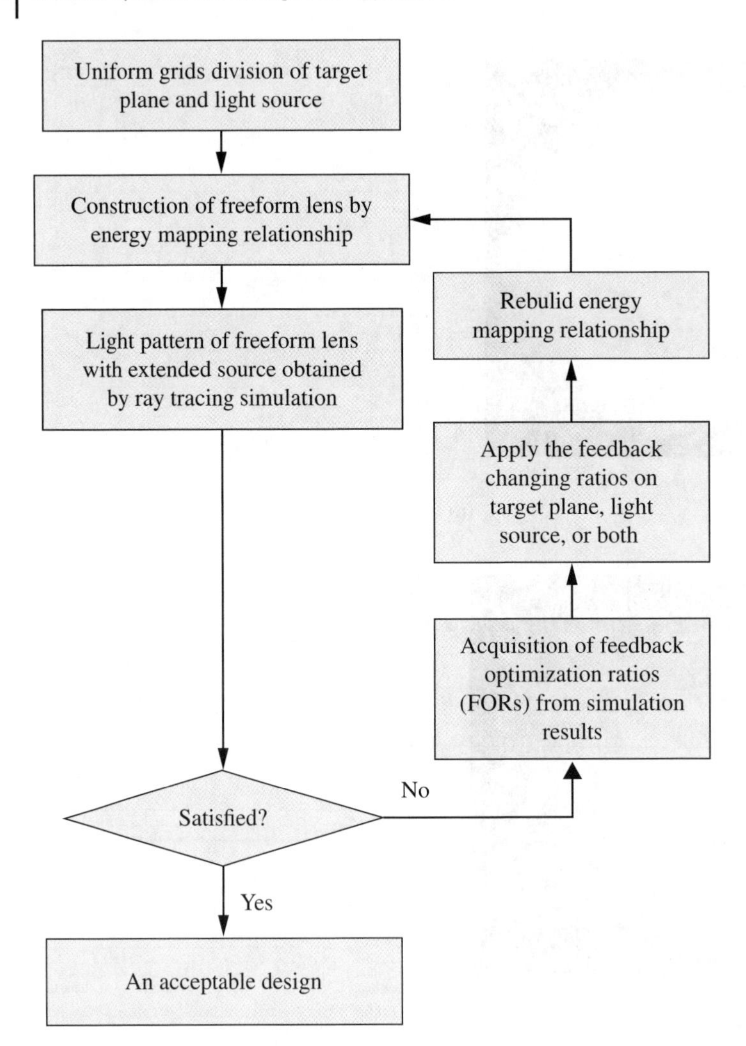

Figure 3.18 Flow chart of a feedback reversing optimization design method of a circular-symmetrical freeform lens for an extended source.

1. First of all, we build a freeform lens for a point source with a circular-symmetrical uniform light pattern based on the method mentioned in Section 3.3.2.
2. In Step 2, we apply the freeform lens designed based on a point source algorithm on the extended source and trace light transportation by optical software. As we expected, illumination results show that the combination of an extended source with a point source freeform lens will lead to serious light pattern deterioration. We compare the deteriorated illuminance distribution with the anticipated uniform illuminance to obtain feedback ratios.
3. For Step 3, the feedback optimization ratios (FORs) are applied on target plane grids division and/or light source grids division, and therefore derive three different algorithms. In this method, grids division will be adjusted according to feedback ratios and is not equal division like in the point source freeform lens design method. Based

on the optimized grids division, we rebuild the energy relationship between a light source and target plane and reconstruct the freeform lens.

4. In Step 4, we evaluate the simulation results the same way as in Step 2. If the uniformity of illuminance on the target plane satisfies our criterion, we complete the feedback optimization; otherwise, we continue our design until the freeform lens meets our requirement. In this design method, the feedback ratios acquisition, redivision of grids, and energy relationship establishment are three key steps, and the other steps are similar with the point source design method (which we do not discuss in detail).

In the remainder of this section, we specifically describe the extended source freeform lens construction procedures.

3.3.1.1 Step 1. Construction of a Point Source Freeform Lens

The design method is the same as the algorithm mentioned in Section 3.3.1.

3.3.1.2 Step 2. Calculation of Feedback Optimization Ratios

After construction of a point source freeform lens, we adopt a ray-tracing method to simulate a point source freeform lens with an extended source. Since light source intensity distribution, the shape of the lens, and the target plane area are all circular-symmetrical, illuminance distribution along the x-axis direction is able to reflect the whole light pattern illuminance distribution on the target plane. In the following steps, we only discuss the illuminance distribution along the x-axis. We set the simulation results of illuminance with an extended source as E_i. According to the simulation results, we can calculate the total flux of the target plane Φ_{total_target} as:

$$\Phi_{total\_t\arg et} = \eta \times \Phi_{total\_source} \tag{3.21}$$

Among them, η is the efficiency of freeform lens light emitting, and Φ_{total_target} is the total flux of the target plane. The target light pattern area is $S = \pi R^2$, where R is the radius of the circular light pattern. Therefore, the average illuminance of the target plane is E_{av}:

$$E_{av} = \frac{\Phi_{total\_t\arg et}}{S_{t\arg et}} \tag{3.22}$$

We use E_{av} as the expected average illuminance of each grid. When the division number of grids is large enough, we can realize uniform illuminance distribution of the whole lighting area. With an extended source, the simulation result shows that the illuminance values on the target plane at different positions are usually quite different from the average illuminance value E_{av}. It is expected that by changing the target plane or light source grids division, the average illuminance of each grid could gradually turn into E_{av} and a uniform illuminance of the whole target plane can be achieved. The FOR for each grid is set as k_i, which reflects the difference between the simulation result and the expectation illuminance value.

$$k_i = \frac{E_i}{E_{av}} \tag{3.23}$$

Since we can only obtain limited illuminance values at the target plane, they are probably not in accordance with target plane grid division positions. In order to obtain k_i at each desired point on the target plane, we adopt the least squares method to fit k_i, and the relationship between coordinates r_i and k_i is set as $k_i = f(r_i)$. By this method, we can get FORs at any position located at the target plane along the x-axis direction.

3.3.1.3 Step 3. Grids Redivision of the Target Plane and Light Source

From the illuminance definition in Eq. 3.24, illuminance of each grid can be changed by varying the area of each grid at the target plane, the flux of each grid of the light source, or both.

$$E = \frac{\Phi}{S} \tag{3.24}$$

We set Φ_i as the flux of each grid of the light source, and S_i as the area of each grid at the target plane. In the point source design method, the light source and target plane are divided into equal grids, which means $\Phi_i = \Phi_0$ and $S_i = S_0$. We also set a_i and b_i as the changing ratios of Φ_i and S_i. Then, the flux of each changed grid will be $\Phi_i' = a_i\Phi_i$ and the area of each changed grid at the target plane will be $S_i' = b_iS_i$. Therefore, the illuminance of each changed grid at the target plane will be:

$$E_i' = \frac{\Phi_i'}{S_i'} \tag{3.25}$$

In order to realize uniform illumination, the illuminance value at each grid will approach E_{av} and the ideal situation is that all illuminance values are equal, as shown in Eq. 3.26:

$$E_i' = E_{av} \tag{3.26}$$

Combination Eqs. 3.24 and 3.25, we can obtain Eq. 3.27:

$$k_i = \frac{E_i}{E_i'} = \frac{\Phi_i/S_i}{\Phi_i'/S_i'} = \frac{\Phi_i/S_i}{a_i\Phi_i/b_iS_i} = \frac{b_i}{a_i} \tag{3.27}$$

We know that k_i can be obtained from simulation results, and the problem can be divided into three different situations:

1. When $a_i = 1$ and $b_i = k_i$, the algorithm shows that the light source still uses the point source uniform division method, and we only optimize the target plane grids division by k_i.
2. When $b_i = 1$ and $a_i = 1/k_i$, the algorithm shows that the target plane still uses the uniform division method, and we only optimize the light source grids division by $1/k_i$.
3. When $a_i \neq 1$, $b_i \neq 1$, and $k_i = b_i/a_i$, the algorithm shows that both the target plane and light source are optimized by changing ratios at the same time. In this chapter, as an example to this situation, we set $a_i = (1/k_i)^{0.5}$ and $b_i = k_i^{0.5}$.

3.3.1.4 Step 4. Rebuild the Energy Relationship between the Light Source and Target Plane

In the point source design method, the light source and target plane are both divided into N equal parts. Each grid of the target plane has the same area S_0, and each grid of the light source has an equal luminous flux Φ_0. In this method, after optimization of the grids' division, each grid's luminous flux of light source and each grid area of target plane are changed into Φ_i and S_i, respectively. It is important to notice that during the first time of feedback optimization, each grid of the light source will be changed into $\Phi_i = a_i\Phi_0$, and each grid of the target plane will be changed into $S_i = b_iS_0$. However, during the second-time optimization, the changing ratios of a_i and b_i possess different values from the first time. Meanwhile, each grid of the light source and target plane are optimized based on the first-time optimized result instead of the original point source grids' equal division method. To clearly illustrate, we introduce subscript j to represent the times of optimization. Light source grids, target plane grids, and corresponding feedback changing ratios are expressed as follows: Φ_{ji}, S_{ji}, a_{ji}, and b_{ji}. For example, we use a_{23} to represent the changing ratios of the second-time optimization applied to the third piece of the light source.

According to the light source energy division method, we can obtain emitting angles θ_{ji} of an optimized light source grid:

$$2\pi \int_0^{\theta_{ji}} I(\theta)\sin\theta d\theta = \sum_{i=0}^{N} \Phi_{ji} = \sum_{i=0}^{N} a_{ji}\Phi_{j-1i} \tag{3.28}$$

Among them, Φ_{0i} means that before optimization, the luminous flux of each light source grid (i.e., Φ_0); θ_{j0} represents the first emitting light ray direction that is vertically pointed, and the value is *0*.

In the same way, we can calculate each grid's area after optimization and the corresponding radius of each grid.

$$\pi r_{ji}^{\ 2} = \sum_{i=0}^{N} S_{ji} = \sum_{i=0}^{N} b_{ji}S_{j-1i} \tag{3.29}$$

Among them, S_{0i} denotes before optimization the area of each grid, which is S_0; r_{j0} means the first emitting ray arrives at the target plane and the value is *0*. By combining Eqs. 3.28 and 3.29, we can calculate a light ray emitting angle θ_{ji} after the j_{th} optimization and a corresponding emitting point at the target plane with radius r_{ji}. In this way, we can build an energy mapping relationship between the latest divided light source and the target plane.

3.3.1.5 Step 5. Construction of a Freeform Lens for an Extended Source

According to the new energy mapping relationship between the light source and target plane, we employ the seed curve method to build the circular-symmetrical freeform lens.

3.3.1.6 Step 6. Ray-Tracing Simulation and Feedback Reversing Optimization

We analyze the light pattern illuminance of an extended source with the new designed freeform lens. If the evaluation results satisfy our preset standards, we accept the design; otherwise, we circulate more times to optimize the design until the results satisfy the illuminance requirements.

3.4 Noncircularly Symmetrical Freeform Lens – Point Source

In Sections 3.2 and 3.3, we discussed a series of algorithms to achieve circular-symmetrical freeform lenses for point sources and extended sources. In LED lighting, except the circular-symmetrical light pattern, noncircular-symmetrical light patterns also have many applications. Currently, LED street and road lighting and LED tunnel lighting are being promoted throughout China for energy saving. If a circular light pattern is adopted in road lighting, this will decrease both the illuminance and luminance uniformities of road lighting, decrease visual comfort, and cause poor lighting performance, as shown in Figure 3.19. At the same time, a large part of light will illuminate the area behind the LED road lamp, resulting in unnecessary energy waste. Therefore, a rectangular light pattern is required for LED road and tunnel lighting. Moreover, in addition to LED road and tunnel lighting, there still are many applications requiring a noncircular- symmetrical light pattern, such as LED pico-projectors, LED automotive headlamps, and so on. In this section, we will introduce freeform optics algorithms for noncircular-symmetrical light patterns in LED lighting, extending the application range of LEDs.

In this section and Section 3.5, we will illustrate noncircular-symmetrical freeform lens design methods for point sources and extended sources, respectively. A rectangular light pattern is the most widely used pattern in noncircular-symmetrical light patterns. Thus, in this section, we will mainly illustrate freeform lens algorithms for rectangular light patterns, and light patterns with other shapes (e.g., hexagon) may be designed conveniently based on the algorithms described here. The freeform lens design methods based on the point source, according to the difference in algorithms, can be sorted as

Figure 3.19 Poor lighting performance of an LED lamp with a circular-symmetrical light pattern.

discontinuous and continuous freeform lens design methods. We will discuss the discontinuous freeform lens design process first.

3.4.1 Discontinuous Freeform Lens Algorithm

In this section, a modified discontinuous freeform lens design method in 3D is presented, with advantages of a flexible energy mapping relationship, accurate light irradiation control, and easy manufacture. During the prescribed illumination design, in most cases, optical systems are designed based on the LIDC of light sources and the expected illuminance distributions on the target plane. As shown in Figure 3.20, the freeform lens refracts the incident ray **I**, represented by spherical coordinates ($\gamma,\ \theta,\ \rho$), into output ray **O**. Then **O** will irradiate at corresponding point $Q(x,y,z)$ on the target plane. According to the energy mapping relationship, the edge ray principle, and Snell's law, the coordinates and normal vector of point $P(x,y,z)$ on the surface of the freeform lens is able to be calculated. For the design and manufacturing convenience, the inner surface of the lens is designed as an inner concave spherical surface, which will little change the transmission directions of the incident lights. We will focus on the construction of the outside surface of the lens in this study. This chapter will provide an effective way to deal with rectangularly prescribed illumination problems.

Since the light patterns of most LEDs are rotationally symmetric, we will only discuss this type of LED in this study. The design method includes three main parts: establishing the light energy mapping relationship between the light source and the target, constructing the lens, and validating the lens design by ray-tracing simulation. The flowchart of an algorithm of a discontinuous freeform lens for a rectangular light pattern is shown in Figure 3.21.

3.4.1.1 Step 1. Establishment of a Light Energy Mapping Relationship

In this design method, both light energy distribution of the light source and illumination target plane are divided into several grids with equal luminous flux Φ_0 and area S_0, respectively. Then a mapping relationship is established between a couple of light source

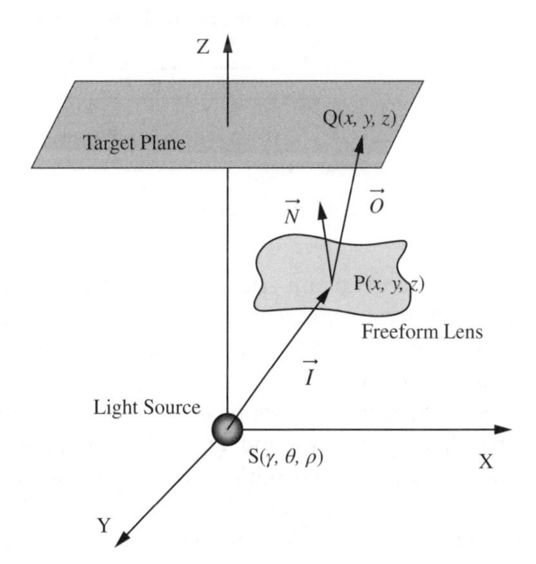

Figure 3.20 Schematic of the rectangularly prescribed illumination problem.

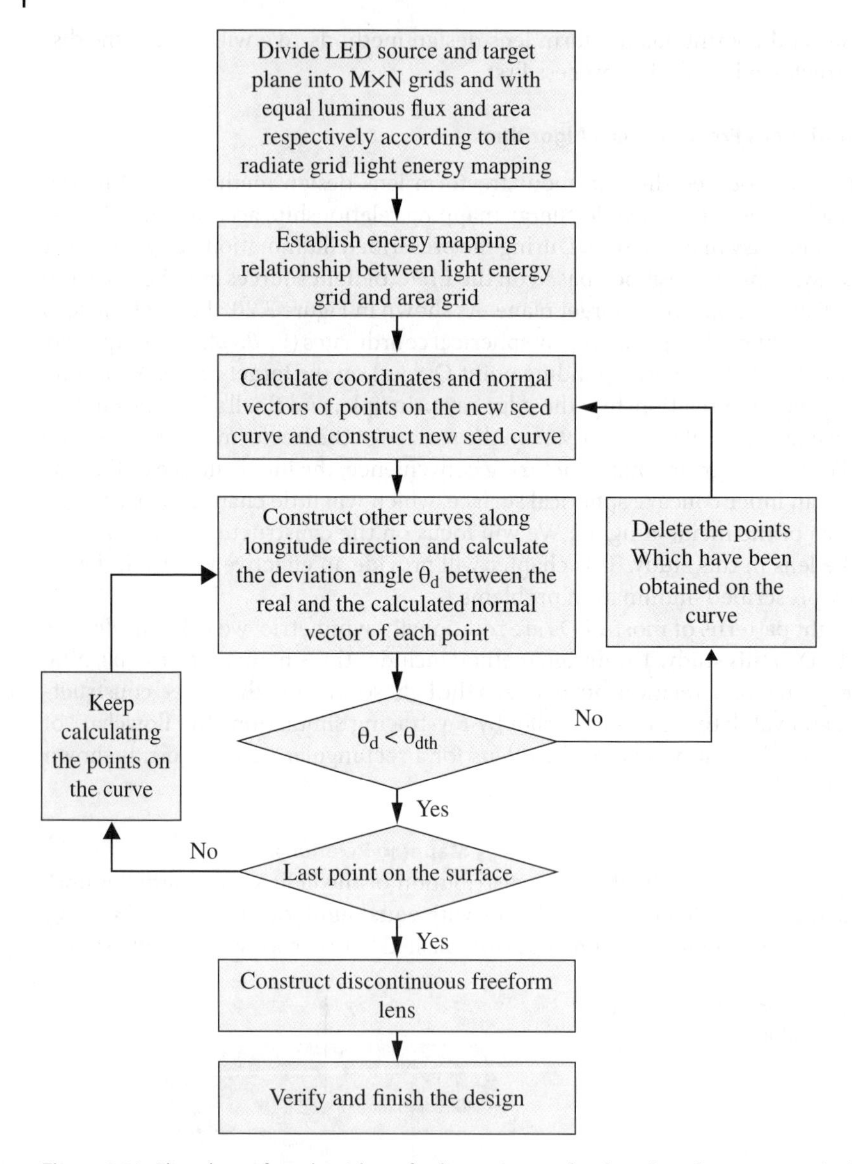

Figure 3.21 Flowchart of an algorithm of a discontinuous freeform lens for a rectangular light pattern.

grid and target plane grid. Therefore, the average illuminance $E_0 = \eta\Phi_0/S_0$, where η is the light output efficiency of the freeform lens, of each target plane grid is the same, and a uniform light pattern can be obtained when the grid is quite small compared with the whole target plane.

Since both the light source and target plane are of axial symmetry, only one-quarter of the whole light source and target plane are to be considered in this discussion. First of all, the light energy distribution of the light source is divided into $M \times N$ grids with equal luminous flux. As shown in Figure 3.22, the light source's light energy distribution Ω could be regarded as composed of a number of unit conical object Ω_0, which represents

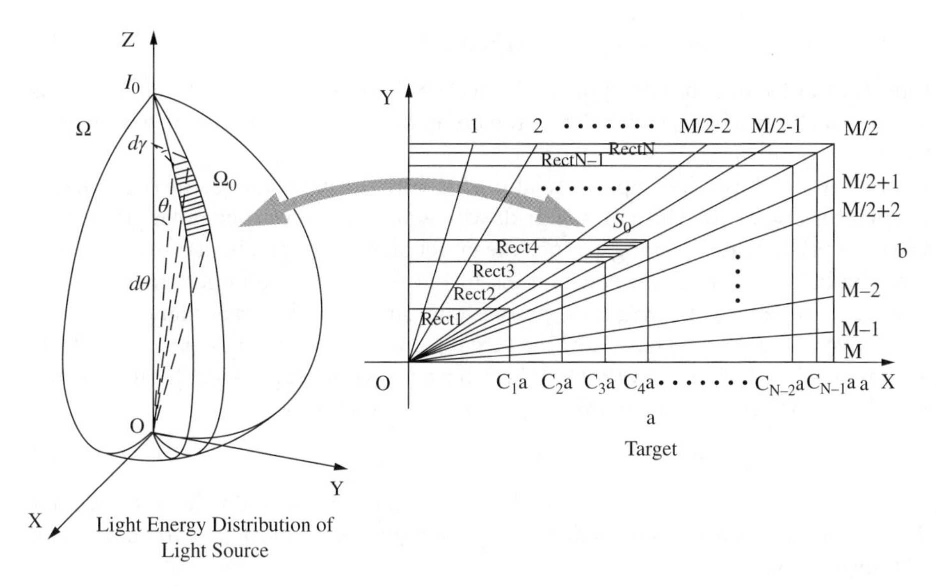

Figure 3.22 Schematic of light energy mapping between the light source and rectangular target.

the luminous flux within the angular range with field angle $d\gamma$ in the latitudinal direction and $d\theta$ in the longitudinal direction. The luminous flux of Φ_0 can be expressed as follows:

$$\Phi_0 = \int I(\theta)d\omega = \int_{\gamma_1}^{\gamma_2} d\gamma \int_{\theta_1}^{\theta_2} I(\theta)\sin\theta d\theta \tag{3.30}$$

where $I(\theta)$ is the light intensity distribution of the light source and $d\omega$ is the solid angle of one unit conical object. Least-squares curve fitting in the form of a polynomial is one of the most often used curve-fitting methods in numerical analysis. Thus, most $I(\theta)$, expressed in the form of LIDC, can be fitted by a polynomial of θ as follows:

$$I(\theta) = I_0 \sum_{i=0}^{m} a_i \theta^i \tag{3.31}$$

where I_0 is the unit light intensity; a is the polynomial coefficient; and m is the order of this polynomial. The larger m is, the more accurate the curve fitting will be, but the corresponding computational amount will increase too. For most LIDCs, m should not be less than 9. The total luminous flux of this one-quarter light source can be expressed as follows:

$$\Phi_{total} = \int_0^{\pi/2} d\gamma \int_0^{\pi/2} I(\theta)\sin\theta d\theta = \frac{\pi}{2}\int_0^{\pi/2} I(\theta)\sin\theta d\theta \tag{3.32}$$

We divide the Ω into M fan-shaped plates along the latitudinal direction with an equal angle of $\Delta\gamma = \pi/2M$ and equal luminous flux of Φ_{total}/M (as shown in Figure 3.22). Then each fan-shaped plate is divided into N parts equally along the longitudinal direction. The field angle $\Delta\theta_{j+1}$, which is different for every j, of each separate conical object along the longitudinal direction can be obtained by iterative calculation as follows:

$$\frac{\pi}{2M}\int_{\theta_j}^{\theta_{j+1}} I(\theta)\sin\theta d\theta = \frac{\Phi_{total}}{MN} \quad (j = 0, 1, 2, \ldots\ldots, N-1,\ \theta_0 = 0) \tag{3.33}$$

$$\Delta\theta_{j+1} = \theta_{j+1} - \theta_j \quad (j = 0, 1, 2, \ldots\ldots, N-1) \tag{3.34}$$

Thus, the light source has been divided into M × N subsources with equal luminous flux. The directions of rays, which define the boundary of one subsource, also have been defined.

Secondly, to establish the mapping relationship with the light source, the one-quarter target plane is also divided into $M \times N$ grids with equal area. A warped polar grid topology is appropriate to fit a rectangle. Since the target plane is perpendicular to the central axis of the light source, z is a constant Z_0 for all points on the target. As shown in Figure 3.22, the length and width of the one-quarter rectangle target plane are a and b, respectively, and the area is $S = ab$. The target plane is divided into N parts equally by subrectangles *Rect1, Rect2, …, RectN*, which have the same length–width ratios as the whole plane. The relationship of area S_{Rect_k} of each subrectangle is:

$$S_{\mathrm{Rect}_k} = kS_{\mathrm{Rect}_1} \quad (k = 1, 2, \ldots\ldots, N) \tag{3.35}$$

where $S_{\mathrm{Rect}_N} = S$. Therefore, the rectangular target plane has been divided into N parts with the same area of S/N. The coefficient C_q of length of $C_q\, a$ of each subrectangle can be obtained as follows:

$$C_q = \left(\frac{q}{N}\right)^{1/2} \quad (q = 1, 2, \ldots\ldots, N) \tag{3.36}$$

Then the plane is divided into M parts equally along the central radiation directions by $M-1$ radial lines. The endpoints of these lines equally divide the edges of the target plane, which ensures that the plane has been divided into M parts with equal area S/M. The edges of subrectangles and radial lines construct the warped polar grids. Therefore, the plane has been divided into $M \times N$ grids S_0 with equal area of S/MN.

According to the edge ray principle, rays from the edge of the source should strike the edge of the target. This principle is true in 2D, and in 3D the skew invariant will lead to loss, but this could be partly recovered by increasing the number of grids. Therefore, if we desire to map the light energy in Ω_0 into the target grid S_0, we should ensure that four rays, which construct the Ω_0 as boundary, irradiate at the four corresponding end points of the target grid S_0 after being refracted by the freeform lens. Since the light source and target plane are both divided into $M \times N$ grids, each ray from the light source could has and only has one corresponding irradiate point on the target plane. Thus, the light energy mapping relationship between the light source and target plane has been established. When dealing with nonrectangular illumination problems, it only needs to change the sizes and shapes of the grids on the target plane, which makes it more easy and flexible to reestablish the light energy mapping relationships.

3.4.1.2 Step 2. Construction of the Lens

In this section, we will find out which lens can realize the mapping between the light source and the target plane. There are four main steps to construct the outside surface of the freeform lens:

Substep 1. Construction of the seed curve The seed curve is the first curve to generate other lens curves, and we will construct curves along the longitudinal direction. As shown in Figure 3.23, we fix a point P_0 as the vertex of the seed curve. The first subscript of a point designates the sequence number (1, 2, …, M+1) of the longitude curve, and the second

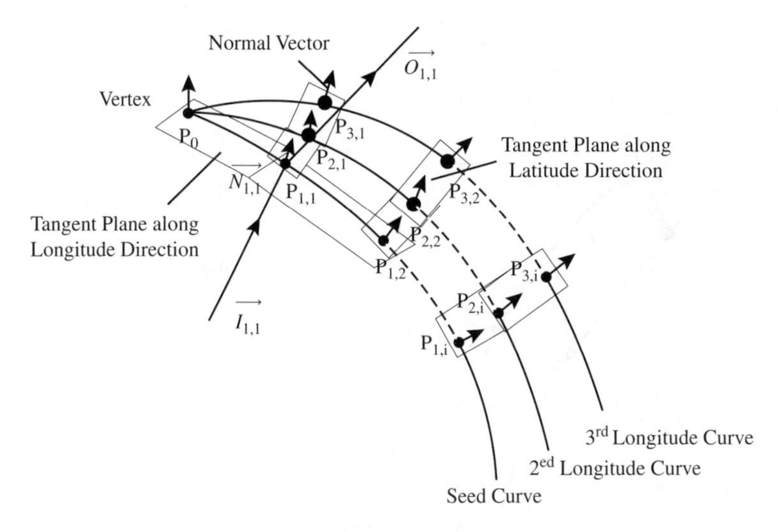

Figure 3.23 Schematic of points generation on the outside surface of the lens.

designates the sequence number (1, 2, ..., N) of the point on a longitude curve except the vertex. The second point $P_{1,1}$ is calculated by the intersection of incident ray $\boldsymbol{I}_{1,1}$ and the tangent plane of the previous point. The direction of the refracted ray $\boldsymbol{O}_{1,1}$ can be also obtained as $\boldsymbol{Q}_1 - \boldsymbol{P}_{1,1}$, where Q_1 is the corresponding point of incident ray $\boldsymbol{I}_{1,1}$ on the target plane. Then we can obtain the normal vector $\mathbf{N}_{1,1}$ of the second point according to Snell's law, expressed as follows:

$$[1 + n^2 - 2n(\mathbf{O} \cdot \mathbf{I})]^{1/2}\mathbf{N} = \mathbf{O} - n\mathbf{I} \tag{3.37}$$

where $\boldsymbol{I}$ and $\boldsymbol{O}$ are the unit vectors of incident and refracted rays; $\boldsymbol{N}$ is the unit normal vector on the refracted point; and n is the refraction index of the lens. Based on this algorithm, we can obtain all other points and their normal vectors on the seed curve. This method can guarantee that the tangent vectors of the seed curve are perpendicular to its calculated normal vectors.

Substep 2. Generation of other longitude curves Since one-quarter of the light source is divided into M parts along the latitudinal direction, there are M corresponding longitude curves to be calculated except the first seed curve. First of all, we calculate the second curve whose vertex coincides with that of the seed curve. As shown in Figure 3.23, different from the seed curve algorithm, point $P_{2,i}$ on the second longitude curve is calculated by the intersection of the incident ray and the tangent plane of point $P_{1,i}$ on the previous curve. Then the following longitude curves, such as the third curve, fourth curve, and so on, are easy to obtain based on this algorithm.

Substep 3. Error control The unit normal vector ($\boldsymbol{N}$) of each point is calculated based on the corresponding incident ray and exit ray at that point. However, from Fig. 3.23 and Fig. 3.24 we can find that, according to the surface construction algorithm, the real unit normal vector ($\boldsymbol{N}'$) is orthogonal to both $\overrightarrow{t1}$ and $\overrightarrow{t2}$, while the calculated unit normal vector ($\boldsymbol{N}$) is only orthogonal to $\overrightarrow{t2}$ and has no relationship with $\overrightarrow{t1}$ except the seed curve. Thus, we cannot guarantee the real unit normal vector ($\boldsymbol{N}'$) of the point of the

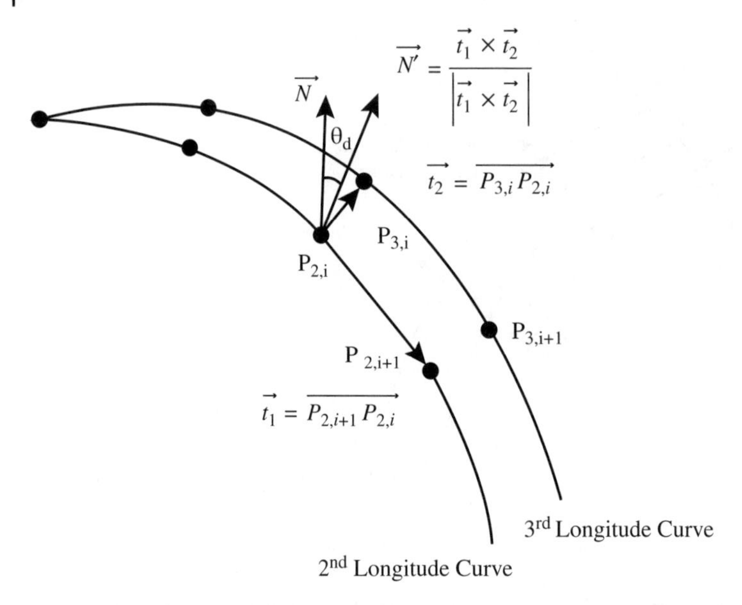

Figure 3.24 Deviation between the real unit normal vector (N′) and the calculated unit normal vector (N) of one point on the surface.

lens surface is still the same as the calculated one (N). Consequently, the direction of the exit ray will deviate from the expected one, which will result in poor illumination performance. As shown in Figure 3.24, starting from the second longitude curve, every point (e.g., $P_{2,i}$) on the curve only has one adjacent point (e.g., $P_{3,i}$) on its tangent plane in the latitudinal direction. Therefore, deviations between the real normal vectors and calculated ones generate on these longitude curves. The deviation can be estimated by the deviation angle θ_d:

$$\theta_d = \cos^{-1}\left(\frac{\mathbf{N} \bullet \mathbf{N'}}{|\mathbf{N}||\mathbf{N'}|}\right) \tag{3.38}$$

Since θ_d becomes larger with the increase of the sequence of longitude curves, if θ_d does not confine within a certain range, the deviations between the real exit rays and the expected rays will become larger, resulting in serious deterioration of the lighting performance of the rectangular light pattern eventually. For example, we design a freeform lens to realize a 30 ×10 m uniform illumination light pattern on the target plane that is 8 m away. According to method described here, we calculate longitude curves and construct the corresponding freeform lens. The freeform lens and its lighting performance are shown in Figure 3.25. We can find that the light pattern is a rectangle, but the illuminance distribution is very nonuniform. There is a very bright pattern along the diagonal, and the whole pattern looks like an "X", which is quite different from our initial expectation. The reason for such serious deterioration is the deviation angle θ_d. The largest deviation angle reaches above 30°, so the light exit direction is totally deviated from the expected one. Therefore, a threshold θ_{dth}, for example 6°, is needed to confine the deviation. If the maximum deviation of one longitude curve is larger than θ_{dth}, we go back to Substep 1 and calculate another seed curve to replace this longitude curve.

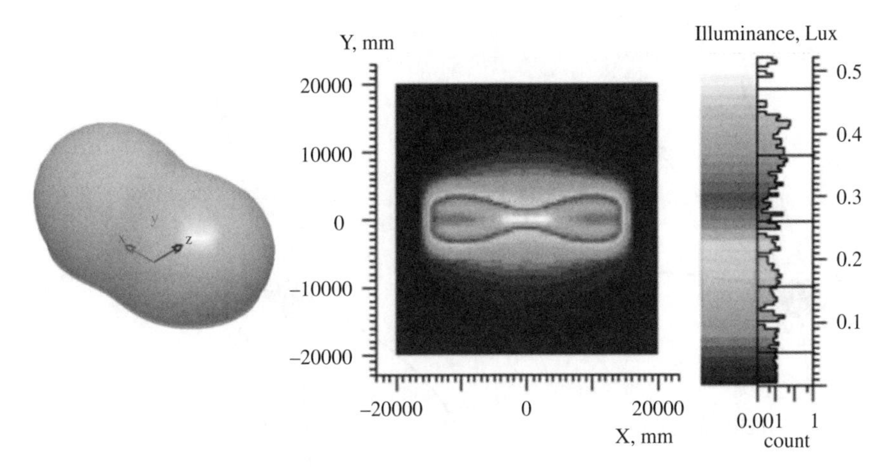

Figure 3.25 Freeform lens without error control and its lighting performance.

Substep 4. Construction of the surface The lofting method is utilized to construct a smooth surface between these longitude curves. Since the newly generated seed curves are discontinuous with the longitude curves before them, the surface of the lens becomes discontinuous.

3.4.1.3 Step 3. Validation of Lens Design

Since it is costly to manufacture a real discontinuous freeform lens, a numerical simulation based on the Monte Carlo ray-tracing method is an efficient way to validate the lens design. According to the simulation results, slight modification by trial and error is needed to improve the illumination performance. For instance, the construction of a transition surface between two subsurfaces is not contained in this method, and the shape of the transition surface should be modified according to the simulation results and manufacturing process.

3.4.2 Continuous Freeform Lens Algorithm

Since it is inevitable to induce manufacturing error, in this algorithm we propose a new design concept that inversely utilizes manufacturing errors to realize the required lighting performance. Firstly, we introduce three concepts of light exit directions: expected light exit direction, designed light exit direction, and real light exit direction. The expected light exit direction is the refracted light direction through the lens surface according to the energy mapping relationship. The designed light exit direction is the direction that is used in the calculation of the point coordinates and unit normal vector in the process of constructing lenses. The real light exit direction is the real direction of incident light refracted by the constructed lens.

As shown in Figure 3.26a, in the design algorithm, the designed light exit direction is consistent with the expected one. But due to the existence of manufacturing error, the real light exit direction deviates from the designed one (i.e., the expected one) to some extent. As a result, the real light pattern on the target plane deviates from the expected pattern greatly, and the illumination uniformity is deteriorated. The novel idea based on the compensation method is shown in Figure 3.26b. The expected light exit direction

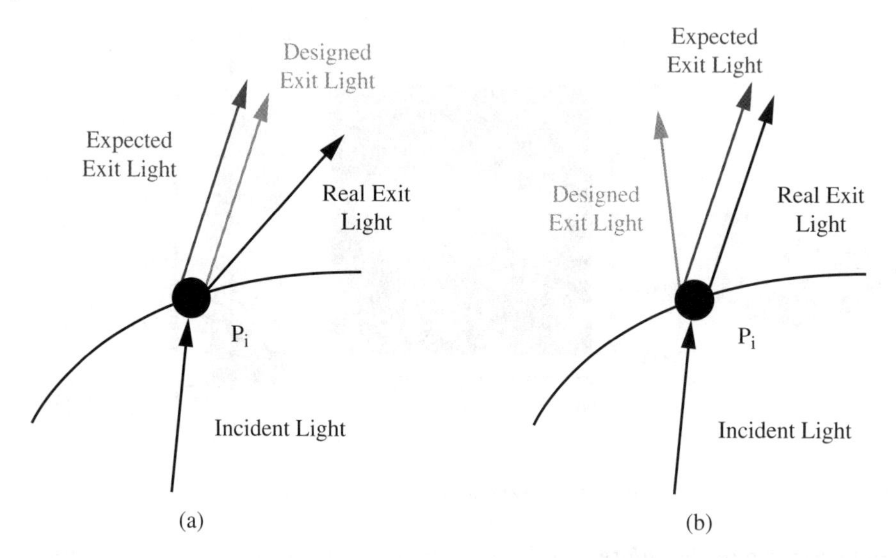

Figure 3.26 Schematic of exit lights of a (a) discontinuous freeform lens and (b) continuous freeform lens.

is quantified based on the energy mapping relationship between the light source and the target plane; therefore, it can't be changed. In the novel algorithm, we change the designed light exit direction and made some deviation with the expected light exit direction. It is expected that the designed deviation could compensate for the deviation between the real light exit direction and the designed one, so that the real light exit direction is consistent with the expected one. In this way, the real lighting performance of the lens could be consistent with the expected one, resulting in good lighting performance.

In this algorithm, we change the designed light exit direction by changing the division of grids on the target plane. Note that the light energy mapping relationship is quantified to realize uniform illumination, so the purpose of changing the gridding on the target plane here is not to change the expected light energy mapping relationship, but to change the real light exit direction through the lens construction so as to make the real light exit direction consistent with the expected one in the initial establishment of the light energy mapping relationship. The design flowchart is shown in Figure 3.27. Firstly, we divide the spatial energy distribution of light source and the area of target plane equally, and establish the mapping relationship. Secondly, we construct the freeform lens according to the incident direction and the designed light exit direction of each ray. Then, we build the optical model and simulate the lighting performance of the constructed lens on the receiver plane by the ray-tracing method. The lighting performance would be evaluated to judge whether the result meets the requirement. If this is the case, the design of the freeform lens is done; otherwise, we have to modify the division of the grids, adjust the grid size, and reconstruct the lens with the application of these modifications. In this case, the designed light exit direction deviates from the expected one. In the second-round simulation and evaluation, if the result still cannot meet the requirement, multiple optimizations will be requested until it is met.

In Step 2, proper optimization of the division of the grids on the target plane helps decrease the times of optimization and accelerate the design speed. In the design

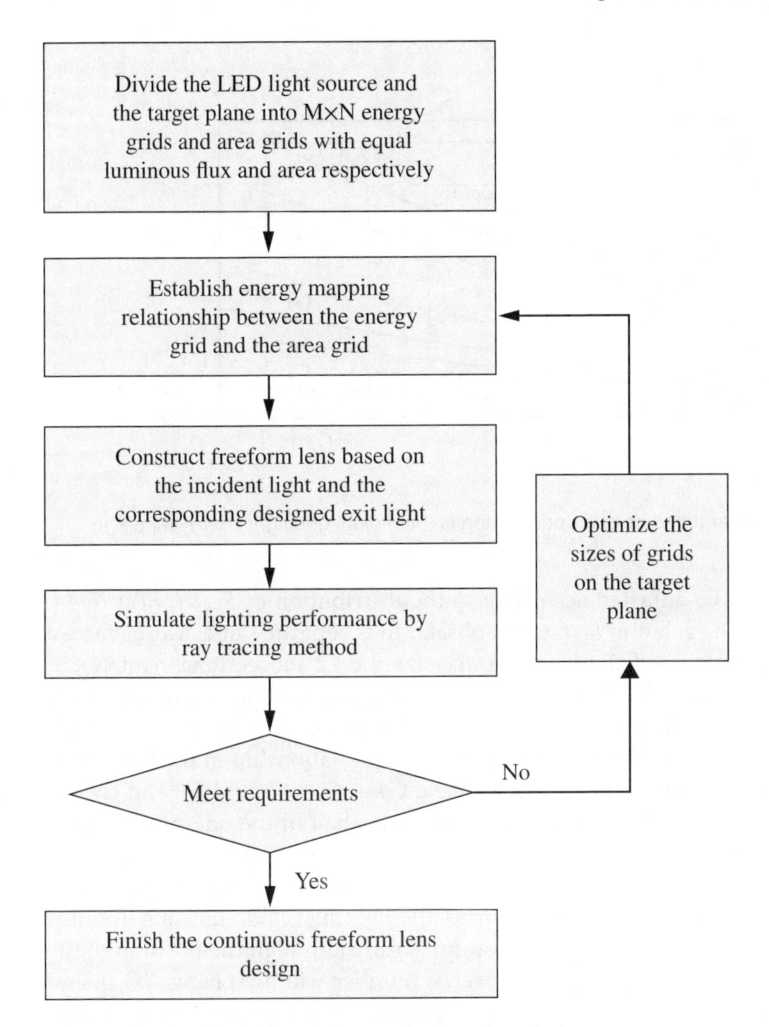

Figure 3.27 Flowchart of a continuous freeform lens design.

process, we propose the compensation idea, where the grid size of some certain region is increased when the simulated illuminance is higher than the expected, and vice versa. Using this method, we could conveniently and quickly optimize the grid division and realize the uniform illumination in the end.

3.4.2.1 Radiate Grid Light Energy Mapping

According to different shapes of light energy grids and area grids, there usually are two kinds of mapping relationships, radiate grid and rectangular grid light energy mapping. In this section, we take the radiate-type division of the target plane to design the continuous freeform lens.

As shown in Figure 3.28, based on the grid division in Figure 3.22, we divide the one-quarter rectangular target plane unequally. We introduce two adjustable rectangular divisions *Rect I* and *Rect II*, and three adjustable radiation lines *LI*, *LII*, and *LIII*. These divide the whole one-quarter rectangular region into 12 subsections, and the area

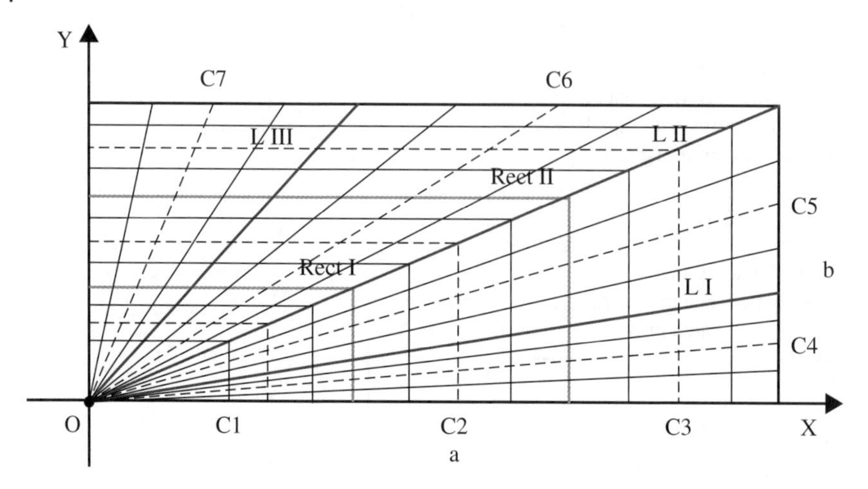

Figure 3.28 Schematic of target plane grid optimization in the radiate grid light energy mapping situation.

of each subsection can be adjusted according to the distribution of *Rect I*, *Rect II*, *LI*, *LII*, and *LIII*. The area of each unit S_i in each subsection is kept the same. Moreover, we also introduce seven ratio coefficients to characterize the 12 subsections, namely *C1*, *C2*, …, *C7*, where *C1*, *C2*, and *C3* denote the area ratio relationship of the adjustable rectangles and the whole rectangle (one-quarter target plane), and we have $C1 + C2 + C3 = 1$; *C4* and *C5* are used to adjust the area division relationship in the lower-half of the one-quarter rectangular region, and we have $C4 + C5 = 1$; and *C6* and *C7* are used to adjust the area division relationship in the upper-half of the one-quarter rectangular region, and we also have $C6 + C7 = 1$. In the continuous freeform lens optimization, we can adjust these seven parameters to decrease the pattern deterioration caused by the surface construction error to make the lighting region become uniform gradually. It is notable that in the parameter optimization, although the position of the grid on the target plane will be changed, the serial number will not change ($Q_{i,j}$), and the incident rays have the same serial number, like the incident ray $I_{i,j}$ still corresponds to $Q_{i,j}$.

3.4.2.2 Rectangular Grid Light Energy Mapping

Here, we will introduce another algorithm for the realization of a continuous freeform lens. The main difference compared to the above-mentioned algorithm is that in this algorithm, we adopt rectangular grids in both the LED light source and target plane. It is expected that this new division method and the energy mapping could weaken the effects of the construction error, and realize relatively good lighting performance in a continuous freeform lens.

Since both the LED source and target plane are of axial symmetry, only one-quarter of them are to be considered in this discussion. First of all, as shown in Figure 3.29, the one-quarter light energy distribution of the source is divided into $M \times N$ grids with equal luminous flux. The LED source's light energy distribution Ω is specified by coordinates (u, v) rather than coordinates (γ, θ, ρ), where u is the angle between the ray and the x-axis, and v is the angle between the x-axis and the plane containing the ray and x-axis. The total luminous flux Φ_{total} of the light energy spatial distribution Ω is:

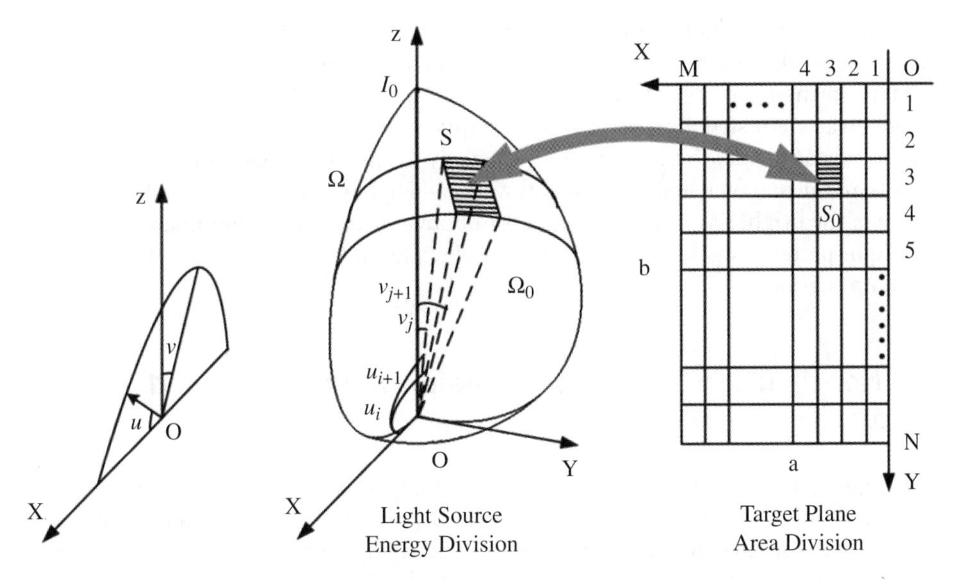

Figure 3.29 Schematic of an energy mapping relationship based on rectangular grids.

$$\Phi_{total} = \int_0^{\pi/2} \int_0^{\pi/2} I(u, v) \sin u du dv \tag{3.39}$$

where $I(u, v)$ is the light intensity distribution of the LED source in the coordinate system of (u, v). The luminous flux $\Phi_0(u, v)$ of the unit object Ω_0 could be expressed as:

$$\Phi_0(u, v) = \int_{v_j}^{v_{j+1}} \int_{u_i}^{u_{i+1}} I(u, v) \sin u du dv \tag{3.40}$$

As shown in Figure 3.29, we divide Ω into M parts along the u direction and N parts along the v direction equally, and then each division point on the light source could be obtained through solving Eqs. 3.41 and 3.42 as follows:

$$\int_0^{\pi/2} \int_0^{u_i} I(u, v) \sin u du dv = \frac{i\Phi_{total}}{M} \quad (i = 0,1, \ldots M) \tag{3.41}$$

$$\int_0^{v_j} \int_0^{\pi/2} I(u, v) \sin u du dv = \frac{j\Phi_{total}}{N} \quad (j = 0,1, \ldots N) \tag{3.42}$$

For a rectangular target plane, we evenly divide the target plane into $M \times N$ rectangular grids. As shown in Figure 3.29, in the (x, y, z) coordinate system, the target plane is divided evenly as N grids along the length direction and M grids along the width direction. The total one-quarter target plane is divided evenly as an $M \times N$ rectangular area grid, and the area of each grid is $ab/M/N$. When establishing the light energy mapping relationship, we make a corresponding relationship between the light exit ray at $u = \pi/2$ and x-axis or y-axis in the target plane, regarded as the initial freeform curve. We can build the one-to-one corresponding relationship of the LED light exit direction and the grids on the target plane, resulting in light energy mapping. Meanwhile, we don't introduce the error control in the lens construction, so as to maintain the continuous and smooth lens surface.

We adopt the rectangular grid in the division of the target plane, and each light energy unit corresponds to a rectangular region rounded by two lines parallel to the x-axis or y-axis on the target plane. Therefore, the light pattern deterioration caused by the curved surface construction error mainly happens on the directions that are parallel to the x-axis or y-axis, so the bright strips usually form at the edges of the light pattern rather than the diagonal lines (an obvious X-shaped light pattern). Meanwhile, the overlapping of the bright strips and the edges is not obvious due to the illuminance decrease at the light pattern edges. As a result, the whole light pattern can realize relatively uniform illumination.

3.5 Noncircularly Symmetrical Freeform Lens – Extended Source

An extended source is a key issue challenging compact LED packaging freeform lens design. The size of a high- power LED chip is usually around 1 × 1 mm, while the distance between the chip and outside surface of the packaging lens is about 2.5 mm or even smaller. Since the distance between the source and lens is less than five times the source diameter, the LED chip could not be regarded as a point source during packaging lens design according to the far-field conditions of LED.[20] Thus, the compact freeform lens should be designed based on an extended source. However, if we design a noncircularly freeform lens according to the assumption of a point source while using the lens for an extended source, lighting performance will deteriorate significantly. Figure 3.30a shows a freeform lens for LED tunnel lighting and its uniform rectangular radiation pattern when using a point source. However, as shown in Figure 3.30b, the shape and uniformity of the radiation pattern deteriorate extremely when the source extends to an extended source with an emitting area of 4 × 4 mm. Therefore, the design method of the freeform lens should be modified to improve lighting performance when using extended sources.

The design method includes three parts: establishing a light energy mapping relationship between the source and target, constructing the lens, and validating the lens design

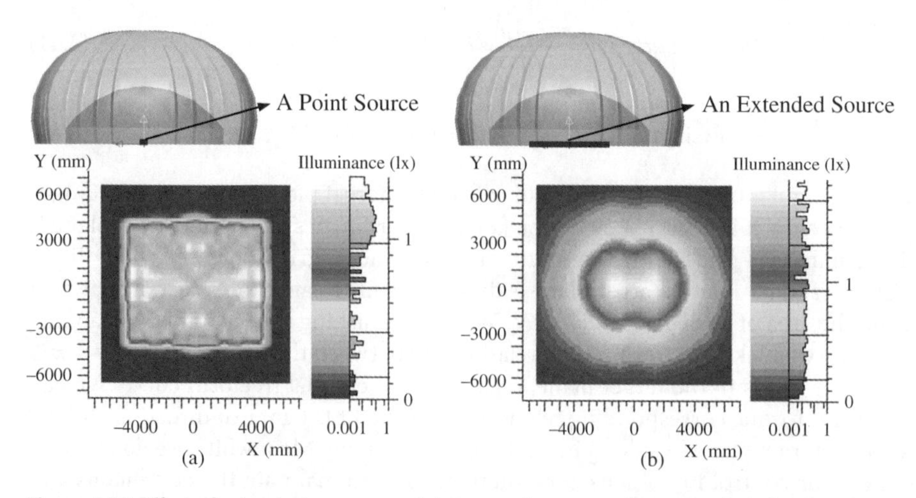

Figure 3.30 Effect of extended sources on lighting performance of non-circularly freeform lens.

by numerical simulation. A rectangular target plane is adopted as an example in the following design. Normally, refractive indexes of encapsulants and lenses of LED packaging are usually different from each other. For example, the refractive index is 1.586 for polycarbonate and is from 1.4 to 1.6 for silicone. It is hard to find an interface that will not deflect rays emitted from an extended source when rays transmit through the interface of two materials with different refractive indexes. In this method, optimization of overall lighting performance is to be achieved by adjusting the optimization coefficients of the target grids and reconstructing the outside surface of the lens. Thus, there is no strict restriction of the shape of the inner surface of the lens, and it is to be designed as an inner concave spherical surface in this method. We will focus on the construction of the outside surface of the lens in this study.

3.5.1.1 Step 1. Establishment of the Light Energy Mapping Relationship

Establishment of the light energy mapping relationship between the LED light source and target plane for the extended source is quite similar with that of rectangular grid light energy mapping as described in Section 3.4.2.2. However, there is one important difference. To overcome the problem of lighting performance deterioration caused by extended sources, rectangular grids with unequal areas are adopted in this method. The width (W_{grid}) and length (L_{grid}) of each grid on the target plane can be expressed as Eq. 3.43 and Eq. 3.44:

$$W_{grid,i} = \frac{C_{Wi}a}{M} \quad (i = 0,1,\ldots M) \tag{3.43}$$

$$L_{grid,j} = \frac{C_{Lj}b}{N} \quad (j = 0,1,\ldots N) \tag{3.44}$$

where C_{Wi} and C_{Lj} are optimization coefficients of grids that can optimize the light energy distribution on the target. Both C_{Wi} and C_{Lj} are equal to 1 when dealing with the point source problem. During design for an extended source, we adjust these two coefficients to make the trend of illuminance distribution on the target plane to be inverse of the trend of lighting performance deterioration caused by the extended source. Then, the optimized illuminance distribution will compensate for the deterioration caused by the lights irradiating from the edge area of the LED chip, and much better lighting performance would be obtained.

According to the edge ray principle, rays from the edge of the source should strike the edge of the target. This principle is true in 2D, and in 3D the skew invariant will lead to loss, but this could be partly recovered by increasing the number of grids. Therefore, if we desire to map the light energy in $d\Omega$ into the target grid dS, we should ensure that four rays, which construct the $d\Omega$ as boundary, irradiate at the four corresponding end points of the target grid dS after being refracted by the lens. Thus, the light energy mapping relationship between the source and the target plane has been established.

3.5.1.2 Step 2. Construction of a Freeform Lens

In this section, we will find out the lens that can realize the mapping between the source and the target plane. There are three main substeps:

Substep 1. Construction of the seed curve. The seed curve is the first curve to generate other lens curves. As shown in Figure 3.31, we fix a point P_0 as the vertex of the seed curve.

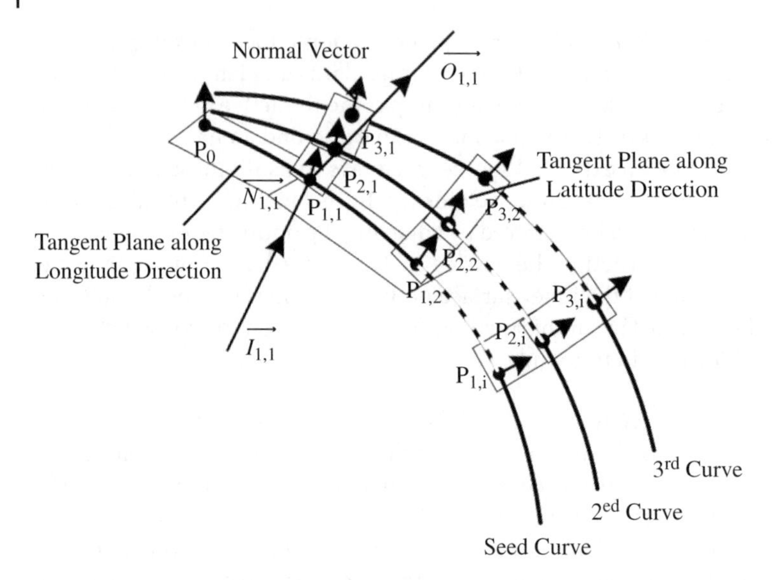

Figure 3.31 Schematic of generation of points on the outside surface of freeform lens.

The second point $P_{1,1}$ on the seed curve is calculated by the intersection of incident ray $\mathbf{I}_{1,1}$ and the tangent plane of the previous point. Then we can obtain the normal vector $\mathbf{N}_{1,1}$ of the second point according to Snell's law. Based on this algorithm, we can obtain all other points and their normal vectors on the seed curve.

Substep 2. Generation of other curves. First of all, we calculate the second curve. As shown in Figure 3.31, different from the seed curve algorithm, point i on the second longitude curve is calculated by the intersection of the incident ray and the tangent plane of point i on the previous curve. Then the following curves, such as the 3rd curve, 4th curve, and so on, are easy to obtain based on this algorithm.

Substep 3. Construction of the surface. The lofting method is utilized to construct a smooth surface of the lens between these curves.

3.5.1.3 Step 3. Validation of Lens Design

Since it is costly to manufacture a real freeform lens, numerical simulation based on the most widely used Monte Carlo ray-tracing method is an efficient way to validate the lens design. The Monte Carlo ray-tracing method traces the desired number of rays from randomly selected points on the surface of, or within the volume of, the sources and into randomly selected angles in space. The selection of starting points and ray direction is based on probabilistic functions that describe emissive characteristics of light sources. Each ray starts with a specific amount of power, determined by each source's characteristics; this power is then modified by the various surfaces hit by the ray in its path through the system. These rays are then collected on specified receiver surfaces for statistical analysis and graphical display. An extended source (e.g., a 1×1 mm LED chip) is adopted during simulation. According to simulation results, we optimize the coefficients of C_{Wi} and C_{Lj} until the lighting performance of the lens meets the requirements of some specific applications. Then we obtain the final design of the compact freeform lens. The coefficients of C_{Wi} and C_{Lj} are usually within range from 0.2 to 2.

3.6 Reversing the Design Method for Uniform Illumination of LED Arrays

These freeform optics algorithms demonstrated in Sections 3.2 to 3.5 are all focused on achieving a uniform illuminated single light pattern by a single freeform lens. However, in practical applications, uniform illumination of an LED array or an LED luminaire array is crucial and at the same time is required by most applications, such as LED direct-lit backlighting for LCDs, LED office lighting, LED plant lighting, and so on. Moreover, within these applications, the distance–height ratio (DHR) is always given, and the LIDC of the LED source is required to design and to optimize. To solve this problem, a new reversing design method for LED uniform illumination is proposed in this section, including design and optimization of LIDC to achieve high uniform illumination and a new algorithm of a freeform lens to generate the required LIDC by an LED light source. This method is practical and simple. Moreover, complicated LIDCs are also able to be realized at will by the new algorithm of a freeform lens, which makes it possible to design new LED optical components for uniform illumination.

In LED lighting, illumination in terms of a square LED array is the most widely applied, for example LED backlighting, interior lighting, and so on. In this study, the design method will be discussed based on the square LED array. Analyses of other arrangements of LED arrays, like linear or circular rings, are quite similar and not included in this discussion. As shown in Figure 3.32, a $M \times N$ (even-number) square LED array is set at the xy plane. The distance between the LED array and the target plane is $z0$. In this

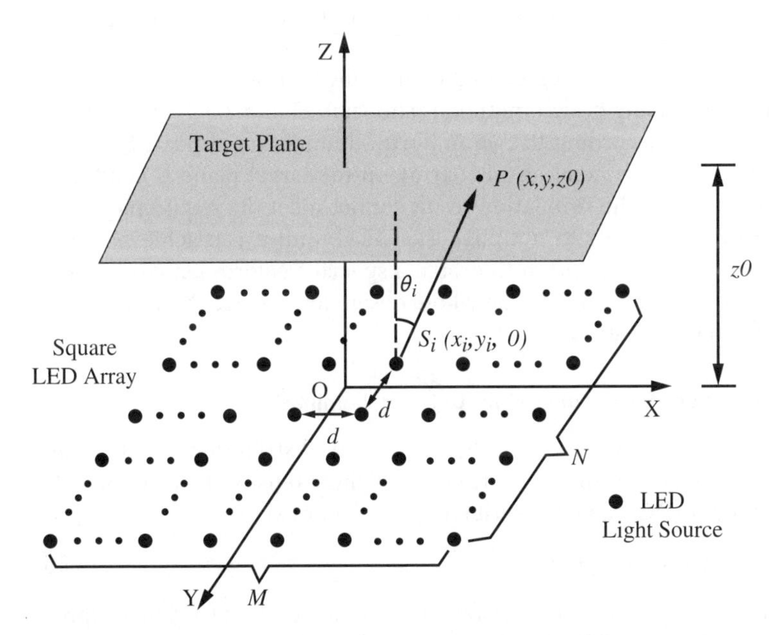

Figure 3.32 Schematic of a square LED light source array and the target plane.

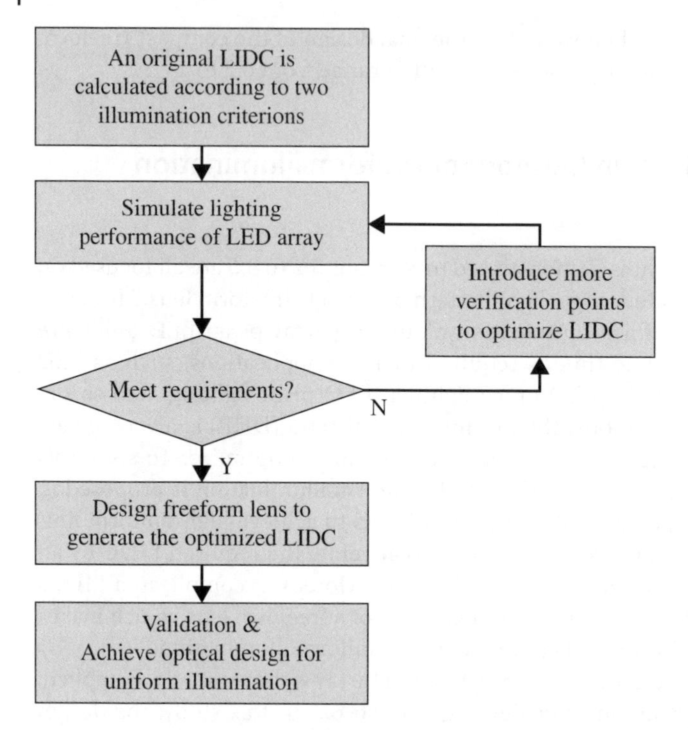

Figure 3.33 Flowchart of this new reversing design method for LED uniform illumination.

design, values of d, $z0$, and DHR $= d/z0$ are given by customer design requirements, and our work is to optimize the LIDC of the LED light source to achieve uniform illumination on the target plane. There are two key issues of concern. One is how to optimize the LIDC, and the other is how to design a freeform lens to realize the required LIDC. A detailed design method will be discussed in the following section.

The flowchart of this reversing design method is shown in Figure 3.33. First of all, an original LIDC is calculated according to two uniform illumination criteria. Secondly, simulation of lighting performance of an LED array on the target plane is conducted by Monte Carlo ray tracing. If the simulation result cannot meet the requirements, we introduce more verification points to optimize the LIDC until it can achieve a good performance. Then we use the algorithm to design a special freeform lens to generate the optimized LIDC. Finally, we validate the whole design and realize the LED optical component for uniform illumination.

3.6.1 Reversing the Design Method of LIDC for Uniform Illumination

Considering the circular symmetry of the light intensity distribution of LED light sources as well as convenience for calculation, the light intensity distribution $I(\theta)$, expressed in the form of LIDC, can be described by a polynomial of θ as follows:

$$I(\theta) = a_0 + a_1\theta^2 + a_2\theta^4 + a_3\theta^6 + a_4\theta^8 \tag{3.45}$$

where θ is the emitting angle between the ray and the z-axis. By using this expression of $I(\theta)$, different kinds of LIDCs including complicated ones can be expressed and calculated. Next we will calculate the illuminance of an arbitrary point $P(x, y, z0)$ on the target plane as shown in Figure 3.32. Firstly, the illuminance $E_i(x, y, z0)$ generated by an LED light source $S_i(x_i, y_i, 0)$ at point $P(x, y, z0)$ can be calculated as follows:

$$E_i(x, y, z0) = \frac{I(\theta_i)\cos\theta_i}{r_i^2} = \frac{I(\theta_i)z0}{r_i^3} = \frac{I(\theta_i)z0}{[(x - x_i)^2 + (y - y_i)^2 + z0^2]^{3/2}} \tag{3.46}$$

where $\theta_i = arctan\{[(x{-}x_i)^2{+}(y{-}y_i)^2]^{1/2}/z0\}$. Therefore the total illuminance $E(x, y, z0)$ generated by all $M \times N$ LED light sources at point $P(x, y, z0)$ can be expressed as follows:

$$E(x, y, z0) = \sum_{i=1}^{M \times N} E_i(x, y, z0) \tag{3.47}$$

To obtain LIDCs satisfying the uniformity requirement, we introduce two uniform illumination criteria in this study. One criterion is Sparrow's criterion to achieve a maximally flat central region, and the other is small illuminance variation between the side point and the central point on the target plane.

For the first criterion, by differentiating $E(x, y, z0)$ twice and setting $x = 0$ and $y = 0$, a function $f(a_0, a_1, a_2, a_3, a_4) = \partial^2 E/\partial x^2|_{x=0,y=0}$ is acquired with variables of a_0, a_1, a_2, a_3, and a_4. According to Sparrow's criterion, $f(a_0, a_1, a_2, a_3, a_4) = 0$ will lead to uniform illuminance at the region across the central point.[21,22] By solving the equation of $f(a_0, a_1, a_2, a_3, a_4) = 0$, four independent variables and one dependent variable are obtained, for example, $a_0 = a_0$, $a_1 = a_1$, $a_2 = a_2$, $a_3 = a_3$, and $a_4 = g(a_0, a_1, a_2, a_3)$. Obviously, other criterion is needed to confine the values of these variables further.

Although using Sparrow's criterion enables achieving uniform illuminance at the central region, it cannot guarantee uniform distribution of illuminance across the whole target plane. Therefore, another criterion of small illuminance variation is needed. CV(RMSE) (coefficient of variation of root mean square error) is a useful concept to evaluate the uniformity of illuminance.[13,23] A group of points on the target plane are sampled, and the value of CV(RMSE) is the ratio of RMSE to the mean value of the samples:

$$CV(RMSE) = RMSE/\overline{x} \tag{3.48}$$

where $\overline{x}$ is the mean value. Unfortunately, few standards adopt CV(RMSE) as an evaluation index for uniformity of illuminance. Thus, CV(RMSE) can evaluate lighting performance but cannot be regarded as one criterion. In this design, we introduce a ratio of $R(x_j, y_j, z0)$ as another criterion:

$$R(x_j, y_j, z0) = \frac{E(x_j, y_j, z0)}{E(0,0,z0)} \tag{3.49}$$

where $E(x_j, y_j, z0)$ is the illuminance of point $Q_j(x_j, y_j, z0)$ except the central point on the target plane. Ratio $R(x_j, y_j, z0)$ reflects the illuminance variance between the side point and the central point. To obtain high uniformity, the range of $R(x_j, y_j, z0)$ is set as $0.85 \leq R(x_j, y_j, z0) \leq 1.15$, which further confines the ranges of variations of the four independent variables. During design, point Q_j is regarded as the verification point, and proper selection of these points is a key issue. For example, if all Q_j points are around the central point, illuminance uniformity of side points becomes unpredictable. Therefore, suitable verification points that can reflect the whole illuminance distribution across the target plane should be selected as references, like $(M/2, N/2, z0)$, $(M/2-d, N/2-d, z0)$, and so on. More verification points could be introduced to optimize the LIDC to meet requirements.

According to the analysis here, it is clear that values of $a_0, a_1, a_2, a_3,$ and a_4 are adjusted to meet criteria of $f(a_0, a_1, a_2, a_3, a_4) = 0$and $0.85 \leq R(x_j, y_j, z0) \leq 1.15$ and at the same time to have a reasonable $I(\theta)$ during optimization. The designed LIDC will be validated by ray-tracing simulation and optimized further if necessary. Based on this reversing method, the optimized LIDC can be acquired easily to achieve uniform illumination. We should note that the calculated $I(\theta)$ just reflects the characteristic of light intensity distribution of light source and its values are meaningless. The calculated $I(\theta)$ will be normalized if necessary during the following design.

3.6.2 Algorithm of a Freeform Lens for the Required LIDC

In this section, an algorithm of a freeform lens will be introduced by which special freeform lenses could be designed to generate the optimized LIDC. This algorithm includes three main steps: establishing a light energy mapping relationship between the light source and the required light intensity distribution, constructing the lens, and validating the lens design by ray-tracing simulation. The first step is the key issue in this design.

In this design, light emitted from the light source (an LED module) is regarded as the input light, and light exiting from the lens (secondary optics or primary optics) is regarded as the output light. During Step 1, firstly the light energy distributions of both the light source and the optimized LIDC are divided into B parts equally. As shown in Figure 3.34, the light energy distribution Ω of the light source could be regarded as composed of a number of unit conical objects Ω_0, which represents the luminous flux within the angular range with a field angle $d\gamma$ in the latitudinal direction and $d\theta$ in the longitudinal direction. The luminous flux Φ_{input_0} of Ω_0 can be expressed as follows:

$$\phi_{input\_0} = \int I_{input}(\theta) d\omega = \int_{\gamma_1}^{\gamma_2} d\gamma \int_{\theta_1}^{\theta_2} I_{input}(\theta) \sin\theta d\theta \tag{3.50}$$

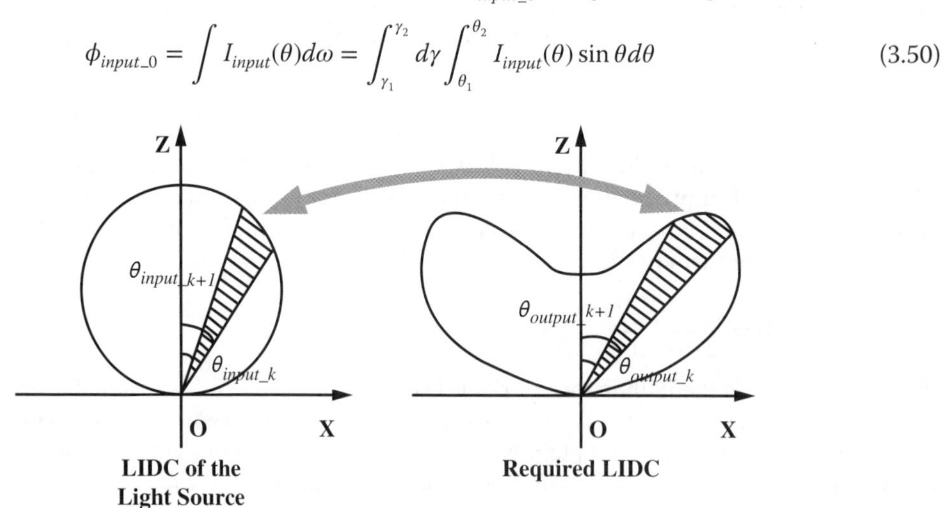

Figure 3.34 Schematic of a light energy mapping relationship between the light source and the required LIDC.

where $d\omega$ is the solid angle of one unit conical object. Since the light intensity distribution is circular symmetry, $\gamma_1 = 0$ and $\gamma_2 = 2\pi$ are set and the 3D design problem is changed into a much simpler 2D problem as shown in Figure 3.34. Then, Φ_{input_0} and the total luminous flux of the light source can be expressed as follows:

$$\phi_{input\_0} = 2\pi \int_{\theta_1}^{\theta_2} I_{input}(\theta) \sin\theta d\theta \tag{3.51}$$

$$\phi_{input\_total} = 2\pi \int_{0}^{\pi/2} I_{input}(\theta) \sin\theta d\theta \tag{3.52}$$

Since the total luminous flux of the light source is divided into B parts, then the field angle $\Delta\theta_{input\_k}$ of each subsource along the longitudinal direction can be obtained by iterative calculation as follows:

$$2\pi \int_{\theta_{input\_k}}^{\theta_{input\_k+1}} I_{input}(\theta) \sin\theta d\theta = \frac{\phi_{input\_total}}{B} \ (k = 1,2,\ldots B,\ \theta_{input\_1} = 0) \tag{3.53}$$

$$\Delta\theta_{input\_k+1} = \theta_{input\_k+1} - \theta_{input\_k} \ (k = 1,2,\ldots B) \tag{3.54}$$

Thus, the light source has been divided into B subsources with equal luminous flux. The directions of rays, which define the boundary of one subsource Ω_{input_k}, have been also calculated out as θ_{input_k}. Therefore, using the same calculation method and setting the optimized LIDC as $I_{output}(\theta)$, θ_{output_k} can be also obtained, which divides the light energy of output light into B parts equally.

Then, we establish the light energy mapping relationship between Ω_{input_k} and the same order of Ω_{output_k}. As shown in Figure 3.34, since light energy of the light source and optimized LIDC both are divided into B parts equally, if light energy within the range of θ_{input_k} to θ_{input_k+1} is able to be mapped to and redistributed within the range of θ_{output_k} to θ_{output_k+1}, then the light intensity distribution of the light source could be converted to the required light intensity distribution of output light, by which obtaining the optimized LIDC. According to the edge ray principle, rays from the edge of the source should strike the edge of the target. In this design, the target means the region constructed by output rays. Therefore, if we desire to map the light energy in Ω_{input_k} into the output region of Ω_{output_k}, we should ensure that two rays, which construct the Ω_{input_k} as boundary, exit as the boundary of Ω_{output_k} after being refracted by the freeform lens. In other words, an input ray with emitting angle of θ_{input_k} establishes a mapping relationship with the output ray witn h aexiting angle of θ_{output_k}. Thus, the light energy mapping relationship between the light source and the required LIDC has been established.

Since we have obtained the emitting directions of input rays and the corresponding exiting directions of output rays, it is easy to design a freeform lens to meet this mapping relationship according to Snell's law and the surface lofting method. Moreover, due to a detailed construction process of freeform lens and validation being the same as the methods proposed in this chapter, Step 2 and Step 3 will not be discussed in detail in this study. In addition, sometimes the LIDC generated by a freeform lens can be deteriorated by an extended source. This problem also can be solved by the feedback optimal freeform optics design method presented in Ref. [24]. Therefore, by using the algorithm mentioned in this chapter, it is able to design a freeform lens to realize the required LIDC and thus achieve uniform illumination.

References

1 Parkyn, W.A. Design of illumination lenses via extrinsic differential geometry. *Proc. SPIE* **3428**, 154–162 (1998).
2 Parkyn, W.A. Segmented illumination lenses for step lighting and wall washing. *Proc. SPIE* **3779**, 363–370 (1999).
3 Wang, L., Qian, K., and Luo, Y. Discontinuous free-form lens design for prescribed irradiance. *Appl. Opt.* **46**, 3716–3723 (2007).
4 Ding, Y., Liu, X., Zheng, Z., and Gu, P. Freeform LED lens for uniform illumination. *Opt. Expr.* **16**, 12958–12966 (2008).
5 Chen, F., Liu, S., Wang, K., Liu, Z., and Luo, X. Free-form lenses for high illumination quality light-emitting diode MR16 lamps. *Opt. Eng.* **48**, 123002–123002–7 (2009).
6 Wang, K., Liu, S., Chen, F., Qin, Z., Liu, Z., and Luo, X. Freeform LED lens for rectangularly prescribed illumination. *J. Opt. A: Pure Appl. Opt.* **11**, 105501 (2009).
7 Wang, K., Chen, F., Liu, Z., Luo, X., and Liu, S. Design of compact freeform lens for application specific Light-Emitting Diode packaging. *Opt Expr.* **18**, 413–425 (2010).
8 Wang, S., Wang, K., Chen, F., and Liu, S. Design of primary optics for LED chip array in road lighting application. *Opt. Expr.* **19** Suppl 4, A716–A724 (2011).
9 Zhao, S., Wang, K., Chen, F., Wu, D., and Liu, S. Lens design of LED searchlight of high brightness and distant spot. *J. Opt. Soc. Am. A* **28**, 815–820 (2011).
10 Ries, H. and Rabl, A. Edge-ray principle of nonimaging optics. *J. Opt. Soc. Am. A* **11**(10), 2627–2632 (1994).
11 Bass, M., Stryland, E.W.V., Williams, D.R., and Wolfe, W.L. *Handbook of Optics: Fundamentals, techniques, and design (McGraw-Hill)* (1994).
12 Wang, K., Wu, D., Qin, Z., Chen, F., Luo, X., and Liu, S. New reversing design method for LED uniform illumination. *Opt. Expr.* **19** Suppl 4, A830–A840 (2011).
13 Qin, Z., Wang, K., Chen, F., Luo, X., and Liu, S. Analysis of condition for uniform lighting generated by array of light emitting diodes with large view angle. *Opt. Expr.* **18**, 17460–17476 (2010).
14 Qin, Z., Ji, C., Wang, K., and Liu, S. Analysis of light emitting diode array lighting system based on human vision: normal and abnormal uniformity condition. *Opt. Expr.* **20**, 23927–23943 (2012).
15 Chen, F., Wang, K., Qin, Z., Wu, D., Luo, X., and Liu, S. Design method of high-efficient LED headlamp lens. *Opt. Expr.* **18**, 20926–20938 (2010).
16 Zhao, S., Wang, K., Chen, F., Qin, Z., and Liu, S. Integral freeform illumination lens design of LED based pico-projector. *Appl. Opt.* **52**, 2985–2993 (2013).
17 Moreno, I., and Sun, C.-C. Modeling the radiation pattern of LEDs. *Opt. Expr.* **16**, 1808–1819 (2008).
18 Hu, R., and Luo, X. Adding an extra condition: a general method to design double freeform-surface lens for LED uniform illumination. *J. Sol. State Light* **1**, 1–9 (2014).
19 Hu, R., Gan, Z., and Luo, X. Design of double freeform-surface lens by distributing the deviation angle for light-emitting diode uniform illumination. In 2013 14th International Conference on Electronic Packaging Technology (ICEPT), pp. 1150–1153 (2013).
20 Moreno, I., and Sun, C.-C. *LED array: Where does far-field begin? Proc. SPIE.* 70580R–70580R–9 (2008).

21 Moreno, I. Configurations of LED arrays for uniform illumination. *Proc. SPIE* **5622**, 713–718 (2004).

22 Moreno, I., Avendaño-Alejo, M., and Tzonchev, R.I. Designing light-emitting diode arrays for uniform near-field irradiance. *Appl. Opt.* **45**, 2265–2272 (2006).

23 Yang, H., Bergmans, J.W.M., Schenk, T.C.W., Linnartz, J.-P.M.G., and Rietman, R. Uniform Illumination Rendering Using an Array of LEDs: A Signal Processing Perspective. *IEEE Trans. Signal Proc.* **57**, 1044–1057 (2009).

24 Wu, D., Wang, K., and Chigrinov, V.G. Feedback Reversing Design Method for Uniform Illumination in LED Backlighting With Extended Source. *J. Disp. Tech.* **10**, 43–48 (2014).

21 [illegible] CD [illegible] 32, [illegible] 2022.
[illegible]

22 [illegible]

23 [illegible]

24 [illegible] Kitchen [illegible] 12 [illegible]

4

Application-Specific LED Package Integrated with a Freeform Lens

4.1 Application-Specific LED Package (ASLP) Design Concept

Light-emitting diodes (LEDs), with increasing luminous flux and lumen efficiency in recent years, have more and more applications in our daily life, such as road lighting, backlighting for LCD display, automotive headlamps, interior and exterior lighting, and so on.[1–3] Brighter, smaller, smarter, and cheaper are development trends of both LED packaging modules and luminaires.[3] For example, smaller LED packaging using less material will decrease the cost of LED modules. In LED applications, secondary optics are essential for most LED luminaires because radiation patterns of most LEDs are circular-symmetrical with nonuniform illuminance distribution, which cannot directly meet the requirements of specific applications (e.g., the subrectangular radiation pattern required in road lighting). The freeform lens is an emerging optical technology with advantages of high design freedom and precise light irradiation control, which are useful for LED lighting design.[4–8] The traditional freeform lens, however, due to the category of secondary optics in nature, has limitations of large size in some space-confined applications, possibly resulting in large volume and weight of both the thermal management part and the luminaires, relatively low system optical efficiency, and possible inconvenience for customers to assembly.

In addition, the optical performance of many LED luminaires existing in the market is relatively poor, mainly due to a lack of technology of secondary optics for many new LED companies. In contrast, thousands of traditional lighting companies, which have been main bodies of employment and marketing and sales, are suffering from even poorer understanding of LEDs, facing significant technical and societal challenges of transferring from traditional lighting to the new LED business, proposing a strong need for easy-to-use LED technology. Therefore, if we integrate secondary optics with traditional LED packaging, achieving new ASLP,[9] as shown in Figure 4.1. it will not only decrease the size of LED modules and systems, and increase the efficiencies of LED luminaires, but also provide a convincing solution to LED lighting for old and new LED application companies.

In ASLP, since the secondary freeform lens has to be designed more compactly to integrate with an LED package module, the distance between source and lens will be less than five times the source diameter, and the LED chip could not be regarded as a point source during packaging lens design according to the far-field conditions of LED, which results in the major challenge in ASLP design – the extended source problem. As the description in Section 3.3 and 3.5 explained, a freeform lens designed based on a

Freeform Optics for LED Packages and Applications, First Edition. Kai Wang, Sheng Liu, Xiaobing Luo and Dan Wu.

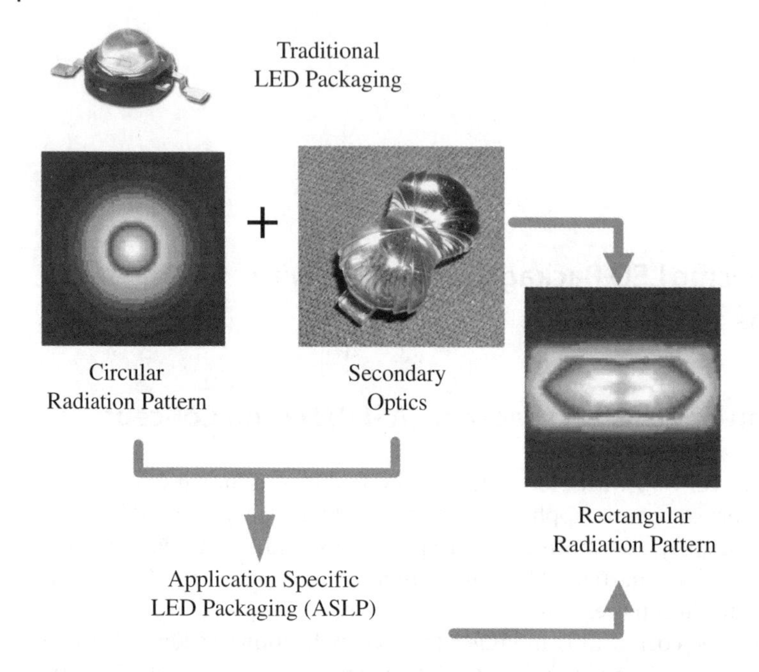

Figure 4.1 Schematic of the concept of application-specific LED packaging.

point source assumption mounted on an LED extended source will lead to a deteriorated lighting performance.

In this chapter, a polycarbonate (PC) compact freeform lens for a new ASLP used in road lighting will be designed based on the circularly and noncircularly symmetrical freeform lens design methods for extended sources mentioned in Sections 3.3 and 3.5, which will eliminate deteriorations caused by extended sources. Optical performance of the ASLP also will be studied both numerically and experimentally. Results demonstrate that the ASLP has high system optical efficiency and good lighting performance with a subrectangular radiation pattern, which meet the requirements of road lighting. Besides ASLP for a single LED chip, ASLP for an LED chip array integrated with multiple functions also will be introduced.

4.2 ASLP Single Module

4.2.1 Design Method of a Compact Freeform Lens

Details of the design method are in Sections 3.3 and 3.5. There are three main parts: establishing a light energy mapping relationship between the source and target, constructing the lens, and validating the lens design by numerical simulation. In our case, the refractive index is 1.586 for PC and ranges from 1.4 to 1.6 for silicone. The inner surface of the freeform lens is designed as a concave spherical surface in this method. Optimization of overall lighting performance is to be achieved by adjusting optimization coefficients of target grids and reconstructing the outside surface of the lens. We are able to build a freeform lens that satisfies our requirements.

4.2.2 Design of the ASLP Module

Due to the significance of road lights in the LED community, demonstrated by a recent Chinese government program and other programs around the world, an ASLP for road lighting is to be designed as an example in this study. A PC compact freeform lens with a refraction index of 1.586 will be designed to form a 32 m long and 12 m wide rectangular radiation pattern at a height of 8 m, as shown in Figure 4.2, which can meet the requirements of road lighting.

4.2.2.1 Optical Modeling

Light intensity distribution curves (LIDCs) of LED chips and LED chips covered by phosphor layer are two key issues for accurate LED packaging optical modeling. This is because most white LEDs are obtained either by integrating blue LED chips with yellow phosphor or by integrating RGB LED chips in one packaging module. Vertical electrode LED chips with size of $1 \times 1 \times 0.1$ mm were adopted for the lens design, and a phosphor layer with a thickness of about 70 μm was conformally coated on the chip. Figure 4.3 depicts that the measured LIDCs of the chip and the chip coated by phosphor are quite similar with the standard Lambert distribution. Therefore, a Lambert source with an emitting area of 1×1 mm would be used as an equivalent source during optical design and simulation.

4.2.2.2 Design of a Compact Freeform Lens

As shown in Figure 4.4, a PC compact freeform lens is designed according to the method mentioned here. Firstly, we divide both the one-quarter source and target plane into $200\ (M) \times 100\ (N) = 20{,}000$ grids with equal luminous flux and unequal area, respectively, and establish a light energy mapping relationship between these grids. Secondly, we calculate coordinates and a normal vector of each point on the freeform surface and construct the lens utilizing these points. Finally, we validate the lighting performance of the lens through simulation and obtained the final design after optimizing coefficients of the target.

The distance between the LED chip and the central point at the outside surface of the lens decides the size of the compact freeform lens. To provide enough space for the LED

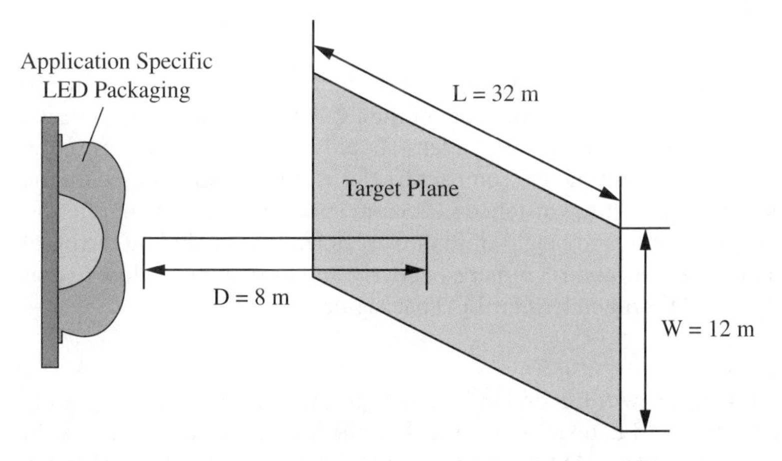

Figure 4.2 Schematic of a design target of an ASLP for road lighting.

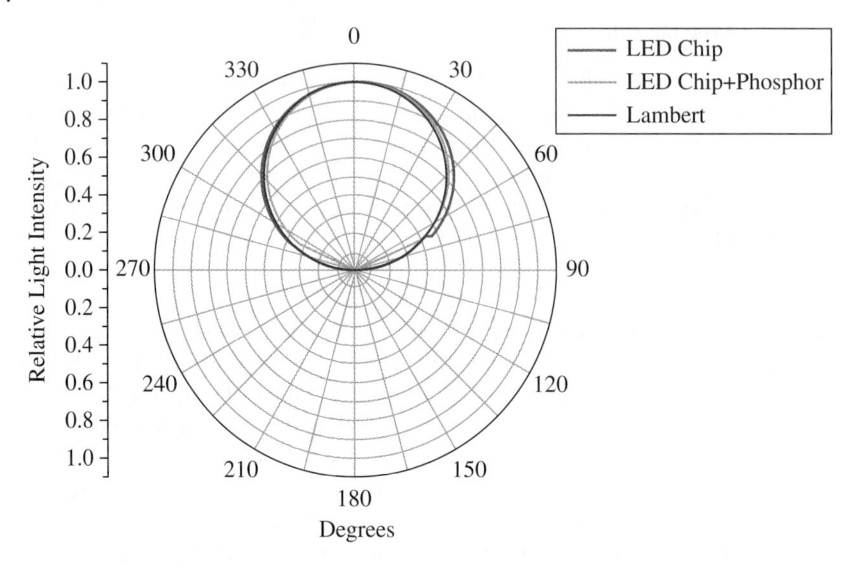

Figure 4.3 Light intensity distribution curves of an LED chip and an LED chip coated by phosphor layer.

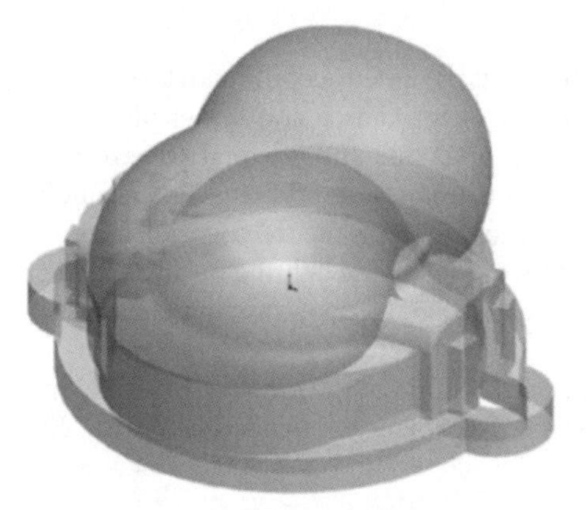

Figure 4.4 A PC compact freeform lens for ASLPs for road lighting.

packaging process (e.g., wire bonding), the radius of the inner concave spherical surface is set as 1.8 mm. Considering the requirements of the PC lens manufacturing process, for example the thinnest thickness of the PC lens should be larger than 0.3 mm, the limit of the distance is 2.5 mm of the PC compact freeform lens for ASLP based on the lead frame and heat sink. We set the distance as 2.5 mm in our design. The volume and largest values of length, width, and height of the PC freeform lens (including the base) are 45.5 mm^3, 6.4 mm, 5.8 mm, and 3.8 mm, respectively, and the volume is close to that of the most widely used PC domed lens for LED packaging.

4.2.2.3 ASLP Module

As shown in Figure 4.5a, a new ASLP design for road lighting is achieved by integrating this PC lens with traditional LED packaging based on the lead frame and heat sink. In detail, the optical structure of the ASLP includes a LED chip, a phosphor layer, silicone, and the PC freeform lens. As shown in the detailed optical structure of traditional LED

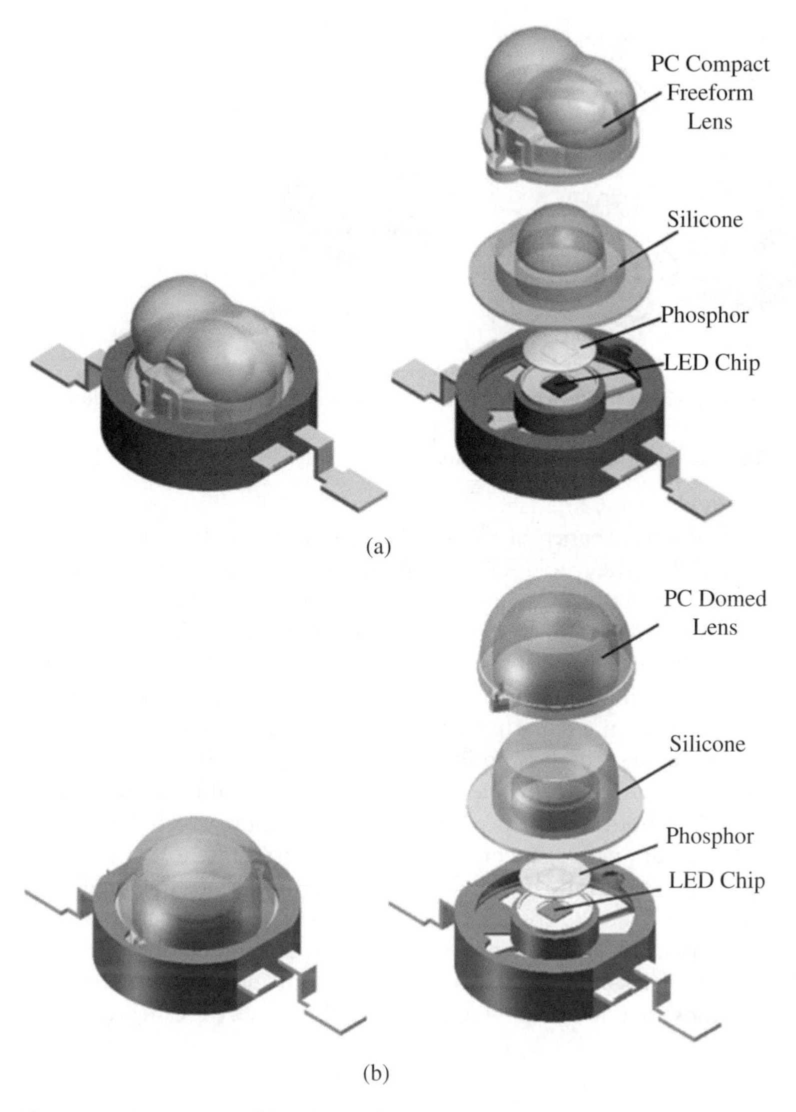

Figure 4.5 Comparison of detail optical structures between (a) an ASLP and (b) traditional LED packaging.

packaging in Figure 4.5b, we can find that during the ASLP packaging process, the only change we need to make is replacing the traditional PC domed lens with the new PC freeform lens, and that is quite compatible with current LED packaging processes, which makes it easier for LED manufacturers to adopt this new technology with little change to the existing process.

Figure 4.6 depicts an LED module for road lighting consisting of a traditional LED and a secondary optical element with a kind of freeform lens. From the comparison shown in Figure 4.6, we can find that the height and volume of the ASLP are only about one-half and one-eighth of those of the traditional LED module, which provides an effective way for designing compact-sized LED lighting systems and also provides more design freedom for new concepts in LED luminaires.

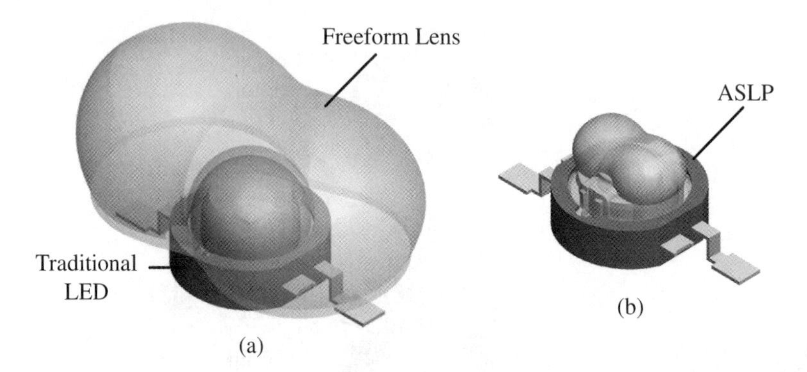

Figure 4.6 Comparison of size between (a) a traditional LED module and (b) an ASLP for road lighting.

4.2.3 Numerical Analyses and Tolerance Analyses

4.2.3.1 Numerical Simulation and Analyses

We simulate the optical performance of the ASLP numerically by the Monte Carlo ray-tracing method, and then the whole ASLP is simulated by 1 million rays. As shown in Figure 4.7, the ASLP forms a subrectangular radiation pattern with a length of 33.4 m and width of 13.6 m at the height of 8 m, which is in agreement with the expected shape. The utilization ratio of the ASLP reaches as high as 95.1%, which means more than 95% of light energy exiting from the ASLP is delivered into the desired lighting area. This utilization ratio is quite close to that of a good designed freeform lens and even higher than that of some secondary optics with poor design (~80%).

Since ASLP has good lighting performance and can replace the traditional LED lighting module with secondary optics directly, system optical efficiency (which decides the final performance of LED luminaires) becomes the most important issue. Light output efficiency (LOE) is defined as the ratio of light energy exiting from a lens to light energy incident into a lens. LOE of the PC compact freeform lens is 94.2%, which is slightly lower than that of traditional PC domed lenses of 96.0%. This phenomenon is caused by the reason that incident angles of most incident rays at the outside surface of freeform lenses are larger than those of domed lenses, and that results in more Fresnel reflection

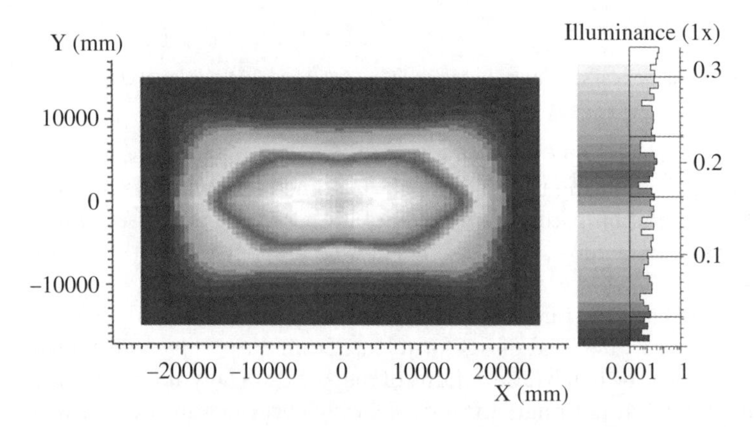

Figure 4.7 Simulation lighting performance of the ASLP at a height of 8 m.

Table 4.1 Simulation comparison of system optical efficiencies between a traditional LED module and an ASLP for road lighting.

	LED module integrated with secondary optics, %	**ASLP, %**
Light output efficiency of LED packaging lens (η_{LOE})	96.0 (Domed lens)	94.2 (Freeform lens)
Light output efficiency of secondary optics (η_{SO})	90.0	–
Utilization ratio (η_{UR})	95.1	95.1
System optical efficiency ($\eta_{SOE} = \eta_{LOE} \times \eta_{SO} \times \eta_{UR}$)	82.2	89.6
Enhancement	–	7.4

loss occurring in the compact freeform lens. Considering the utilization ratio of 95.1%, the system optical efficiency of the ASLP reaches as high as 89.6%. The LOE of traditional freeform lenses is always at a low level of about 90% because of Fresnel reflection loss from two lens surfaces and absorption of material. As shown in Table 4.1, considering the light loss (~10%) of secondary optics as well as the utilization ratio (assuming the same with ASLP), the system optical efficiency of the LED module for road lighting is only about 82.2%, which is 7.4% lower than that of ASLP. Therefore, the ASLP has good lighting performance and higher system optical efficiency and could be used directly for road lighting so that no secondary optics are needed, which makes it convenient for old and new LED luminaire designers and manufacturers to use and also will further reduce the cost of LED luminaires.

4.2.3.2 Tolerance Analyses

Tolerance is an important issue for freeform lenses. Since the same scale installation or manufacturing error (e.g., 0.1 mm) has more effect on the lighting performance of small-size freeform lenses, tolerance is especially important for the compact freeform lenses for ASLP. In this section, tolerance analyses will be focused on installation errors, including deviations in horizontal (d_H) and vertical up (d_V) directions and rotational deviation (θ_R) of the lens (as shown in Figure 4.8). To evaluate deterioration of lighting performance, the coefficient of an effective lighting area (η_{ELA}) is adopted in this study. *Effective lighting area* (A_{EL}) is defined as the area on the target plane containing more than 95% light energy exiting from the lens. We use A_{ELE} and A_{ELO} to express the effective lighting area of an ASLP with and without errors, respectively. Then η_{ELA} can be obtained as Eq. 4.1:

$$\eta_{ELA} = \frac{A_{ELE}}{A_{ELO}} \tag{4.1}$$

We define the value of η_{ELA} as 1 when there are no errors. The worse the lighting performance is, the smaller the η_{ELA} will be. Thus, η_{ELA} is able to reflect the effects of various errors on lighting performance of the ASLP.

The effects of installation errors on lighting performance of the ASLP are shown in Figure 4.8. Horizontal deviation d_H and vertical deviation d_V of the lens have significant

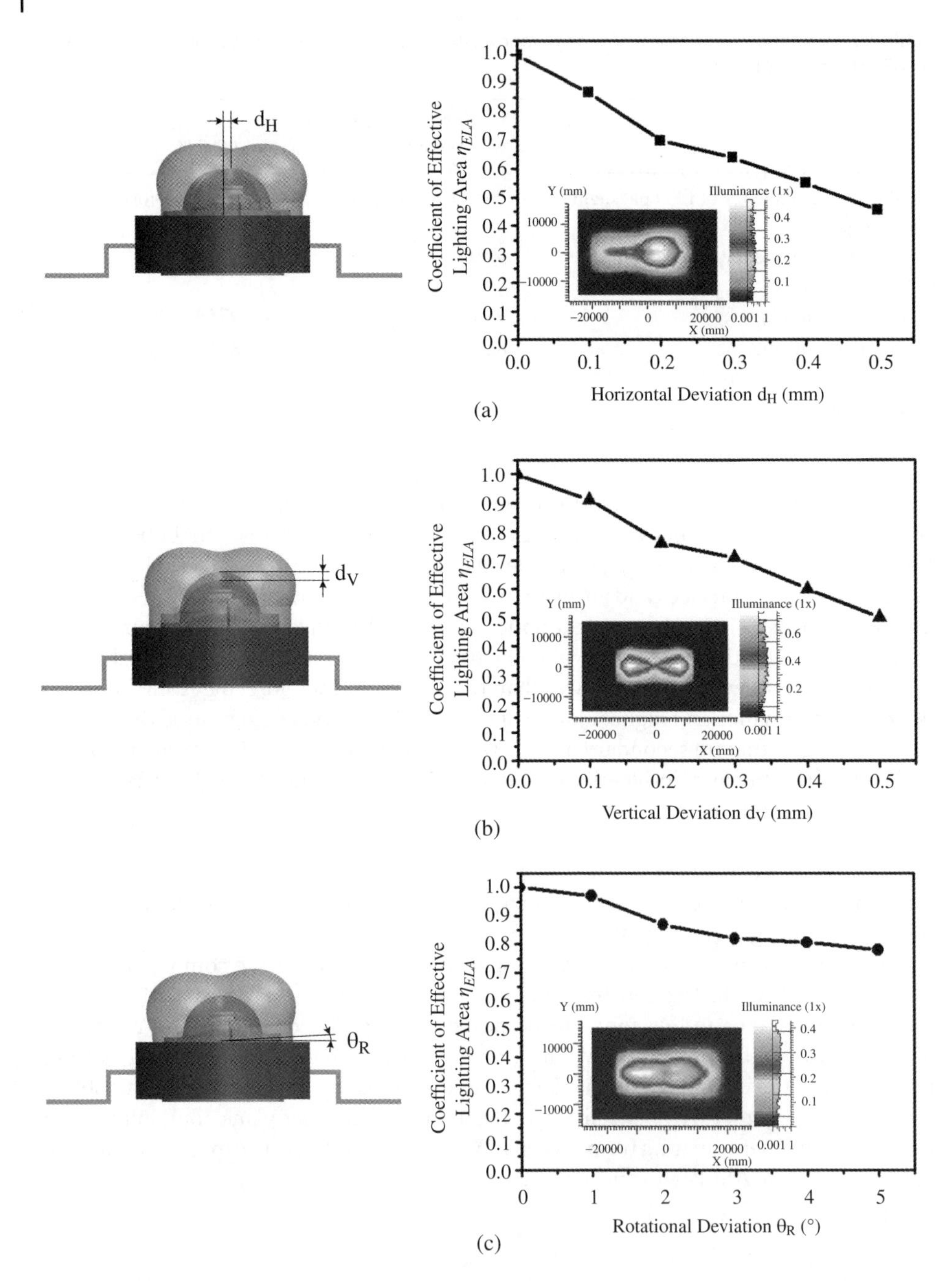

Figure 4.8 Effects of installation errors on lighting performance of the ASLP: (a) horizontal deviation d_H; (b) vertical deviation d_V; and (c) rotational deviation θ_R.

effects on the shape and uniformity of the radiation pattern. When d_H increases from 0.1 to 0.5 mm, the radiation pattern becomes asymmetric and the right pattern is much brighter than the left one, which results in a decline of η_{ELA} from 0.87 to 0.46. Moreover, when d_V increases, the size of the radiation pattern becomes small and light energy concentrates on the left and right ends of the pattern, causing low uniformity. η_{ELA} also sharply declines from 0.91 to 0.50 when d_V increases from 0.1 to 0.5 mm. However, we can find that rotational deviation θ_R has less effect on the shape and uniformity of the radiation pattern compared with d_H and d_V. η_{ELA} only declines from 0.97 to 0.78 when θ_R increases from 1° to 5°. Therefore, it seems that lighting performance is more sensitive to the installation errors in the horizontal and vertical directions.

The deteriorated radiation pattern, whose η_{ELA} is larger than 0.85, is still acceptable in engineering. Therefore, the limits on the tolerance of horizontal, vertical, and rotational deviation of the freeform lens are 0.11 mm, 0.14 mm, and 2.4°, respectively, which are acceptable for mass production. Since the two glue-injection holes at the base of the PC compact freeform lens confine deviation of the lens in the horizontal direction, we should pay more attention to minimizing or even avoiding vertical and rotational deviations during the LED packaging process.

Besides the above-mentioned lead frame ASLP, we also designed a surface-mount device (SMD) ASLP based on the ceramic board.[9–11] In this design, we use silicone with a refractive index of 1.54 as the ASLP freeform lens material. As shown in Figure 4.9, there is a conventional SMD LED package with a hemispherical silicone lens structural schematic and its illumination performance. This SMD LED is mainly composed of a ceramic integrated circuit board, LED chip, phosphor layer, and hemispherical silicone lens. Its LIDC is Lambertian type, and the light pattern is circular with nonuniform illuminance distribution. For the SMD ASLP, as shown in Figure 4.10, similarly as in traditional SMDs, in the package process, we only need to change the molding module for the hemispherical silicone lens into the freeform lens. The rest of the packaging

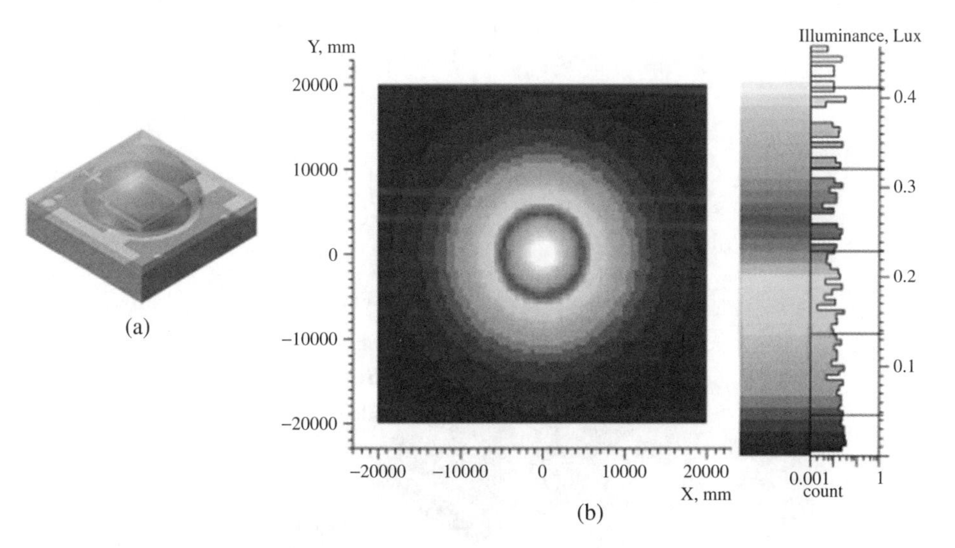

Figure 4.9 (a) Traditional SMD LED package based on a ceramic board, (b) its optical performance, and (c) its detailed optical structure.

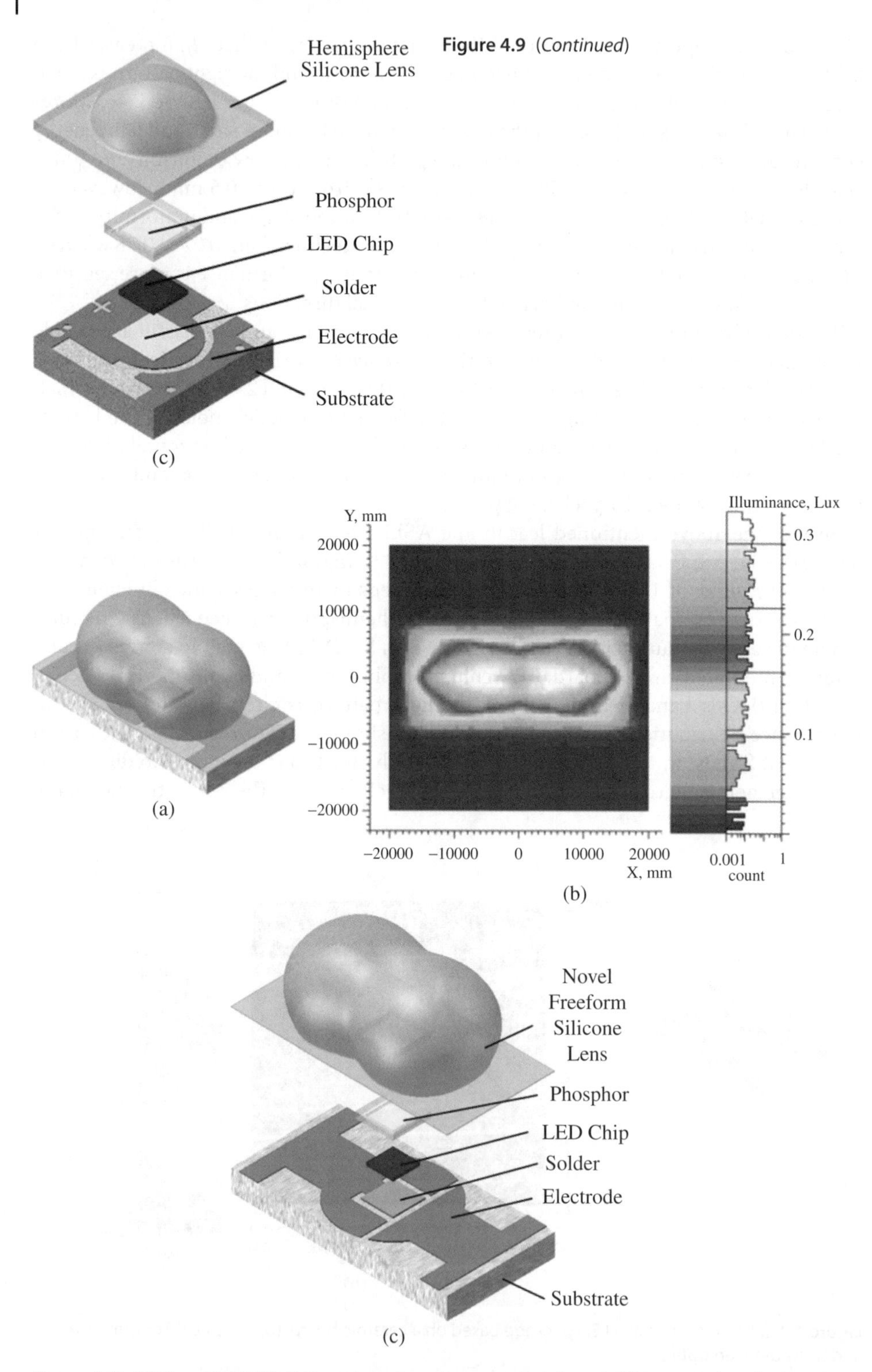

Figure 4.9 (*Continued*)

Figure 4.10 (a) New ASLP SMD LED package based on a ceramic board, (b) its optical performance, and (c) its detailed optical structure.

process for manufacturing remains the same. It is flexible to change from one type of lens to another. The light pattern of the SMD ASLP is quite similar with that of the lead frame ASLP and presents a rectangular distribution.

4.2.3.3 Experiments

As shown in Figure 4.11 and Figure 4.12, the PC compact freeform lens is manufactured by injection molding method, and a white-light ASLP for road lighting is also manufactured by integrating the lens with traditional LED packaging. As shown in a comparison in Figure 4.13, the size of the ASLP is quite close to that of the traditional LED, but it is much smaller than that of an LED module for road lighting. Thus, the ASLP will reduce the size of LED luminaires and make them easier for customers to assemble.

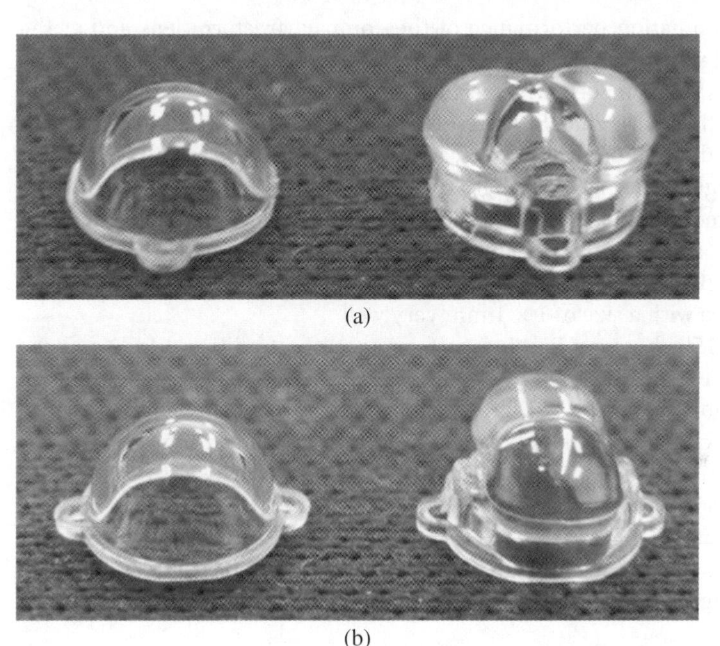

Figure 4.11 (a) Front view of and (b) left view of a PC domed lens (left) and a PC compact freeform lens (right).

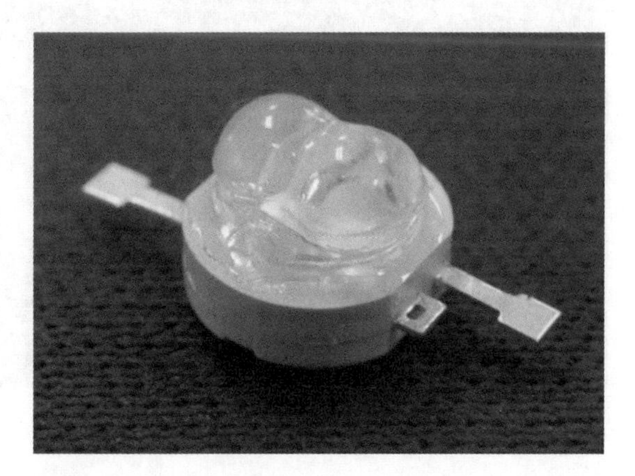

Figure 4.12 White-light ASLP for road lighting.

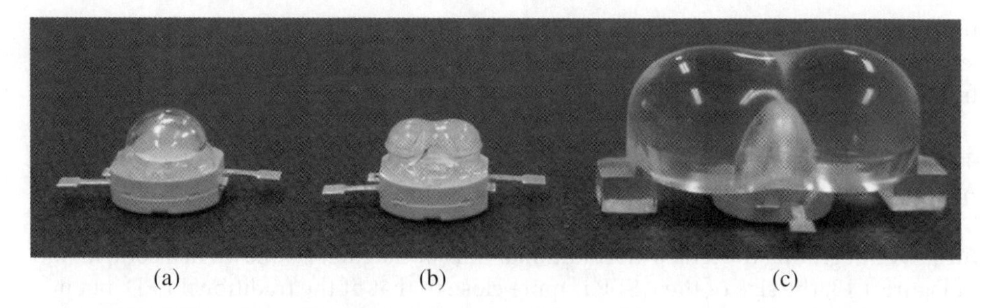

Figure 4.13 (a) Traditional LED packaging, (b) ASLP, and (c) LED module for road lighting.

To evaluate the illumination performance of the compact freeform lens and at the same time avoid the light pattern deterioration caused by a phosphor freely dispensing process, a blue-light ASLP is made with a vertical electrode LED chip of SemiLeds. As shown in Figure 4.14, the blue-light ASLP presents a rectangular blue-light pattern with a length of 236 cm and width of 88 cm on a test plane 58.5 cm away from the ASLP. It equals a rectangular light pattern with a length of 32.3 m and width of 12.1 m at 8 m away, which meets design requirements. The blue-light pattern has clear edges and the light energy uniformly distributed in the rectangular region. This experiment demonstrates that the designed ASLP lens is able to overcome the illumination deterioration caused by the extended source with a size of 1×1 mm very well.

Figure 4.15a and 4.15b show the lighting performance of a traditional LED and the ASLP, respectively. The light pattern of the traditional LED is circular with nonuniform illuminance distribution, while the ASLP redistributes the LED's light energy distribution and forms a subrectangular radiation pattern on the target plane, which is more uniform than the circular radiation pattern. The target plane is 58.5 cm away from the LED. Experimental results demonstrate that 92.8% light energy of the ASLP is distributed in a subrectangular area with a length of 240 cm and width of 90 cm, and it will be enlarged to 32.8 m long and 12.3 m wide at the height of 8 m, which is very close to the simulation results and is also in agreement with the expected performance.

Figure 4.14 Light pattern of the blue-light ASLP.

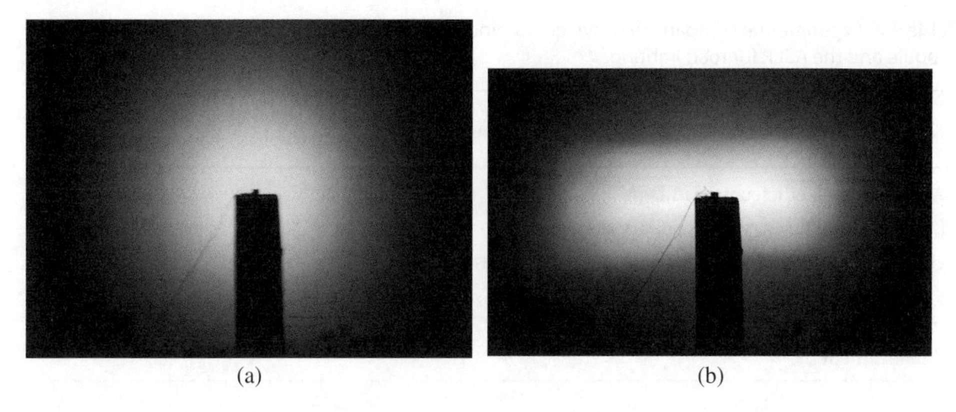

Figure 4.15 Lighting performance of (a) a traditional LED packaging and (b) the ASLP.

Unit: lx

Y \ X	120 cm	100 cm	80 cm	60 cm	40 cm	20 cm	0	20 cm	40 cm	60 cm	80 cm	100 cm	120 cm
50 cm	5.4	7.0	8.3	9.7	11.1	11.6	11.9	11.7	11.2	9.8	8.2	6.8	5.2
40 cm	6.2	12.7	16.6	16.8	17.9	18.3	17.8	18.4	17.8	16.4	16.1	12.5	6.0
30 cm	6.9	15.4	21.3	30.4	34.5	33.6	31.4	33.2	34.0	30.7	21.1	15.6	7.0
20 cm	7.3	16.5	27.8	40.9	47.5	48.9	44.2	48.6	47.3	41.2	27.5	16.9	7.4
10 cm	7.8	18.8	35.1	53.0	60.7	61.6	59.5	61.1	60.2	52.8	34.8	18.6	7.7
0	8.5	20.9	41.8	64.1	68.7	67.3	62.1	67.5	69.1	63.9	41.2	20.1	8.1
10 cm	7.8	18.3	34.9	53.3	61.5	62.0	58.7	62.3	61.3	53.1	35.2	17.9	7.8
20 cm	7.6	17.2	28.3	41.3	47.8	49.4	44.7	49.1	46.9	41.8	28.1	16.7	7.5
30 cm	7.3	15.9	21.4	31.1	34.2	33.8	30.8	34.0	34.3	30.1	20.8	15.1	7.1
40 cm	6.5	13.0	16.3	17.0	17.6	18.4	17.2	18.9	18.0	17.1	16.5	13.1	6.8
50 cm	5.7	7.3	8.4	10.1	10.9	11.4	11.5	11.3	11.1	10.0	8.6	7.1	5.8

Figure 4.16 Illuminance distribution of a radiation pattern of the ASLP.

Illuminance distribution of a radiation pattern of the ASLP is shown in Figure 4.16. We can find that illuminance distribution in the central area is more uniform than in the edge area of the radiation pattern. Most light energy is uniformly distributed within the central area (about 160 cm long and 60 cm wide), and illuminance declines sharply out of this area. Although illuminance distribution is not uniform enough within the whole subrectangular radiation pattern, it is already acceptable for road lighting in engineering.

System luminous efficiencies of the traditional LED module and ASLP were also investigated as a key issue. The experimental utilization ratio of ASLP is slightly lower than the simulation result. As shown in Table 4.2, although the lumen efficiency and utilization ratio of traditional LEDs are slightly higher than those of ASLPs, without light loss caused by secondary optics, the system lumen efficiency of the ASLP reaches as high as 94.7 lm/W, which is 8.1% higher than that of the LED module integrated with secondary optics. Therefore, the ASLP provides a solution to LED lighting systems with higher system lumen efficiency.

Based on the ASLP module discussed here, we develop a new 108 W high-power LED street lamp as shown in Figure 4.17. An ASLP module array distributes uniformly on the metal core printed circuit board (MCPCB), and there is no need to assemble a secondary lens. Without a secondary lens, the light generated from the ASLP will pass through the

Table 4.2 Experimental comparison of system luminous efficiencies between the traditional LED module and the ASLP for road lighting.

	LED module integrated with secondary optics	**ASLP**
Average lumen efficiency of LED (η_{LE})	103 lm/W	102 lm/W
Light output efficiency of secondary optics (η_{SO})	90.0%	–
Utilization ratio (η_{UR})	94.5%	92.8%
System lumen efficiency of LED module ($\eta_{SLE} = \eta_{LE} \times \eta_{SO} \times \eta_{UR}$)	87.6 lm/W	94.7 lm/W
Enhancement	–	8.1%

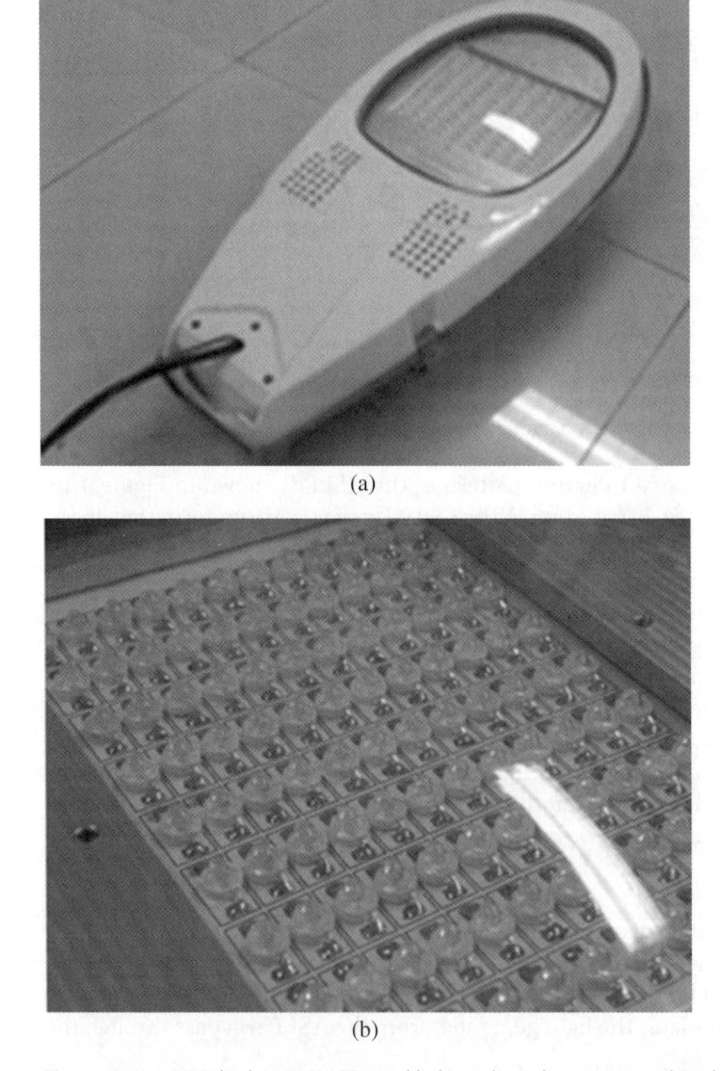

(a)

(b)

Figure 4.17 108 W high-power LED road lighting based on a warm white-light ASLP array.

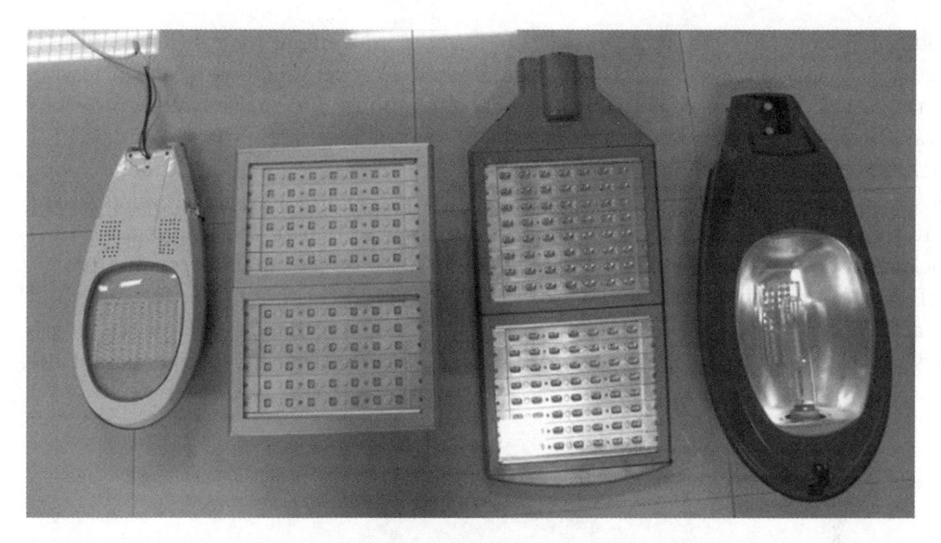

Figure 4.18 Various types of lamps: LED road lamp, LED tunnel lamp, and conventional HPS lamp.

glass cover and illuminate on the road surface. In the meantime, with elimination of the secondary lens, the ASLP module array gap can be reduced. Compared with a conventional LED road lamp, the light source area is dramatically reduced. Integrated with the active heat dissipation method, which is based on fans and fins for heat dissipation, the whole road lamp volume is greatly reduced. This compact ASLP lamp design can be fitted into a small road lamp housing, including the optics, heat dissipation, and driving and control parts. From Figure 4.18, we can find that the size of a 108 W ASLP LED road lamp is much smaller than the 112 W LED road lamp design that uses a secondary lens. The size of the 108 W LED road lamp is also smaller than the 250 W HPS (high-pressure sodium) lamp and 84 W LED tunnel lighting. With the volume continuing to reduce, the structure material and the cost are reduced. The weight of LED lighting is reduced from 0.14 kg/W to 0.07 kg/W, and we realize the design of low-weight and high-power LED lighting.

From these analyses, we can find that the ASLP integrated with a small freeform lens has the ability to achieve a rectangular light pattern with uniform illuminance distribution, which can satisfy road lighting application requirements. There is no need for other secondary optical systems, and it is convenient for usage. At the same time, compared with the solution that includes the conventional LED module and the secondary lens, this new ASLP module has advantages such as compact size (~1/8), higher lighting efficiency (higher than 8.1%), low cost (without secondary optics, reduced 17%), convenient assembly, and great potential for applications. Besides, the design concept that combines secondary optics and LED package modules will be the developing LED optics design trend.

4.3 ASLP Array Module

Two application-specific LED array modules for street lighting, with chip-on-board (CoB) packaging, are also designed by integrating 3 × 3 freeform lens arrays with traditional CoB LED modules to make them more convenient for customers to use. Figure 4.19 shows a 3 × 3 LED array module based on ceramic board, and it mainly consists of a ceramic board with circuits, LED chips, phosphor, and a novel silicone lens array. The LOE of this module is 95.1%, which is a little lower than that of a single LED package because some lights with large emergence angles irradiate into other silicone lenses and are absorbed by the material. As shown in Figure 4.19, the simulated light pattern of this LED array module is quite similar with that of the single novel LED package and also could be used in street lighting directly. The light source of a 108 W LED road lamp, which is one of the most used types of LED lamps in the market, could be achieved easily only by integrating 12 of this type of LED module. Since the length and width of this module are only 34 mm and 30 mm, respectively, the size of the

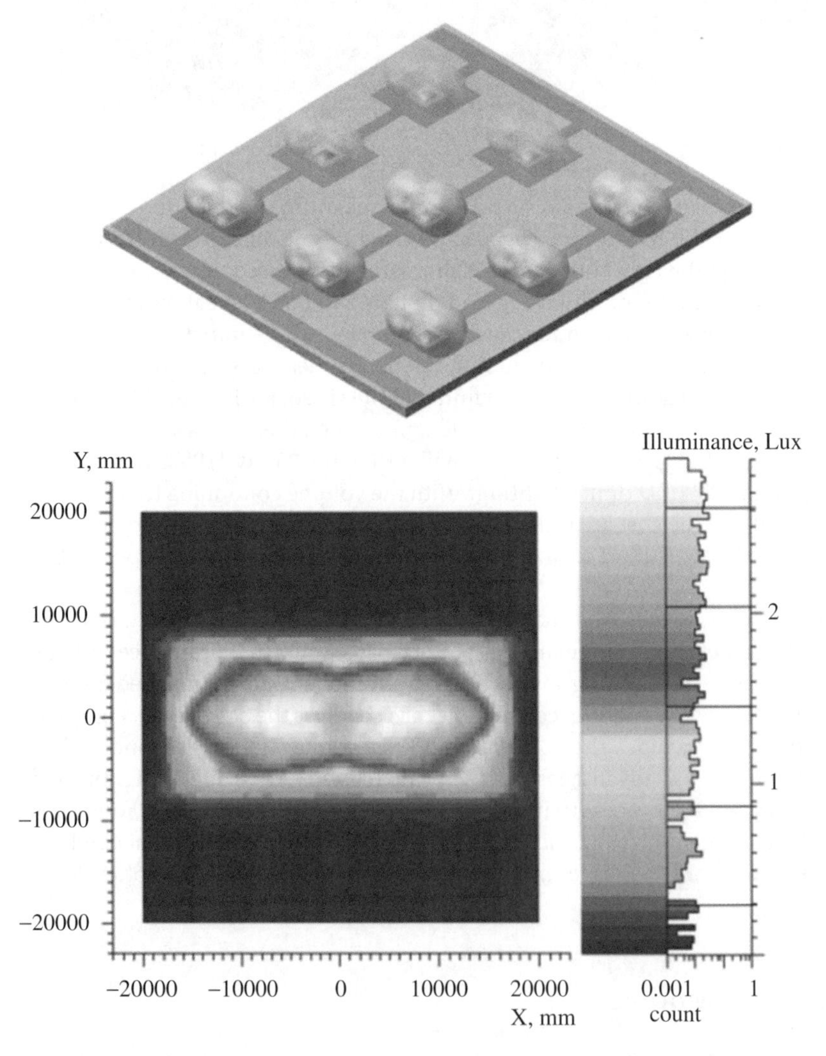

Figure 4.19 A novel 3 × 3 ASLP LED array module based on a ceramic board for road lighting and its optical performance.

light source of the 108 W LED road lamp could be smaller than 120 mm × 105 mm, which will considerably reduce the size and cost of LED street lamps and fling down a challenge to heat dissipation technologies for LED fixtures in the future.

In most LED applications, since the LED array module with a ceramic board will be bonded on the MCPCB before connecting with heat sinks, an application-specific LED array module directly based on MCPCB will provide a more effective solution with the advantages of low thermal resistance and cost. Figure 4.20 shows a 3 × 3 ASLP LED array module based on an MCPCB, and it mainly consists of an MCPCB with circuits, LED chips, phosphor, and a novel freeform PC lens array. This module can be bonded directly on the heat sink of LED road lamps and reduce one thermal interface between the ceramic board and MCPCB, which will decrease the thermal resistance from LED chips to the heat sink. Since the solder mask existing on the surface of MCPCB will reduce the bonding strength between the silicone lens and MCPCB, a PC freeform lens array is adopted in this LED module packaging, and it could be fixed on the MCPCB by bonding or mechanical fastening. Then, silicone will be injected into the cavities of the PC lens through the injection holes and fill the cavities. Moreover, a PC lens array will reduce the cost of application-specific LED array modules further and make it more convenient to assemble.

By comparing with the traditional LED illumination module consisting of a LED and a secondary optical element, the novel ASLPs have the advantages of low profile, small volume, high LOE, low cost, and customer convenience. Therefore, ASLPs provide effective solutions to high-performance and low-cost LED fixtures and will probably become a trend in LED packaging, providing a convincing solution to general lighting.

4.4 ASLP System Integrated with Multiple Functions

Optical design, thermal management, and powering supply are three key issues in the design of LED luminaires. However, the overall performance of many LED luminaires in the market is quite poor, mainly due to a lack of technology integration of optics, cooling, and powering at many new LED companies. In addition, hundreds of traditional manufacturers of lighting are hesitating to enter LED markets due to lack of secondary optics, thermal management, and powering. Therefore, LED packaging modules integrated with multiple functions that are easy to install are essential to these manufacturers for both the survival of their businesses and fast time to market.

In this section, we will present a 16 W replaceable ASLP module integrated with optical, cooling, and powering functions for road lighting. The structure of the LED module was designed for the PAR 38 lamp, which is widely used in general lighting. A borosilicate glass freeform lens with high light extraction efficiency was designed according to a lens design method for an extended source. Lighting performance of the LED module could meet requirements of road lighting well. The heat dissipation structure for the module was designed under several requirements, and the thermal performance was checked and verified by thermal calculation and simulation.[12] The designed heat dissipation structure could keep the junction temperature of LEDs at a relatively low level. A powering unit was also embedded in the bottom of the module. Compared to traditional LED road lamps, this ASLP module for road lighting had advantages of high

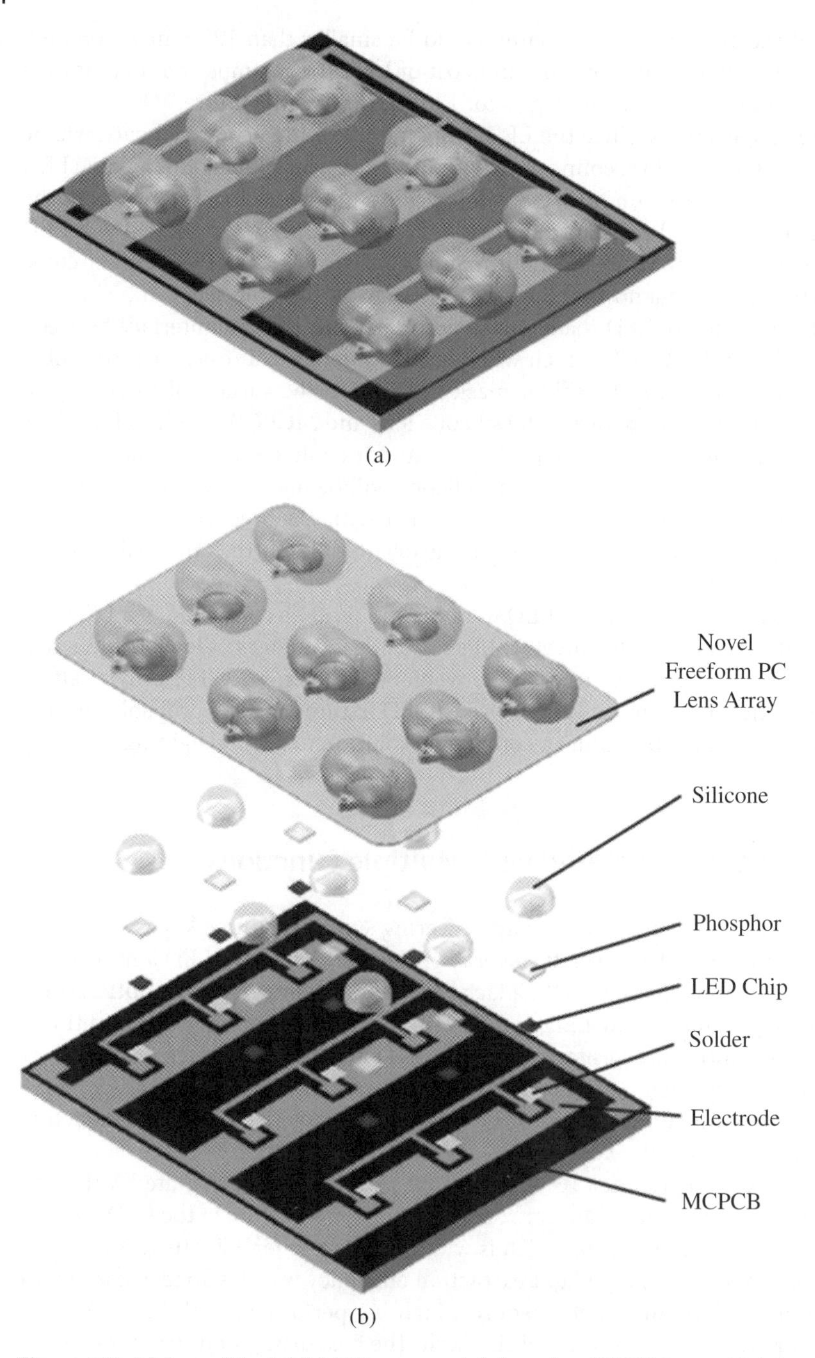

Figure 4.20 (a) A novel 3 × 3 ASLP LED array module based on MCPCB for road lighting and (b) its detailed optical structure.

system optical efficiency, relatively low junction temperature, replaceability, and easy assembly. The ASLP module provided a new concept for LED road lamp design.

4.4.1 Optical Design

4.4.1.1 Problem Statement

In this section, we try to design a freeform lens for a 16 W replaceable ASLP module for road lighting. Blue LED chips integrated with yellow phosphor are the most common way to obtain white LEDs. Blue light emits from the LED chips. The phosphor layer first absorbs the blue light and then re-emits yellow light. As shown in Figure 4.21, light sources of the 16 W ASLP module consist of a 4×4 LED chip array and a phosphor layer. Sixteen LED chips are bonded at the center of a circular MCPCB. Dimensions of each LED chip are $1 \times 1 \times 0.1$ mm. The space between the centers of two chips is 2.4 mm for the convenience of the chip-bonding and wire-bonding processes. Therefore, the size of the light source of blue light reaches as large as 8.2×8.2 mm. Moreover, a phosphor layer is coated on the chips. The base of the phosphor is constrained by the edges of an inner square groove in the MCPCB. The side length of the phosphor layer is 10 mm, so that the light source of yellow light takes an area of 100 mm^2.

The length of the lens is limited within 70 mm when considering the size of a PAR 38 lamp, heat dissipation structure, mechanical structure, and so on. The longest distance between the light sources and lens is 35 mm, which is less than five times the light sources' diameters. Therefore, neither the chip array nor the phosphor layer could be regarded as a point source during lens design according to the far-field conditions of LED.[13] The lens should be designed based on an extended source.

4.4.1.2 Optical Modeling

Since the freeform lens is designed by modifying optimization coefficients of the target according to simulation results, it is essential to build precise optical models. The vertical electrode LED chip is lifted off from sapphire and bonded on Si. The top surface of Si is coated with Ag to reflect lights. The size of the chip is 1×1 mm. The thickness of each layer is: N-GaN 4 μm, MQW 100 nm, P-GaN 300 nm, and Si 100 μm. The refractive indexes and absorption coefficients for N-GaN, MQW, and P-GaN are 2.42, 2.54, 2.45 and 5 mm^{-1}, 8 mm^{-1}, 5 mm^{-1}, respectively.[14] Blue light (465 nm) is isotropically emitted from the top and bottom surfaces of MQW with uniform distribution.

The phosphor layer is a bulk scattering material, which is fabricated by mixing the transparent silicone with phosphor particles. The height of the phosphor layer is 0.4 mm, while the side length of the base is 10 mm. The phosphor concentration is 0.35 g/cm^3.

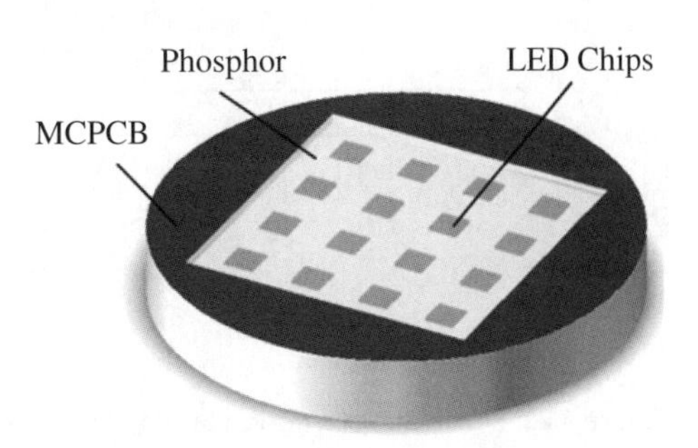

Figure 4.21 Light sources of the 16 W ASLP module.

Based on a Mie scattering model, the absorption and reduced scattering coefficients of phosphor are 3.18 mm^{-1} and 5.35 mm^{-1} for blue light and 0.06 mm^{-1} and 7.44 mm^{-1} for yellow light.[14] The refractive index of silicone is 1.50.

Blue light (465 nm) and yellow light (555 nm) are separately calculated with the Monte Carlo ray-tracing method. The phosphor layer first absorbs blue light and then re-emits yellow light from the top and bottom surfaces. Experiments show that the radiation pattern of yellow light is similar to Lambertian when blue light passes through the phosphor layer. In those LEDs tested, *blue light* covers wavelengths of 380 to 490 nm and *yellow light* covers wavelengths of 490 to 780 nm. In this analysis, single wavelengths of blue light (465 nm) and yellow light (555 nm) are adopted in simulation, and this simplification has been verified as effective for white LED simulations.[15]

4.4.1.3 Design of a Freeform Lens

Borosilicate glass with a refractive index of 1.47 is adopted as the material for the lens. The refractive index of the silicone filled in the cavity between the phosphor and the lens is 1.50, which is close to that of the lens. Therefore, the inner surface of the lens is to be designed as an inner concave spherical surface, which will have little change in the transmission directions of incident lights from LED chips and phosphor. We will focus on the construction of the outside surface of the lens in this study. A borosilicate glass freeform lens would be designed to form a 40 m long and 18 m wide rectangular radiation pattern at a height of 10 m, as shown in Figure 4.22, which can meet the requirements of road lighting.

As shown in Figure 4.23, a borosilicate glass freeform lens with a refractive index of 1.47 is designed according to the method discussed here. Firstly, we divide both the one-quarter source and target plane into 20,000 grids with equal luminous flux and unequal area, respectively, and establish a light energy mapping relationship between these grids. Secondly, we calculate coordinates and the normal vector of each point on the freeform surface and construct the lens utilizing these points. Finally, we validate the lighting performance of the lens through simulation and obtain the final design after optimizing coefficients of the target. The largest values of the length, width, and height of the glass freeform lens are 64.0, 35.5, and 24.1 mm, respectively. To provide

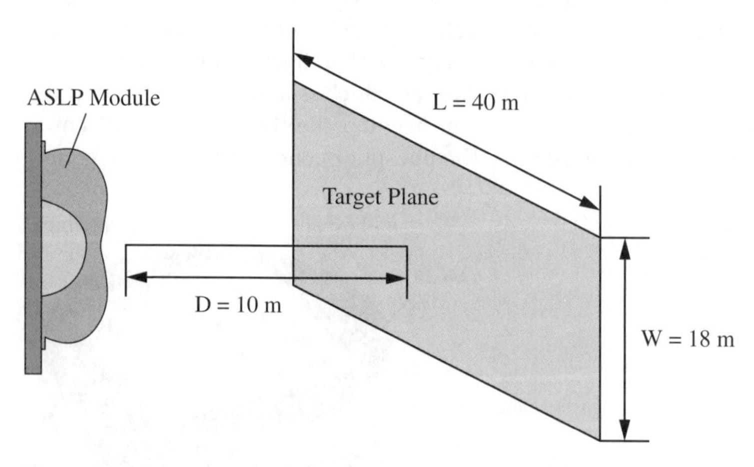

Figure 4.22 Schematic of a design target of the ASLP module for road lighting.

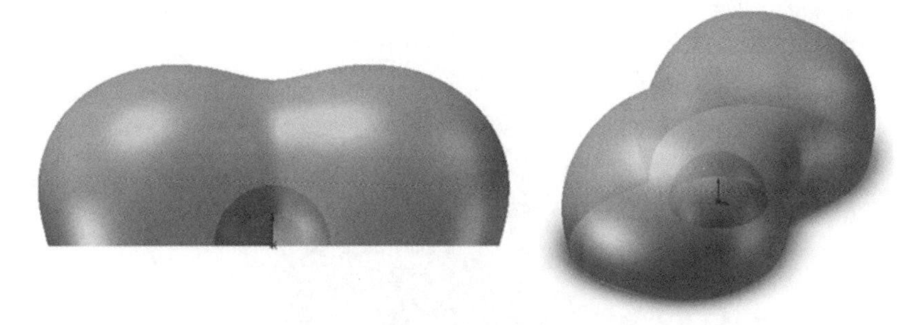

Figure 4.23 A borosilicate glass freeform lens for the ASLP module.

space for LED packaging, the radius of the inner concave spherical surface is set as 8.0 mm.

4.4.1.4 Simulation of Lighting Performance

As shown in Figure 4.24, optical components of this 16 W ASLP module include an LED chip array, a phosphor layer, a glass freeform lens, and the silicone filled in the cavity of the lens. We simulate the lighting performance of the ASLP module numerically by the Monte Carlo ray-tracing method. The lighting performance is the combination results of blue light and yellow light. Figure 4.24 shows that the ASLP module forms a subrectangular radiation pattern with a length of 43 m and width of 20 m at a height of 10 m, which is a little larger than the expected shape.

Road-lighting performance of the ASLP module is also evaluated. We design a 144 W LED road lamp consisting of nine pieces of 16 W ASLP modules for major road lighting. DIALux software is adopted to calculate lighting performance on the road. Figure 4.25 shows illuminance distribution on a two-way eight-lane major road. From the comparisons shown in Table 4.3, we can find that the 144 W LED road lamp based on the ASLP modules could meet the requirements of road lighting very well.

In addition, light extraction efficiency of this ASLP module reaches as high as 96.7% of that of the traditional LED module with a hemisphere lens, whose light extraction efficiency is defined as a reference of 100%. As we know, light output efficiencies of traditional secondary optics (e.g., freeform lenses) and lampshades for LED road lamps are always at a low level of about 90% because of Fresnel reflection loss and material absorption. As shown in Table 4.4, since neither secondary optics nor a lampshade is needed for the LED road lamp based on the ASLP modules, the system optical efficiency of the new road lamp is 19.4% higher than that of the traditional LED road lamp. Therefore, the ASLP module has good lighting performance and higher system optical efficiency. The module could be directly used for road lighting, and no secondary optics and lampshade are needed; this makes it convenient for old and new LED luminaire designers and manufacturers to use, and also it will further reduce the cost of LED luminaires.

4.4.2 Thermal Management

In the ASLP module, LED chips are placed close to each other, which will increase the heat dissipation per area and create challenges for thermal management of the new module. Details of thermal design readers are in Ref. [12]. A simulation made by software

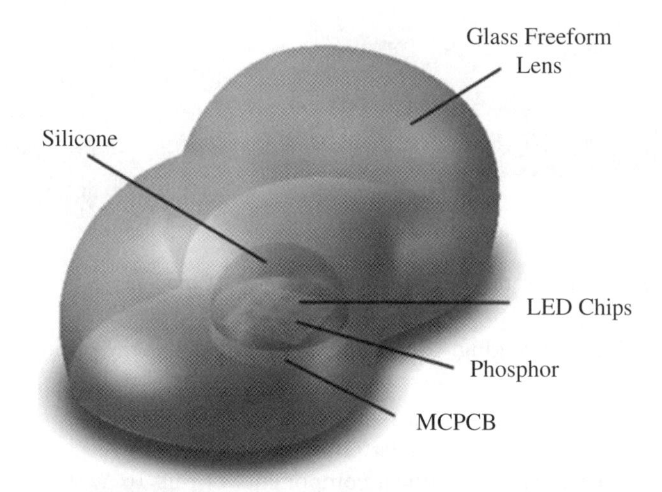

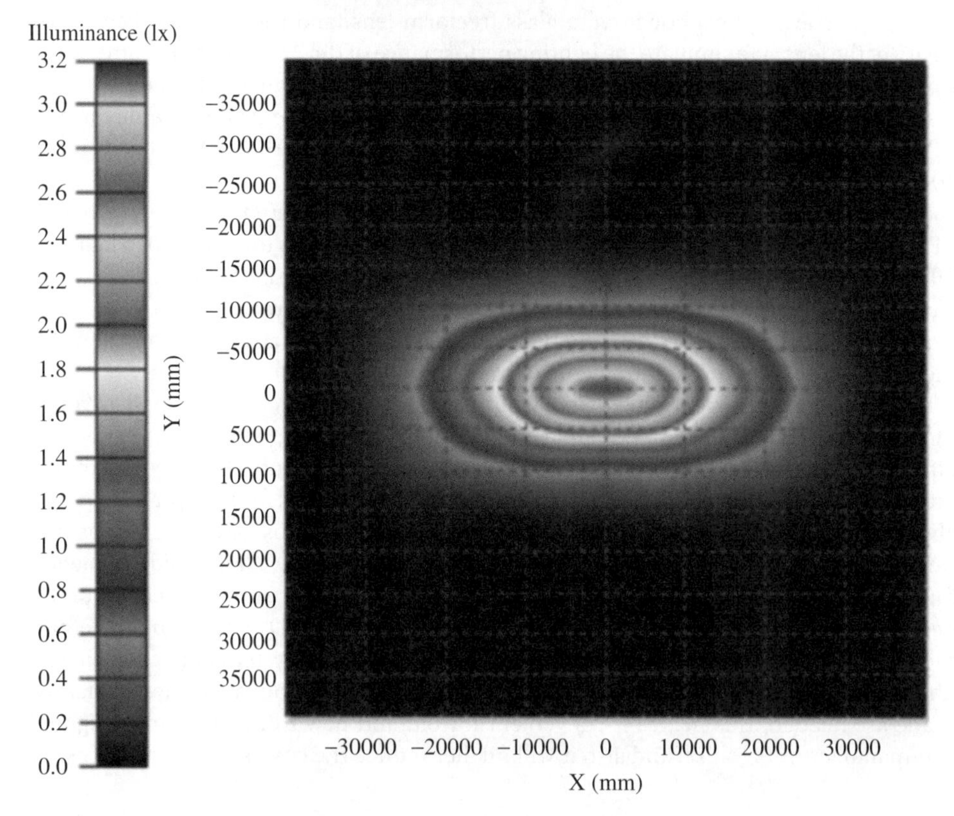

Figure 4.24 Optical components of the 16 W ASLP module and its lighting performance where illuminance decreases gradually from the center to the side of the light pattern.

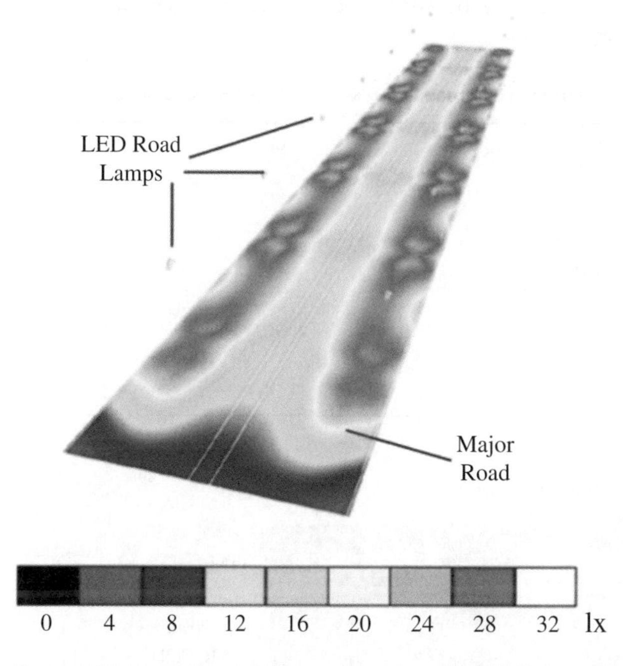

Figure 4.25 Simulation lighting performance of 144 W LED road lamps based on the 16 W ASLP modules where illuminance decreases gradually from the area under the LED road lamp to the center of the road.

Table 4.3 Comparison of the road-lighting performance of the 144 W LED road lamp with the national standards.

			Road luminance		
Grade	**Road type**		**Average luminance L_{av}(cd/m^2)**	**Overall uniformity U_o**	**Longitudinal uniformity U_L**
I	Major road	National standard	1.5/2.0	0.4	0.7
		144 W LED road lamp	1.6	0.5	0.7

			Road illuminance		Glare limitation threshold increment
Grade	**Road type**		**Average illuminance E_{av}(lx)**	**Uniformity U_E**	***TI* (%)**
I	Major road	National standard	20/30	0.4	10
		144 W LED road lamp	24.0	0.7	8

Table 4.4 Comparison of system optical efficiencies between a traditional LED road lamp and a new LED road lamp based on the ASLP modules.

	Traditional LED road lamp, %	New LED road lamp, %
Light extraction efficiency of LED module (η_{LEE})	100 (Reference)	96.7
Light output efficiency of secondary optics (η_{SO})	90.0	–
Light output efficiency of lampshade (η_L)	90.0	–
System optical efficiency ($\eta_{SOE} = \eta_{LEE} \times \eta_{SO} \times \eta_L$)	81.0	96.7
Enhancement	–	19.4

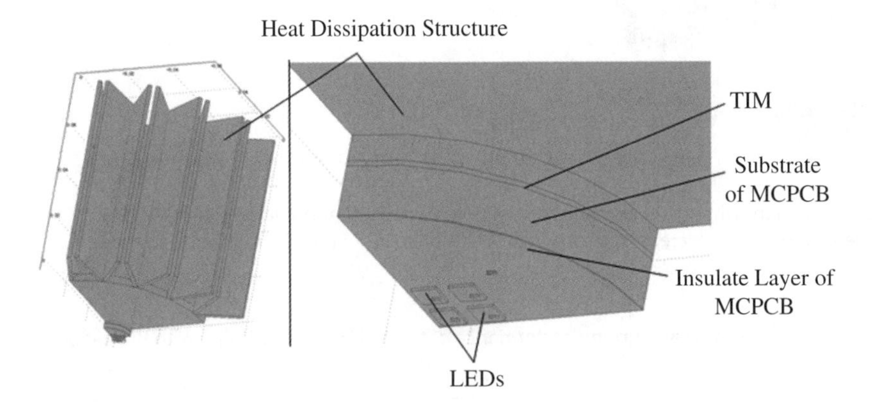

Figure 4.26 Simulation model of a heat dissipation structure.

COMSOL is used to verify thermal performance of the whole module. As the module is axi-symmetric, only one-fourth of the module is simulated. Figure 4.26 is the model for the simulation. Thicknesses of thermal interface material (TIM), substrate, and the insulate layer of MCPCB are 0.2 mm, 1.9 mm, and 0.05 mm, respectively. Dimensions of the LED are $1 \times 1 \times 0.104$ mm. Heat dissipation structure is made of aluminum. Material used for substrate of MCPCB is aluminum 6063-T83. The LED is supposed to be silicon.

The simulation obtained the temperature field of the whole module, and it is shown in Figure 4.27. The maximum temperature is 86.6 °C found at the LEDs. The minimum temperature is 48.9 °C found at the heat dissipation structure surface, so that the temperature difference between the fins and ambient temperature is less than 15 °C. The result shows that the designed heat dissipation structure is capable of maintaining a junction temperature of LEDs in the range of <90 °C when ambient temperature was 35°C.

4.4.3 ASLP Module

Figure 4.28 shows a whole ASLP module for road lighting integrated with multiple functions, including optical, cooling, and powering functions. An AC–DC powering unit is

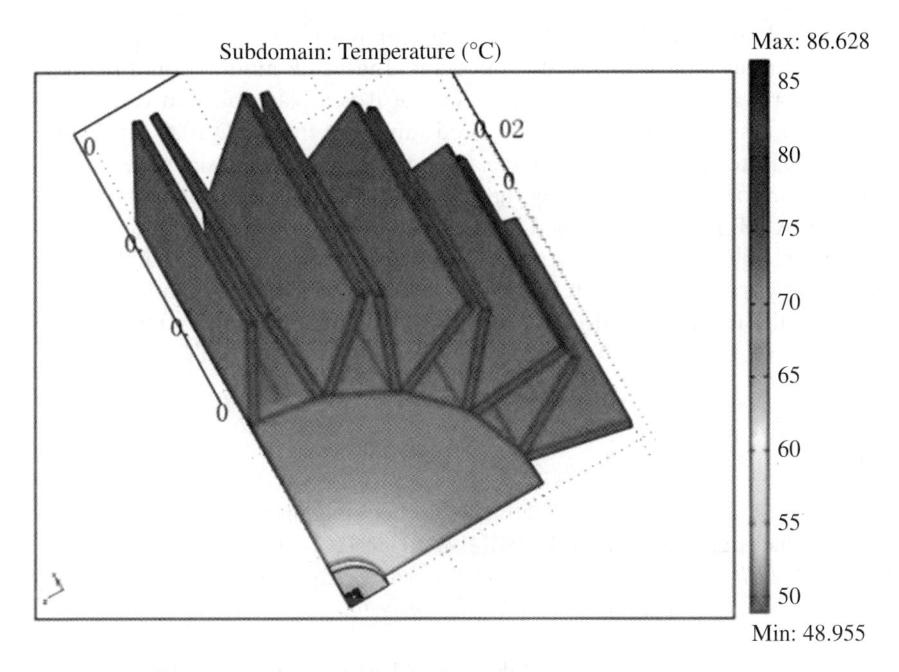

Figure 4.27 Temperature field obtained by simulation.

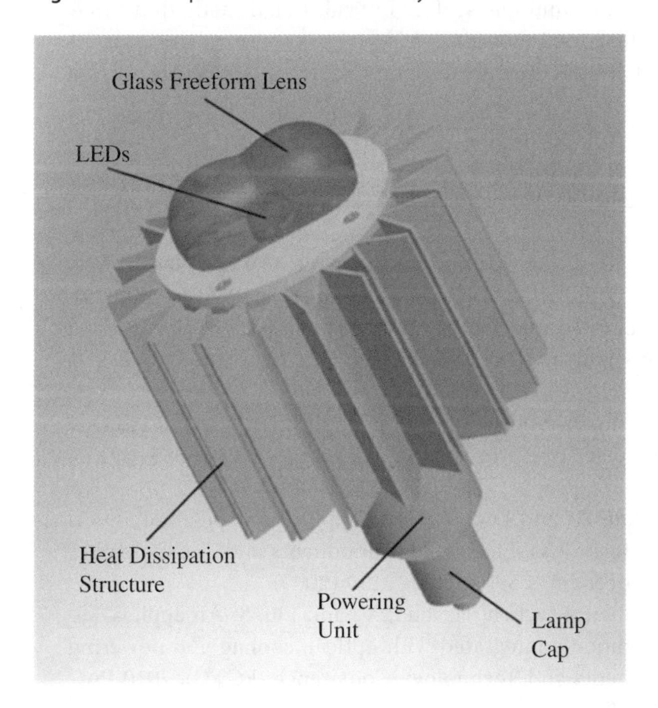

Figure 4.28 An ASLP module for road lighting.

embedded in the bottom of the module. The lamp cap has functions not only of electric connection but also of mechanical connection. The ASLP module could be fixed easily on a LED road lamp by the lamp cap, so that it is replaceable and convenient for customers to assemble. Since most required functions in a LED road lamp have already been integrated in this ASLP module, no more complicated designs are needed, and manufacturers only need to supply the mechanical structure for assembly. With the replaceable ASLP module proposed here, it would be very easy for those companies in traditional lighting to switch to an LED community so as to minimize the pains caused to them by transferring from traditional lighting to solid-state lighting. Therefore, ASLP modules integrated with multiple functions probably lead the trend of LED packaging.

References

1 Craford, M.G. LEDs for solid state lighting and other emerging applications: status, trends, and challenges. Proc. of SPIE 5941, 594101–594101–10 (2005).

2 Schubert, E.F. *Light-Emitting Diodes* (Cambridge University Press, 2006).

3 Liu, S. and Luo, X.B. *Design of LED Packaging for Lighting Applications* (John Wiley and Sons, 2009)

4 Benítez, P., Miñano, J.C., Blen, J., Mohedano, R., Chaves, J., Dross, O., Hernández, M., and Falicoff, W. Simultaneous multiple surface optical design method in three dimensions. *Opt. Eng.* **43**, 1489–1502 (2004).

5 Ries, H., and Muschaweck, J. Tailored freeform optical surfaces. *J. Opt. Soc. Am. A* **19**, 590–595 (2002).

6 Wang, L., Qian, K., and Luo, Y. Discontinuous free-form lens design for prescribed irradiance. *Appl. Opt.* **46**, 3716–3723 (2007).

7 Ding, Y., Liu, X., Zheng, Z., and Gu, P. Freeform LED lens for uniform illumination. *Opt. Expr.* **16**, 12958–12966 (2008).

8 Wang, K., Liu, S., Chen, F., Qin, Z., Liu, Z., and Luo, X. Freeform LED lens for rectangularly prescribed illumination. *J. Opt. A: Pure Appl. Opt.* **11**, 105501 (2009).

9 Wang, K., Chen, F., Liu, Z., Luo, X., and Liu, S. Design of compact freeform lens for application specific Light-Emitting Diode packaging. *Opt Expr.* **18**, 413–425 (2010).

10 Wang, K., Chen, F., Liu, Z., Luo, X., and Liu, S. Novel application-specific LED packages integrated with compact freeform lens. *ASME InterPACK'09*, 685–691 (2009).

11 Wang, K., Liu, S., Chen, F., Liu, Z., and Luo, X. Novel application-specific LED packaging with compact freeform lens. Electronic Components and Technology Conference (ECTC), 2009 Proceedings 59th, 2125–2130 (2009).

12 Wang, K., Mao, Z., Liu, Z., Chen, F., Chen, H., Luo, X., and Liu, S. An application specific LED packaging module integrated with optical, cooling and powering functions. Electronic Components and Technology Conference (ECTC), 2010 Proceedings 60th, pp. 686–692 (2010).

13 Moreno, I. and Sun, C.C. LED array: where does far-field begin. 8th International Conference on Solid State Lighting, San Diego, CA, August. 70580R-1–70580R-9 (2008).

14 Liu, Z.-Y., Liu, S., Wang, K., and Luo, X.-B. Studies on optical consistency of white LEDs affected by phosphor thickness and concentration using optical simulation. *IEEE Trans. Comp. Pack. Tech.* **33**, 680–687 (2010).
15 Liu, Z., Liu, S., Wang, K., and Luo, X. Optical analysis of color distribution in white LEDs with various packaging methods. *IEEE Photon. Tech. Lett.* **20**, 2027–2029 (2008).

5

Freeform Optics for LED Indoor Lighting

5.1 Introduction

With the increased luminous efficiency and decreased costs of LED chip and packaging modules, the LED lighting market is gradually transferring from outdoor lighting, such as LED road lighting that is supported and promoted by government first, to indoor lighting, such as commercial lighting, office lighting, factory lighting, residential lighting, and so on. Within these LED indoor-lighting applications, since the LED commercial, office, and factory lighting have reached tipping points in the market, these three LED indoor-lighting markets have developed much faster than LED residential lighting. At the same time, it is widely believed that LED residential lighting is no farther away from us with the rapid decrease of LED cost.

Like LED outdoor lighting, LED indoor lighting also has strict requirements on the quality and controllability of the light pattern of LED luminaires. Regarding the light pattern, there are mainly two types of optical design concepts for LED indoor lighting. One is focusing on the quality of a single light pattern, including the large emitting angle (≥90°, e.g., an LED ceiling light) and small emitting angle (<90°, e.g., a spotlight for commercial lighting), and another is paying more attention to the whole lighting performance of the LED luminaire array (e.g., an LED high bay lamp array for a factory). Both of these types of concepts require precise optical design methods. In this chapter, we will introduce freeform lens designs for three typical LED indoor-lighting situations: a single light pattern with a large emitting angle, one with a small emitting angle, and whole uniform illumination with an LED lights array, based on the algorithms demonstrated in Chapter 3.

5.2 A Large-Emitting-Angle Freeform Lens with a Small LED Source

In the indoor-lighting luminaires, like LED down lighting and LED panel lighting, people usually choose a diffuser to enhance the illumination uniformity of the light pattern; this diffuses the light and increases the exit angle. However, the light transmittance of the diffuser is usually as low as around 70~85%. Its diffusing ability is also limited. For some collimating light, a conventional diffuser has difficulties realizing uniform illumination. Besides, the price of a diffuser is usually high. Using these diffusers extensively

Freeform Optics for LED Packages and Applications, First Edition. Kai Wang, Sheng Liu, Xiaobing Luo and Dan Wu.

in LED luminaires not only will increase the price but also will decrease the optical efficiency. Therefore, it is not a cost-effective solution. If we choose the rotational symmetrical freeform lens, which has advantages like low volume, high light extraction efficiency, uniform light pattern, and low cost for mass production, it is convenient to realize large-angle uniform illumination and provide a kind of high-cost performance solution for LED uniform illumination.

Now we use the most-frequently used Philip Lumileds K2 LED package module[1] and CREE XLamp XR-E LED package module[2] as the light sources to design the large-emitting-angle rotational symmetrical freeform lenses for the indoor luminaires.

5.2.1 A Freeform Lens for a Philip Lumileds K2 LED

The optical model for a Philip Lumileds K2 LED is shown in Figure 5.1a. When the module is driven by 350 mA, the optical flux was 100 lm in the year 2008, and the spatial

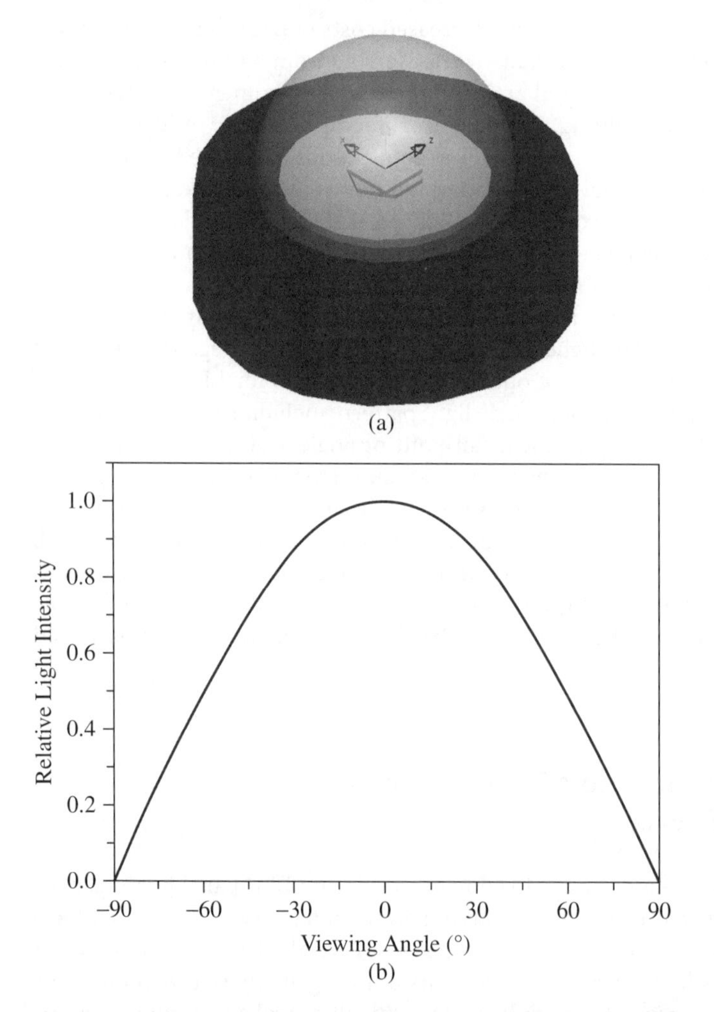

Figure 5.1 (a) Philips Lumileds K2 LED optical model, and (b) its Lambertian light intensity distribution curve (LIDC).

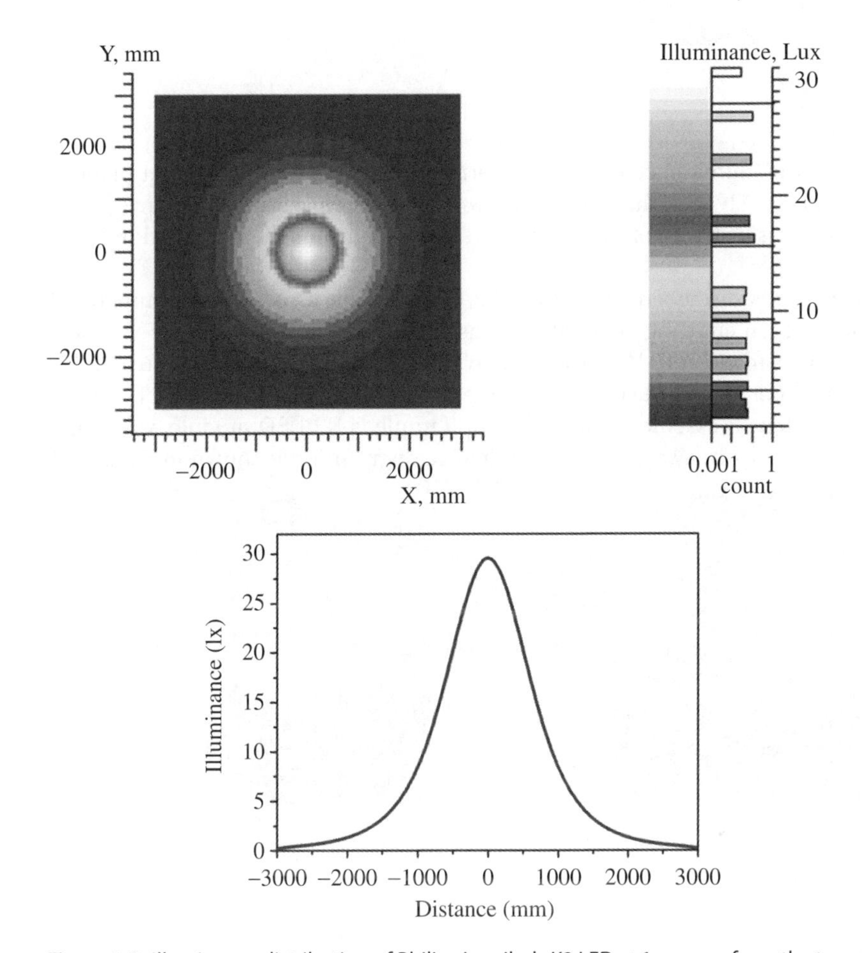

Figure 5.2 Illuminance distribution of Philips Lumileds K2 LED at 1 m away from the target plane.

distribution of the light intensity is close to Lambertian as shown in Figure 5.1b. As shown in Figure 5.2, the circular pattern is not uniform because the central region has larger illuminance, while the edge region has relatively low illuminance. In the simulation, the distance between the LED module and the receiver is 1 m, and 2 million rays are traced. Since there is no obvious illuminated region in the light pattern, it is rather difficult to calculate the illumination uniformity in the target region.

In theory, we can calculate the illuminance distribution of the Lambertian light source. According to photometry, the relationship between the illuminance E and light intensity $I(\theta)$ is:

$$E = \frac{I(\theta)\cos\theta}{l^2} \tag{5.1}$$

where $I(\theta)$ is the light intensity at the test direction, θ is the angle between the ray emitting direction and the center axis of the light source, and l is the distance from the light source to the test point. Assuming the perpendicular distance from the light source to the test plane is d_0, here we have $l = d_0/cos\theta$. With Lambertian light intensity distribution, the relationship between the illuminance E and the ray exit angle θ is as follows:

$$E = \frac{I_0 \cos^4\theta}{d_0{}^2} \tag{5.2}$$

From Eq. 5.2, the illuminance of the Lambertian light source on the target plane is decreasing progressively with the function of $cos^4\theta$. For example, the illuminance at 60° (the ray exit angle) is just 1/16 of that at the center, which means the illumination is not uniform at all.

Based on methods discussed in Section 3.2.1, we design a rotational symmetrical freeform lens with an emitting angle of 120°, as shown in Figure 5.3, where N is 500. The lens material is made of PMMA (polymethyl methacrylate), whose refractive index is 1.49. The initial height and the largest diameter of the lens are 5.0 mm and 11.36 mm, respectively. The lens is assembled on the Philip Lumileds K2 LED module, where the target plane is placed 1 m away, and its illuminance distribution is shown in Figure 5.4

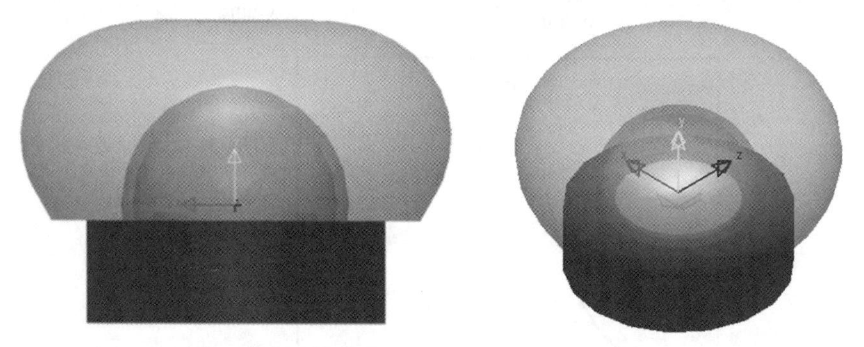

Figure 5.3 Circularly symmetrical PMMA freeform lens with a 120° emitting angle, based on the Philips Lumileds K2 LED.

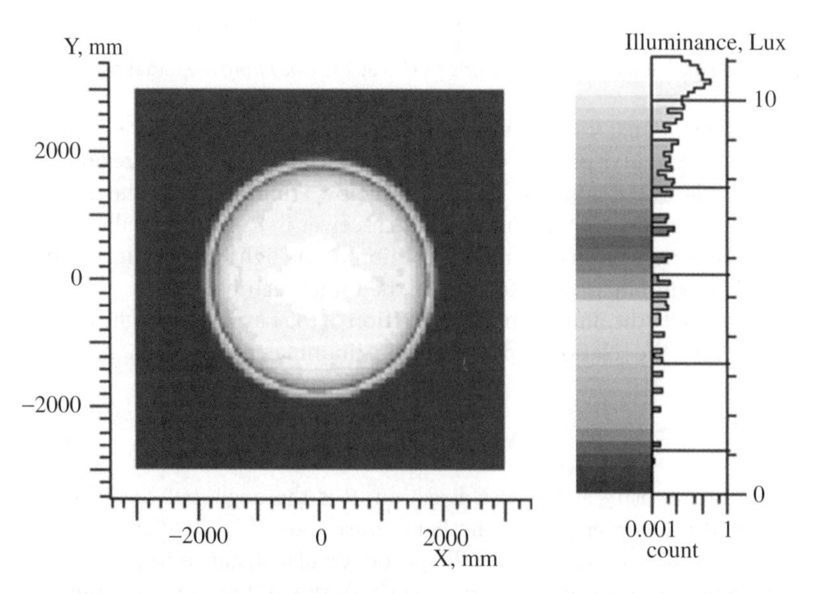

Figure 5.4 Illuminance distribution at 1 m away from the target plane for a circularly symmetric freeform lens with a 120° emitting angle.

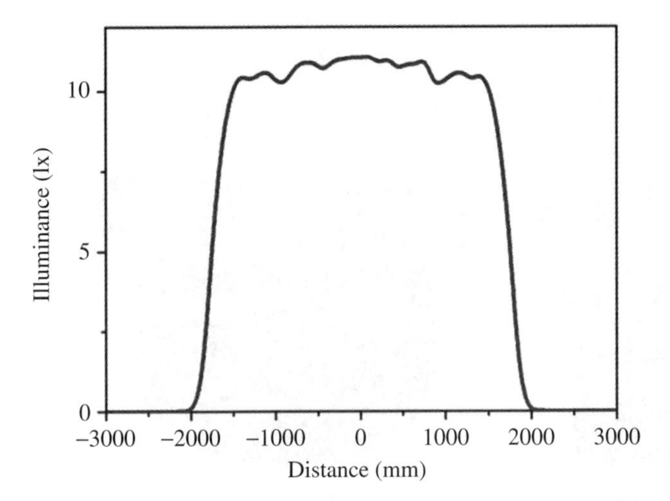

Figure 5.4 *(Continued)*

after tracing 2 million rays. It is seen that most light energy (~98%) is uniformly distributed on the targeted illumination region. We introduce the illuminance uniformity U_E to describe the illuminance distribution on the target plane, which is usually defined as the ratio of the minimum illuminance E_{min} and the maximum illuminance E_{max}:

$$U_E = E_{\min}/E_{\max} \tag{5.3}$$

Based on our simulations, in the uniform illumination region, E_{min} = 10.25 lx, E_{max} = 11.10 lx, and U_E = 0.923. Therefore, this freeform lens could realize large-emitting-angle (120°) uniform illumination. In addition, when the light source is placed at the center of a semispherical lens, almost all the light emitted from the light source could transmit through the lens. There is no total internal reflection loss, but material absorption loss and Fresnel loss, so the extraction efficiency is the largest. If we assume the extraction efficiency of a semispherical lens is 1, then the relative extraction efficiency of our freeform lens is 99.9%.

As seen in Figure 5.5 and Figure 5.6, we also design another rotational symmetrical freeform lens with an emitting angle of 90°. The lens material is also PMMA, the height is 7.0 mm, the largest diameter is 9.98 mm, and the relative light extraction efficiency is 99.8%. Through simulations, we found that this lens also could realize uniform illumination on the target plane, U_E =0.877. As a short conclusion, this algorithm can effectively design a freeform lens to realize large-view-angle uniform illumination.

5.2.2 Freeform Lens for a CREE XLamp XR-E LED

The CREE XLamp XR-E LED package module is another extensively used LED light source, and it output 110 lm at 350 mA in the year 2008. Here, we discuss its application as the light source. The sample and the optical model are shown in Figure 5.7. There is a deep reflector structure when a silicone lens is placed at the top side. The light paths can be categorized into two parts: one part is large-viewing-angle rays, which are

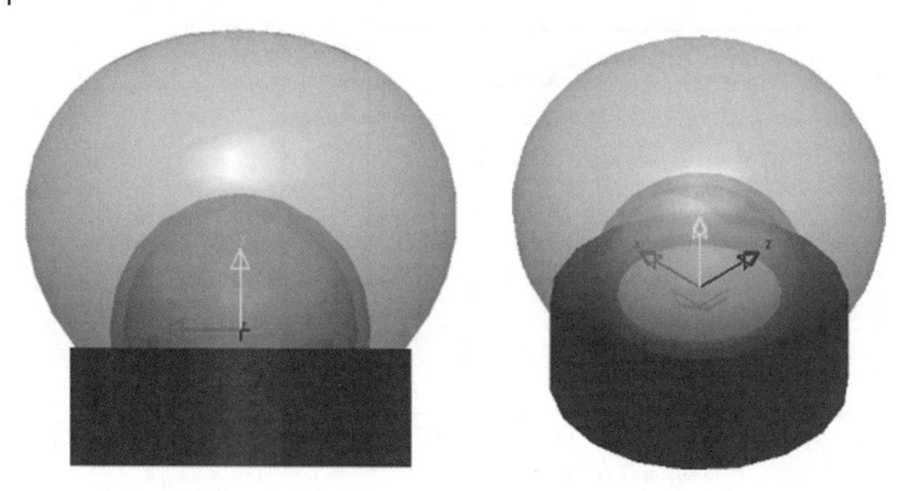

Figure 5.5 Circularly symmetrical PMMA freeform lens with a 90° emitting angle, based on the Philips Lumileds K2 LED.

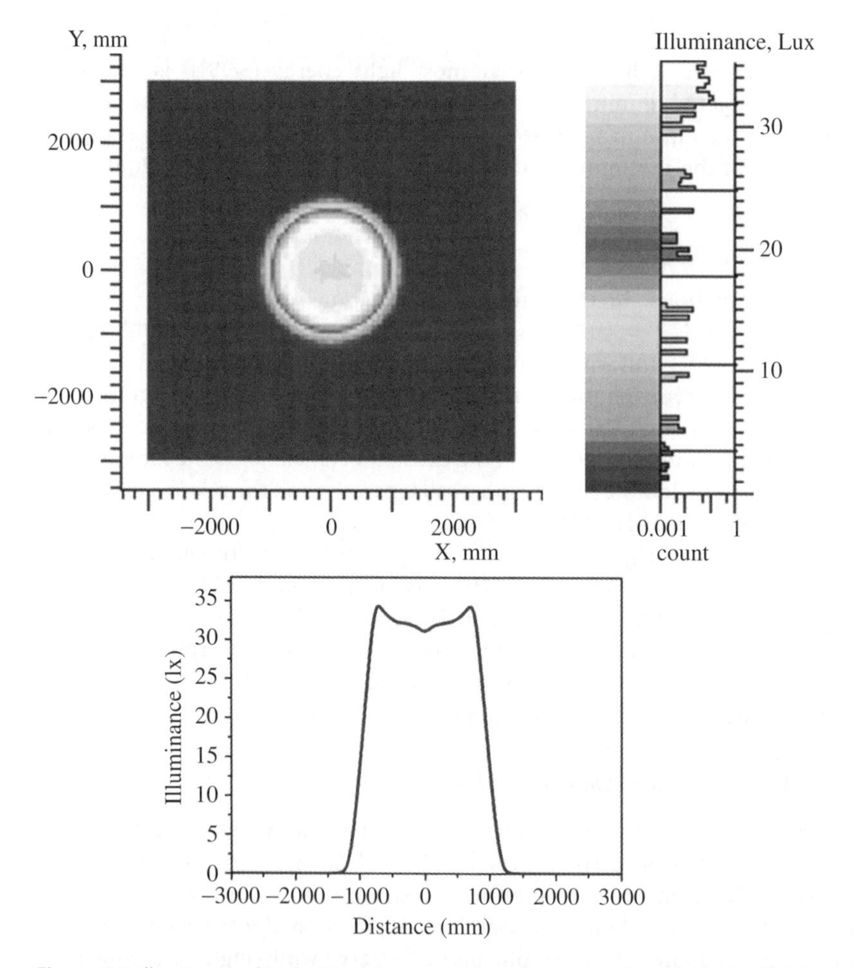

Figure 5.6 Illuminance distribution at 1 m away from the target plane for a circularly symmetrical freeform lens with a 90° emitting angle.

(a)

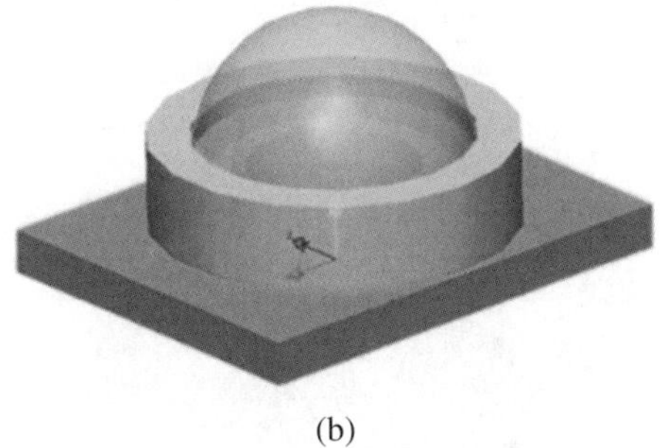

(b)

Figure 5.7 (a) CREE XLamp XR-E LED real object, and (b) optical model.

Figure 5.8 CREE XLamp XR-E LED's LIDC.

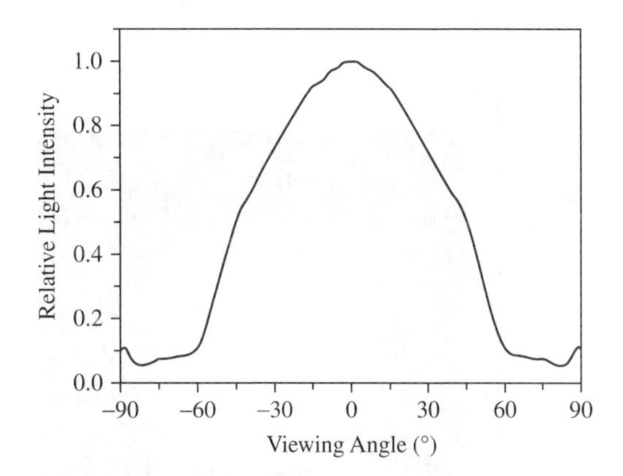

multiply reflected by the reflector and then transmit through the lens, and the other is small-viewing-angle rays, which will directly transmit through the lens. Therefore, the light intensity distribution curve (LIDC) is different from a conventional Lambertian light source. We use an LIDC measurement system (EVERFINE GO1900) to test the LED module. Figure 5.8 shows the LIDC for a CREE XLamp XR-E LED module. Here, the intensity distribution tends to be closer to the center axis than a Lambertian light source, falling in the category of *collimating type*. Also, when the angle increases, the light intensity increases. This LED module has a relatively complex intensity distribution.

We design two large-emitting-angle rotational symmetrical freeform lenses (120° and 90°) based on a CREE XLamp XR-E LED light source, as shown in Figures 5.9 and 5.11. For the 120° freeform lens, the material is PMMA, the height is 3.7 mm, the largest diameter is 14.74 mm, and the relative light extraction efficiency is 98.9% with $U_E = 0.833$. The illuminance performance is shown in Figure 5.10. For the 90° freeform lens in Figure 5.11, the material is PMMA, the height is 3.7 mm, the largest diameter is 9.06 mm, and the relative light extraction efficiency is 98.5% with $U_E = 0.774$. Therefore, for the complicated LIDC LED modules, this algorithm is also suitable for the effective design of a freeform lens to realize large-view-angle uniform illumination. The illuminance performance is shown in Figure 5.12.

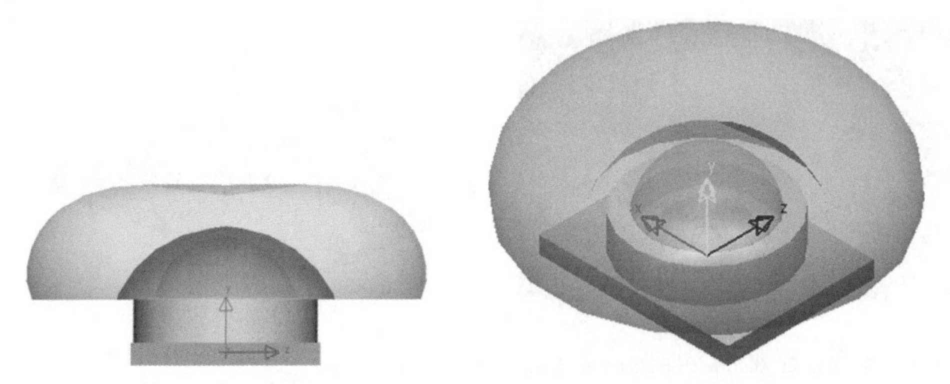

Figure 5.9 Circular-symmetrical PMMA freeform lens with a 120° emitting angle based on the CREE XLamp XR-E LED.

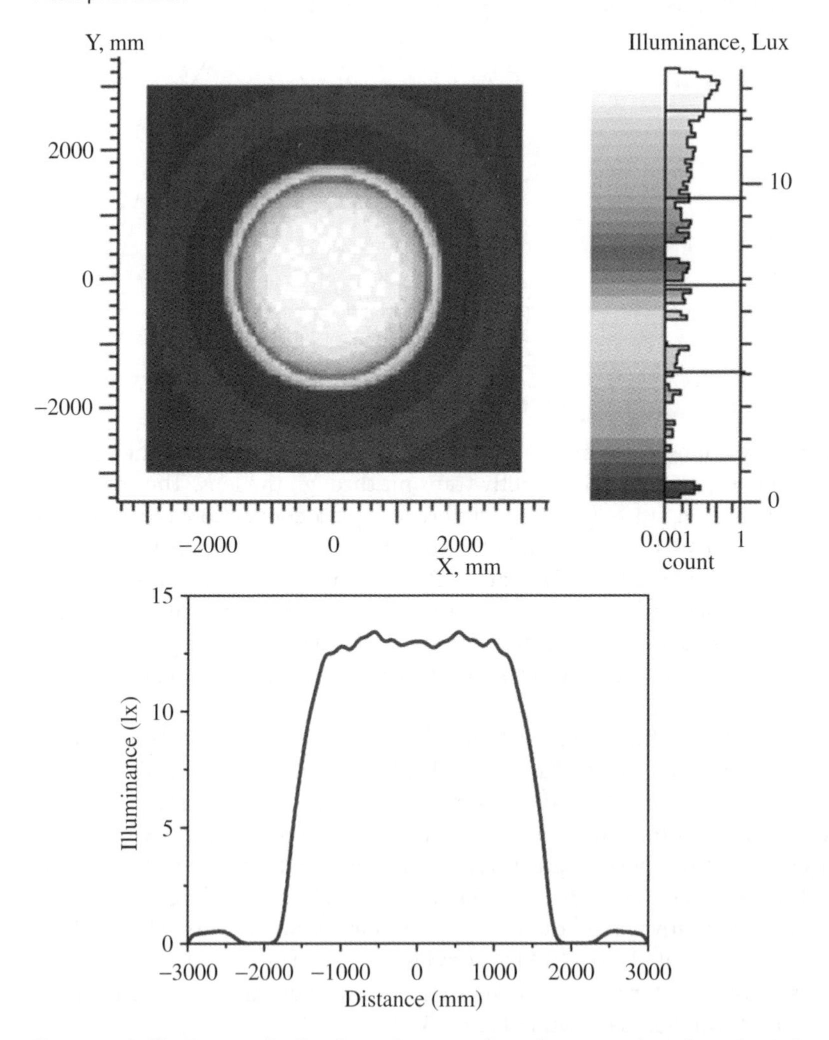

Figure 5.10 Illuminance distribution at 1 m away from the target plane for a circularly symmetrical freeform lens with a 120° emitting angle based on the CREE XLamp XR-E LED.

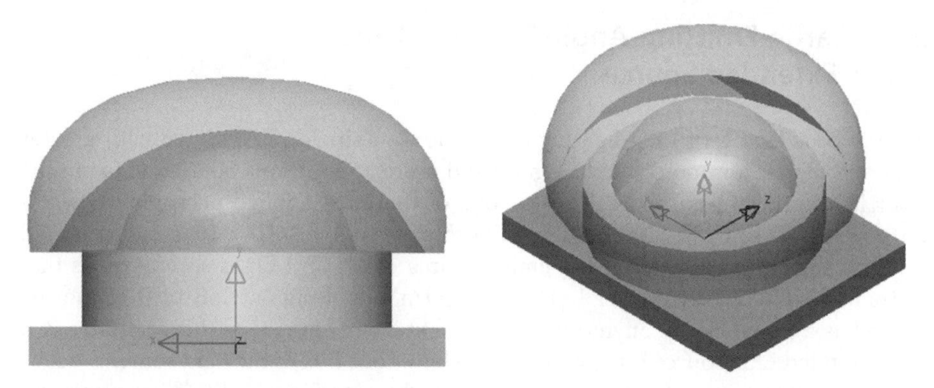

Figure 5.11 Circularly symmetrical PMMA freeform lens with a 90° emitting angle based on the CREE XLamp XR-E LED.

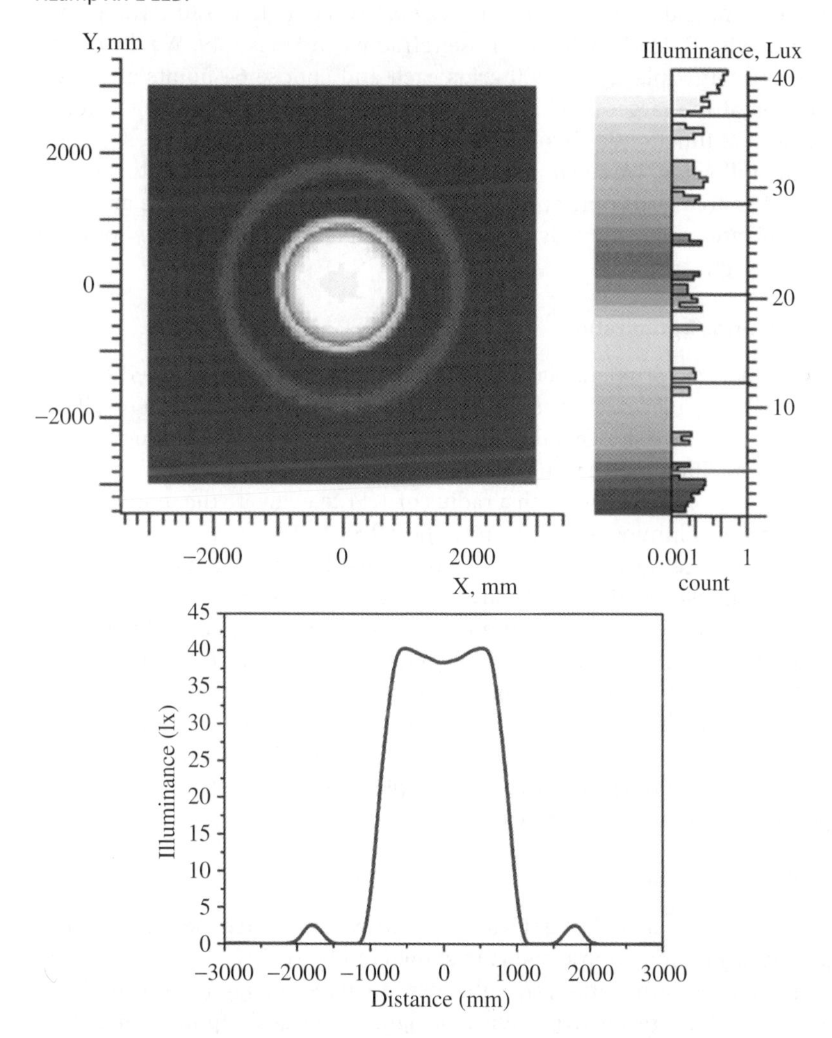

Figure 5.12 Illuminance distribution at 1 m away from the target plane for a circularly symmetrical freeform lens with a 90° emitting angle based on the CREE XLamp XR-E LED.

5.3 A Large-Emitting-Angle Freeform Lens with an Extended Source

With the broadening of LED lighting applications, a single LED source with greater power and higher luminous flux has more and more application requirements in LED indoor lighting. Chip-on-board (CoB) is one of the most applied packaging types of high-power LED sources with multiple LED chips (e.g., 3 × 3, 10 × 10, etc.) integrated in one packaging module. Since the emitting area of a CoB LED is several times that of a traditional 1 W high-power LED, in freeform lens design a CoB LED cannot be regarded as a point source anymore, but is an extended source. In this section, a 3 × 3 mm extended LED source is introduced. The luminous flux of the light source is set as 100 lm, just as a design reference. Our goal is to add a freeform lens to form a uniform circular light pattern with a radius of 1.73 m. The distance between the light source and the target plane is 1 m, and therefore the edge ray emitting angle is 120°. The height of the lens is 7 mm with PMMA material whose refractive index is 1.49. We divide the light source and the target plane into 500 grids each and choose 64 points along the horizontal direction at the target plane. Based on these test points' changing ratios, a 10th-order polynomial function is introduced to fit the values and obtain the function of $k_i = f(r_i)$. In the following, we separately use the (1) target plane grids optimization algorithm, (2) light source grids optimization algorithm, and (3) target plane and light source grids optimization algorithm, as described in detail in Section 3.3, to optimize the freeform lens design.

5.3.1 Target Plane Grids Optimization

First of all, we design a freeform lens according to the point source design method. We use a 3 × 3 mm LED extended source as the light source, and the simulation results of the horizontal central axis are shown in Figure 5.13. The central region has a higher illuminance value, and the illuminance value decreases gradually as it approaches the edge. We mainly consider a circular region with a radius of 1.44 to evaluate the uniformity at the target plane, and the illuminance uniformity U_E is 0.53.

We only adjust the target grids division, which means that $a_i = 1$, and $b_i = k_i$; the light pattern illuminance simulation results are shown in Figure 5.13. Simulation results show that after the first time of optimization U_E increases to 0.6, and after the second time of optimization U_E increases to 0.9. At this point, we realize a very uniform illuminance. After the third time of optimization, the illuminance distribution curve shows little change compared with the second time, and U_E is still 0.9. Before and after our optimization algorithm, the size of the lens is shown in Figure 5.14; we can find that the shapes of these lenses are similar except that after optimization, the sizes of the lenses are larger, especially at the top of the lenses.

5.3.2 Light Source Grids Optimization

In this algorithm, we only change the light source grids division, which means $b_i = 1$ and $a_i = 1/k_i$. Light pattern illumination results are shown in Figure 5.15. The results show that after the first time of optimization, U_E enhances into 0.58, and after the second and third time it is 0.76 and 0.77, respectively. With the light source grids optimization algorithm, we can enhance illuminance uniformity of the extended source at the target plane,

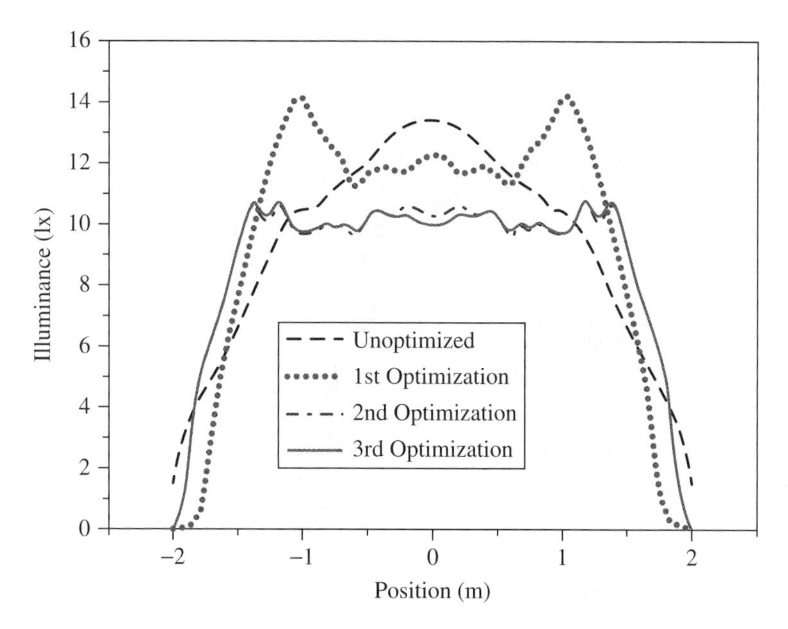

Figure 5.13 Simulation results of the target plane grids optimization algorithm.

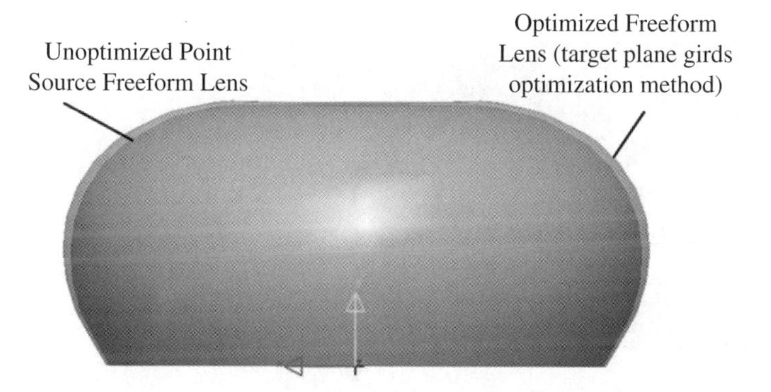

Figure 5.14 Shapes of freeform lenses with the target plane grids optimization algorithm.

but it is not as uniform as the plane grids optimization algorithm. From Figure 5.16, we can find the shapes of two lenses, and the size change is not very obvious. It varies mainly at the bottom area.

5.3.3 Target Plane and Light Source Grids Coupling Optimization

We also study the effect of optimizing the target plane and the light source grids division at the same time. In this section, we adopt $a_i = (1/k_i)^{0.5}$, $b_i = k_i^{0.5}$, as shown in Figure 5.17; after the first time of optimization, U_E increases to 0.64, and after the second time of optimization it increases to 0.88, through which we obtain uniform illumination. As shown in Figure 5.18, the optimized shapes of the lenses are very similar with that in the target plane optimization algorithm.

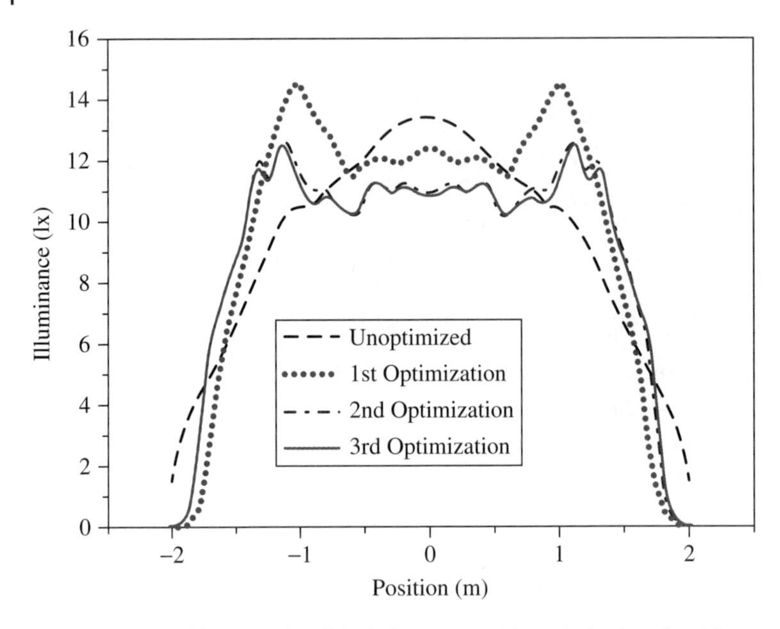

Figure 5.15 Simulation results of the light source grids optimization algorithm.

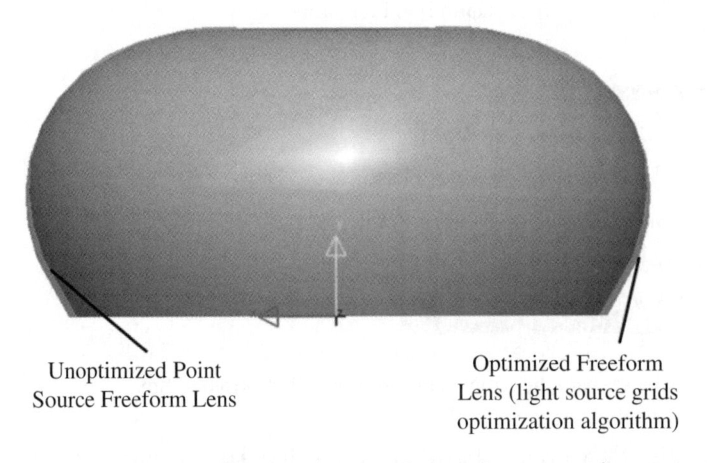

Figure 5.16 Shapes of freeform lenses with the light source grids optimization algorithm.

5.4 A Small-Emitting-Angle Freeform Lens with a Small LED Source

A freeform total internal reflection (TIR) lens is a flexible lens design method to realize small-view-angle project lighting, and it can be applied in LED general indoor lighting, such as LED spotlights, down lights, and stage lamps; it also can be applied in LED microprojectors. A TIR lens can collimate and control the light beam, enhancing the utilization efficiency of projecting the light beam to the display chips.

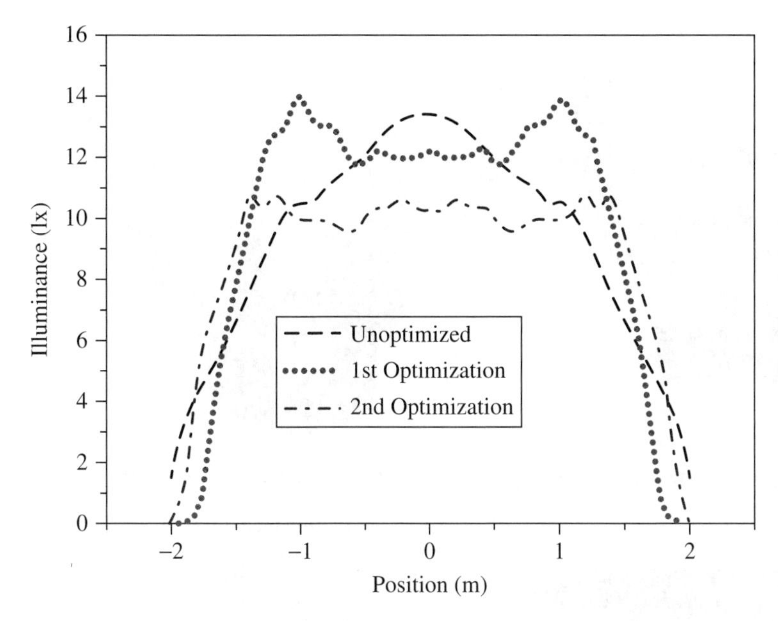

Figure 5.17 Simulation results of the target plane and light source grids optimization algorithm.

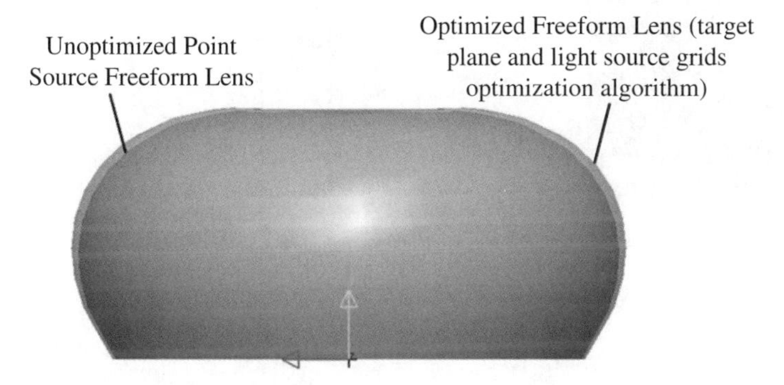

Figure 5.18 Shapes of freeform lenses with the target plane and light source grids coupling optimization algorithm.

According to the TIR lens design methods mentioned in Section 3.2.2, we design freeform TIR lenses based on a Philips Lumileds K2 LED light source with two divergence angles (60° and 38°), as seen in Figures 5.19, 5.20, and 5.21. For the freeform lens with a divergence angle of 60°, the height is 11.14 mm, the largest diameter is 19.64 mm, and the relative light extraction efficiency is 99.6%. It is seen from the lighting performance in Figure 5.20 that the light beam can be well controlled in the target plane after using our freeform lens, and the light pattern is uniformly distributed with an illuminance uniformity of $U_E = 0.722$, which is obviously improved compared with the lens without such a design in Chapter 3 ($U_E = 0.417$). For the freeform lens with a divergence angle of 38°, the height is 10.40 mm, the largest diameter is 19.64 mm, and the relative light extraction efficiency is 99.2%. Similarly, we can see from Figure 5.22 that

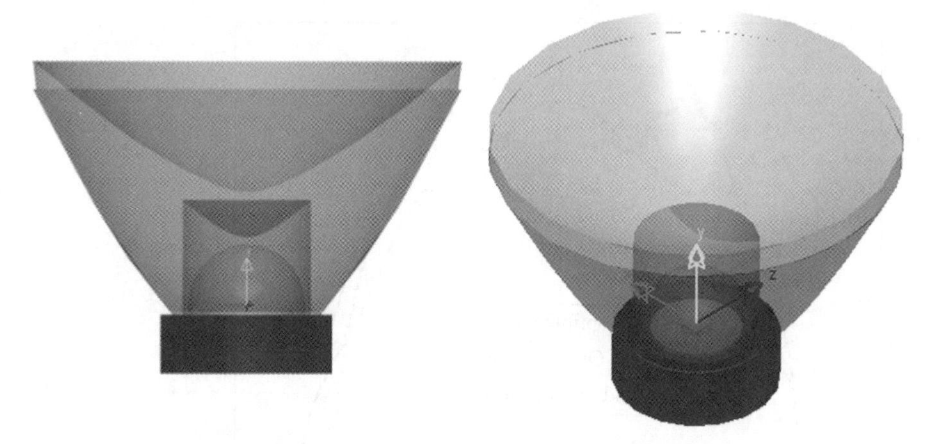

Figure 5.19 TIR freeform lens with a 60° emitting angle for the Philips Lumileds K2 LED.

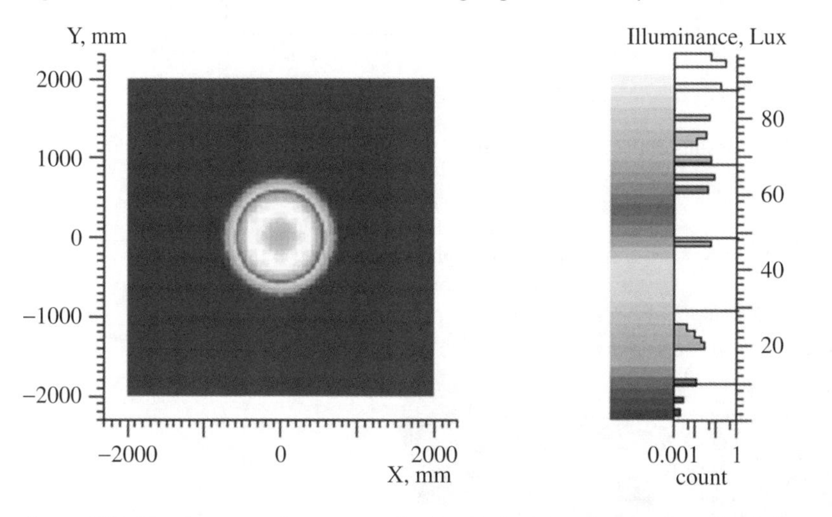

Figure 5.20 Illuminance performance at the target plane locating 1 m for a circularly symmetrical TIR freeform lens with a 60° emitting angle.

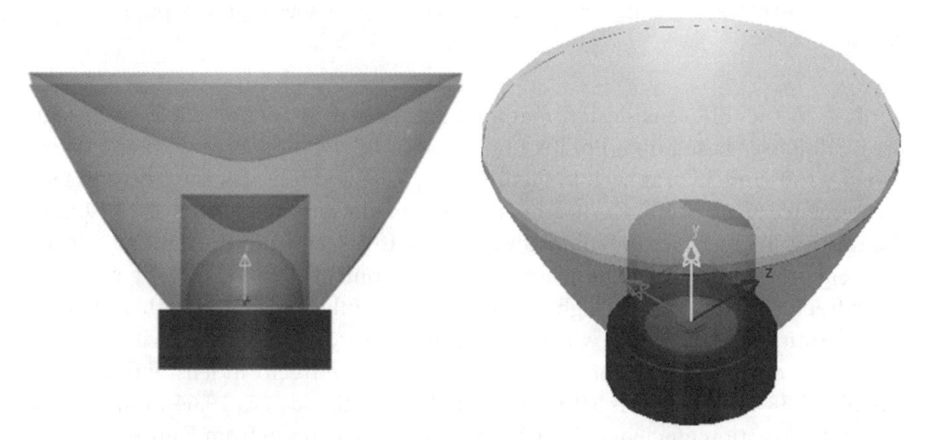

Figure 5.21 TIR freeform lens with a 38° emitting angle for the Philips Lumileds K2 LED.

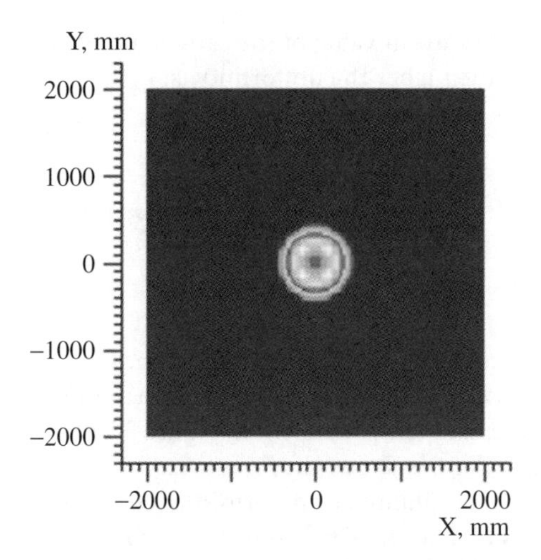

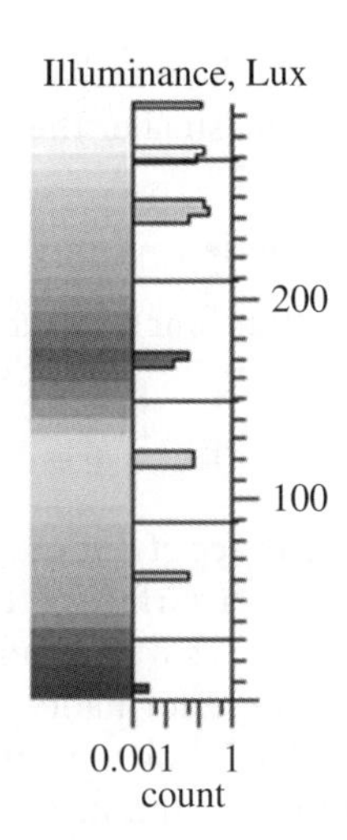

Figure 5.22 Illuminance performance at the target plane locating 1 m for a circularly symmetrical TIR freeform lens with a 38° emitting angle.

the modified freeform lens could control the light-emitting direction better and realize illumination with higher uniformity. Therefore, for a small-emitting-angle case, we could use the freeform TIR lens to obviously enhance the controlling ability of the exit beam and realize uniform illumination.

In this design, the light ray incident direction of freeform Surface 4, as shown in Figure 3.11, is vertical up, which is actually a special case. Due to the flexibility of the freeform surface design, we could adjust the incident direction on Surface 4 to some extent by improving the design of Surface 2 and Surface 3 to meet more application requirements. More details about the algorithms can be found in Section 3.2.2. In addition, we have discussed the rotational symmetrical freeform lens in this chapter. Based on the nonrotational symmetrical freeform lens design methods mentioned in Chapter 3, we could design Surface 4 as nonrotational-symmetrical and realize other shapes of uniform light patterns (e.g., a rectangular light pattern will be presented in Chapter 9), extending the application limits of this freeform TIR lens.

5.5 A Double-Surface Freeform Lens for Uniform Illumination

To validate the double-surface freeform lens design method mentioned in Section 3.2.3, we design several examples by providing them with different extra conditions.[3] We also build the contrastive lenses where we replace the freeform surfaces with hemispherical surfaces. Monte Carlo ray-tracing simulations are conducted to demonstrate the illumination performance. In all the simulations, the distance between the light source and the target plane is 50 mm, and the radius of the target plane is 100 mm. The material of the lenses is selected as PMMA. To evaluate the illumination uniformity, the variation coefficient of the root mean square error (CV(RMSE)) is calculated. It is defined as follows:

$$CV(RMSE) = RMSE/\overline{x} \quad (5.4)$$

where the *RMSE* is the standard error and $\bar{x}$ is the mean value of the sample points of the target plane. The smaller the CV(RMSE) is, the higher the uniformity is.

5.5.1 Design Example 1

The function of the inner surface of the freeform lens in the first design sample is given as:

$$f_1(x, y, z) = x^2 + y^2 - 1 = 0, z \in [0, 1] \tag{5.5}$$

where the inner surface of the lens is a cylinder and the outer surface is freeform. The initial height of the outer surface of the lens is set as 2 mm. With the present design method, we calculate the outer surface and build the freeform lens, as shown in Figure 5.23a.

The illumination performance on the target plane is shown in Figure 5.23b, and its CV(RMSE) is 0.3967. For the contrastive case, the illumination performance is illustrated in Figure 5.24b, and its CV(RMSE) is as high as 3.2996. From the comparisons, we could see that the present freeform lens could enhance the illumination uniformity

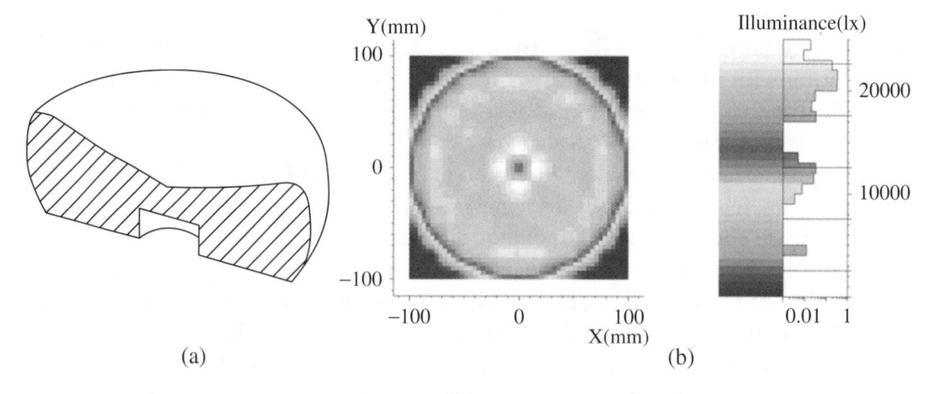

Figure 5.23 (a) Schematic of a freeform lens with a cylindrical inner surface and freeform outer surface, and (b) its illumination performance on the target plane.

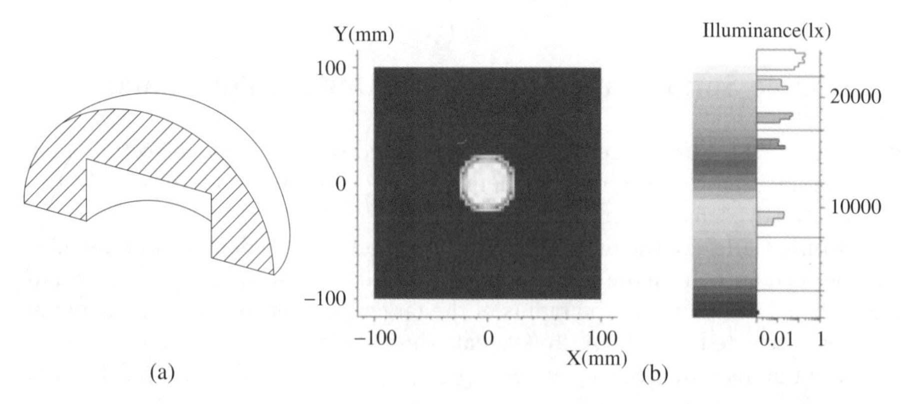

Figure 5.24 (a) Schematic of a contrastive lens with a cylindrical inner surface and hemispherical outer surface, and (b) its illumination performance on the target plane.

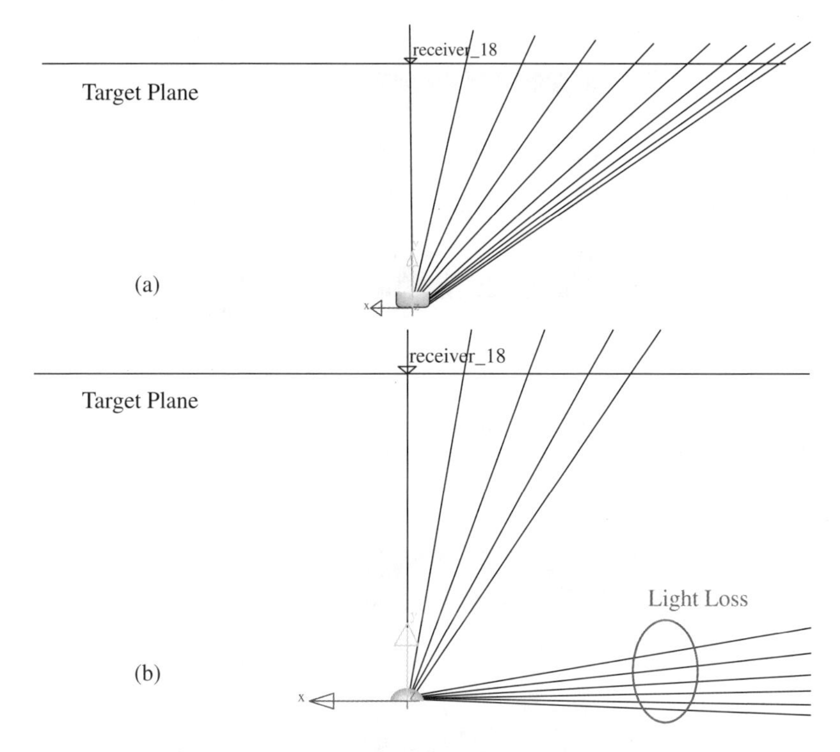

Figure 5.25 Light propagation paths from the light source to the target plane when crossing (a) the present freeform lens, and (b) the contrastive lens.

87.98% and decrease the light loss. To figure out the reasons, we examine the light propagation path when transmitting across the lens. As shown in Figure 5.25, it is found that most light eradiated from the light source could reach the target plane when crossing the freeform lens, whereas only a small fraction of the light could reach the target plane when crossing the contrastive lens, and most light is lost due to the refraction and total inner reflection. That's the exact reason for the light spot formation on the target plane in Figure 5.24b. From the comparisons in Figures 5.23, 5.24, and 5.25, we can see that the freeform lens could increase the illumination uniformity and decrease the light loss.

Inspired by the flat inner surface of such a double freeform lens, we design a novel lens to realize uniform illumination and conformal phosphor coating simultaneously. The flat inner surface can realize the uniform thickness of the phosphor layer, and then we designed the outer surface to achieve uniform illumination by controlling the direction of the emergent ray from the preset inner surface. The detailed design process and validation can be referred to in Chapter 10.

5.5.2 Design Example 2

In the second design example, the function of the outer surface of the freeform lens is given as:

$$f_2(x, y, z) = x^2 + y^2 - 4 = 0,\ z \in [0, 2] \tag{5.6}$$

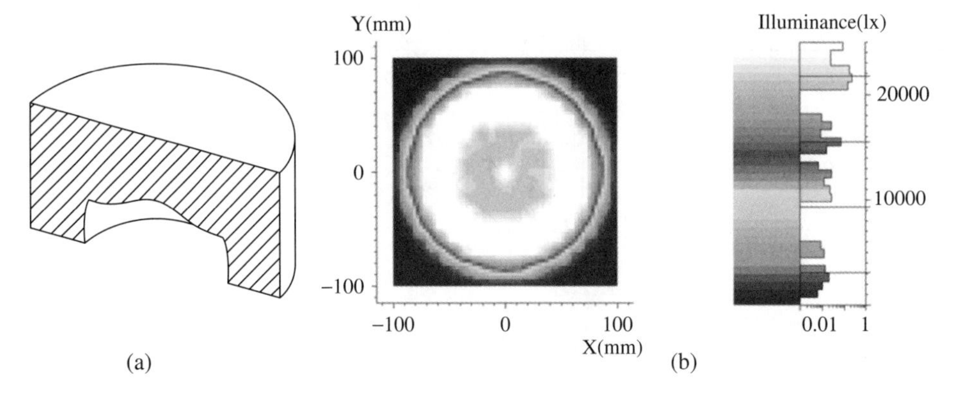

Figure 5.26 (a) Schematic of the second freeform lens with a freeform inner surface and cylindrical outer surface, and (b) its illumination performance on the target plane.

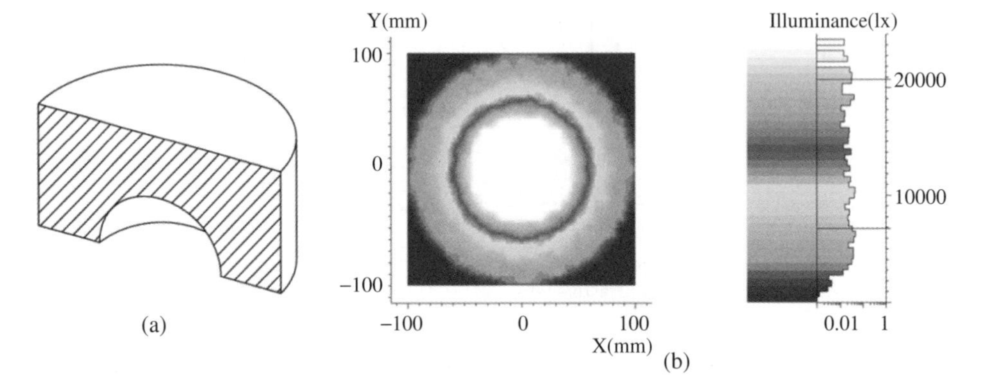

Figure 5.27 (a) Schematic of the second contrastive lens with a spherical inner surface and cylindrical outer surface, and (b) its illumination performance on the target plane.

where the outer surface of the lens is a cylinder and the inner surface is freeform. The initial height of the inner surface of the lens is set as 1 mm. With the same method, we calculate the coordinates of the inner freeform surface and build the freeform lens as shown in Figure 5.26a. Figure 5.27a shows the contrastive lens, whose inner surface is spherical and outer surface is cylindrical. Their illumination performance is illustrated in Figure 5.26b and Figure 5.27b. The CV(RMSE) of the freeform lens is 0.7381, and the CV(RMSE) of the contrastive lens is 1.266. From the comparison, we are able to see that the illumination uniformity of the freeform lens designed by the present method is enhanced 41.70%.

5.5.3 Design Example 3

Figure 5.28 shows another kind of double freeform-surface lens, which is designed by controlling the deviation angle on the inner and outer surfaces.[3] We distribute the total deviation angle into two parts corresponding to the two refractions on the inner and outer surfaces of the lens. The incident ray deviates for the first time when transmitting through the inner surface, and then deviates for the second time when transmitting

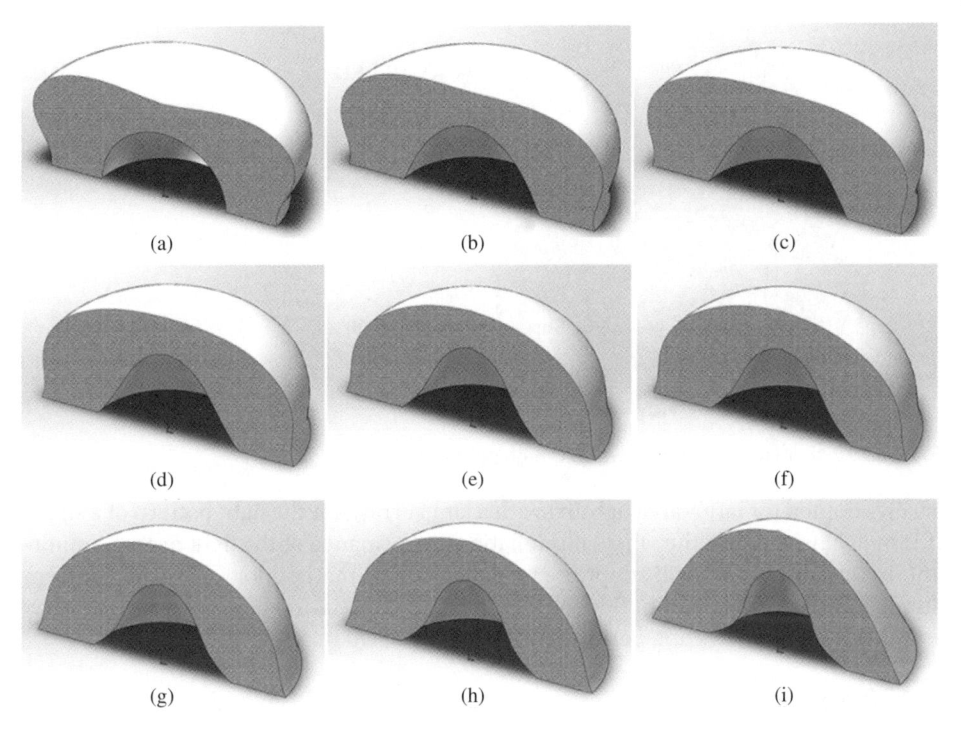

Figure 5.28 A double freeform-surface lens designed (Lens II) by providing the distribution of the deviation angle. The ratios of the first and the total deviation angles are (a) 0, (b) 0.25, (c) 0.33, (d) 0.50, (e) 0.66, (f) 0.70, (g) 0.75, (h) 0.80, and (i) 1.0.

through the outer surface. With this method, illustrated in Figure 5.28, nine cases are designed. Monte Carlo ray-tracing simulations and Fresnel loss calculations are conducted. It was proven that the present double freeform-surface lens can realize uniform illumination as well as minimize the Fresnel loss.

5.6 A Freeform Lens for Uniform Illumination of an LED High Bay Lamp Array

5.6.1 Design Concept

High-power LED high bay lamps, due to their significant energy-saving efficiencies, have been widely applied in many indoor-lighting applications with large lighting areas, such as factories, warehouses, gyms, and so on. A schematic of a typical LED high bay lamp is shown in Figure 5.29, which mainly consists of a CoB LED source, lens, reflector, and fins. Traditional high bay lamps usually adopt high-intensity discharge (HID) as a light source, whose light emitted at 360° full angles, and the reflector is always used to reshape the light intensity distribution to make it meet requirements. However, unlike traditional high bay lamps, due to the Lambert LIDC of the LED source, the reflector has little effect on the redistribution of light energy of the LED source, but the lens does. Therefore, it's important to optimize the lens design. Moreover, since LED high bay lamps are

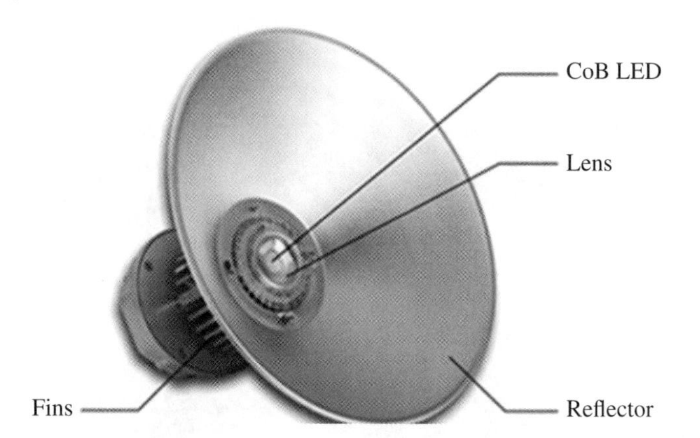

Figure 5.29 Schematic of a high-power LED high bay lamp.

usually adopted for large-area lighting with a lamp array, not the light pattern of a single LED high bay lamp but the whole illumination performance of the light array is important. In this section, we will adopt the reversing design method for LED array uniform illumination as mentioned in Section 3.6 to design a freeform lens for LED high bay lamps with uniform lighting performance when installed in arrays.

5.6.2 Design Case

5.6.2.1 Algorithms and Design Procedure

Considering a CoB LED light source whose light intensity is circular-symmetrical, we express the LED LIDC in the form of a high-order cosine function.

$$I(\theta) = I_0 \cos^m \theta \qquad (m > 0) \tag{5.7}$$

In Eq. 5.7, I_0 is the light intensity of LED in the central axis, θ is the angle between the emitting ray and the central axis, and m is the control parameter of the LIDC that is to be calculated. Through this LIDC expression, we are able to describe more convergent LIDC than Lambertian LIDC.

We select the installation height to be $h = 8.2$ m and the installation spacing to be $d = 6$ m to optimize the LIDC of the light source. From the design algorithms mentioned in Chapter 3, we can obtain the LIDC control parameter to be $m = 5.6$, and an initial LIDC will be achieved. Moreover, due to the extended light source effect, the initial LIDC (which is $I(\theta) = I_0 \cos^{5.6}\theta$) has an illumination performance on the target plane that is not the best case. After optimizing the control parameter, we are able to obtain the optimized LIDC as follows:

$$I(\theta) = I_0 \cos^4 \theta \tag{5.8}$$

This LIDC, as shown in Figure 5.30, is more convergent when compared with the Lambertian type.

When light sources with these LIDCs are installed as an array, the illuminance distribution in the target area is shown in Figure 5.31. When the array is 3×3, we calculate the 6×6 m size target area. The illuminance uniformity can reach as high as 0.894, which can meet the overall illuminance uniformity design target and therefore confirm that the LIDC can satisfy the requirements.

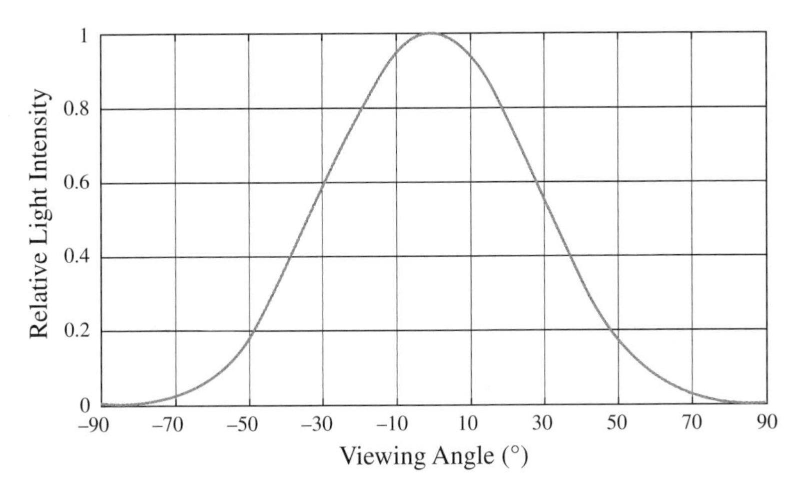

Figure 5.30 Optimized LIDC.

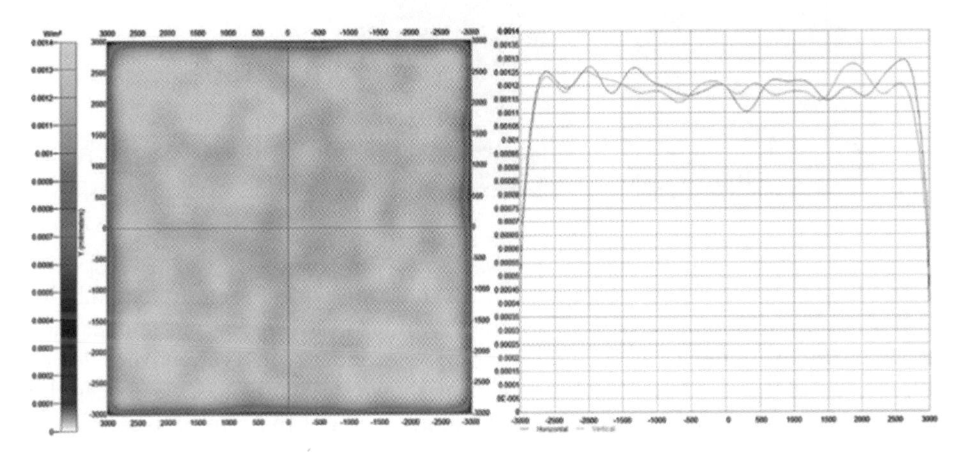

Figure 5.31 Illuminance distribution on the target plane with optimized LIDC.

5.6.2.2 Optical Structures

In order to realize high-power and high-brightness LED high bay illumination, we use four pieces of CXA2520 CoB LEDs to fabricate a 2 × 2 CoB LED array light source and the total power is 160 W, as shown in Figure 5.32. In a CoB LED array arrangement, we need to design the spacing between each pair of LED chips, which will be discussed in the lens design further.

Since the light source of each LED high bay lamp is a 2 × 2 CoB LED array, we need to design a secondary freeform lens for each CoB LED module. According to the lens design algorithms mentioned in Section 3.6, we are able to obtain the required LIDC for uniform illumination with an LED array first and then use the reversing design method to calculate each point on the freeform curve of the lens. For each single lens, the outer shape is shown in Figure 5.33; the height of the lens is 25 mm, and the widest is 31 mm.

When considering that the optical loss is mainly caused by the TIR generated at the interface of CoB LED phosphor silicone and air existing between the LED source and

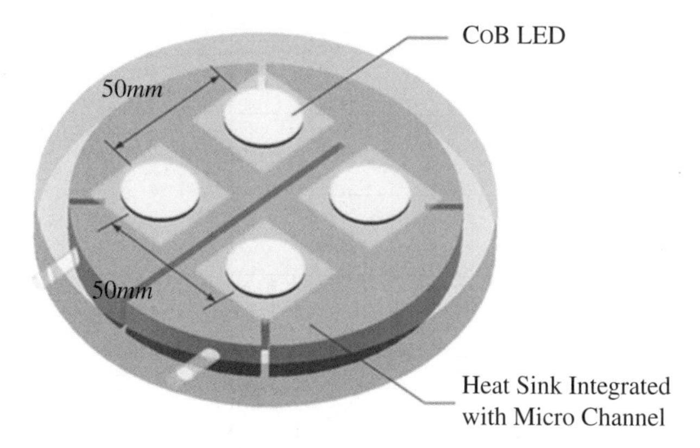

Figure 5.32 Schematic of a CoB LED array module.

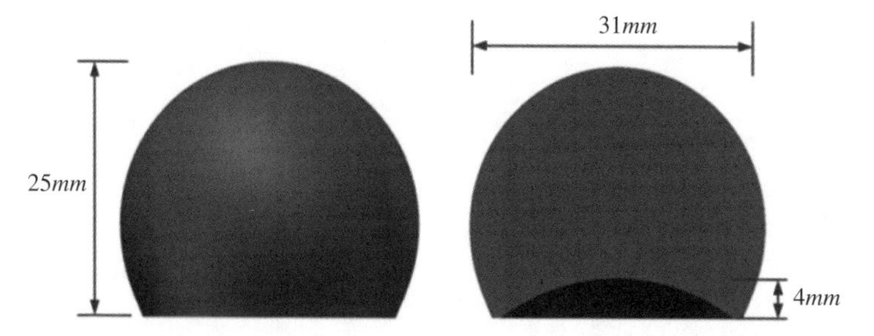

Figure 5.33 Freeform lens outer shape (left) and the cross-sectional view (right).

the lens, the solution to reduce the optical loss is to fill the gap between the lens and the emitting surface with a material with a refractive index between that of the phosphor silicone (refractive index of 1.54) and the PMMA secondary lens (refractive index of 1.49) with good transmittance. In our design, as shown in Figure 5.34, we choose the silicone with a refractive index of 1.53 as the filling material that can both satisfy the refractive index requirement and have good transmittance.

In our design, in order to save the silicone consumption, we design the lens inside to be a spherical cap that is 4 mm in height. The lens array could be fabricated as a whole, and it is easy to machine and assemble with the whole LED light source. The lens array

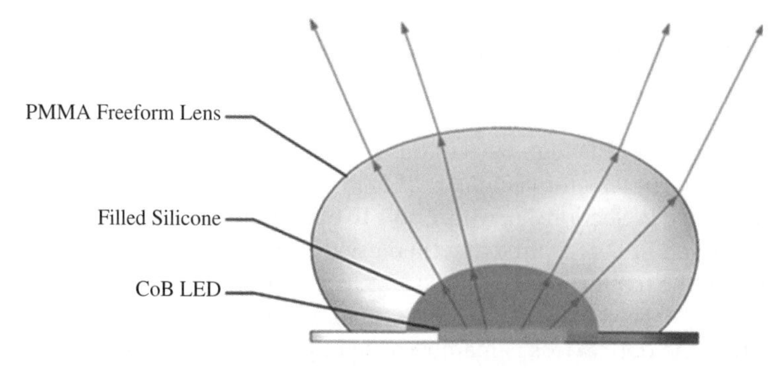

Figure 5.34 Silicone filled the gap between the LED light source and the secondary freeform lens.

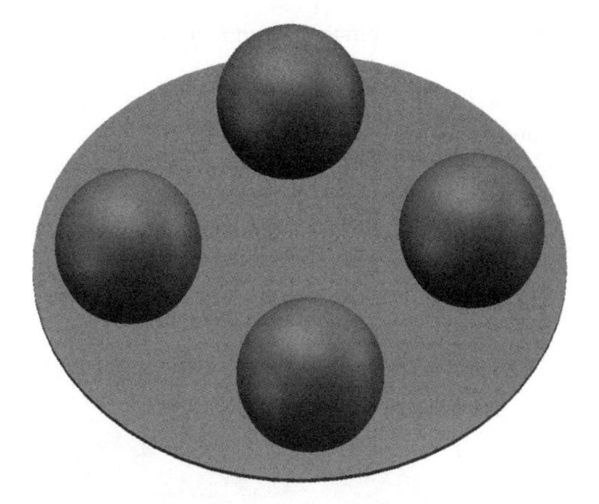

Figure 5.35 Outer shape of the designed freeform lens array.

with the size of 110 × 110 × 27 mm is as shown in Figure 5.35. Besides, the distance between the centers of each adjacent lens is determined based on two principles: the size of the overall lens array should not be too large, and the emitting rays should not interfere with each other. We choose the distance to be 50 mm, which can meet both requirements.

A high bay lamp must satisfy the glaring prevention requirement: when the viewer looks beyond a certain angle at the lamp, there should be no light entering the eye directly from the source. Therefore, we add a 30° lampshade over the whole light source, as shown in Figure 5.36.

5.6.2.3 Monte Carlo Optical Simulation

At the last step of the whole LED high bay lamp design, we need to verify the optical performance of the whole lamp by Monte Carlo optical simulation.

Output *efficiency simulation verification* First of all, we verify the output efficiency of the whole system by optical simulation. From the output efficiency simulation, we discuss the different structures and their effects on the whole lamp's output efficiency.

From the simulation results shown in Table 5.1, when the optical structure is composed of only one single LED chip, the whole module output efficiency is set as 100%, which means that all the light rays emitting from the LED chip will arrive at the receiving plane. When the optical module is a CoB LED packaging module, only 41.7% of light rays will exit from the module, and most of the light rays will have TIR at the interface of the phosphor silicone and the air, resulting in the reflected light rays being reabsorbed

Figure 5.36 30° lampshade for the LED high bay lamp.

Table 5.1 Output efficiency simulation results of an LED source and LED high bay lamp.

Optical structure	Output efficiency, %
Single LED chip	100
CoB LED module	41.7
CoB LED module + silicone filled	85.4
CoB LED module + silicone filled + freeform lens	96.4

Figure 5.37 Whole optical model of the LED high bay lamp.

by the LED chip and the packaging board. When silicone is mounted on the CoB LED module as a 4 mm high spherical cap, the overall output efficiency will be as high as 85.4%. Moreover, when there is a silicone-filled air gap as well as a freeform lens, the TIR effect can be eliminated as much as possible, and more light rays will arrive at the receiving plane with an output efficiency of 96.4%. Therefore, the design concept of filling in the air gap between the CoB LED and the freeform lens with silicone can greatly improve the whole lamp's efficiency.

Here, we will analyze the whole optical performance of the LED high bay lamp integrated with a freeform lens array, as shown in Figure 5.37. We use 2 million light rays to simulate the optical performance and achieve whole-lamp output efficiency as high as 96.1%, which is quite similar with the silicone-filled and freeform-lens-added case. This result demonstrates that there is little light exit from the freeform lens, which is affected by adjacent lenses. In the meantime, we obtain the LIDC of the whole lamp as shown in Figure 5.38. Its shape is very similar with the LIDC in the initial stage of our design, which demonstrates that although four CoB LEDs and freeform lenses are adopted in the lamp, the whole lamp still could be conducted as one LED module with a freeform lens in situations of large installation height and spacing.

Illumination area uniformity evaluation To evaluate the illuminance uniformity of the LED high bay lamps array, we adopt DIALux software to simulate the whole performance. The optical model is built as shown in Figure 5.39. There is an 8 × 8 LED high bay lamps array. The whole illuminating space is 48 × 48 × 9 m.

In order to reduce the effects of absorbance and reflectance from the surfaces of ceiling and the surrounding wall, we select two areas for illuminance calculation. The individual sizes are 24 × 24 m for region 1 and 36 × 36 m for region 2. The final calculation results are shown in Figure 5.40. From the illumination area's pseudo-color map, we can find

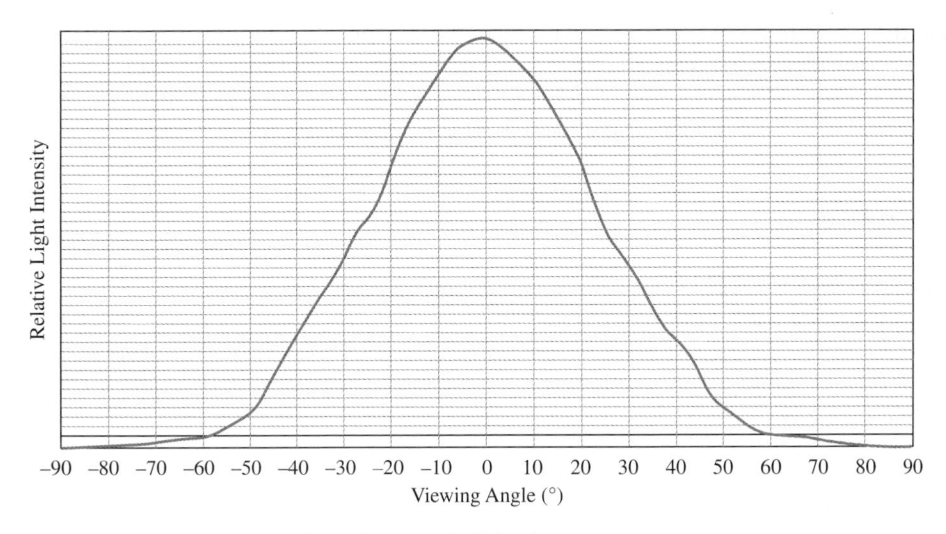

Figure 5.38 Simulated LIDC of the whole LED high bay lamp.

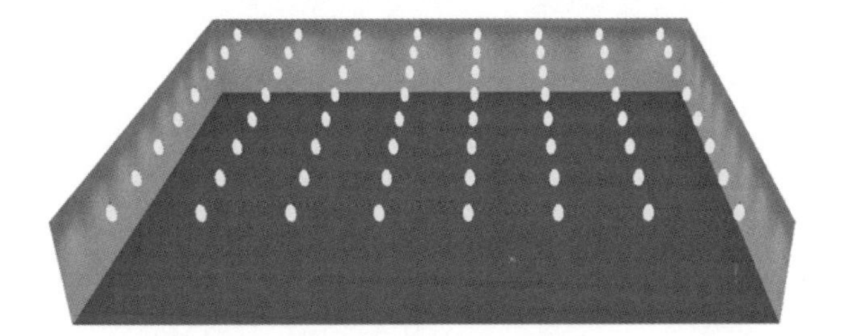

Figure 5.39 Optical model of the 8×8 LED high bay lamp array.

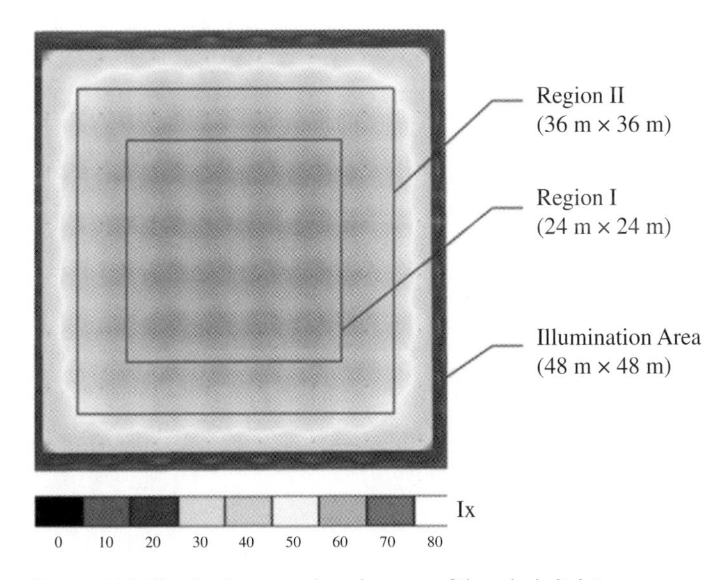

Figure 5.40 Illumination pseudo-color map of the whole lighting area.

that the whole illumination area has uniform illuminance distribution. The simulation results show that in the selected region 1, the illuminance uniformity is as high as 0.936, and in region 2 it is 0.883. Therefore, the designed high-efficiency LED high bay lamp integrated with freeform lenses is able to meet the illumination requirements very well.

References

1 Philip Lumileds K2 LED. http://www.philipslumileds.com
2 CREE XLamp XR-E LED. http://www.CREE.com.
3 Hu, R., Gan, Z., Luo, X., Zheng, H., and Liu, S. Design of double freeform-surface lens for LED uniform illumination with minimum Fresnel losses. *Optik – Intl. J. Light Elect. Opt.* **124**, 3895–3897 (2013).

6

Freeform Optics for LED Road Lighting

6.1 Introduction

Within the total electricity consumption of city lighting, street and road lighting mostly dominates. Decreasing the electricity consumption of street and road lighting will bring significant energy saving for city lighting. Intelligent lighting control is one of the most highly adopted and effective ways to reduce electricity consumption of road lighting when using traditional light sources of high-pressure sodium (HPS). With the development of light-emitting diode (LED) epitaxy growth. and chip-manufacturing and chip-packaging technologies, the luminous efficiency of LED modules has increased dramatically in recent years, and it reached more than 300 lm/W in the lab[1] and 150 lm/W in the market in 2014, providing a new way to reduce the power consumption of road lighting – LED road lighting. The lighting performance of a 100 W high-power LED road lamp was even better than that of a traditional 250 W HPS road lamp in 2014, and the energy saving efficiency is more than 50%. Therefore, LED road lighting has been strongly supported by countries around the world and also is the earliest mass application of high-power LEDs. In 2005, several cities demonstrated LED road-lighting projects in China. Since then, especially supported by the Chinese government's 10 City 10,000 Lamps project in 2009, both LED road-lighting technology and its market have developed very quickly in China, and more and more cities are adopting LED road lamps in terms of the Energy Management Contract (EMC). In 2014 3.3 million LED road lamps were sold in China, which is much higher than 2.56 million in 2013.[2] Not only in China, but also in other countries and areas around the world (e.g., the United States and European Union), LED road-lighting projects have been promoted to save energy. Recently, US President Barack Obama announced acceleration of the LED road lamps replacing project, as part of a Better Buildings program of the US Department of Energy (DOE), and the number of LED road lamps in the United States was projected to reach 1.5 million in 2016.

High luminous efficiency and energy saving are the major advantages of LED road lamps compared with traditional road lamps. To achieve higher efficiency, LED lamps require not only higher luminous efficiency but also a higher light energy utilization rate. Around 2009, the luminous efficiency of a commercial LED packaging module was about 100 lm/W, which is quite similar with that of HPS. However, less than 20%

Freeform Optics for LED Packages and Applications, First Edition. Kai Wang, Sheng Liu, Xiaobing Luo and Dan Wu.

(a) (b)

Figure 6.1 Poor lighting performance of LED road lamps with (a) zebra stripes, and (b) anti-zebra stripes on the road.

power consumption is required for an LED road lamp compared with a HPS road lamp when achieving the same average illuminance on the road. The major different is that the LED road lamp adopts freeform lenses, which can precisely control and redistribute the light energy emitted from an LED module and make most light energy illuminate on the road when meeting the requirements of the surrounding ratio (SR). Besides efficiency, uniform illumination on the road is very important, not only for driving comfort but also for driving safety. With improper optical design, light spots, such as a circular light pattern (as shown in Figure 3.19 in Chapter 3), zebra stripes (Figure 6.1a), and anti-zebra stripes (Figure 6.1b), will occur on the road, reducing both the luminance and illuminance uniformities of road lighting. Therefore, advanced and precise freeform lens design methods are required for high-performance LED road lighting. In this chapter, design concepts, discontinuous freeform lenses (DFLs), continuous freeform lenses (CFLs), and symmetrical and unsymmetrical freeform lenses for LED road lighting all will be discussed in detail. Moreover, an emerging modularized LED road lamp design concept also will be introduced.

6.2 The Optical Design Concept of LED Road Lighting

Firstly, we analyze the lighting demand on driveways according to the latest national road-lighting standard, *City Road Lighting Design Standard CJJ 45-2006*.[3] Note that although the Chinese road-lighting design standard had been adopted as an example in this book, the analysis method and freeform optics design method are quite similar to other countries' road-lighting standards. Readers are able to design new freeform lenses to meet their own requirements according to the algorithms and methods mentioned in this book. As shown in Table 6.1, there are mainly four lighting standards for different kinds of roads: luminance, illuminance, glare restriction threshold increment, and surround ratio. *Luminance* in roads standards consists of average luminance, total luminance uniformity, and longitudinal luminance uniformity. *Illuminance* consists of average illuminance and illuminance uniformity. The average illuminance shown

Table 6.1 A city road-lighting standard.

Level	Road type	Luminance: Average luminance L_{av} (cd/m²) (Mini)	Luminance: Total uniformity U_o (Mini)	Luminance: Longitudinal uniformity U_L (Mini)
I	Highway, arterial road	1.5/2.0	0.4	0.7
II	Subarterial road	0.75/1.0	0.4	0.5
III	Branch	0.5/0.75	0.4	—

Level	Road type	Illuminance: Average illuminance E_{av}(lx) (Mini)	Illuminance: Uniformity U_E (Mini)	Glare restriction threshold increment (TI), % (Max)	Surrounding ratio (SR) (Mini)
I	Highway, arterial road	20/30	0.4	10	0.5
II	Subarterial road	10/15	0.35	10	0.5
III	Branch	8/10	0.3	15	—

in Table 6.1 is suitable for asphalt roadbeds only. As for cement concrete roadbeds, the value of average illuminance is allowed to decrease by 30%. Then the calculating methods of all the indicators are analyzed as shown in the remainder of this section.

6.2.1 Illuminance

Illuminance is defined as the luminous flux on the unit area, which is measured by an illuminometer located on the road vertically. Therefore, the luminous flux measured is given out by the road lamp directly, instead of the reflection from the road; in other words, the materials and reflection coefficient of the road are not involved. However, human vision and feelings about brightness on the road are decided by the light that is reflected by the road surface. That is to say, illuminance alone is not able to express the brightness seen by human beings on the road.

$M \times N$ test points are conducted on the road, with a distance of 1 or 2 m between every two points. After testing the illuminance E_i of every single test point, the average illuminance on the road E_{av} can be obtained by the expression

$$E_{av} = \frac{1}{4MN}\left(\sum E_{\Delta} + 2\sum E_{O} + 4\sum E_{\Theta}\right) \tag{6.1}$$

where E_{Δ} is the illuminance value on the four vertexes of the testing rectangle; E_O is the illuminance value on the four edges except the four vertexes; and E_{Θ} is the illuminance value of other test points. The illuminance uniformity on road U_E is defined as the ratio of the minimum value E_{min} to the maximum value E_{av} of illuminance.

$$U_E = \frac{U_{\min}}{U_{av}} \tag{6.2}$$

6.2.2 Luminance

Luminance is measured by an illuminometer imitating drivers' eyes of common cars. The observation point is set at a distance of 60 m from the observation area and a height of 1.5 m on the road. The light measured from the observation point is the light reflected by the road surface, so the luminance is able to reflect the brightness degree and uniformity of the road viewed by people.

Luminance is related to the materials, the reflection coefficient at a given observation point, and the illuminance of the road. When the observation point and materials of the road are confirmed, the reflection coefficient can be obtained. Hence, a luminance coefficient is proposed to indicate that *luminance value = illuminance value × luminance coefficient*. For different roads, there are different luminance coefficients. For simplification, a table luminance coefficient for different road materials (mainly asphalt and cement concrete roads) is established. Then the luminance coefficient can be obtained expediently from the table. As shown in Figure 6.2, the luminance value can be calculated by Eq. 6.3:

$$L = q(\beta, \gamma)E(c, \gamma) \tag{6.3}$$

where $q(\beta, \gamma)$ is the luminance coefficient; $E(c, \gamma)$ is the illuminance value; β is the angle between the incident plane and observation plane; γ is the angle between the emergent ray and center vertical of the lamp; and c is the angle between the emergent ray and road edge. Meanwhile, there is a transformational relation between the illuminance $E(c, \gamma)$ and the light intensity $I(c, \gamma)$.

$$E(c, \gamma) = \frac{I(c, \gamma)\cos\gamma}{l^2} = \frac{I(c, \gamma)\cos\gamma}{(H/\cos\gamma)^2} = \frac{I(c, \gamma)\cos^3\gamma}{H^2} \tag{6.4}$$

where H is the height of the road lamp. By substituting Eq. 6.4 into Eq. 6.3, we could obtain the luminance L as:

$$L = \frac{q(\beta, \gamma)\cos^3\gamma I(c, \gamma)}{H^2} \tag{6.5}$$

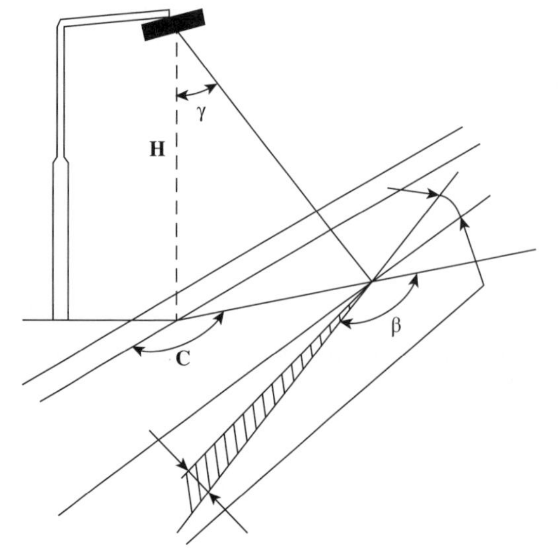

Figure 6.2 Schematic of a road luminance calculation.

A simplified luminance coefficient is proposed as:

$$r(\beta,\gamma) = q(\beta,\gamma)\cos^3\gamma \tag{6.6}$$

The simplified luminance coefficients for different roads, such as asphalt and cement concrete roads, can be examined in tables in Ref. [3]. By substituting Eq. 6.6 into Eq. 6.5, we can obtain the width of the road:

$$L = \frac{r(\beta,\gamma)I(c,\gamma)}{H^2} \tag{6.7}$$

Thereby, if the light intensity distribution curve (LIDC) and the mounting height H of one LED road lamp are known, the brightness distribution on the road can be obtained from the simplified luminance coefficient table. Then the average luminance L_{av} can be calculated by a method similar to that used to calculate the average illuminance. The total luminance uniformity on the road is defined as the ratio of the minimum luminance value L_{min} on the road to the average luminance value L_{av}:

$$U_0 = \frac{L_{\min}}{L_{av}} \tag{6.8}$$

The longitudinal luminance uniformity on the road is defined as the ratio of the minimum luminance value L_{min} to the maximum luminance value L_{max} along the center line of one lane:

$$U_L = \frac{L_{\min}}{L_{\max}} \tag{6.9}$$

What is worth noticing is that the minimum luminance value of longitudinal luminance uniformity is the minimum value on the center line instead of the whole surface of the road.

In addition, by analyzing Eq. 6.3 and Eq. 6.6, we can obtain:

$$L = \frac{r(\beta,\gamma)}{\cos^3\gamma}E(c,\gamma) \tag{6.10}$$

where $r(\beta, \gamma)/cos^3\gamma$ is defined as the equivalent luminance coefficient of luminance to illuminance, by which the relationship between luminance and illuminance can be made known.

6.2.3 Glare Restriction Threshold Increment

Glare is a visual phenomenon that occurs when the brightness contrast is too extreme or the brightness distribution in the visual field is not suitable for view; in such cases, people will feel uncomfortable or the observation capacity will reduce, which affects the illumination quality strongly. Glare can be divided into *discomfort glare* and *disable glare*, which are indicated by glare index G and glare restriction threshold increment TI, respectively. Generally speaking, if disable glare can be acceptable, the discomfort glare can be acceptable, too. Hence, more and more lighting standards use TI as an index of glare, which is expressed as:

$$TI = \frac{65k}{L_{\text{av}}^{\ 0.8}}\sum_{i=1}^{n}\frac{E_{eyei}}{\theta_i^{\ 2}} \tag{6.11}$$

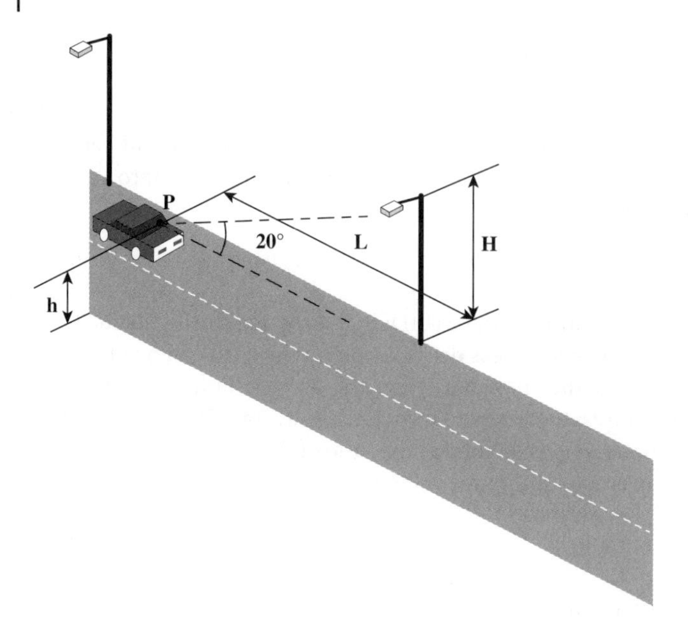

Figure 6.3 Schematic of the observed point of the maximum glare threshold increment TI_{max}.

where k is the age factor, generally 10, suitable for the observers aged 23; L_{av} is average luminance on road (cd/m^2); E_{eyei} is the illuminance on the retinas of people from the ith vertical glare source (lx); θ_i is the angle between the ith incident ray from the glare source to the eyes and the line of sight (°); and n is the number of glare sources.

During driving, with the relative position between the driver (observer) and the lamp position changing, the TI changes accordingly. Thus, a maximum glare threshold increment TI_{max} is needed as an index. For road lighting, the expected TI_{max} is less than 10. The position of TI_{max} on the driving direction is decided by shielding the windshield's top angle. For the demand of road-lighting design, the angle has been standardized to 20° upward horizontal. Generally speaking, the observation point of the TI_{max} appears exactly at that angle, as shown in Figure 6.3, where the point P is the driver's position; h is the height of the driver's eye from the ground, 1.5 m generally; L is the distance between the position of the driver's maximum glare increment threshold and the glare source; and H is the height of the road lamp. Also, the TI_{max} can be calculated by DIALux software during road-lighting evaluation.

6.2.4 Surrounding Ratio

For drivers, the visibility state depends mainly on the average luminance of the road, but the brightness of surroundings can influence the eyes' general adaptive state. When the surroundings are dark and the road surface is bright, the eyes of drivers adapt to the bright road surface. Then the surrounding objects in the dark are unlikely to capture the attention of drivers, which will enhance the probability of traffic accident. The surrounding ratio SR is defined as the ratio of average horizontal illuminance on the region

5 m out of the roadway to the average horizontal illuminance on the roadway within the range of 5 m; it is generally no less than 0.5.

6.3 Discontinuous Freeform Lenses (DFLs) for LED Road Lighting

6.3.1 Design of DFLs for Rectangular Radiation Patterns

In this section, LED tunnel lighting, as one kind of LED road lighting, will be discussed as a design example due to its importance regarding the urgent replacement of many tunnels in planning and construction projects in China.[4] To meet the requirements of tunnel illumination, according to the discontinuous freeform lens design method proposed in Chapter 3, one discontinuous freeform lens will be designed as an example of one LED forming one 10 m long and 8 m wide uniform rectangular illumination area at the height of 5.5 m (as shown in Figure 6.4). The material of this lens is polymethyl methacrylate (PMMA) with a refraction index of 1.49. It is important to note that this method also could be used to design rectangular light patterns with different length–width ratios for other applications.

6.3.1.1 Step 1. Optical Modeling for an LED

In this design case, as shown in Figure 6.6, we used the Cree XLamp XR-E LED as the light source because it is one of the most popular LEDs in the market and its LIDC is special and more complicated than others, which makes it a good example for evaluating this algorithm. A point source optical model of the Cree XLamp XR-E LED is to be established to evaluate this freeform lens design algorithm. However, illumination performance probably will deteriorate in practical application because the size of the LED cannot be ignored. Thus, a practical optical model of the Cree XLamp XR-E LED is also to be established to evaluate the actual illumination performance of the designed lens.

Substep 1.1. Optical modeling for a point light source. The LIDC of the Cree XLamp XR-E LED is shown in Figure 6.5, and it will be adopted as the light energy distribution of a

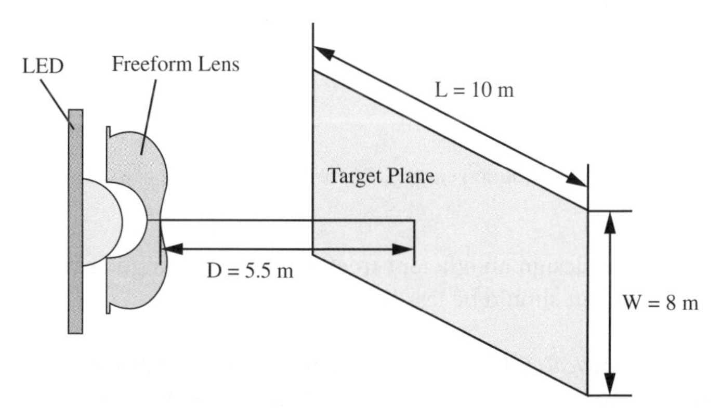

Figure 6.4 Schematic of the design target of LED tunnel lighting.

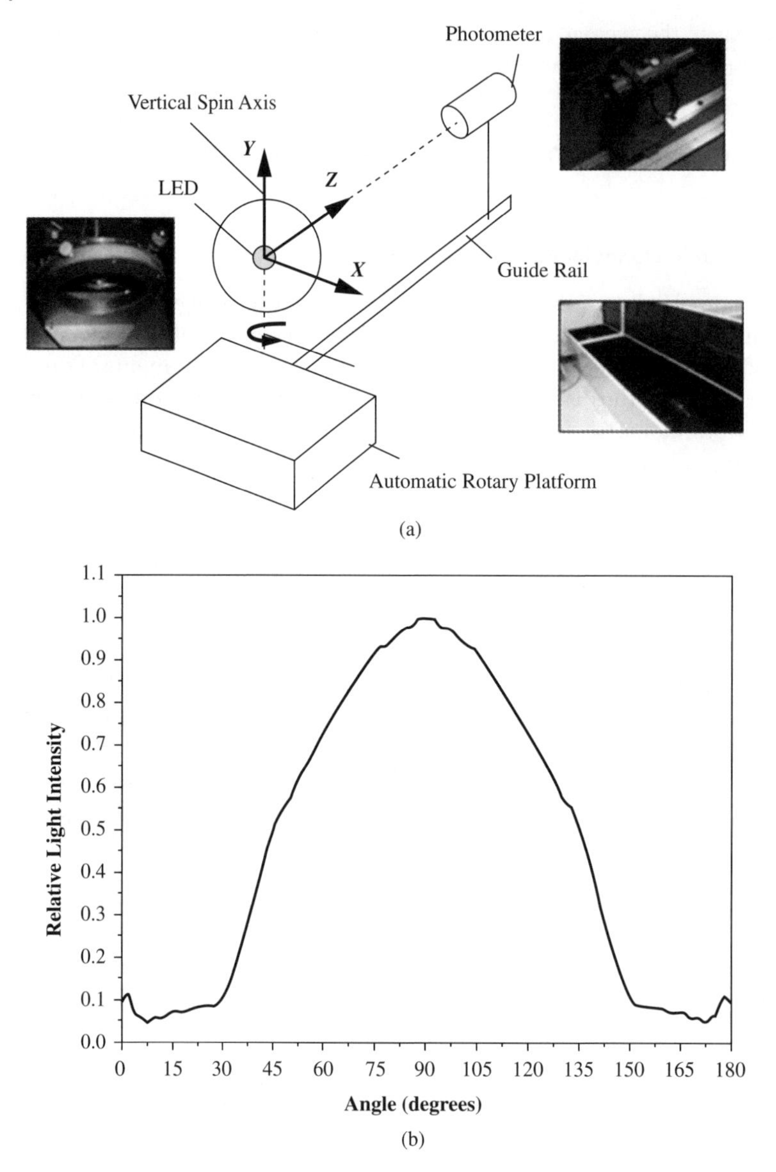

Figure 6.5 (a) Schematic of a light intensity distribution curve (LIDC) tester and (b) LIDC of a Cree XLamp XR-E LED.

point source during simulation. To design an efficient freeform lens, the angular resolution of the LED LIDC measurement should be less than 1°.

Substep 1.2. Optical modeling for a practical LED. First of all, we made some measurements to determine the geometrical parameters and established a practical structural model of the LED. We obtained the LIDC of this LED model via the widely used Monte Carlo ray-tracing method, and the LED was simulated by 1 million rays. Then, a precise LED optical model was established by comparing the similarity between the simulation LIDC

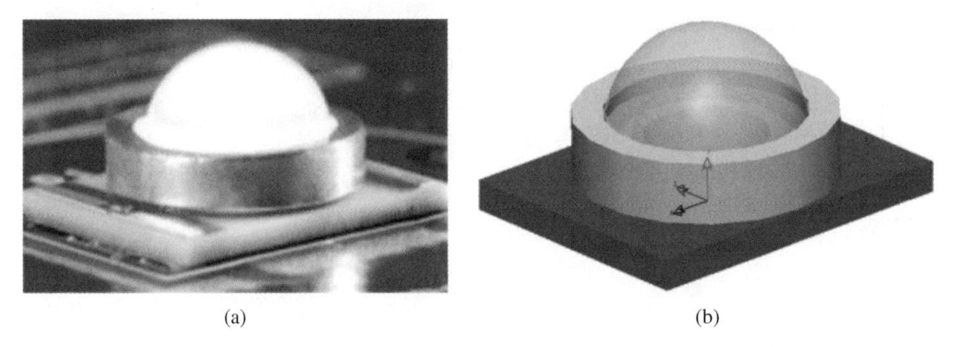

(a) (b)

Figure 6.6 Cree XLamp XR-E LED: (a) material object and (b) practical optical model.

and the experimental LIDC, which was quantified by the normalized cross correlation (NCC).[5] The NCC is written as in Eq. 6.12:

$$NCC = \frac{\sum_x \sum_y (A_{xy} - \overline{A})(B_{xy} - \overline{B})}{\left[\sum_x \sum_y (A_{xy} - \overline{A})^2 \sum_x \sum_y (B_{xy} - \overline{B})^2\right]^{1/2}} \tag{6.12}$$

where A_{xy} and B_{xy} are the intensity or irradiance of the simulation value (A) and experimental value (B); and $\overline{A}$ ($\overline{B}$) is the mean value of A (B) across the x–y plane shown in Figure 6.5. For the modeling algorithm for an LED model mentioned in Ref. [6], we adjusted the scattering parameters and refraction indexes of some packaging materials used in LEDs, such as phosphor, polymer, and silicone, until the NCC reached as high as 97.6% (as shown in Figure 6.7a). Thus, the precise optical modeling for the LED was finished.

6.3.1.2 Step 2. Freeform Lens Design

One LED tunnel illumination freeform lens was designed according to the method mentioned in Chapter 3. Firstly, we calculated a 10-order polynomial to fit the tested LIDC according to the least-squares curve-fitting method. Figure 6.7b depicts that the polynomial fits the test LIDC very well and the NCC reaches as high as 99.9%. Secondly, we divided the one-quarter light energy distribution of both the light source and target plane into 64 (M) × 500 (N) = 32,000 grids with equal luminous flux and area, respectively, and established a light energy mapping relationship between these grids. Finally, we calculated the coordinates and normal vector of each point on the freeform surface and constructed the discontinuous freeform lens utilizing these points. We set a threshold θ_{dth} as 6° in our calculation, and nine seed curves were constructed to reduce deviation of exit rays. Figure 6.8 shows these seed curves, and the sequence numbers of them are 1, 16, 22, 27, 31, 38, 42, 47, and 53, respectively. As shown in Figure 6.9a, the whole discontinuous freeform lens was constructed by only 32 subsurfaces, which makes it easier to manufacture, and fewer manufacturing defects will be induced compared with 450 subsurfaces used in Refs. [7,8]. The largest value of length, width, and height of the lens are 10.6 mm, 9.7 mm, and 5.1 mm, respectively.

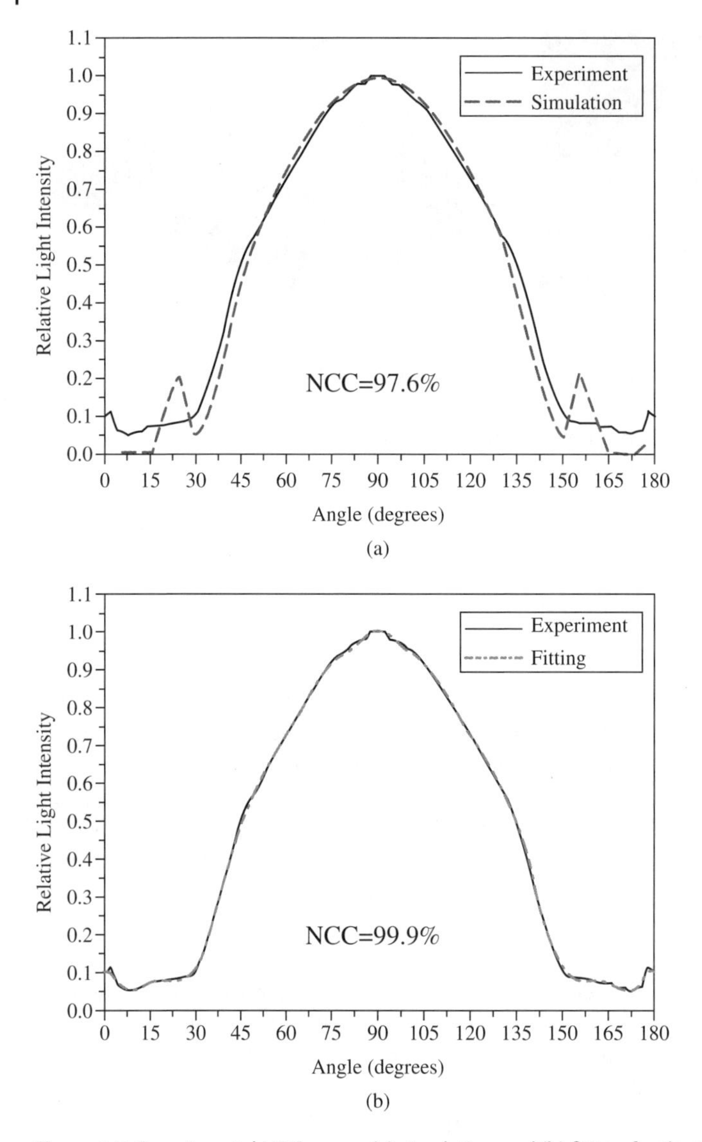

Figure 6.7 Experimental LIDC versus (a) simulation and (b) fitting for the Cree LED.

6.3.2 Simulation Illumination Performance and Tolerance Analyses

The numerical illumination performance of the freeform lens using the point light source is shown in Figure 6.10. This figure depicts simulation illuminance distributions on a test area with a size of 13 m long and 13 m wide that is 5.5 m away from the LED. From the simulation results, we can find that the lens forms a rectangular light pattern with a length of about 9.6 m and width of about 8.0 m, which is quite consistent with the design target. The light output efficiency (LOE) of the freeform lens is 97.4%. Fresnel loss is the main reason for light loss. The other light loss is mainly due to lens material absorption and total internal reflection (TIR) at the interface of lens and air.

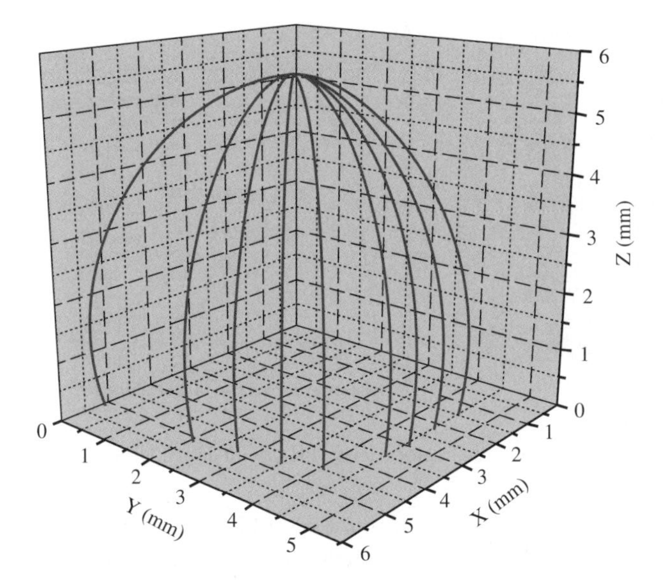

Figure 6.8 Seed curves of a one-quarter discontinuous freeform lens.

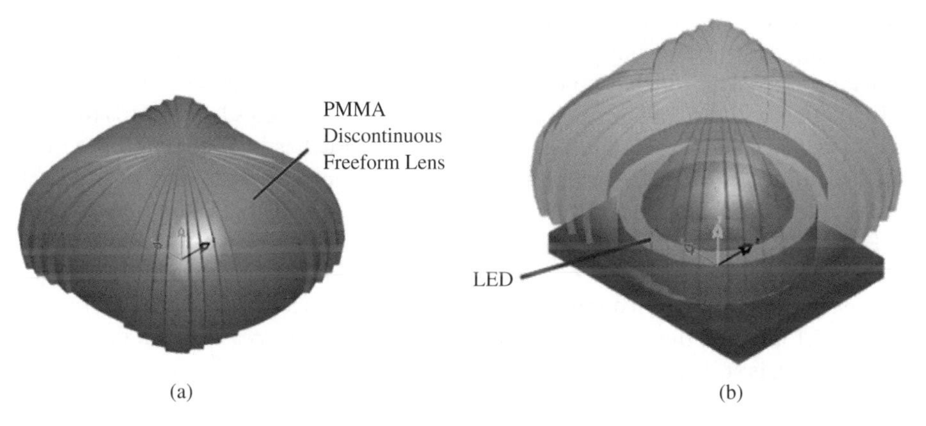

Figure 6.9 (a) A PMMA discontinuous freeform tunnel lens and (b) a numerical model for a Cree LED module with this lens.

The uniformity, defined as $U = E_{min}/E_{max}$, across the center of the target plane along the x-axis and y-axis is $U_x = 82.0\%$ and $U_Y = 81.3\%$, respectively. Since the shape of the light pattern will deteriorate sharply when there are some installation errors, we will calculate the uniformity within the area that is surrounded by the sharp edges of a light pattern, for example 9.6 m long and 8 m wide in Figure 6.10.

Tolerance is an important issue for freeform lenses. In this study, the tolerance analysis will be focused on increasing the size of the light source and on installation errors, including migration and rotation of the lens.

The practical optical model for the Cree XLamp XR-E LED with this freeform lens is shown in Figure 6.9b. The Cree EZ1000 LED chip is adopted in this LED module, and its area is 0.98 × 0.98 mm. Phosphor is conformal coated on the chip with a thickness of about 50 μm. Considering the converted light, yellow light, irradiating

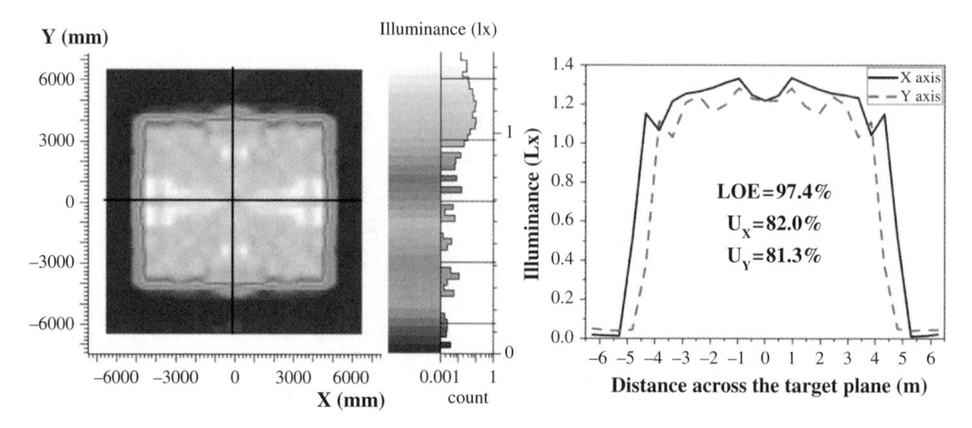

Figure 6.10 Numerical illumination performance of the freeform lens with a point light source.

from the phosphor layer, the area of light source in this Cree LED optical model is 1.08 × 1.08 mm. Since the distance from the chip to the outside surface of the lens is less than five times the chip size, this practical Cree LED could not be regarded as a point light source according to the far-field conditions of LED.[9] From simulation results shown in Figure 6.11, we can find that due to the increasing size of the light source, the edge of the light pattern becomes dim and the effective illumination area is decreased to 8.4 m long and 6.8 m, which is much smaller than that of the point light source condition. The LOE is still at a high level of 97.0%, and U_x (U_Y) increases from 82.0% (81.3%) to 85.8% (88.2%). Therefore, increasing the size of the LED has little effect on the LOE of this discontinuous freeform lens, but it significantly decreases the area of the light pattern. Large LED chips are a development trend due to their high luminous flux of single LEDs. Thus, to limit the effect of extended light sources maximally, the size of the freeform lens should be enlarged correspondingly according to the far-field condition.

The effects of lens migration on the illumination performance are shown in Figure 6.12 and Figure 6.13. When the vertical deviation d_V of the lens increases from 0.2 mm to 0.6 mm, the LOE is still at a high level of more than 97.0%, but the light pattern becomes

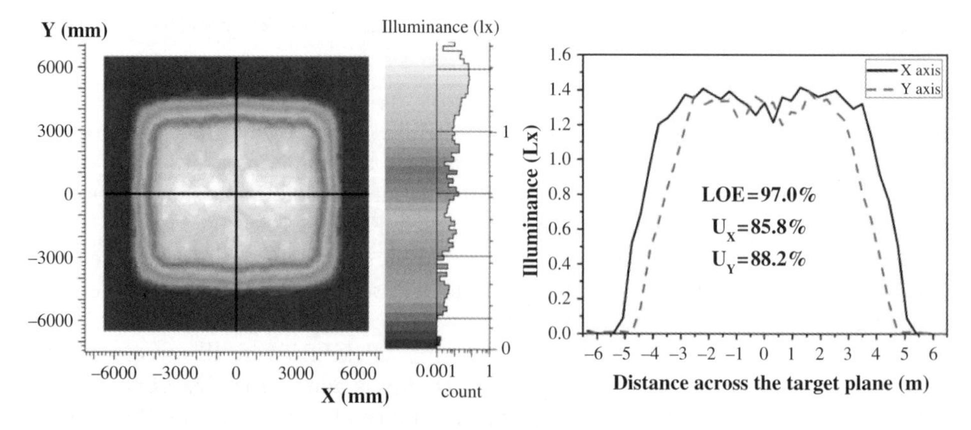

Figure 6.11 Numerical illumination performance of the freeform lens with a practical Cree LED.

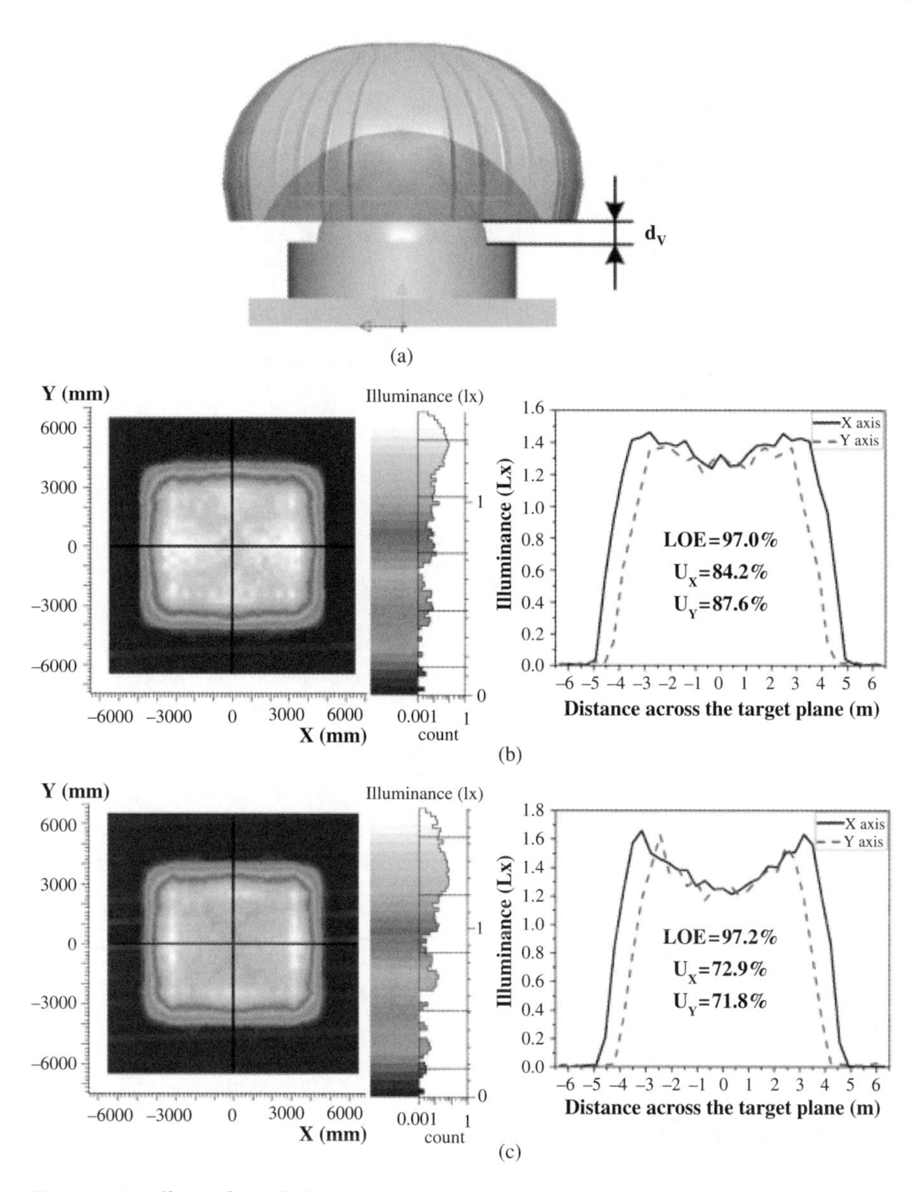

Figure 6.12 Effects of installation errors in the vertical up direction: (a) schematic of vertical deviation d_V; and (b) numerical illumination performance when $d_V = 0.2$ mm, (c) when $d_V = 0.4$ mm, and (d) when $d_V = 0.6$ mm.

smaller and U_x (U_Y) declines from 85.8% (88.2%) to 61.7% (61.4%). Moreover, as shown in Figure 6.13, the shape and uniformity of the light pattern both deteriorate sharply when the freeform lens deviates in the horizontal direction, except the LOE. Although the variation of U_Y is small, U_x declines from 85.8% to 60.2% as the horizontal deviation d_H increases up to 0.6 mm. Furthermore, the light pattern also becomes asymmetrical and larger when d_H increases. Therefore, it seems that the illumination performance is more sensitive to the installation errors in the horizontal direction.

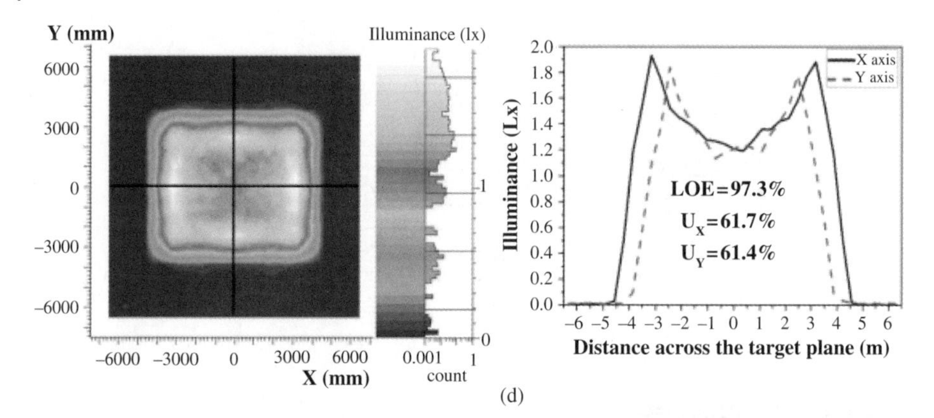

Figure 6.12 (*Continued*)

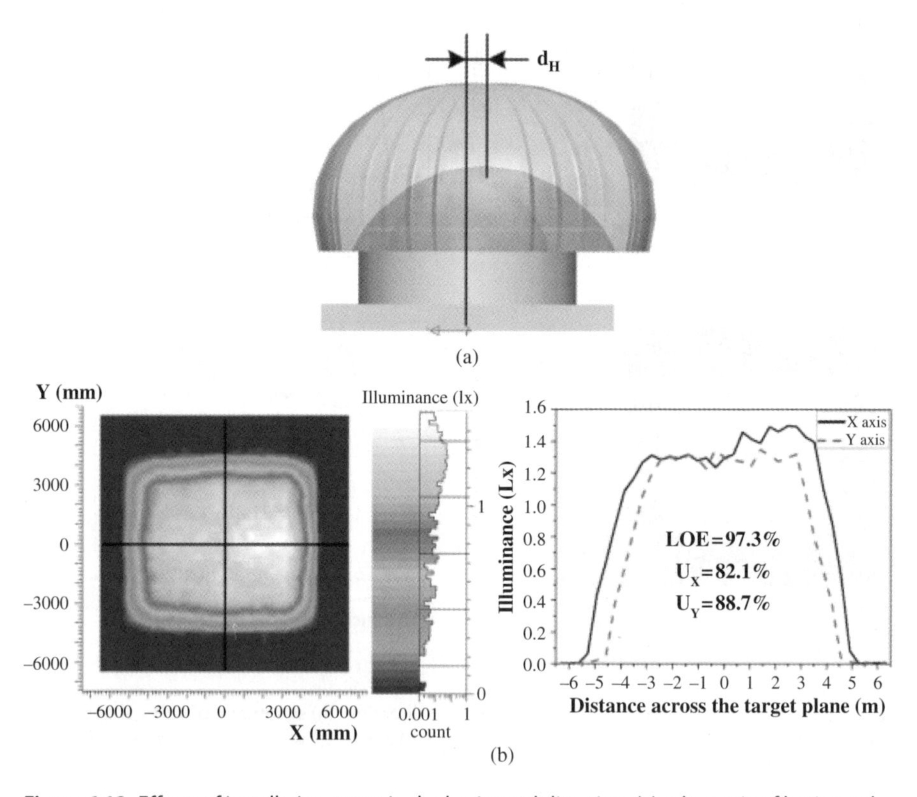

Figure 6.13 Effects of installation errors in the horizontal direction: (a) schematic of horizontal deviation d_H; and (b) numerical illumination performance when $d_H = 0.2$ mm, (c) when $d_H = 0.4$ mm, and (d) when $d_H = 0.6$ mm.

The effects of lens rotation errors were calculated as shown in Figure 6.14. The illumination performance also deteriorates sharply when the rotation error of freeform lens increases from 2° to 6°. The light pattern becomes larger and U_x declines from 85.8% to 53.7%, which is similar to the effects of installation errors in the horizontal direction.

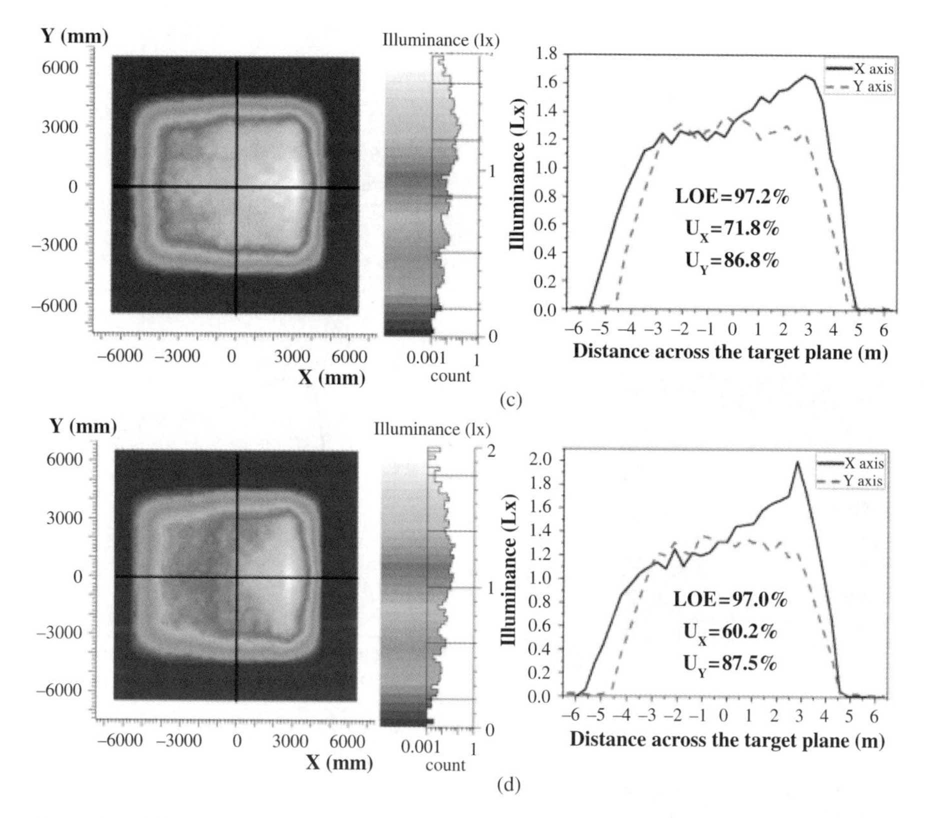

Figure 6.13 (*Continued*)

The effects of installation errors on illumination performance are concluded in Figure 6.15. Humans are distracted with illuminance variation of about 33% across the target; thus, the illumination performance of a light pattern cannot be acceptable until the uniformity is larger than 67% in either the X or Y direction. From comparisons as shown in Figure 6.15, we can find that the limits on the tolerance of vertical deviation, horizontal deviation, and rotation error are 0.4 mm, 0.4 mm, and 2°, respectively.

6.3.3 Experimental Analyses

The PMMA discontinuous freeform lens was manufactured by a molding process as shown in Figure 6.16. The lighting performance of the lens on the test plane is shown in Figure 6.17a. We can see that the light pattern is rectangular with a clear cutoff line, and the illuminance distribution of the light pattern is uniform, which is close to the result of simulation. Based on the discontinuous freeform lens, an 84 W high-power LED tunnel lamp, as shown in Figure 6.17b, was developed with cooperation from the Guangdong Real Faith Enterprises Group.

6.3.4 Effects of Manufacturing Defects on the Lighting Performance

Since the surface of this kind of freeform lens is constructed by a series of discrete sub-surfaces, microsized ultraprecise multi-axis diamond machining systems are needed

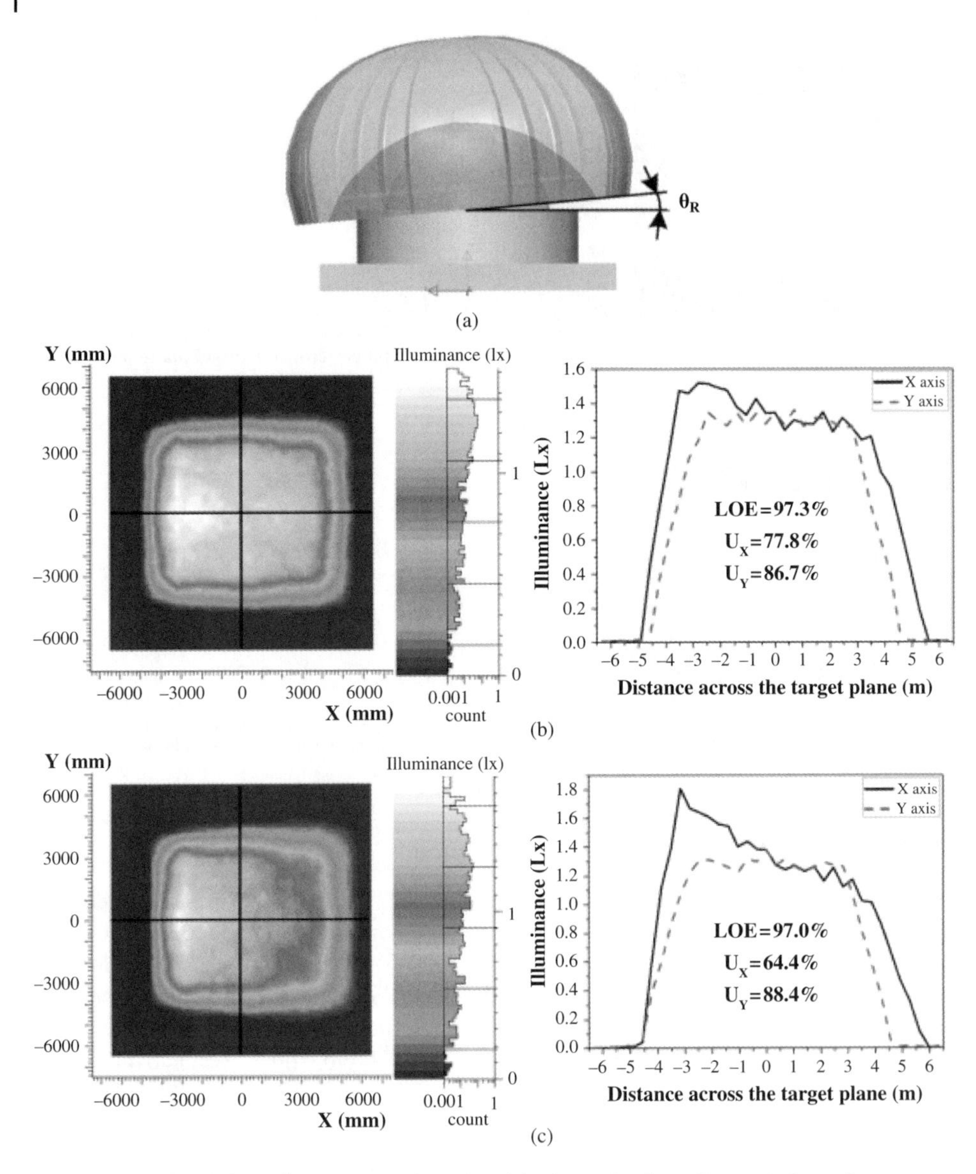

Figure 6.14 Effects of installation errors of rotation: (a) schematic of rotation angle θ_R; and (b) numerical illumination performance when $\theta_R = 2°$, (c) when $\theta_R = 4°$, and (d) when $\theta_R = 6°$.

for the lens or the lens mold fabrication. However, during mass production (i.e., injection molding), many manufacturing factors (e.g., surface morphology of mold, injection molding temperature and pressure, viscosity of liquid, etc.) will affect the surface morphology of the discontinuous freeform lens and thereby affect the optical performance of the lens. Surface roughness and smooth transition between discrete subsurfaces are two of the most common manufacturing defects existing in discontinuous freeform lenses. This section focuses on the effects of these two manufacturing defects on the light output efficiency and the shape and uniformity of the light pattern of the kind of

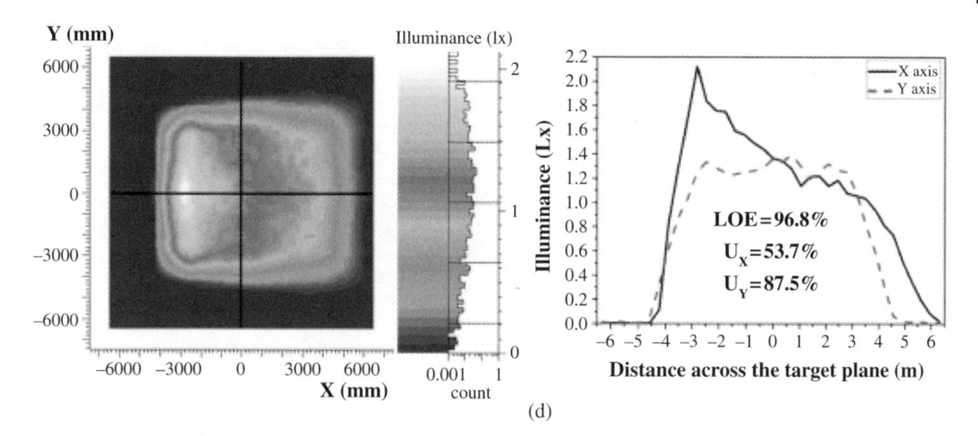

Figure 6.14 (*Continued*)

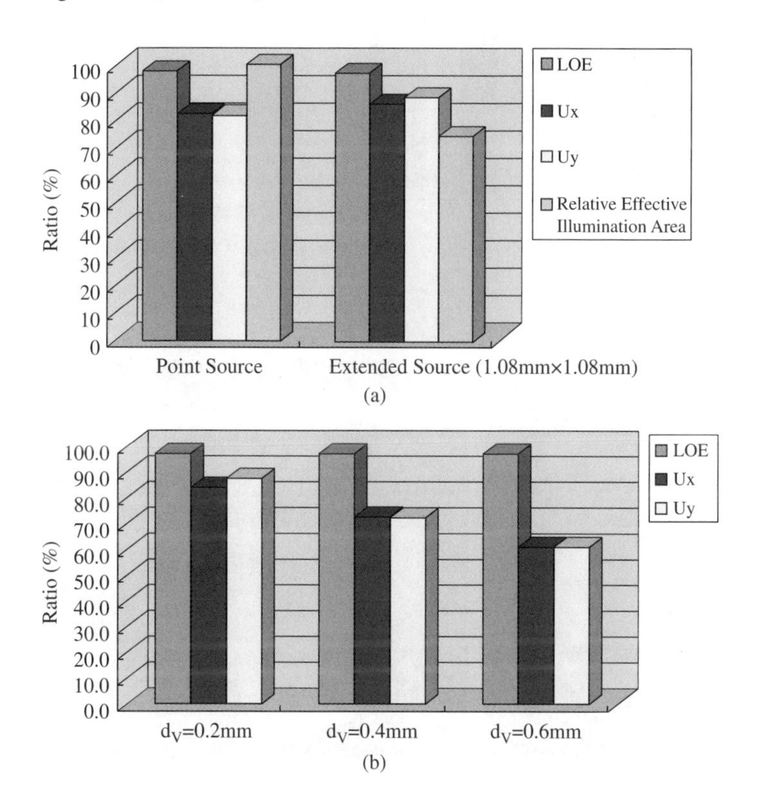

Figure 6.15 Comparisons of the effects of increasing size of light source and installation errors on illumination performance.

discontinuous freeform lens. The objectives of this section are to find out the effects of manufacturing defects on the optical performance and to come up with possible improvements.

To meet the requirements of road illumination, according to the design method mentioned in Chapter 7, two discontinuous freeform lenses were designed to form a 30 m

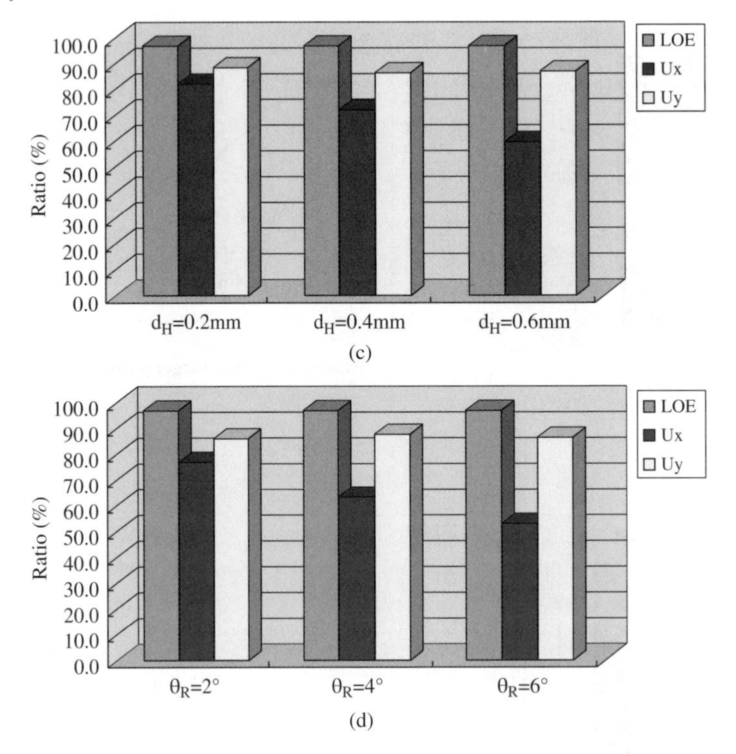

Figure 6.15 (*Continued*)

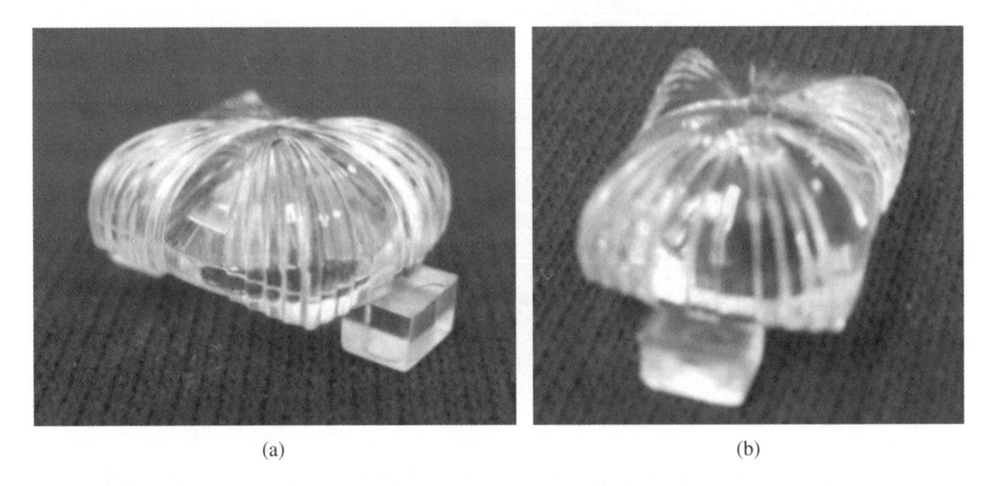

Figure 6.16 A PMMA discontinuous freeform lens for LED tunnel lighting.

long and 10 m wide uniform rectangle illumination area at a height of 8 m, which is acceptable for LED road lighting. The materials of these two lenses are PMMA and BK7 (named K9 in China) optical glass, which have refraction indexes of 1.49 and 1.59, respectively. The point clouds of the freeform surface and the lens model of the PMMA lens are shown in Figure 6.18. The numerical models for the LED module with these

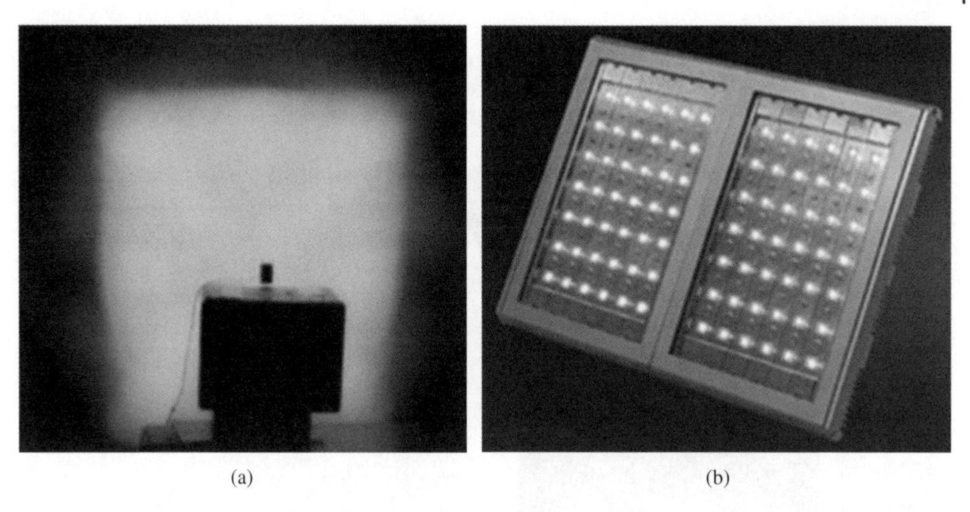

Figure 6.17 (a) Rectangular light pattern of the discontinuous freeform lens, and (b) an 84 W LED tunnel lamp integrated with these freeform lenses.

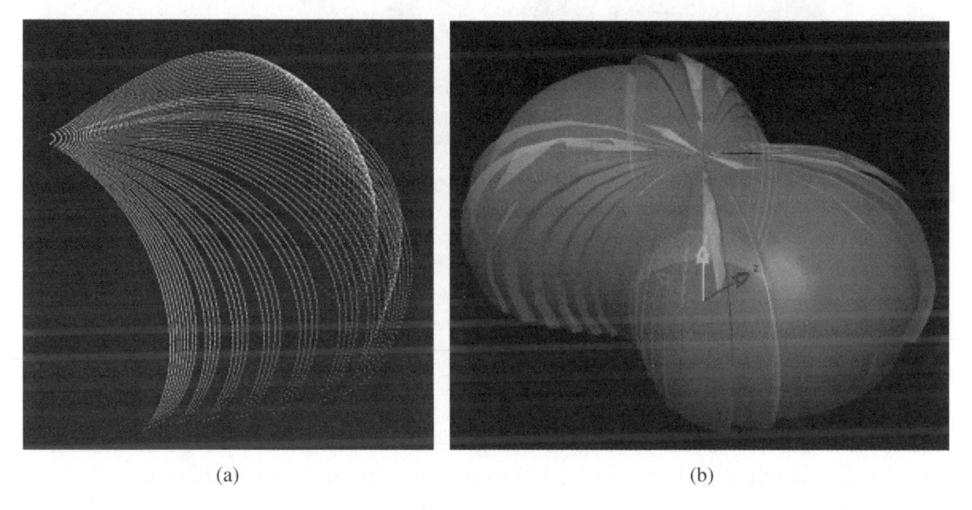

Figure 6.18 (a) Point clouds, and (b) a model of the PMMA discontinuous freeform lens.

discontinuous freeform lenses are shown in Figure 6.19a and Figure 6.20a. Then, two whole LED modules with different freeform lenses were simulated by 1 million rays. Figure 6.19b and Figure 6.20b show simulation illuminance distributions on two test areas with the same size of 42 m long and 42 m wide that are 8 m away from LED modules. In both simulation results, we can find that more than 95% light energy is uniformly distributed in the central area of about 28 m long and 10 m wide, which is consistent with the design target. During the central area, the minimal and average illuminances in simulation are 0.172 lx and 0.267 lx, respectively, for the PMMA lens; and 0.180 lx and 0.274 lx, respectively, for the BK7 glass lens. Consequently, the uniformity of the light patterns with a PMMA lens and BK7 glass lens is 0.644 and 0.657, respectively. The light output efficiency of the PMMA lens is 93.2%, and the efficiency of the BK7 glass lens reaches as high as 96.9%.

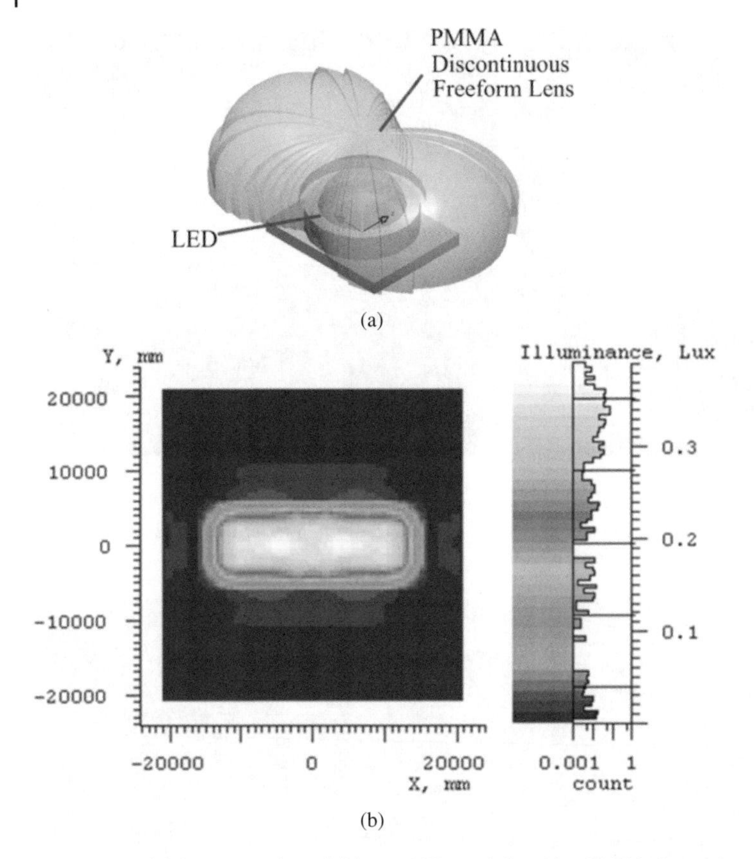

Figure 6.19 (a) A numerical model for an LED module with a PMMA discontinuous freeform lens and (b) its illumination performance in simulation.

6.3.4.1 Surface Morphology

As shown in Figure 6.21, these two discontinuous freeform lenses were also manufactured by the injection molding method. We observed the shape and surface morphology of these two discontinuous freeform lenses by microscopy. To control the irradiation directions of lights accurately, a sharp transition surface between two discrete subsurfaces was designed. The transition surface is nearly perpendicular to the two adjacent subsurfaces, and the deviation angle is less than 5°. By comparing Figure 6.22a with 6.22b, we can find that the shape of the manufactured PMMA lens is quite in agreement with the designed lens. The transition surface of the lens is sharp, and the boundaries between two discrete subsurfaces are clear. However, as shown in Figure 6.23, when we increase the magnification of microscopy, we can clearly find that there are a lot of micron-sized particles distributed on the surface of lens, especially on the transition surfaces, where are supposed to be smooth. Furthermore, some parts of this discontinuous freeform lens are totally composed of numbers of discontinuous particles with the size of tens of micrometers, and the surface morphology of these parts is quite different from the expected one, which will result in severe scattering. These particles were produced during manufacturing processes and were mainly caused by the unpolished surface of

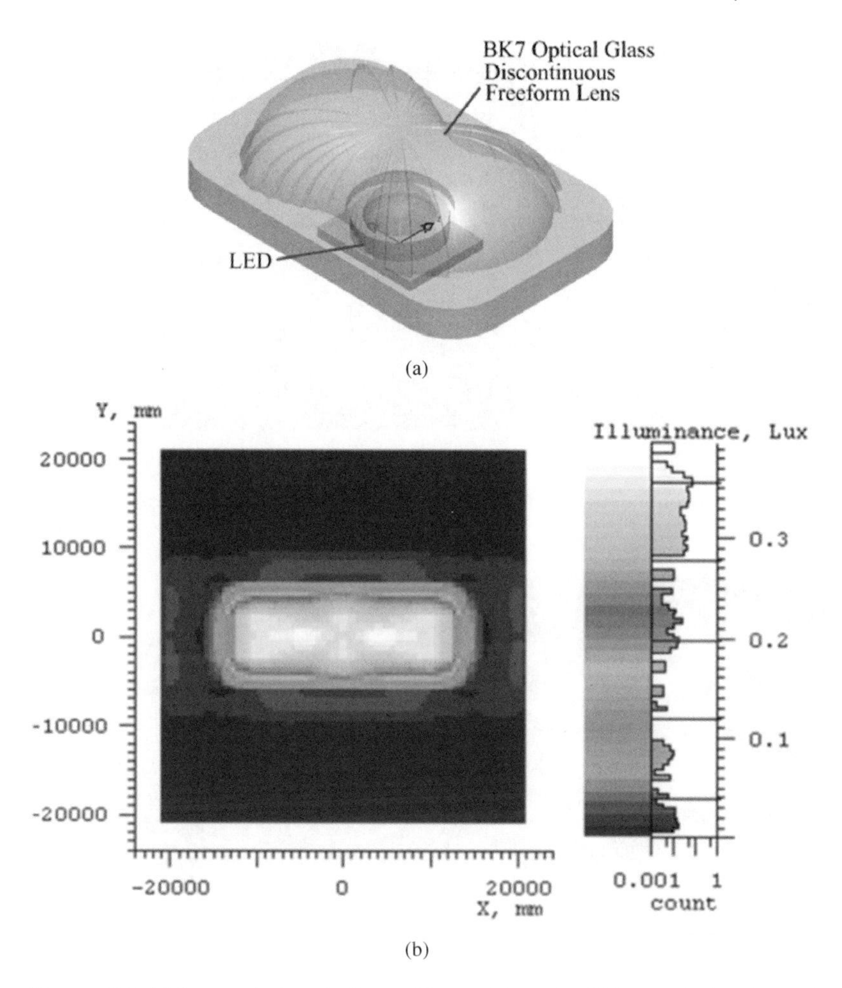

Figure 6.20 (a) A numerical model for an LED module with a BK7 optical glass discontinuous freeform lens and (b) its illumination performance in simulation.

mold. Moreover, unsuitable injection molding temperature, pressure, and viscosity of liquid also could result in these particles.

Figure 6.24 shows the comparison of a manufactured BK7 glass lens with its numerical optical model, from which we can find that although the shape of the manufactured BK7 glass lens is approximately similar with the designed one, significant differences exist at the positions of transition surfaces and the upper side of some discrete subsurfaces. The transition surfaces of the BK7 glass lens were also designed to be sharp; however, as shown in Figure 6.24a, they become quite smooth in the manufactured lens and the boundaries between two discrete subsurfaces also are blurry, especially at the middle part of the lens. Moreover, since some discrete subsurfaces are narrow, the deformation of their edges caused by the smooth transition will significantly affect the whole shape of these subsurfaces, which will result in the irradiation directions of lights having a large deviation from the expect directions. The smooth transition surfaces were mainly caused by the low processing precision of the mold. The same reasons as mentioned in

Figure 6.21 (a) The PMMA discontinuous freeform lens and (b) the BK7 optical glass discontinuous freeform lens.

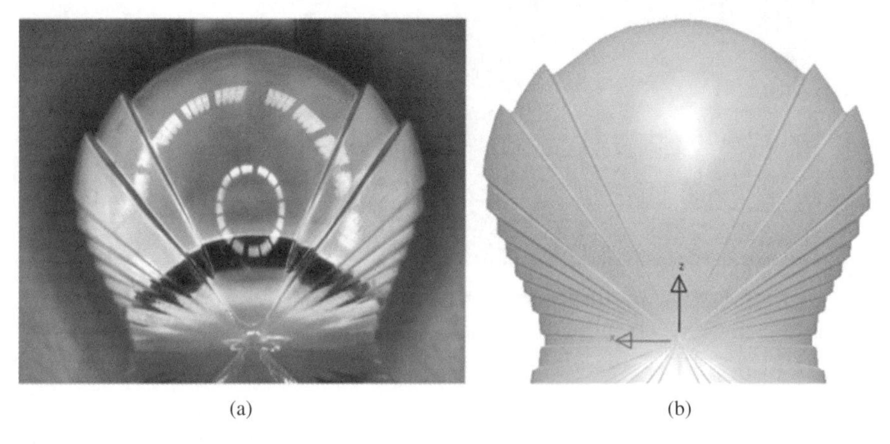

Figure 6.22 (a) Microphotograph of the PMMA lens and (b) a numerical optical model of the PMMA lens.

PMMA manufacturing (e.g., unsuitable injection molding temperature, pressure, and viscosity) could also result in these manufacturing defects. From Figure 6.24a and 6.24c, we can find that there are only little tiny particles distributed on the surface of the BK7 glass lens and the surface is much smoother than the PMMA lens.

6.3.4.2 Optical Performance Testing

To test the optical performance of these two discontinuous lenses, LED freeform lens test modules were built. As shown in Figure 6.25, these are two freeform lens test modules, and each of them is mainly composed of four parts: a 1 W high-power CREE XLamp LED, a discontinuous freeform lens, a lamp frame with a heat sink and fins, and a driving circuit for the power input of the LED.

The total luminous flux of the LED module with and without a freeform lens was measured by ultraviolet-visible (UV-VIS) near-infrared (near-IR) spectrum photo colorimeter measurement and an integrating sphere. The light output efficiency of the

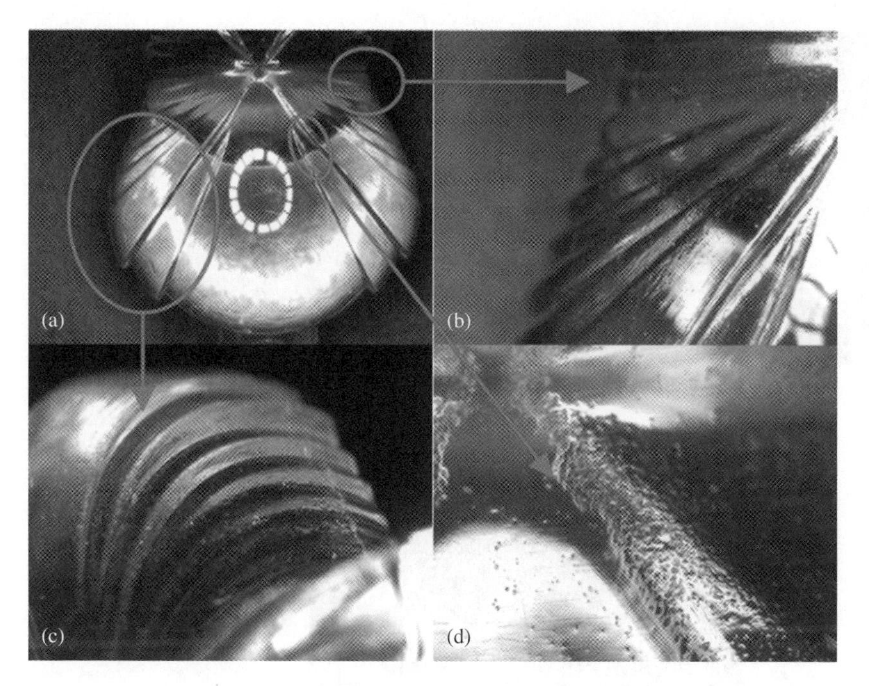

Figure 6.23 Micrographs of different parts of the PMMA discontinuous freeform lens.

PMMA lens reaches as high as 90.5%, which is quite close to the simulation results. Figure 6.26 depicts the light pattern at 70 cm away from the LED. We can find that most of the light energy is distributed in a nearly rectangle area with a length of 240 cm and width of 83 cm, and it will enlarge to 27.5 m long and 9.5 m wide at the height of 8 m according to the light rectilinear propagation principle, which is also quite in agreement with the expected shape. However, obvious dark stripes exist on the light pattern, especially in the middle-upper part of the pattern, which will decrease the uniformity of the pattern and the performance of illumination. We arranged 30 test points equally spaced on the light pattern and measured the illuminance of each point by illuminance meter. The minimal illuminance, appearing at the dark stripe, is 10.7 lx and the calculated average illuminance is 34.2 lx, so the uniformity is only about 0.313, which is much lower than the simulated uniformity. Moreover, we also find that the relative positions of these dark stripes on the light pattern are the same as the relative positions of transition surfaces on the lens surface.

The optical performance of the BK7 glass lens was also tested. The light output efficiency of the lens is 91.0%. Although this efficiency is slightly higher than that of the PMMA lens as the BK7 optical glass has a higher transmittance than PMMA, it is much lower than the simulation results. Compared with the experimental and simulation results, the efficiency decline of the BK7 glass lens is up to 5.9%, which is higher than that of the PMMA lens (2.7%). A light pattern at 70 cm away from the LED is shown in Figure 6.27. We can find that the illumination performance of the BK7 glass lens is quite poor; not only the light energy distribution on the target plane is nonuniform, but also the shape of the light pattern is not rectangular. Obvious dark and bright stripes exist on the light pattern, especially at the central positions and in the diagonal directions.

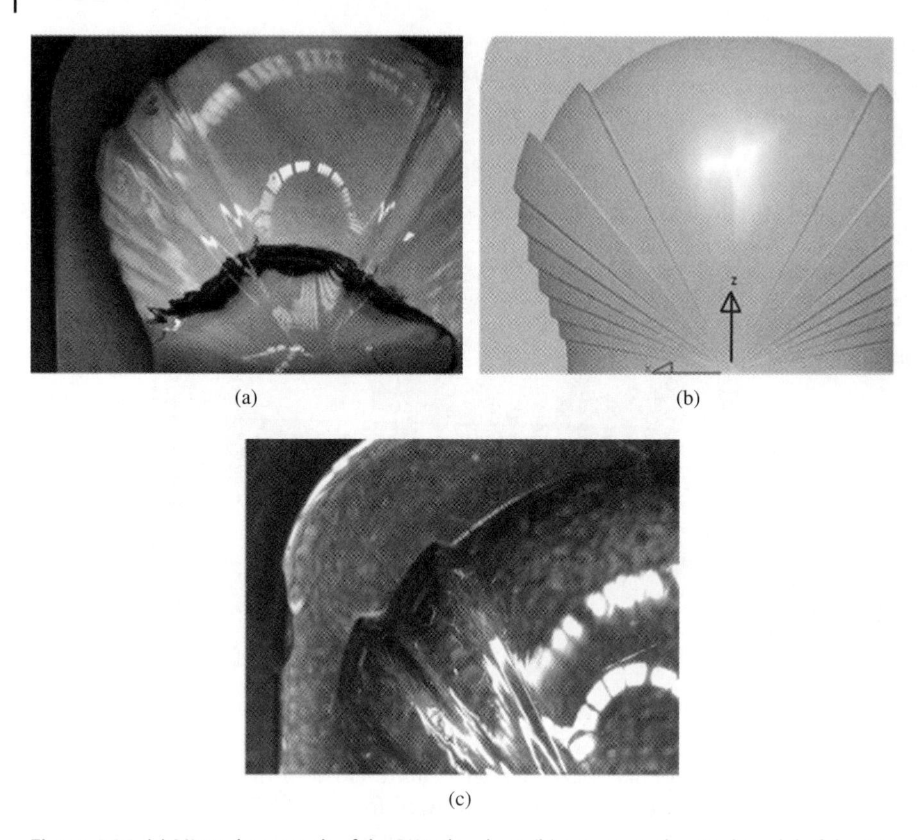

Figure 6.24 (a) Microphotograph of the BK7 glass lens, (b) a numerical optical model of the BK7 glass lens, and (c) a partially enlarged view of the BK7 glass lens.

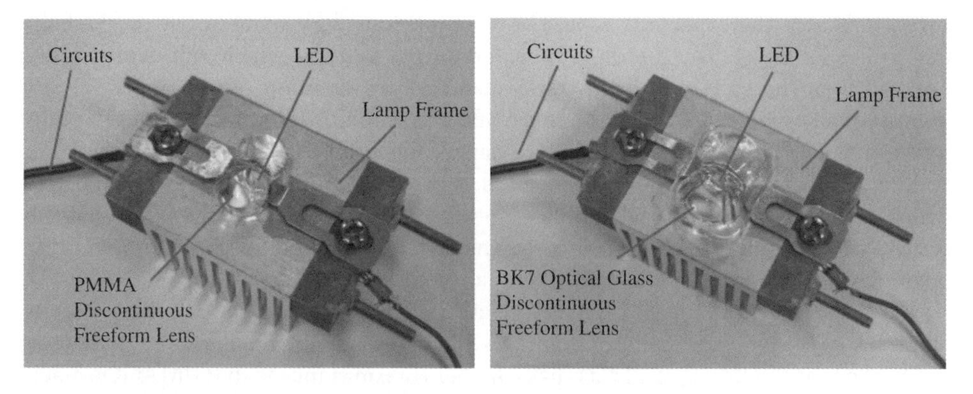

Figure 6.25 The LED discontinuous freeform lenses test modules.

The BK7 glass lens could not control the lights effectively, and some lights irradiate out of the designed rectangular area, which results in fan-shaped light energy distribution at the left and right ends of the light pattern. Since the shape of the light pattern is quite different from the rectangle, it is meaningless to calculate the average illuminance and uniformity.

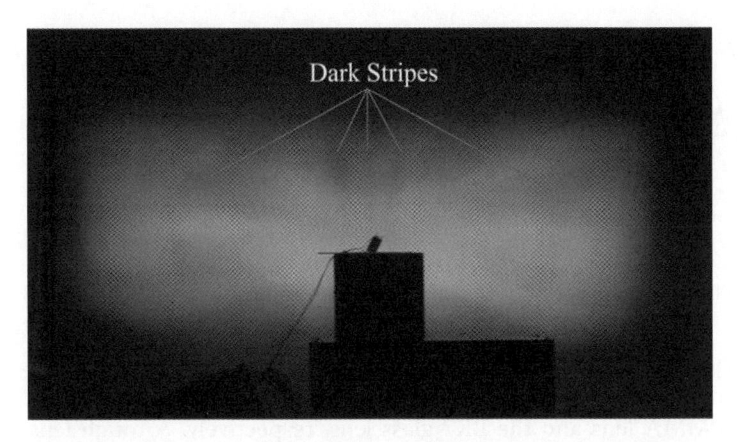

Figure 6.26 Light pattern of the PMMA lens at 70 cm away from the LED.

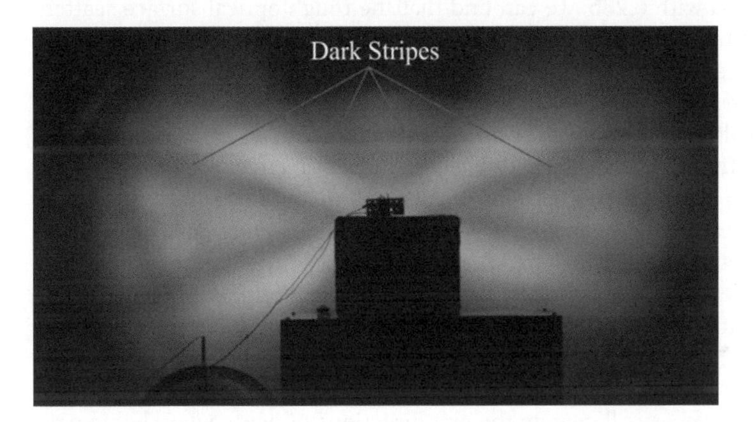

Figure 6.27 Light pattern of the BK7 optical glass lens at 70 cm away from the LED.

Table 6.2 Comparisons between simulation results and experimental results.

		Light output efficiency of the freeform lens, %	Shape of light pattern (8 m high)	Dark stripes	Uniformity
PMMA lens	Simulation	93.2	28 × 10 m	No	0.644
	Experiment	90.5	27.5 × 9.5 m	Yes	0.313
BK7 glass lens	Simulation	96.9	28 × 10 m	No	0.657
	Experiment	91.0	Nonrectangular	Yes	–

From comparisons between the simulation results and the experimental results as shown in Table 6.2, we can find that the experimental light output efficiency and shape of light pattern of the PMMA lens are quite close to the simulation ones, but dark stripes exist, which mainly result in low uniformity of the experimental light pattern. However, the BK7 glass lens has poor illumination performance, and the experimental results are quite different from the design goals.

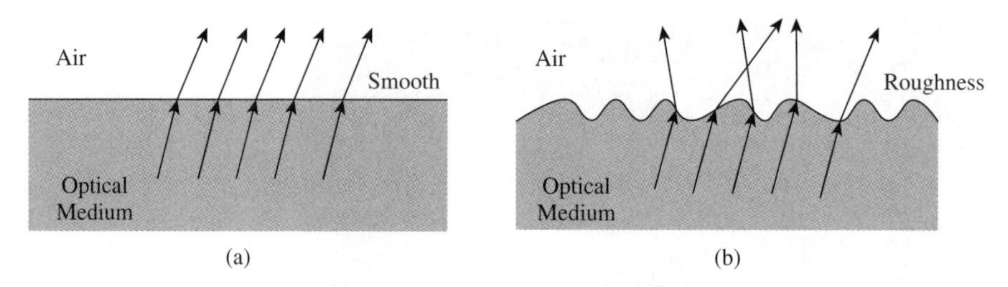

Figure 6.28 Schematic of lights propagation at (a) smooth and (b) rough optical surfaces.

6.3.4.3 Analysis and Discussion

Surface roughness and smooth transition surfaces are two major kinds of manufacturing defects in the PMMA lens and the BK7 glass lens, respectively. A rough lens surface could increase the chance of light scattering at the interface of lens and air. By comparing Figure 6.28a with 6.28b, we can find that the rough optical surface scatters lights randomly, and the directions of exit lights are quite different from those of the designed ones. The scattered lights generated at the roughness transition surface will deviate from the expected irradiation directions and probably will overlap with other lights, which will result in dark stripes appearing in the direction of transition surfaces on the light pattern. The dark stripes not only affect the visual effect but also decrease the illuminance of this illumination area, which will result in low uniformity. Moreover, although lights are scattered by roughness surfaces, most of them just change the irradiation direction and still can exit from the lens surface; this is the reason why defect-induced surface roughness has a small effect on the light output efficiency of the lens. Fortunately, the area of transition surfaces accounts for only <10% of the whole lens surface area; therefore, the dark stripes also have a small effect on the shape of the light pattern.

Smooth transition surfaces will change the exit directions of lights when lights irradiate on these surfaces. As shown in Figure 6.29, the deviation angle θ of the transition surface increases from about 5° to more than 60° when the transition surface becomes smooth. By comparing Figure 6.29a with 6.29b, we can find that most lights, which are supposed to exit from the lens at the sharp transition surface, will have a TIR at the smooth transition surface as the large deviation angle. As shown in Figure 6.29b, the TIR will significantly change the propagation directions of lights in the lens and the exit directions of lights at the lenses' surface. Since the area of the smooth transition surfaces accounts for a high ratio (more than 30%) of the whole BK7 glass lens surface area as shown in Figure 6.24, more than 30% of exiting light deviates seriously from the designed directions, which will induce the generation of dark and bright strips, change the shape of the light pattern, and significantly affect the illumination performance. Furthermore, the shape deformation of some narrow subsurfaces caused by smooth transition surfaces will also change the directions of lights and result in poor illumination performance. Moreover, the TIR increases the propagation length of lights, which will increase the lights absorption by glass and decrease the LOE. Even worse, as the reason of TIR, a small part of lights irradiate to the bottom of the lens; thus, this part of light energy is

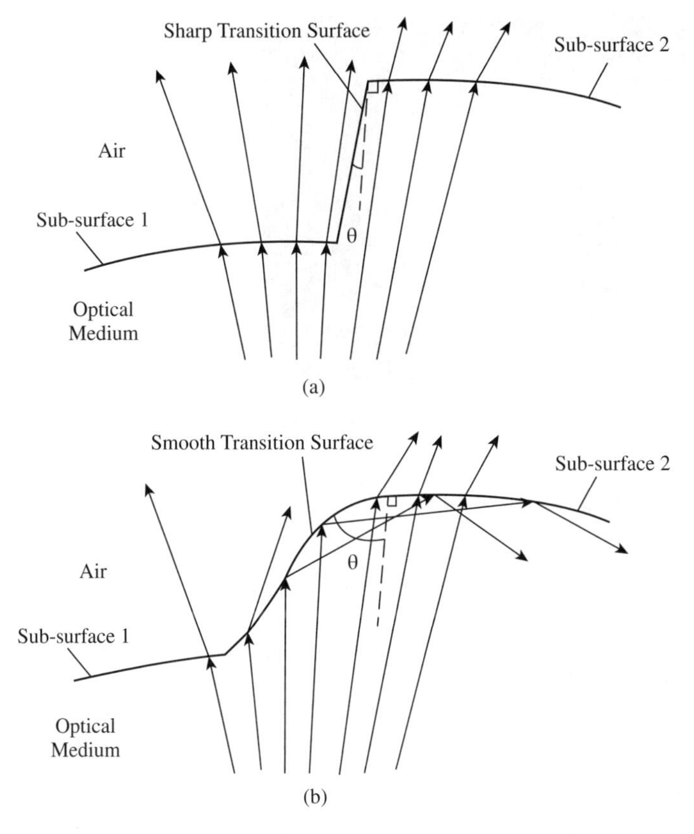

Figure 6.29 Schematic of lights propagation at (a) the sharp transition surface and (b) the smooth transition surface.

totally lost. These are reasons why the efficiency decline of the BK7 glass lens is higher than that of the PMMA lens.

To improve the optical performance of the freeform lens, several suggestions are provided as follows: (1) improving the surface construction algorithm in the discontinuous surface freeform lens design method and ensuring that the surface of the lens is continuous and smooth – this is the radical way to reduce the defects generated during manufacturing; (2) improving and optimizing the manufacturing process control, such as pressure, temperature, viscosity of polymer, and so on, and enhancing the precision of manufacturing; and (3) utilizing the discontinuous surface freeform lenses array while designing LED illumination systems. As the illumination area is overlapped by multilight patterns irradiating from different positions and directions of the illumination system, dark stripes existing on one light pattern might be overlapped by bright illumination areas of other light patterns and we cannot distinguish them anymore. As shown in Figure 6.30, the lighting performance of a LED road lamp with a freeform lenses array at 1 m away is much more uniform than that of a single freeform lens.

Figure 6.30 Lighting performance of an 112 W LED road lamp with a freeform lens array.

6.3.5 Case Study – LED Road Lamps Based on DFLs

Based on the PMMA discontinuous freeform lens mentioned in this chapter, an 112 W LED road lamp integrated with a freeform lenses array was developed by the Guangdong Real Faith Enterprises Group in 2007, as shown in Figure 6.31. It is the first LED road lamp based on the discontinuous freeform lens interiorly. The measured LIDC of the 112 W LED road lamp is shown in Figure 6.32. As we can see, the light shape emanates along the road, but aggregates at the direction perpendicular to the road. Thereby, the energy of light is limited to the range of effective area. The realistic lighting performance of the 112 W LED road lamp is shown in Figure 6.33. As we can see, the road surface is bright and uniform, making people comfortable. Meanwhile, compared with the performance of a 250 W HPS lamp on the same road, the light of the 112 W LED road lamp is brighter and more uniform. Hence, the LED road lamp is able to substitute for the 250 W HPS lamp. It is more energy-efficient: energy use can be cut by more than 50% with better performance of light.

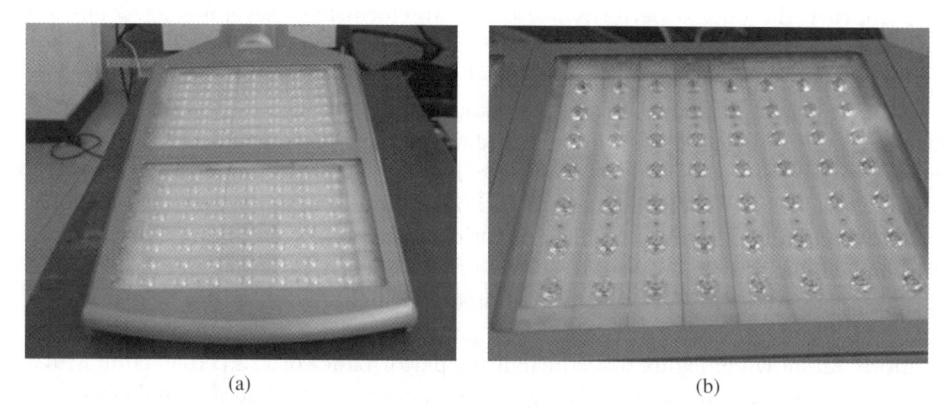

(a) (b)

Figure 6.31 (a) One 112 W LED road lamp based on the PMMA discontinuous freeform lenses, and (b) a PMMA freeform lens array.

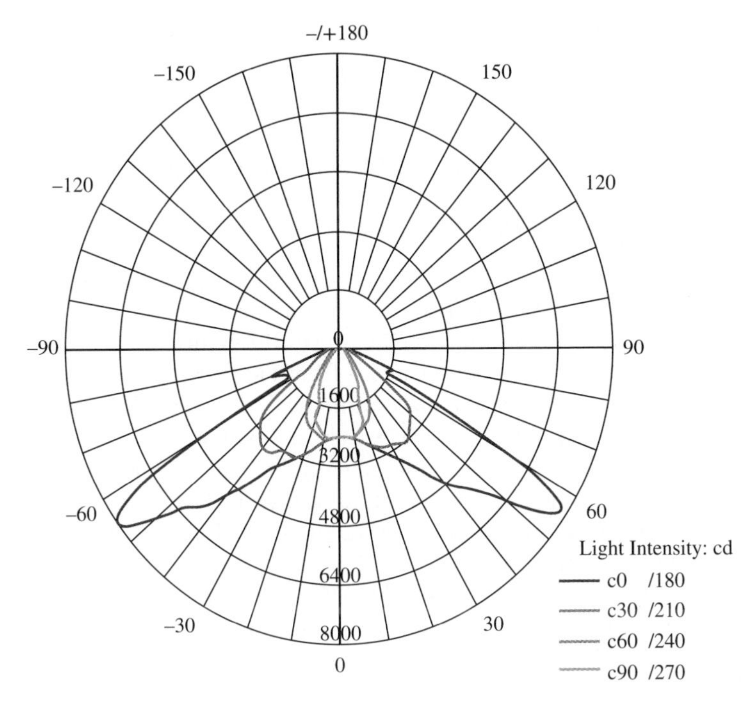

Figure 6.32 LIDC of the 112 W LED road lamp.

Figure 6.33 (a) Good lighting performance of the 112 W LED road lamp integrated with a discontinuous freeform lenses array, and (b) lighting performance comparison between the 112 W LED road lamp and one 250 W HPS lamp. (Supplied by Guangdong Real Faith Enterprises Group, www.gd-realfaith.com)

From the above-mentioned algorithm analysis, simulation verification, experimental test, and practical application, it can be seen that the new algorithm of a discontinuous freeform lens for a rectangular light pattern proposed in this book has the advantage of flexible energy mapping, more accurate control of light shape, greater uniformity, greater ease in manufacturing, and so on. The discontinuous freeform lens is practical and used successfully in reality.

6.4 Continuous Freeform Lens (CFL) for LED Road Lighting

6.4.1 CFL Based on the Radiate Grid Mapping Method

According to the continuous freeform lens algorithm based on radiate grid light energy mapping as mentioned in Section 3.4.2.1, a PMMA continuous freeform lens for road light is designed whose optical model and simulated lighting performance are shown in Figure 6.34 and Figure 6.35b, respectively. As we can see from Figure 6.34, the surface of the lens is continuous and smooth, without a discontinuous fillet surface. The simulation light pattern on the receiving surface is approximately in the shape of a rectangle; in it, the illuminance distribution is uniform. Compared with the lighting performance of the continuous freeform lens without optimization as shown in Figure 6.35a, the uniformity of the newly designed lens is drastically elevated, indicating that the effect of the optimization algorithm is obvious.

Meanwhile, as shown in Figure 6.36a, the lens is manufactured by a molding process, and its surface is very smooth without obvious manufacturing defects. Using Lambertian light distribution as the light source, the efficiency of the Philips Lumileds K2 LED, for example, is 93.1% in the experiment, which is higher than that of the discontinuous freeform lens. The lighting performance of the light pattern is shown in Figure 6.36b. As we can see, the light pattern is approximately in the shape of a rectangle, where the illuminance distribution is uniform. Based on the lens, as shown in Figure 6.37a, a 168 W LED road lamp launching warm white light was developed by Guangdong Real Faith Enterprises Group. The main reason why warm white light was adopted is that people will feel that the surroundings are more comfortable and soft with warm white light at night, instead of the cold and depressive feel of cold white light. Also, uncomfortable glare will be more serious with the increase of the blue component. Consequently, compared with cold white light, there is less glare produced by warm white light. The lighting performance is shown in Figure 6.37b, from which we can see that the road surface is very bright and uniform with great visual effect.

6.4.2 CFL Based on the Rectangular Grid Mapping Method

According to the continuous freeform lens algorithm based on rectangular grid light energy mapping as mentioned in Section 3.4.2.2, we designed a PMMA continuous freeform lens for road lighting with a Lambertian LED light as the light source. The lens' optical model and simulating lighting are shown in Figure 6.38 and Figure 6.39a. We can see that the surface of the lens is continuous and smooth, and the light pattern is nearly a rectangle. The PMMA lens is manufactured by machining (Figure 6.40a), and the optical efficiency is 92.8%. As Figure 6.40b shows, the volume of the freeform lens is much smaller compared with a LED road light lens on the market. Figure 6.39b shows the experimental lighting performance of the lens. We can see that the light pattern is close to a rectangle and the outline is clearly defined. The pattern is overall uniform without clear bright lines or dark lines, achieving a uniform rectangular illumination.

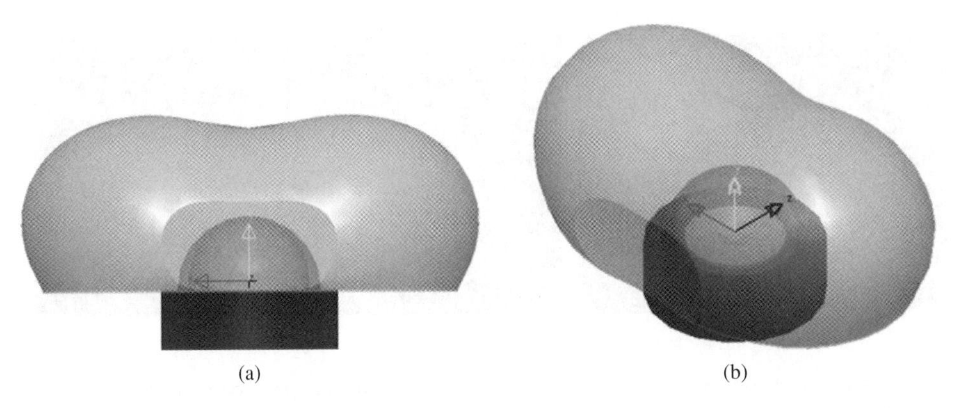

Figure 6.34 Optical model of the PMMA continuous freeform lens (radiate grid mapping) for road lighting.

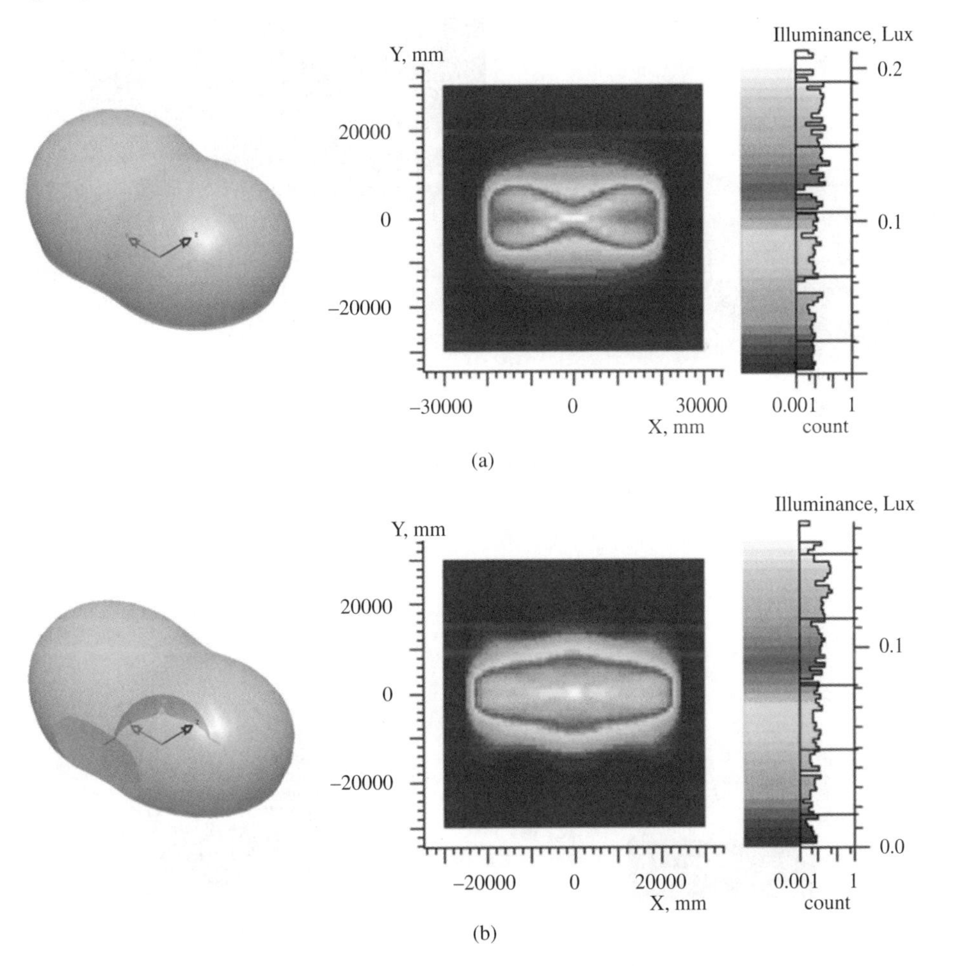

Figure 6.35 Simulation lighting performance of a continuous freeform lens (radiate grid mapping): (a) without optimization and (b) with optimization.

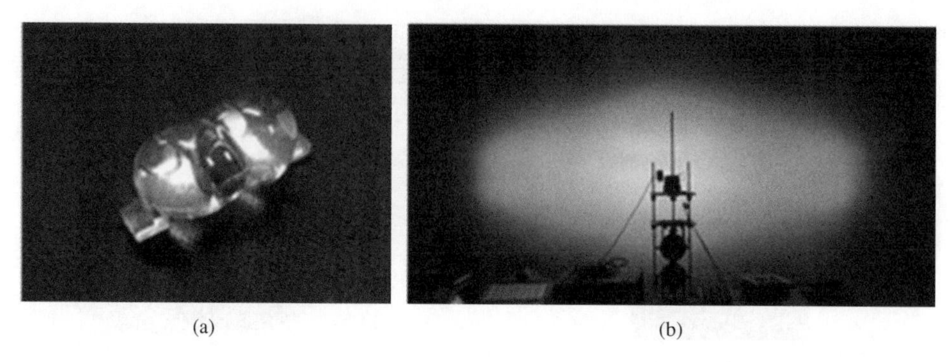

Figure 6.36 (a) A PMMA continuous freeform lens (radiate grid mapping) for road lighting, and (b) its light pattern.

Figure 6.37 (a) A 168 W high-power LED road lamp based on the PMMA continuous freeform lenses (radiate grid mapping), and (b) its road lighting performance. (Supplied by Guangdong Real Faith Enterprises Group, www.gd-realfaith.com)

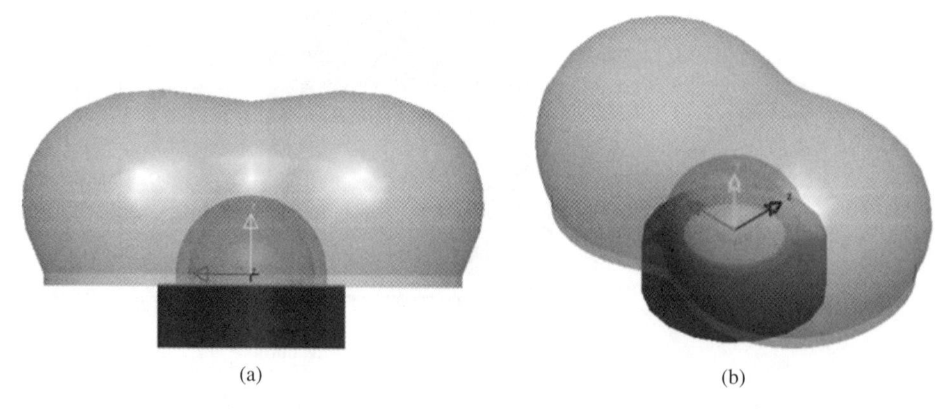

Figure 6.38 Optical model of the PMMA continuous freeform lens (rectangular grid mapping) for road lighting.

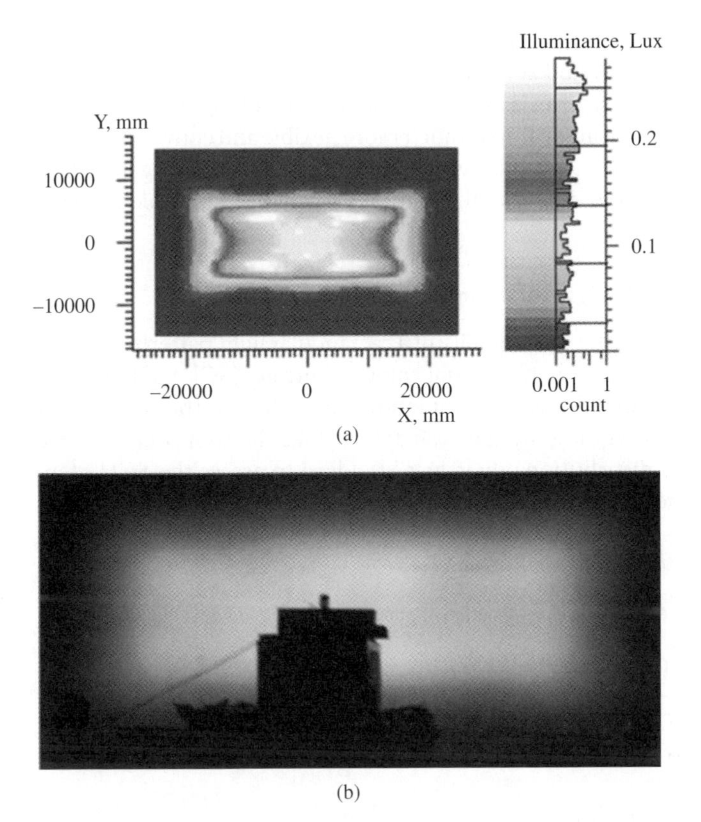

Figure 6.39 (a) Simulation and (b) experimental lighting performance of a single continuous freeform lens (rectangular grid mapping).

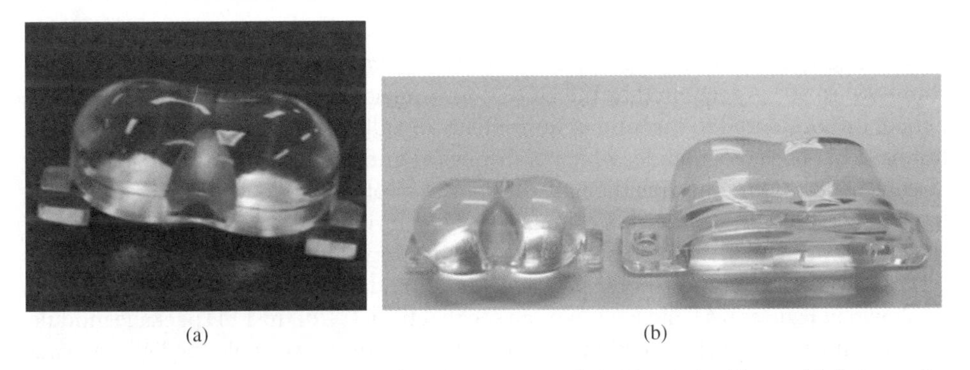

Figure 6.40 (a) A PMMA continuous freeform lens (rectangular grid mapping) for road lighting; and (b) another lens on the market, for comparison.

Through the analysis in this chapter, we can find that a continuous freeform lens with high lighting quality can be designed by the two continuous freeform lens design algorithms proposed in Chapter 3, and a uniform rectangular illumination will be achieved. The lens has the advantages of continuous smooth surface, small size, and high LOE, and the pattern shape can be controlled accurately. Lighting performance can meet the needs of practical application. Therefore, the design of the continuous freeform lens will be based on these two algorithms, and some modifications will be added that depend on different applications. Besides, from a series of applications of algorithms in this section, we can see that this design algorithm is much more flexible and convenient compared to the energy-mapping relationship between the light source and target surface by solving the energy conservation differential equations. It will be applicable in many situations.

6.4.3 Spatial Color Uniformity Analyses of a Continuous Freeform Lens

As with the circular light pattern, the color quality of a rectangular light pattern is also an important factor. Therefore, we analyze the spatial color uniformity (SCU) of the above two designed continuous freeform lenses by simulation. According to the standard of solid-state lighting luminaires required by the DOE,[10] we take the root-mean-square (RMS) value of spatial color distribution $\Delta u'v'$ as a standard to assess the SCU of an LED. For noncircular symmetry light patterns, the $\Delta u'v'$ is:

$$\Delta u'v'(\alpha, \beta) = \sqrt{(u'(\alpha\beta) - u'_{weighted})^2 + (v'(\alpha, \beta) - v'_{weighted})^2} \tag{6.13}$$

where $u'_{weighted}$ and $v'_{weighted}$ are:

$$u'_{weighted} = \frac{\sum\limits_{\beta}\sum\limits_{\alpha} E_N(\alpha, \beta)u'(\alpha, \beta)}{\sum\limits_{\beta}\sum\limits_{\alpha} E_N(\alpha, \beta)} \tag{6.14}$$

$$v'_{weighted} = \frac{\sum\limits_{\beta}\sum\limits_{\alpha} E_N(\alpha, \beta)v'(\alpha, \beta)}{\sum\limits_{\beta}\sum\limits_{\alpha} E_N(\alpha, \beta)} \tag{6.15}$$

where α and β are defined as light output angles: α ranges from 0° to 180°, and β ranges from −90° to 90°. As required by the DOE, $\Delta u'v'$ must be smaller than 0.004, and then it could be considered to meet the requirements of SCU. At this point, we take the traditional LED package module, which is dispensed and coated by a phosphor layer by a freely dispersing method, as the light source. The optical model will be elaborated in detail in Chapter 10. Besides, we calculate these colorimetric parameters according to the color coordinates $u'v'$ and correlated color temperature (CCT).

The CCT of one traditional LED package module and the spatial distribution of $\Delta u'v'$ are shown in Figures 6.41 and 6.42. We can see that the CCT of the LED package module is higher in a small-angle direction, and the CCT will decrease while the light viewing angle increases. The spatial distribution of $\Delta u'v'$ looks like the letter W. As we can see, the value of $\Delta u'v'$ is larger than 0.004 except for the intermediate region with a small angle, and it is up to about 0.022.

We place the PMMA continuous freeform lens (radiate grid mapping) on the LED module, and simulate the blue light emitting from the LED chip and the yellow light emitting from phosphor, respectively. Figure 6.43 shows the optical model of an LED module integrated with the freeform lens, and the blue-light pattern and yellow-light

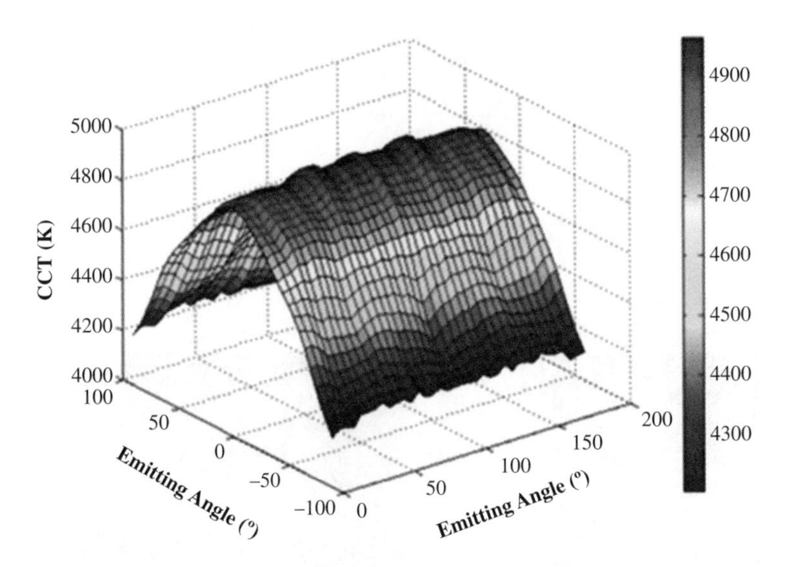

Figure 6.41 CCT spatial distribution of the traditional LED module where CCT decreases gradually from the center to the side.

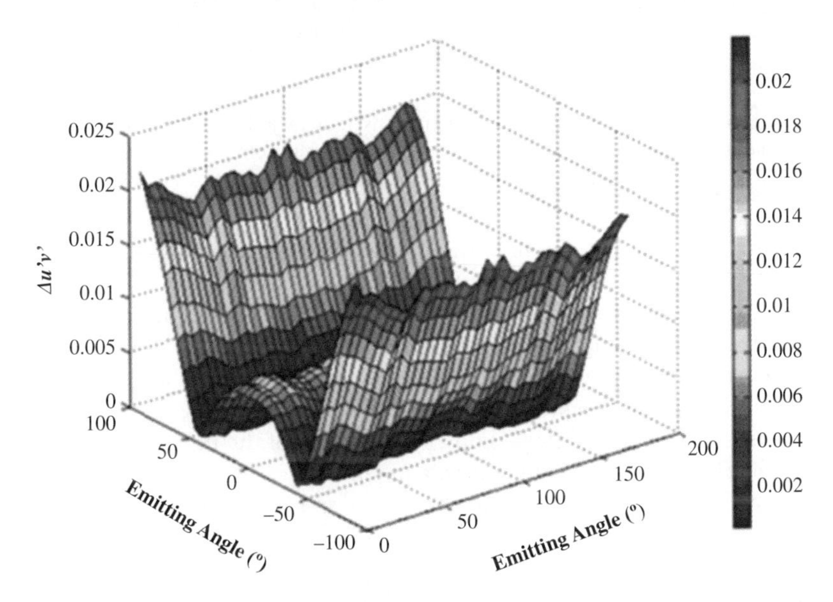

Figure 6.42 $\Delta u'v'$ spatial distribution of the traditional LED module where $\Delta u'v'$ increases from the center to the side.

pattern by simulation. It can be found from the figure that the blue-light pattern is close to the yellow-light pattern, but the yellow-light pattern is slightly longer and wider than the blue-light pattern as a whole. There are two reasons for this phenomenon. First, the blue light and the yellow light have different light-emitting areas. As a result of the freely dispersed method, the phosphor layer reaches a bottom diameter at 3 mm that

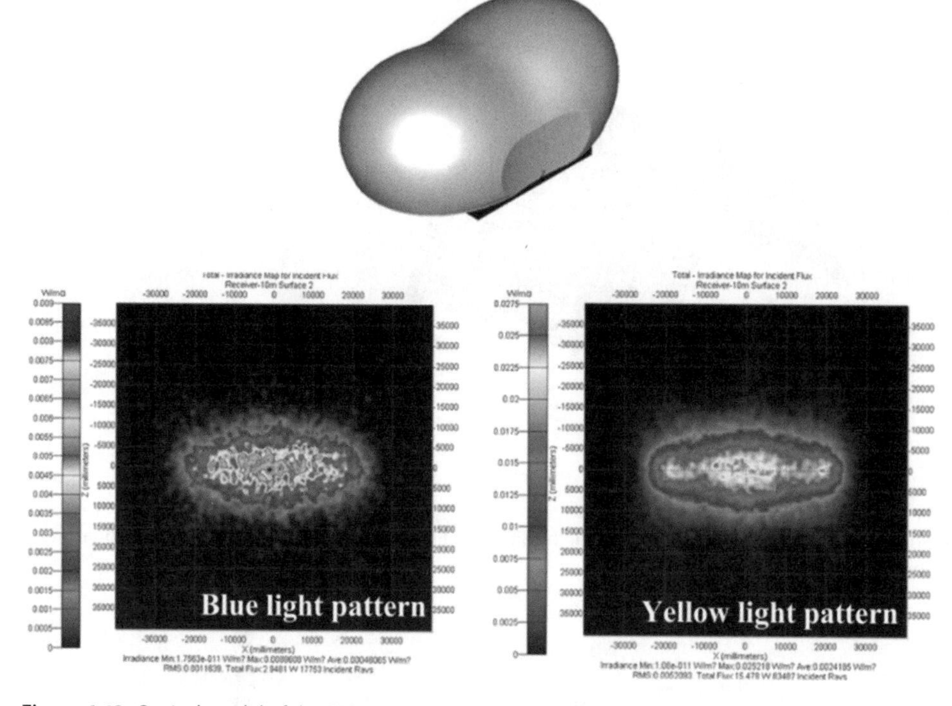

Figure 6.43 Optical model of the PMMA continuous freeform lens (radiate grid mapping) and its blue-light and yellow-light patterns where illuminance decreases gradually from the center to the side of the light pattern.

is much larger than 1 mm of an LED chip. The outputting area of a phosphor layer is larger than an LED chip, resulting in the pattern being relatively diffused. The second is that the emitting areas of blue light and yellow light have a different height. As with the phosphor freely dispersing process, the height of the phosphor layer is about 0.4 mm. The height of the equivalent phosphor layer is between 0.2 mm and 0.3 mm, which is higher than the 0.1 mm height of an LED chip. From the discussion of installation errors on lenses in Section 6.3.2, we know that the size of the light pattern will be smaller while the distance of the light source and lens increases, or vice versa. As the distance between the phosphor layer and the lens is smaller than the spacing of the LED chip and the lens, so the yellow-light pattern is larger than the blue-light pattern. It can be expected that the CCT of the PMMA freeform lens is greater in a bigger light viewing angle than that in a smaller angle, and the $\Delta u'v'$ will rise in the edge part.

The CCT and $\Delta u'v'$ spatial distributions of the LED module integrated with the PMMA continuous freeform lens (radiate grid mapping) are shown in Figure 6.44 and Figure 6.45. We can find from Figure 6.44 that the lens is similar to the LED packaging module without a lens in terms of the spatial distribution of the CCT. The CCT reaches the highest level in the middle region, then decreases as the light viewing angle increases. Also shown in Figure 6.45, the $\Delta u'v'$ distribution of the LED module with a

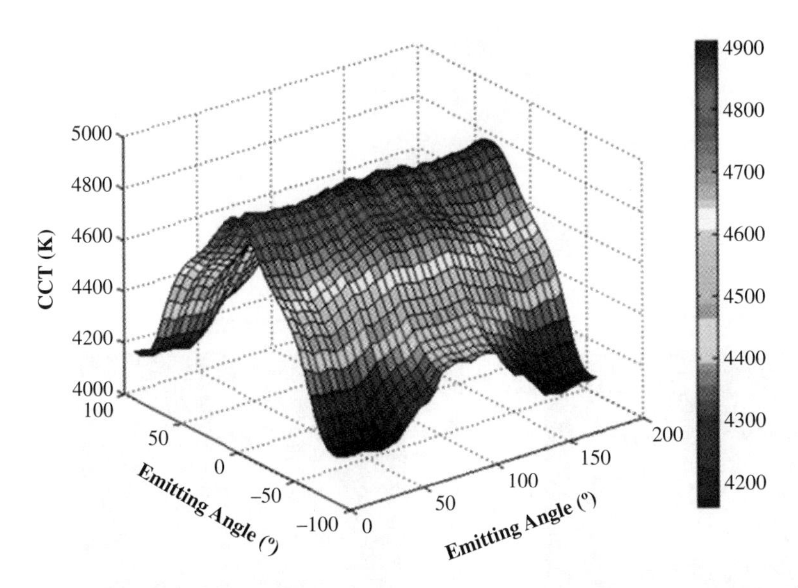

Figure 6.44 CCT spatial distribution of the LED module integrated with the PMMA continuous freeform lens (radiate grid mapping) where CCT decreases gradually from the center to the side.

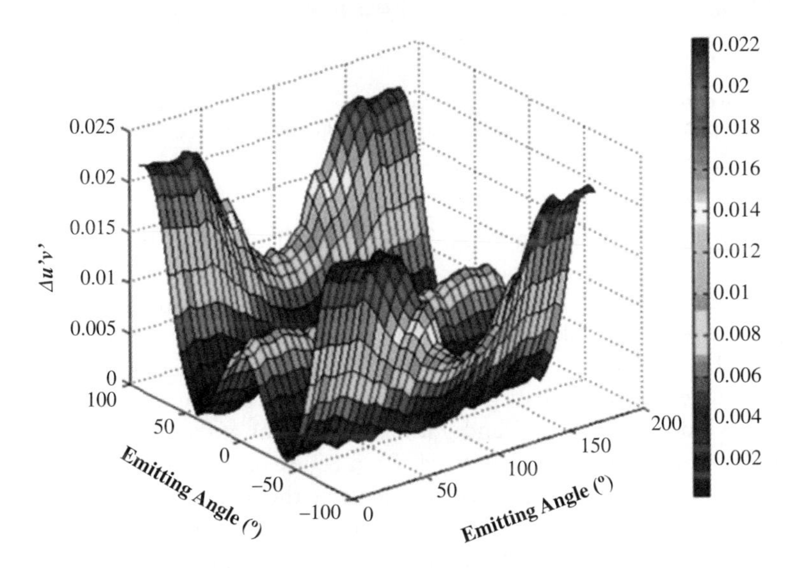

Figure 6.45 $\Delta u'v'$ spatial distribution of the LED module integrated with the PMMA continuous freeform lens (radiate grid mapping) where $\Delta u'v'$ increases from the center to the side.

lens is similar to the distribution without a lens. Both present W-shaped distributions, and $\Delta u'v'$ smaller than 0.004 only occurs at some points within the small-angle region. The main reason for this phenomenon is that the light source and the target plane are divided according to the center radiation method in this design algorithm, which is

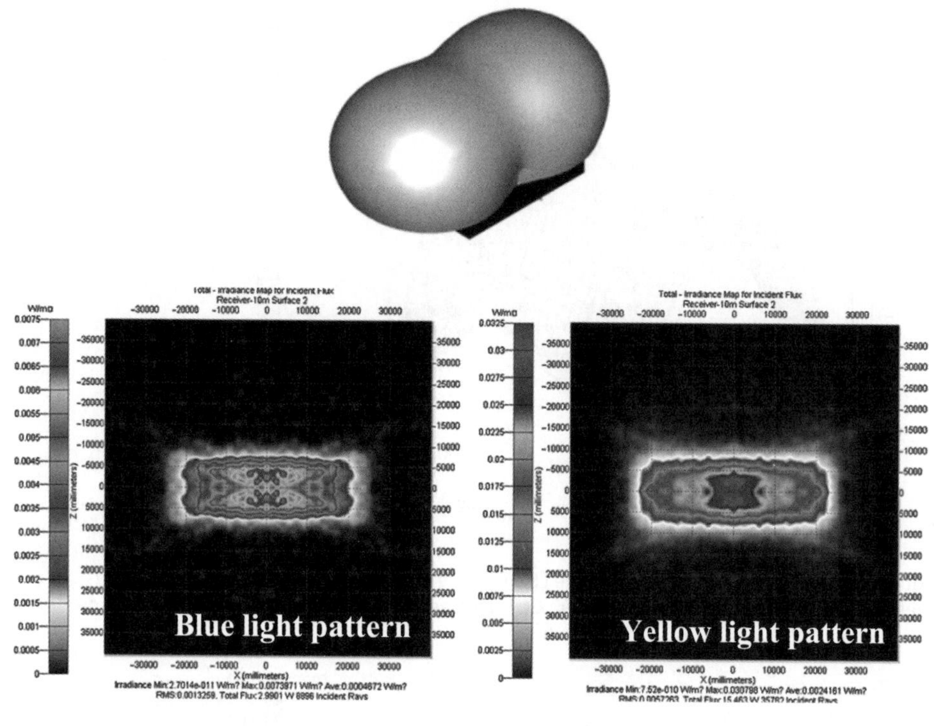

Figure 6.46 Optical model of the PMMA continuous freeform lens (rectangular grid mapping) and its blue-light and yellow-light patterns where illuminance decreases gradually from the center to the side of the light pattern.

in the same direction as LED source outputting. Thus, the spatial color distribution is similar.

The optical model of the LED module integrated with the PMMA continuous freeform lens (rectangular grid mapping) and both of its blue- and yellow-light patterns are shown in Figure 6.46. The yellow-light pattern resembles the blue-light pattern, which is similar to the freeform lens with a radiate grid mapping method, but the yellow one is slightly bigger than the blue one. The CCT and $\Delta u'v'$ distributions of the LED module are shown in Figure 6.47 and Figure 6.48. We can see from Figure 6.47, compared to the distribution of the CCT with the freeform lens designed using the radiate grid mapping method, that the illuminance distribution in the middle area is relatively uniform, almost maintained at 4700 K. However, the CCT at the edge portion of the pattern presents a dramatic decline. This is mainly due to the yellow pattern's shape being larger than that of the blue pattern, resulting in an increase of the proportion of the yellow light at the edge position. It can be seen from Figure 6.48 that the $\Delta u'v'$ is at a lower value in the middle area of the pattern, substantially less than 0.004, which is better than that of the lens designed by the radiate grid mapping method. It can be found from the above analyses that the SCU has been greatly improved at the pattern area after installing a PMMA continuous freeform lens (a rectangular grid mapping). However, due to the phosphor layer of the LED module being coated by a free-flowing process, the SCU is still uneven.

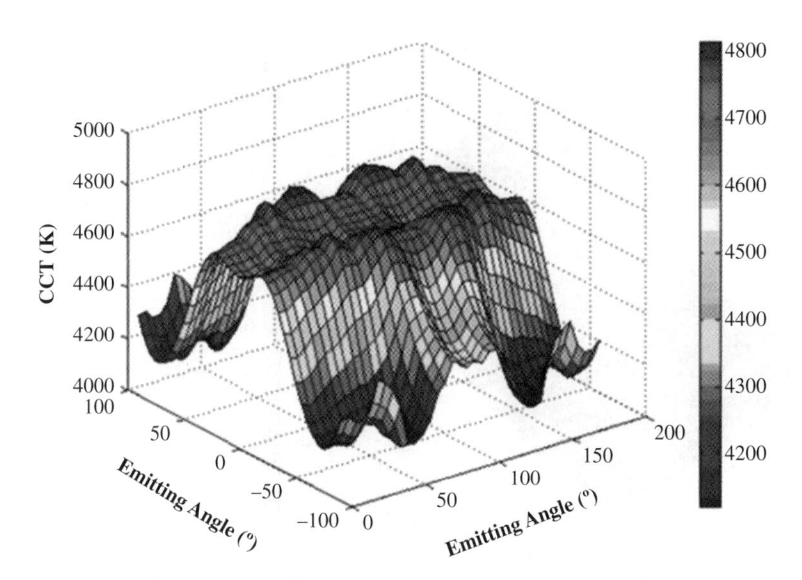

Figure 6.47 CCT spatial distribution of the LED module integrated with the PMMA continuous freeform lens (rectangular grid mapping) where CCT decreases gradually from the center to the side.

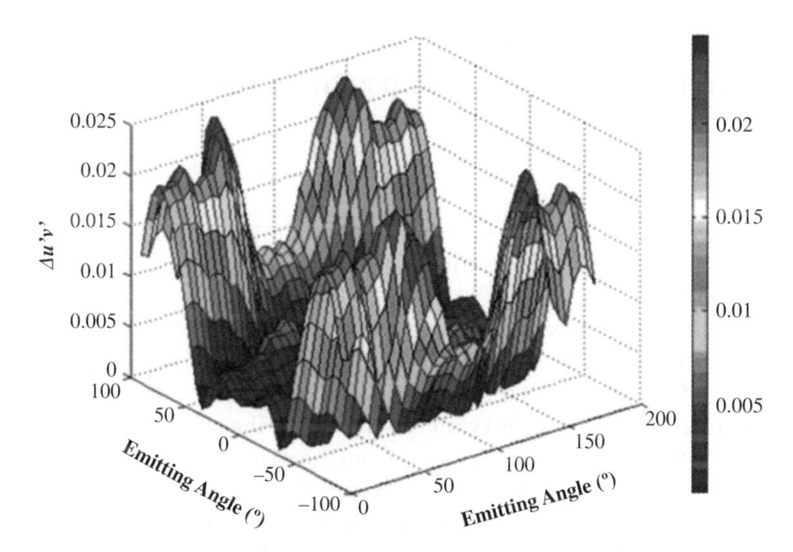

Figure 6.48 $\Delta u'v'$ spatial distribution of the LED module integrated with the PMMA continuous freeform lens (rectangular grid mapping) where $\Delta u'v'$ increases from the center to the side.

From the above analyses, we can find that although both radiate grid mapping and rectangular grid mapping algorithms are able to realize continuous freeform lenses, in terms of SCU, the rectangular grid mapping method is much better than the method of radiate grid mapping. Therefore, in Sections 6.5 through 6.7, we will use the continuous freeform lens algorithm based on rectangular grid mapping to conduct lens designs.

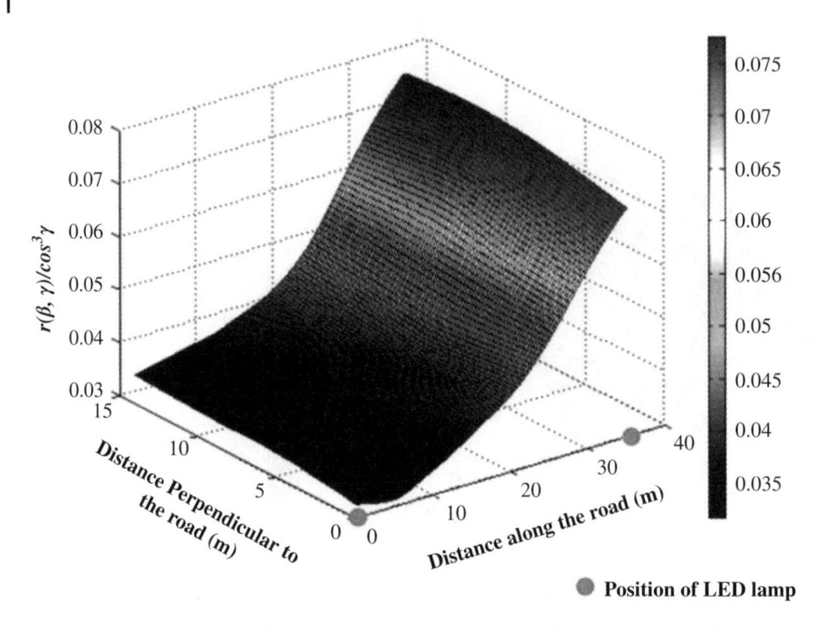

Figure 6.49 Distribution of the equivalent luminance coefficient $r(\beta, \gamma)/cos^3\gamma$ on asphalt pavement.

6.5 Freeform Lens for an LED Road Lamp with Uniform Luminance

6.5.1 Problem Statement

According to Eq. 6.10, $r(\beta, \gamma)/cos^3\gamma$ is defined as the equivalent luminance coefficient of luminance to illuminance, by which the relationship between luminance and illuminance can be made known.

An example of how to calculate an equivalent luminance coefficient is demonstrated as follows. A two-way eight-lane road, with each lane 3.5 m wide, adopts a bilateral road lamp configuration. Therefore, one side of the road with four lanes can be considered to be illuminated by unilateral lamps due to the symmetry situation. The observation point is located in the middle of the road, which is 7 m away from the roadside and 1.5 m high, while the distance from the first lamp is −60 m. The height of the road lamp is 10 m, and the position of the first road lamp is set as 0 m. The second and third road lamps are set at 35 m and 70 m, respectively. We investigate the $r(\beta, \gamma)/cos^3\gamma$ distribution condition in a region of 35 m length and 14 m width (four lanes) that is mainly illuminated by these three lamps. According to a simplified luminance coefficient table on asphalt and concrete pavement given in the *City Road Lighting Design Standard CJJ 45-2006*, the calculated $r(\beta, \gamma)/cos^3\gamma$ distributions of the studied region are shown in Figure 6.49 and Figure 6.50, respectively.

For the asphalt pavement, the coefficient $r(\beta , \gamma)/cos^3\gamma$ is smaller in the area just below the road lamp, and the minimum is found to be 0.033 in this region. As the distance from the road lamp along the road direction increases, the equivalent luminance coefficient increases rapidly and reaches the maximum of 0.077 near the area between the

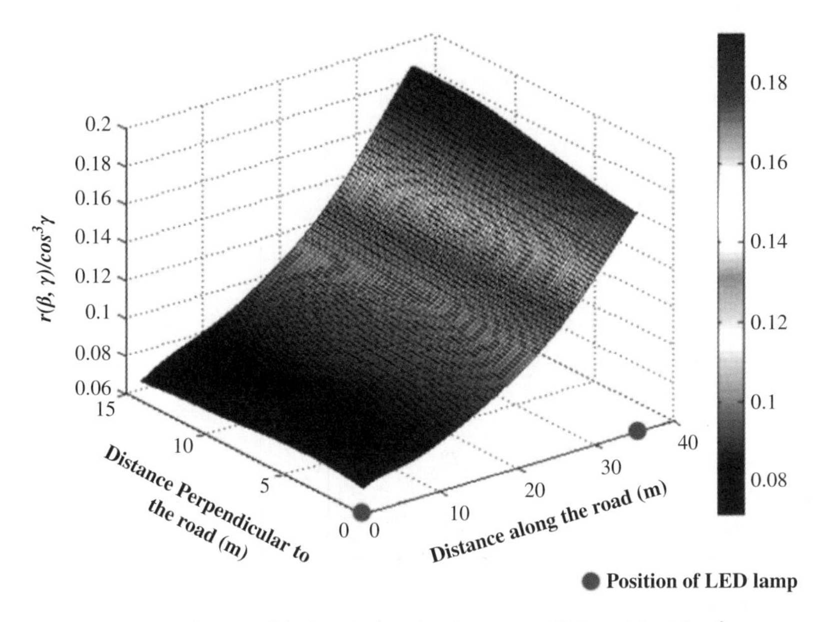

Figure 6.50 Distribution of the equivalent luminance coefficient $r(\beta,\ \gamma)/cos^3\gamma$ on concrete pavement.

two lamps in the middle of the road. Hence, the difference between the maximum and minimum is 2.33 times, meaning that if the illumination on the pavement is kept the same, the luminance distribution would be nonuniform, and the maximum luminance would be 2.33 times than the minimum. For the concrete pavement, the equivalent luminance coefficient has a similar distribution, and there is a 2.78 times difference between its maximum and minimum values, indicating a worse luminance distribution. Thus, the areas between the lamps are brighter and present bright spots, but the areas under the lamps are darker and present dark spots to human vision. Those present an *anti-zebra stripes* phenomenon on the pavement. The so-called anti-zebra stripes in this book are contrasted to the *zebra stripes*. In the conventional sense, road lamps with a non-ideal light distribution effect will present pretty high illumination under the lamp but extremely low illumination between two lamps, although this is compensated by an equivalent coefficient. Therefore, it results in a phenomenon of alternating bright and dark spots, which is the so-called zebra stripes effect.

Here, we will simulate and analyze the road-lighting performance of a 168 W LED road lamp design based on the PMMA continuous freeform lens with a uniform illuminance light pattern mentioned in Section 6.4. The road is a two-way, six-lane, R2 pavement surface, and the total width is 25 m with a 2 m wide middle isolation zone. An *R2* pavement surface means an asphalt road surface with an aggregate composed of a minimum 60% gravel (size greater than 1 cm) and with 10% to 15% artificial brightener in the aggregate mix. The LED road lamps are bilateral-symmetrically arranged. The lamp is 12 m high, the cantilever is 2 m long, and the elevation angle is 15°. Two lamps are spaced 35 m. Using the internationally known lighting software DIALux to simulate and analyze the lighting performance, simulated road illuminance and luminance distributions are shown in Figure 6.51 and Table 6.3. We can find that the distribution of illuminance is very uniform on the road, and illuminance uniformity U_E has a high

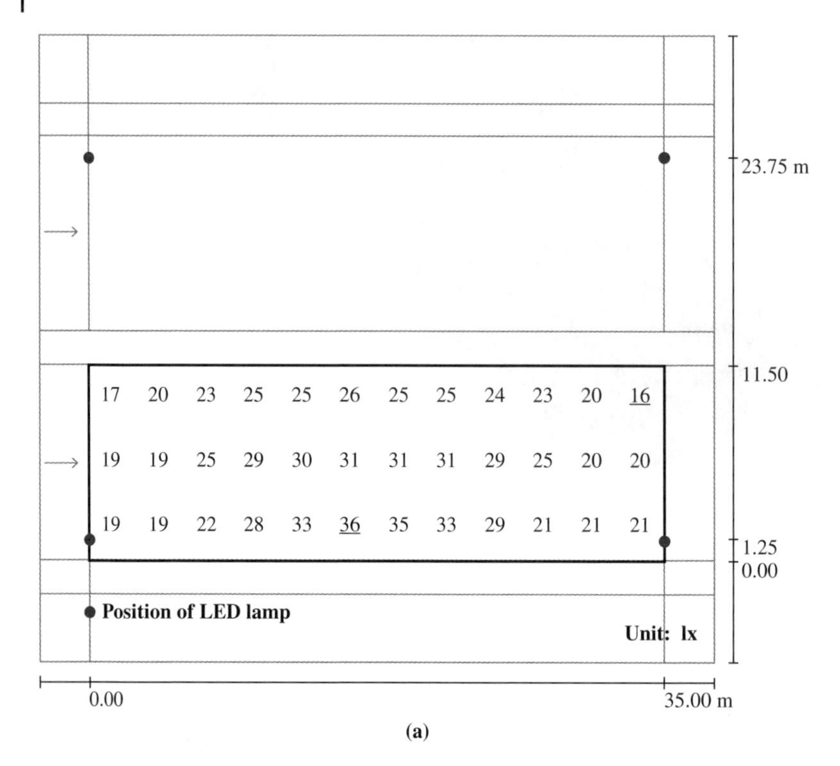

Figure 6.51 Simulated road (a) illuminance distribution and (b) luminance distribution of an LED road lamp based on a freeform lens with a uniform-illuminance light pattern.

value of 0.65. However, the luminance uniformity is low; in particular, the longitudinal luminance uniformity is only 0.30. The major reason is that the minimum luminance under the lamp is only 0.79 cd/m^2, while the maximum luminance between two lamps is 3.91 cd/m^2, which results in *anti-zebra stripes*.

Therefore, it can be found from the analyses in this chapter that a uniform illuminance distribution light pattern can be used in road lighting, resulting in high illuminance uniformity, and that its performance is better than a Lambertian LED road lamp or another traditional road lamp. However, this kind of light pattern is not the best pattern to achieve uniform luminance on roads. It is necessary to improve present algorithms to design a freeform lens to realize uniform luminance.

6.5.2 Combined Design Method for Uniform Luminance in Road Lighting

A rectangular light pattern with uniform illuminance distribution could be achieved by the continuous freeform lens proposed in this chapter. However, due to the light intensity in the small angle being low (and, in the large angle, being high), luminance below the LED road lamp is not large enough, while luminance between two lamps is too high, forming the anti-zebra stripes. At the same time, the illuminance distribution on the plane irradiated by a Lambert-type LED source decreases from the center to the edge in terms of $cos^4\theta$, which results in zebra stripes. Considering there must be a uniform-luminance area between zebra stripes and anti-zebra stripes, we

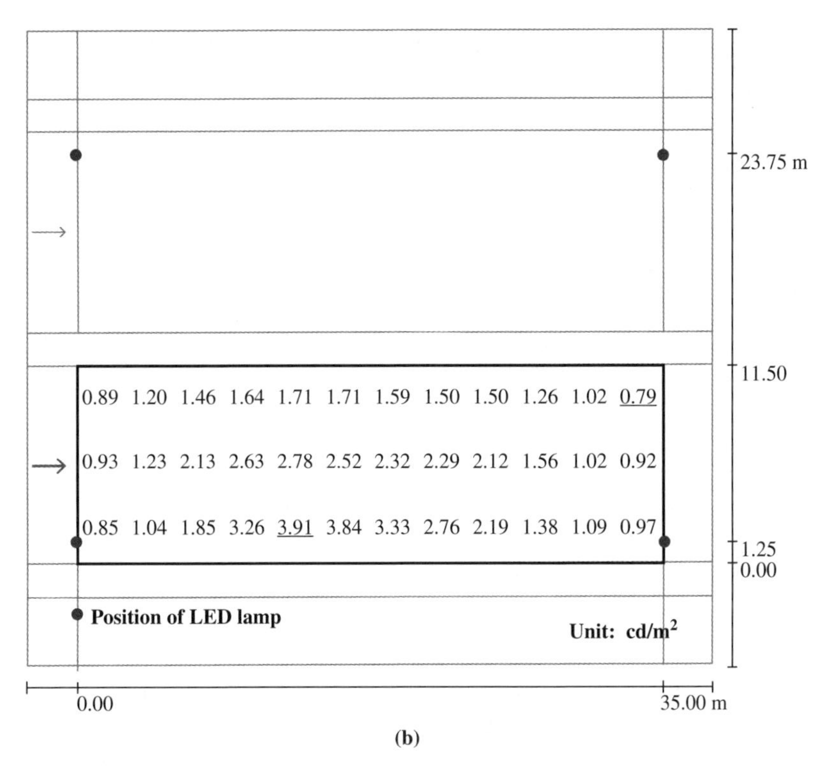

Figure 6.51 (*Continued*)

Table 6.3 Simulated road-lighting performance of a lens with a uniform-illuminance light pattern.

	Luminance		
Road type	**Average luminance L_{av} (cd/m^2)**	**Total uniformity U_o**	**Longitudinal uniformity U_L**
Two-way six-lane	1.81	0.4	0.3

	Illuminance			
Road type	**Average illuminance E_{av}(lx)**	**Uniformity U_E**	**Glare restriction threshold increment (TI), %**	**Surrounding ratio (SR)**
Two-way six-lane	25	0.65	3	0.8

combine these two kinds of optical designs, so as to achieve uniform-luminance road lighting.

Since we mainly consider the luminance uniformity of the road lighting in this design, we can ignore the absolute values of illuminance and luminance for the moment. The design flowchart is shown in Figure 6.52. Firstly, we build the optical models of a

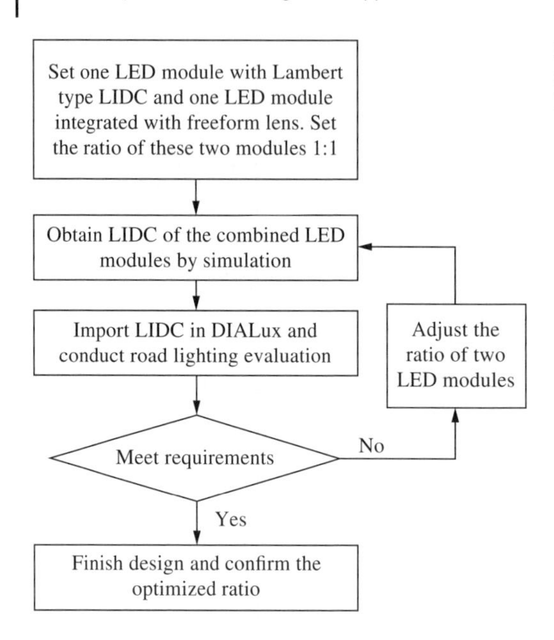

Figure 6.52 Flowchart of the combined design method for uniform-luminance road lighting.

Lambert LED module and a LED module with a freeform lens, and the initial ratio of each luminous flux is set as 1:1. Through ray-tracing simulation, we can get the LIDC of the combined model. Then we import the LIDC to DIALux software to evaluate the performance. When other parameters (e.g., illuminance uniformity and SR) meet the requirements, and if luminance uniformity can meet the requirements, we can quantify the ratio of LED modules with and without a lens as 1:1. If not, we can adjust the ratio between these two modules to recalculate to get a new combined LIDC, and then re-evaluate the performance in the DIALux. Adjust the ratio until it meets the requirements of road lighting, then finish the design and determine the optimal proportion relationship. In addition, because the LIDC of LED package modules is often the Lambert type, in practice, just remove part of the freeform lenses in the LED road lamp to realize the combined light distribution effect.

In this case, we still use a PMMA continuous freeform lens as an example to carry out the design. Figure 6.53a and 6.53b, respectively, show the LIDC of an LED road lamp with full lenses and half lenses with the same power of 168 W. We can see, in the situation of half lenses, that light intensity in the center area of the small angle increased obviously. We use DIALux to simulate the same two-way six-lane road with the LIDC of the half lenses, and the illuminance and luminance distributions are shown in Figure 6.54 and Table 6.4. With half lenses, both total luminance uniformity U_0 and longitudinal luminance uniformity U_L are significantly increased, especially the U_L rapidly increasing from 0.30 to 0.70, which meets the requirements of road lighting. However, with half lenses, the average luminance and average illuminance decrease slightly. The main reason is that the light pattern diffused along the vertical direction of the road, and more light can be projected to the region behind the lamps and less light is projected to the road.

Based on this design method, we improved the existing LED road lamp, and the real road illumination performance is as shown in Figure 6.55. From these figures, we can

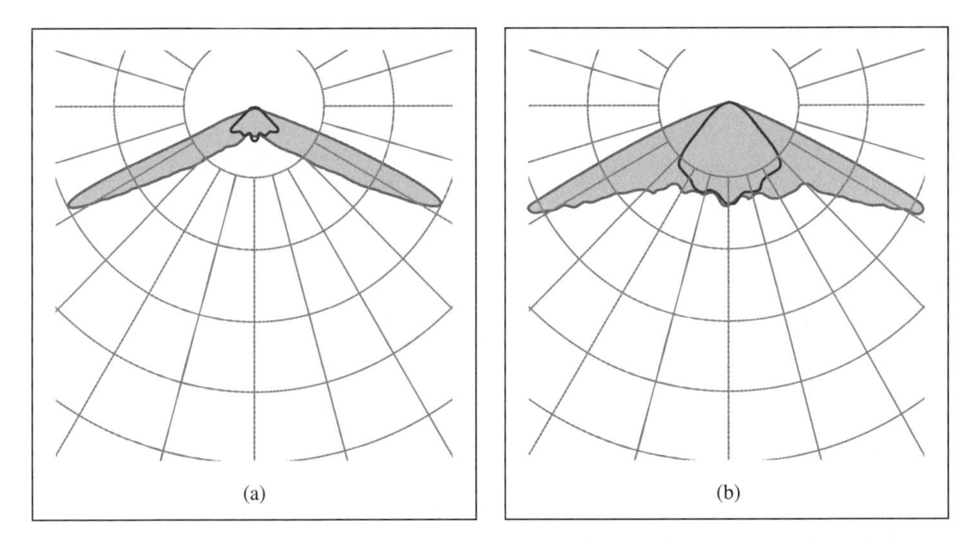

Figure 6.53 (a, b) Flowchart of the combined design method for uniform-luminance road lighting.

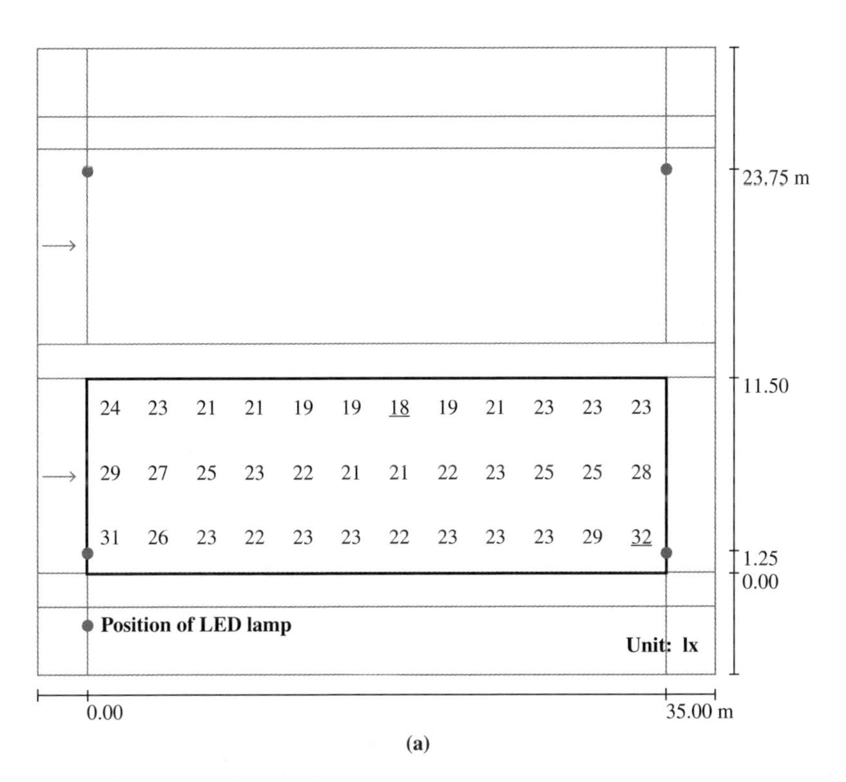

Figure 6.54 Simulated road (a) illuminance distribution and (b) luminance distribution of an LED road lamp based on the combined design method.

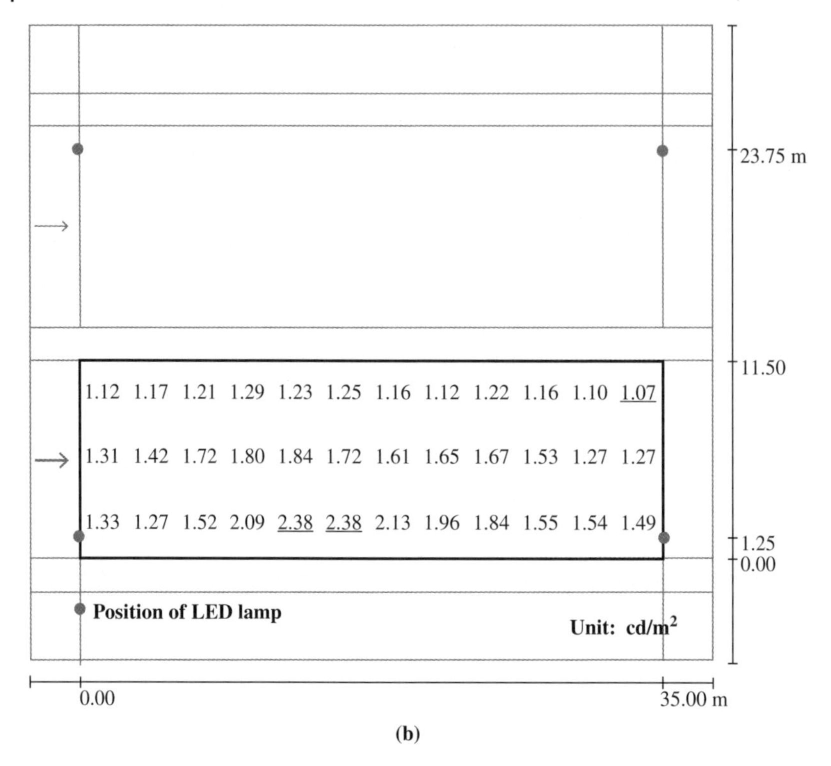

Figure 6.54 *(Continued)*

Table 6.4 Simulated road-lighting performance using the combined design method.

	Luminance		
Road type	**Average luminance L_{av} (cd/m^2)**	**Total uniformity U_o**	**Longitudinal uniformity U_L**
Two-way six-lane	1.51	0.7	0.7

	Illuminance			
Road type	**Average illuminance E_{av}(lx)**	**Uniformity U_E**	**Glare restriction threshold increment (TI), %**	**Surrounding ratio (SR)**
Two-way six-lane	23	0.78	6	0.8

see that luminance distribution on the road is uniform, and no bright or dark spots (i.e., neither zebra stripes nor anti-zebra stripes) can be seen on the road, achieving uniform luminance in road-lighting performance with a comfortable visual feel. Hence, this combined design method is simple, convenient, and practical and can obtain good lighting performance.

(a) (b)

Figure 6.55 Road-lighting performance of an LED lamp based on the optimized LIDC designed by the combined method. (Supplied by Guangdong Real Faith Enterprises Group, www.gd-realfaith.com)

6.5.3 Freeform Lens Design Method for Uniform-Luminance Road Lighting

Although the method described in Section 6.5.2 can easily realize the luminance uniformity of road lighting, it needs to take the lens apart. For different roads and different spaces between road lamps, it may need to have a different lens ratio, and thus may lead to inconvenience in assembly management and also affect the appearance of the LED road lamp. Development of the freeform lens algorithm to achieve uniform luminance is the fundamental way to solve this problem.

As is known from the algorithms mentioned in this chapter, changing the grid division of the target plane can promote the lighting performance of a continuous freeform lens effectively; thus, we can adjust illuminance distribution of the light pattern by changing the grid division of the target plane, so as to adjust the LIDC of the lens, and therefore it can satisfy the requirements of the standard of road lighting. A flowchart for a uniform-luminance freeform lens design method is shown in Figure 6.56. First of all, the target plane is divided according to the uniform meshing method, calculating the free surface lens, and the LIDC is obtained by ray tracing. Then evaluate and analyze the lighting performance of the LED road lamp based on the LIDC in DIALux software. If luminance uniformity cannot meet the requirements, we optimize the target plane meshing and obtain a new freeform lens to re-evaluate. The feedback process will be repeated many times until the requirements of road-lighting standards are met, then we complete the design to achieve a freeform lens with uniform luminance.

Based on this design algorithm, as shown in Figure 6.57, we design a PMMA freeform lens with uniform luminance. From its LIDC, we can find that the redistributed light intensity increases at the direction of the small angle, while it decreases at the large angle. Also, we can find from Figure 6.58 that, although the shape of the light pattern is still close to the rectangle, the illuminance distribution in the illumination area is no longer uniformly distributed, but higher in the center region and lower in the edge region. We evaluate the road-lighting performance of a 210 W LED road lamp based on this freeform lens by simulation. In simulation, a two-way six-lane road with a width of 21 m and an R2 surface is adopted. LED lamps are bilateral-symmetrically arranged with a height of 10 m, a cantilever length of 2 m, an elevation angle of 0°, and spacing of 35 m. Simulation results are demonstrated in Table 6.5. The U_0 reaches 0.6 and U_L reaches 0.7,

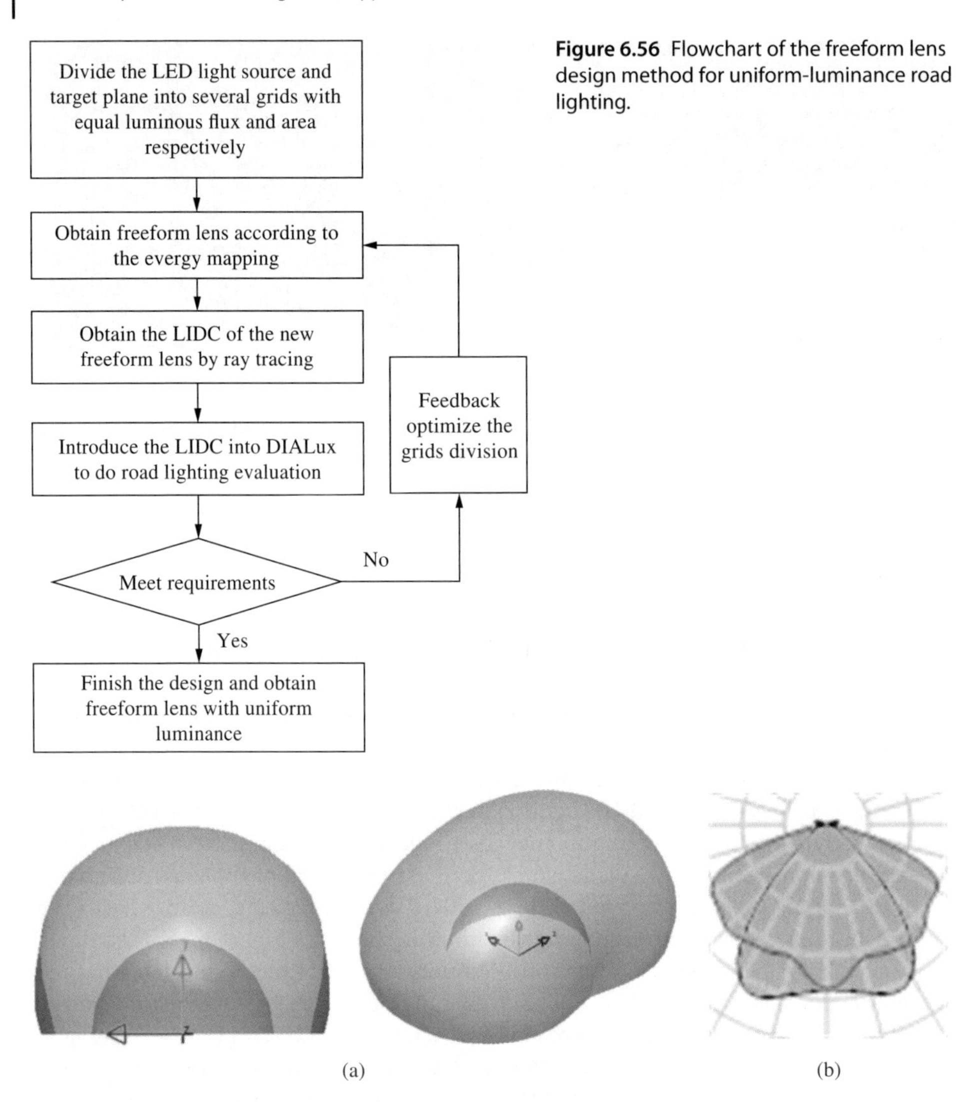
Divide the LED light source and target plane into several grids with equal luminous flux and area respectively
Obtain freeform lens according to the evergy mapping
Obtain the LIDC of the new freeform lens by ray tracing
Introduce the LIDC into DIALux to do road lighting evaluation
Feedback optimize the grids division
Meet requirements
No
Yes
Finish the design and obtain freeform lens with uniform luminance
(a)
(b)

Figure 6.56 Flowchart of the freeform lens design method for uniform-luminance road lighting.

Figure 6.57 (a) A freeform lens for road lighting with uniform luminance, and (b) its LIDC.

which meet the requirements. Therefore, through this design method, only one kind of freeform lens is needed to achieve uniform-luminance road lighting, which has more advantages compared with the combined design method mentioned in Section 6.5.2. The practical road-lighting performance of this freeform lens is shown in Figure 6.59.

Using the same algorithm, we also design a PMMA freeform lens to achieve uniform luminance distribution in tunnel lighting. The freeform lens and its application in tunnel lighting are shown in Figure 6.60. Both the total luminance uniformity and the longitudinal luminance uniformity reach as high as 0.8, which meets the requirements of *Tunnel Design Specifications of JTG D70-2004* in China and helps realize high-performance tunnel lighting with uniform luminance distribution.

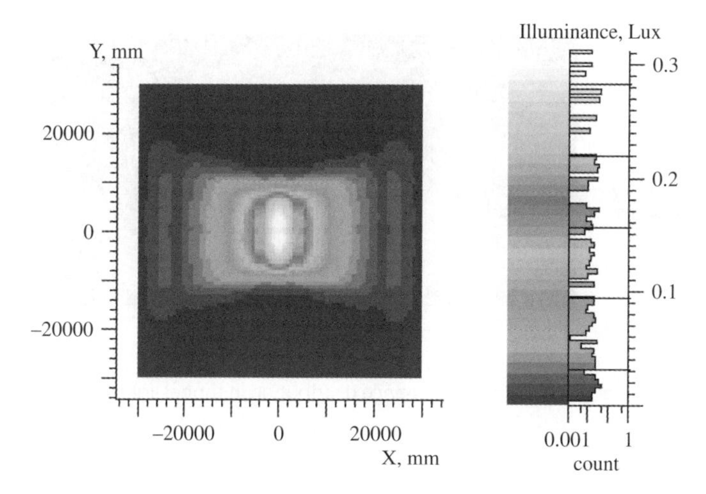

Figure 6.58 Simulated light pattern of the freeform lens for road lighting with uniform luminance.

Table 6.5 Simulated road-lighting performance using a freeform lens with uniform luminance.

	Luminance		
Road type	**Average luminance L_{av} (cd/m^2)**	**Total uniformity U_o**	**Longitudinal uniformity U_L**
Two-way six-lane	1.8	0.6	0.7

	Illuminance			
Road type	**Average illuminance E_{av}(lx)**	**Uniformity U_E**	**Glare restriction threshold increment (TI), %**	**Surrounding ratio (SR)**
Two-way six-lane	29	0.46	7	0.9

(a) (b)

Figure 6.59 Practical road-lighting performance of an LED road lamp with the uniform-luminance freeform lenses. (Supplied by Guangdong Real Faith Enterprises Group, www.gd-realfaith.com)

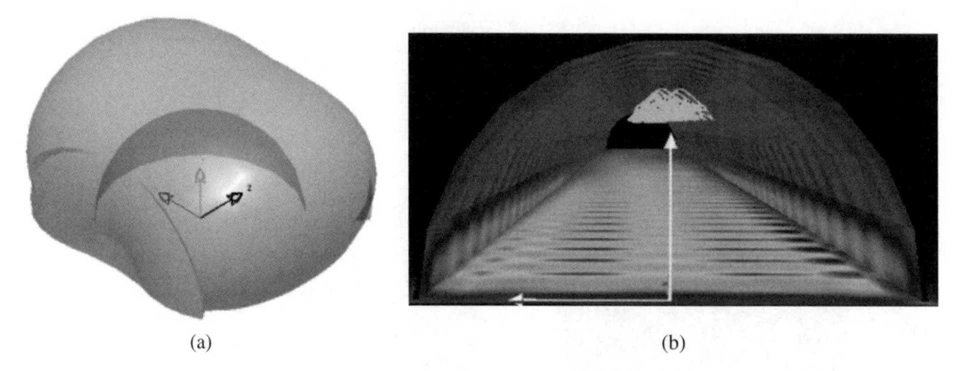

Figure 6.60 A PMMA freeform lens to achieve uniform-luminance distribution in tunnel lighting and its lighting performance.

6.6 Asymmetrical CFLs with a High Light Energy Utilization Ratio

At present, there are three main road lamp placing methods: bilaterally symmetrical distribution, bilaterally staggered distribution, and lane center distribution. Among them, the most common methods are the former two. As shown in Figure 6.61a, LED lamps are placed on the road side. Since the light emitting from the LED road lamp is symmetrical in the vertical direction across the road, when the elevation angle is 0°, nearly 50% of light energy irradiates on the sidewalk behind the lamp. The sidewalk indeed needs light, but only requires that the SR is more than 0.5. Thus, too much light irradiating on the sidewalk causes light energy waste and reduces the light energy utilization ratio of road lighting. Especially because the light of LED road lamp is still very "precious," it is more important to reduce waste and improve the utilization ratio.

In order to let more light irradiate on the road, we always adopt the method of increasing the elevation angle of the road lamp, to make the light emitting from the LED road lamp asymmetrical in the vertical direction and close to the road side, and, as far as possible, to increase the utilization rate of light energy. According to the *City Road Lighting*

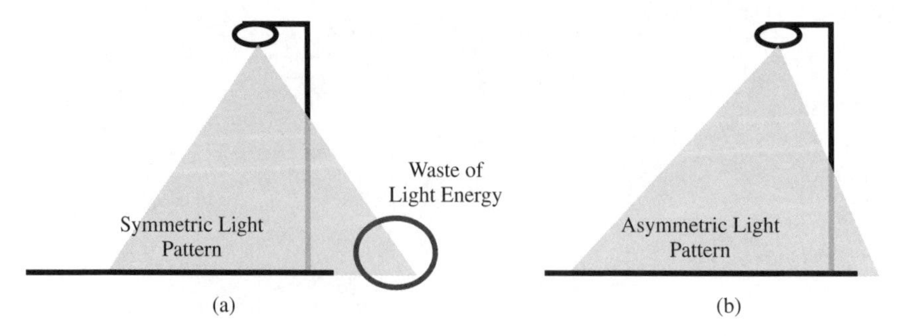

Figure 6.61 Schematic of road lighting with (a) symmetrical and (b) asymmetrical light patterns.

Design Standard CJJ 45-2006, in order to avoid glare, the elevation angle of the road lamp should not be more than 15°. But in the market at present, the LED road lamp elevation angle often is more than 15°, and some even reach 30°. Large elevation angles will cause serious glare and affect drivers' visual comfort, resulting in visual fatigue and traffic accidents.

In this section, we propose an asymmetrical freeform lens. As shown in Figure 6.61b, emergent light from the freeform lens is symmetrical in the direction along the road, but asymmetrical in the vertical direction across the road; it leads to more light irradiated on the road side. Therefore, using this freeform lens creates asymmetrical road lighting with a small elevation angle, improving the utilization ratio of light energy and avoiding glare. An asymmetrical freeform lens algorithm is similar to that of a symmetrical freeform lens (as mentioned in this chapter) except for the establishment of a light energy mapping relationship. In the design of an asymmetrical freeform lens, shift the target plane along the road width direction (vertical direction across the road), and make light emitting from the lens asymmetrical in the road width direction. Note that, since the light pattern is symmetrical only in the length direction, the one-quarter light source and target plane mentioned in Chapter 3 are no longer able to be adopted, and one should use one-half ones as shown in Figure 6.62.

An asymmetrical freeform lens is designed as shown in Figure 6.63a based on the algorithm mentioned here. We can see the lens has obvious asymmetry along the width direction, which will make light deflect to one side on the width direction. The simulated LIDC and light pattern illumination performance of this lens are shown in Figure 6.63b and Figure 6.64, respectively. We can find that the LIDC is asymmetrical along the vertical direction across the road, most light deflects to the road, and a small part of the light deflects to the sidewalk to meet the requirement of the SR. Moreover, the center

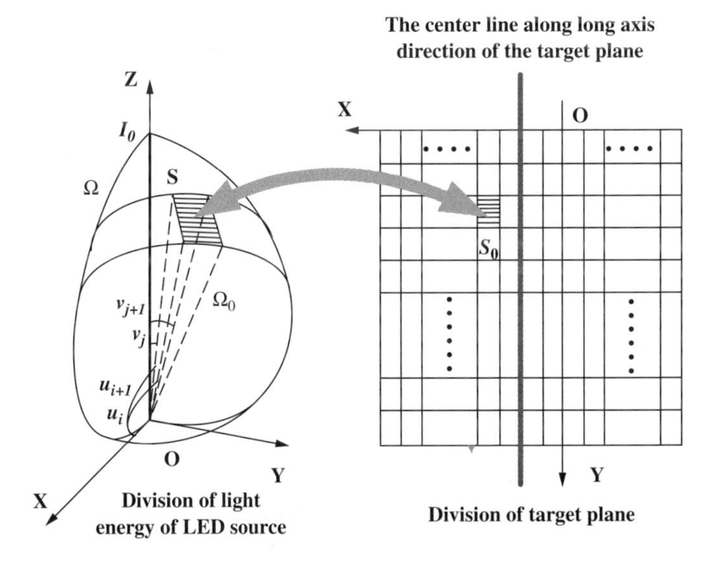

Figure 6.62 Schematic of the light energy mapping relationship of an asymmetrical freeform lens.

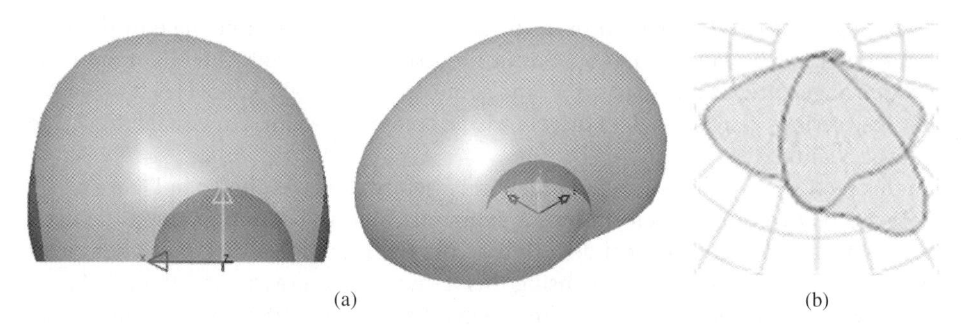

Figure 6.63 (a) An asymmetrical freeform lens for road lighting with uniform luminance, and (b) its LIDC.

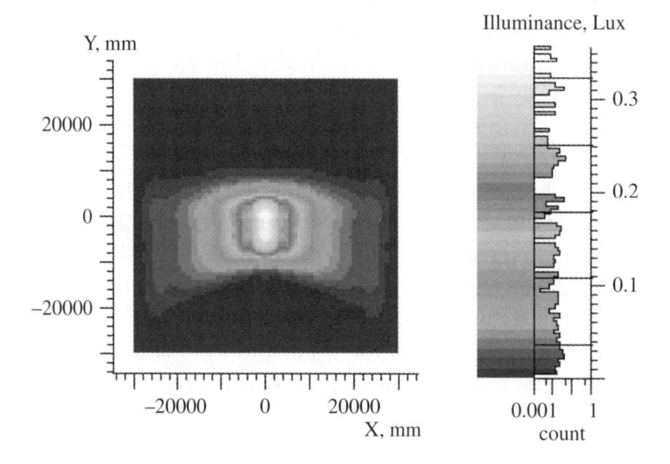

Figure 6.64 Simulated light pattern of the asymmetrical freeform lens for road lighting with uniform luminance.

Table 6.6 Simulated road lighting performance using the asymmetrical freeform lens with uniform luminance.

	Luminance		
Road type	**Average luminance L_{av} (cd/m^2)**	**Total uniformity U_o**	**Longitudinal uniformity U_L**
Two-way six-lane	2.25	0.6	0.7

	Illuminance			
Road type	**Average illuminance E_{av}(lx)**	**Uniformity U_E**	**Glare restriction threshold increment (TI), %**	**Surrounding ratio (SR)**
Two-way six-lane	37	0.45	7	0.7

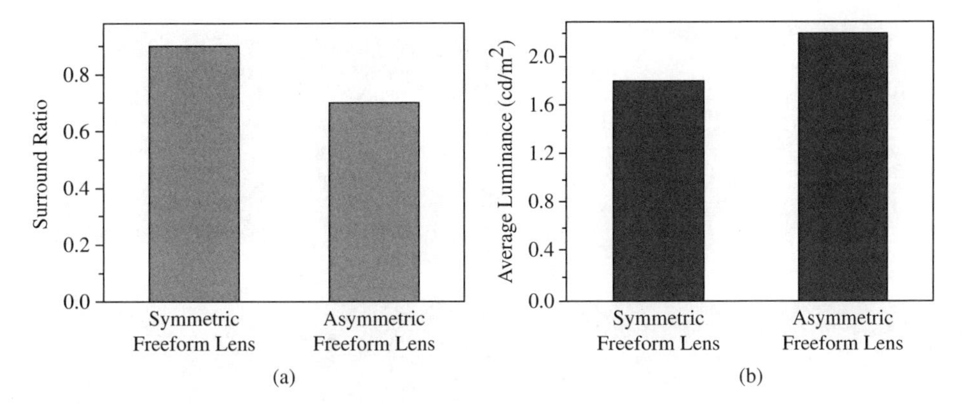

Figure 6.65 Comparison of symmetrical and asymmetrical freeform lenses for LED road lighting: (a) surrounding ratio and (b) average luminance.

of the asymmetrical light pattern shifts to one side of the road, which can increase the light energy utilization ratio.

We evaluate the road-lighting performance of the asymmetrical freeform lens using the same situation mentioned in Section 6.5 (e.g., an elevation angle of 0°), and the illumination performance is shown in Table 6.6 and Figure 6.65. Compared with the symmetrical freeform lens with uniform luminance, the asymmetrical freeform lens' total luminance uniformity and longitudinal luminance uniformity reach high levels of 0.6 and 0.7, respectively, achieving uniform-luminance road lighting. Moreover, after adopting the asymmetrical freeform lens, the SR decreases from 0.9 to 0.7 while the average luminance increases from 1.80 cd/m^2 to 2.25 cd/m^2 with an enhancement of 25%, increasing the light energy utilization ratio of road lighting. The practical application of the designed asymmetrical freeform lens is shown in Figure 6.66. Since an asymmetrical freeform lens has advantages of a high light energy utilization ratio, a small elevation angle, and uniform luminance distribution, it will become a development trend for optical design of LED road lighting.

Figure 6.66 Practical road lighting performance of an LED road lamp with the asymmetrical uniform-luminance freeform lenses. (Supplied by Guangdong Real Faith Enterprises Group, www.gd-realfaith.com)

Figure 6.67 An LED road-lighting engine integrated with multiple functions and various kinds of LED road lamps based on this module. (Supplied by Guangdong Real Faith Enterprises Group, www.gd-realfaith.com)

6.7 Modularized LED Road Lamp Based on Freeform Optics

With the development of LED technologies, an LED road-lighting engine, which integrates multiple functions of freeform optics, heat dissipation, power driving, and intelligent control in one module, as shown in Figure 6.67, has been recognized by more and more people and become another development trend of LED road lighting. As shown in Figure 6.67, different kinds of LED road lamps with different appearances can be achieved by adopting the same LED road-lighting module. One needs to just change the shell of an LED road lamp to meet the aesthetic requirements of different projects, cities, or countries. The modularized LED road lamp is able to overcome disadvantages of the traditional LED road lamp (e.g., poor interchangeability, difficult to maintain, and hard to upgrade in the future) and to realize mass production, so as to reduce the total cost. For example, if one LED package has a problem, one needs not replace the whole road lamp but just one LED road-lighting module, which will decrease maintenance charges. The LED road-lighting engine or module, integrated with the modularized design concept, can be developed into a product platform, realizing the diversification and expansibility of road lamp products. More and more modularized LED road lamps have been adopted in applications, and they have great potential to occupy the market in the near future.

References

1 CREE First to Break 300 Lumens-per-Watt Barrier. http://www.cree.com/
2 China LED Road Lighting Report 2014–2017. http://www.gg-ii.com/

3 China Technical Specification for Detecting and Operation: City Road Lighting Design Standard CJJ 45-2006. Beijing: China Architecture & Building Press, 2006.

4 Wang, K., Liu, S., Chen, F., Qin, Z., Liu, Z., and Luo, X. Freeform LED lens for rectangularly prescribed illumination. *J. Opt. A: Pure Appl. Opt.* **11**, 105501 (2009).

5 Sun, C.-C., Lee, T.-X., Ma, S.-H., Lee, Y.-L., and Huang, S.-M. Precise optical modeling for LED lighting verified by cross correlation in the midfield region. *Opt. Lett.* **31**, 2193–2195 (2006).

6 Wang, K., Luo, X., Liu, Z., Liu, S., Zhou, B., and Gan, Z. Optical analysis of an 80-W light-emitting-diode street lamp. *Opt. Eng*. **47**, 013002–013002 – 13 (2008).

7 Ding, Y., Liu, X., Zheng, Z., and Gu, P. Freeform LED lens for uniform illumination. *Opt. Expr.* **16**, 12958–12966 (2008).

8 Ding, Y., Liu, X., Zheng, Z., and Gu, P. Secondary optical design for LED illumination using freeform lens. Proc. SPIE 7103, 71030K–71030K – 8 (2008).

9 Moreno, I., and Sun, C.-C. LED array: where does far-field begin? Proc. SPIE 70580R–70580R – 9 (2008).

10 ENERGY STAR Program Requirements for SSL Luminaires, Version 1.1 (2008). https://www.energystar.gov/index.cfm?c=new_specs.ssl_luminaires

7

Freeform Optics for a Direct-Lit LED Backlighting Unit

7.1 Introduction

Liquid crystal display (LCD) technology has been widely used in cell phones, pads, laptops, monitors, and TV, and the market is huge. However, liquid crystal doesn't emit light by itself. A backlighting light source is required for all of these LCD applications. Before high-efficiency light-emitting diodes (LEDs) were developed, the cold cathode fluorescent lamp (CCFL) was the major light source for LCD backlighting, but it has disadvantages of bulkiness, poor visual performance, high power consumption, and mercury pollution. Therefore, in recent years, with the fast development of LED technologies, CCFL has been replaced by LED sources in more and more LCD backlighting applications, as shown in Figure 7.1 (according to *DisplaySearch*). LEDs have dominated the backlighting market so far, whether for small displays (e.g., cell phone) or large displays (e.g., a 65-in. TV).

There are two types of LED backlighting, edge-lit and direct-lit. As shown in Figure 7.2, LEDs are arranged at the side of the backlighting unit (BLU) in edge-lit while arranged at the bottom of the BLU in direct-lit. In edge-lit, light emitted from LEDs will be coupled into the light guide plate (LGP) and transmit along the LGP through total internal reflection (TIR). During transmitting, almost all light will be reflected by the dot patterns at the bottom of the LGP and then exit from the LGP upward. However, in direct-lit, light emitted from LEDs will first be mixed by the cavity and then irradiate at the diffuser plate directly. These are two different design concepts and have different applications.

Since edge-lit has many advantages, such as low thickness, high efficiency, less LEDs used, and so on, edge-lit LED BLUs have been widely applied in backlighting for cell phones, laptops, and LCD monitors. At the same time, with the improvement of the quality of people's daily lives, the size of TVs becomes larger and larger, as shown in Figure 7.3, which requires large or even ultra-large LED backlighting. However, with limitations to the light transmit distance in the LGP, LED edge-lit is usually used in backlighting for screens smaller than 40 in., but it is not suitable for large or ultra-large LED backlighting applications where the screen size is usually larger than 40 in., like an LED TV. In addition, local dimming is hard to realize in edge-lit. Direct-lit has advantages

Freeform Optics for LED Packages and Applications, First Edition. Kai Wang, Sheng Liu, Xiaobing Luo and Dan Wu.

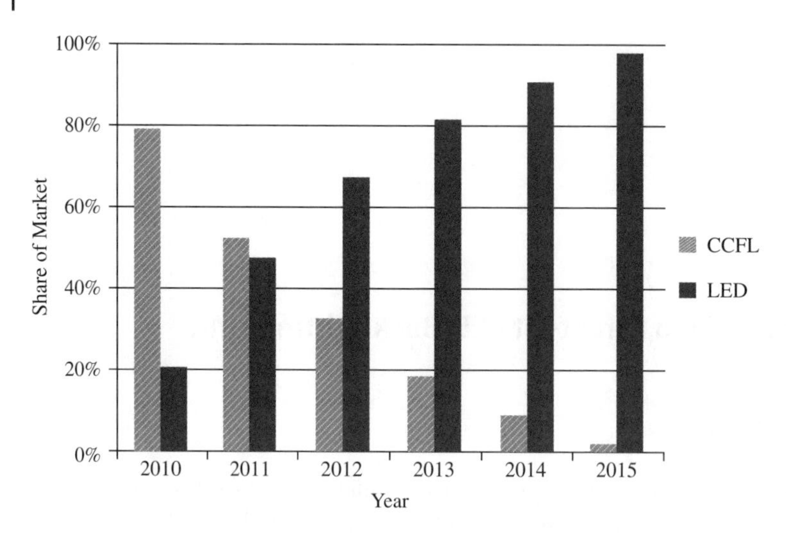

Figure 7.1 Market development of CCFLs and LEDs for LCD backlighting.[1]

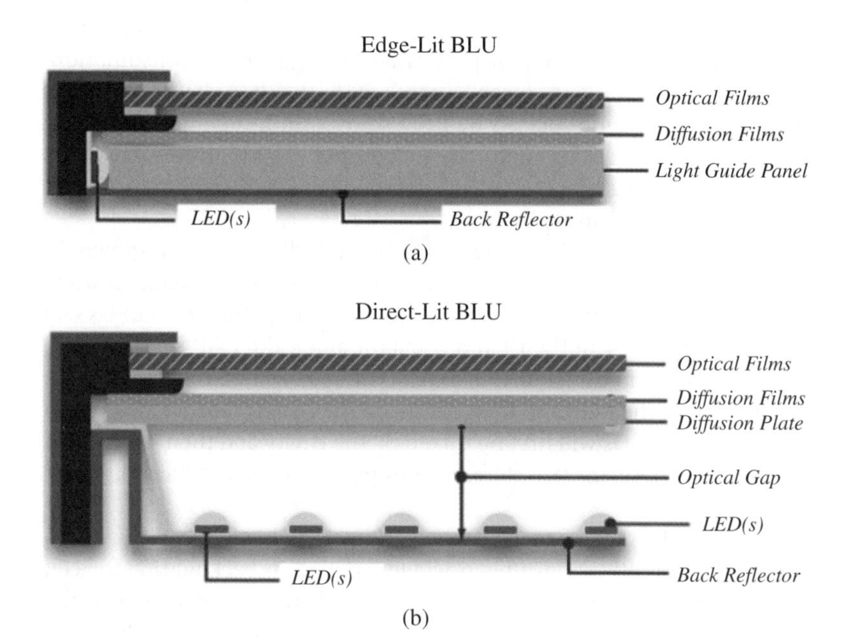

Figure 7.2 Schematic of an LED (a) edge-lit and (b) direct-lit backlighting unit (BLU).

of being not only size-scalable but also capable of local dimming, which could enhance the contrast ratio and decrease power consumption for an LED TV. Therefore, direct-lit probably will become a development trend for ultra-large LED backlighting. In this chapter, we will focus on the freeform optical designs for large-scale direct-lit LED backlighting.

Figure 7.3 Development of size of an LED TV.

7.2 Optical Design Concept of a Direct-Lit LED BLU

In 2003, Lumileds proposed a direct-lit LED BLU based on side-emitting LEDs as shown in Figure 7.4. A special designed primary lens is packaged in the LED module. Light emitting from the LED chip and phosphor will have a TIR and refract exit from the side of the lens with a very large exiting angle, nearly in a horizontal direction. Then the exited light will be reflected many times by side walls of the reflective box to mix light and to achieve high uniform illumination on the diffuser plate. Although high uniformity of illuminance and a small amount of LEDs could be obtained, the BLU optical efficiency is at a low level of only about 60% due to reflection and light loss occurring too many times during reflection. Note that *BLU optical efficiency* here means the ratio of light energy irradiated at the diffuse plate to the light energy emitted from LEDs, without considering the efficiencies of diffuse plate, brightness enhancement film (BEF), and dual-BEF (DBEF). In addition, slim or ultrathin thickness (~10 mm) is the development trend of LED TVs in the future. However, the thickness of this BLU reaches as high as 40 mm, which cannot meet customer requirements. Much research has been conducted to improve the design of the lens since 2003, but few of these efforts can overcome the disadvantages of low efficiency and high thickness.

Another traditional optical design concept for direct-lit LED BLUs is a surface-mount display (SMD) LED array. As shown in Figure 7.5, an SMD LED array is mounted on the bottom of the BLU. The light intensity distribution of the SMD LED module is Lambertian type, and the distance between two adjacent LED modules is usually small (~10 mm) to achieve uniform illumination when the thickness of the BLU is as low as 10 mm. Thin, direct-lit LED BLUs can be obtained by this design. Moreover, since almost all light irradiates at the diffuse plate directly without being reflected by side walls, the BLU

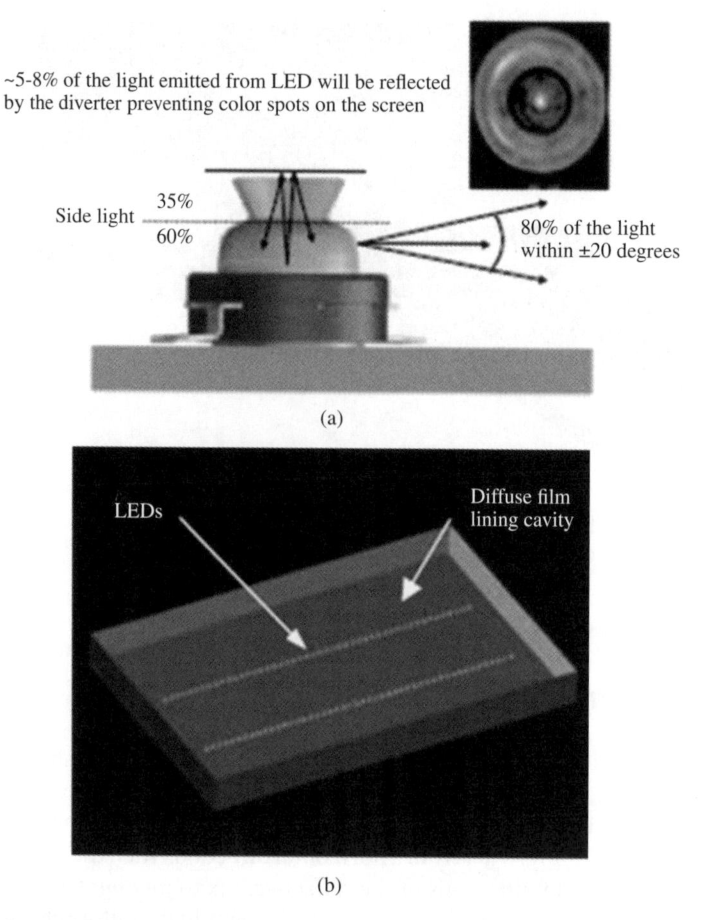

Figure 7.4 (a) A side-emitting LED module and (b) a direct-lit BLU based on this LED module array developed by Lumileds.[2]

Figure 7.5 Direct-lit LED BLU using an SMD LED array.

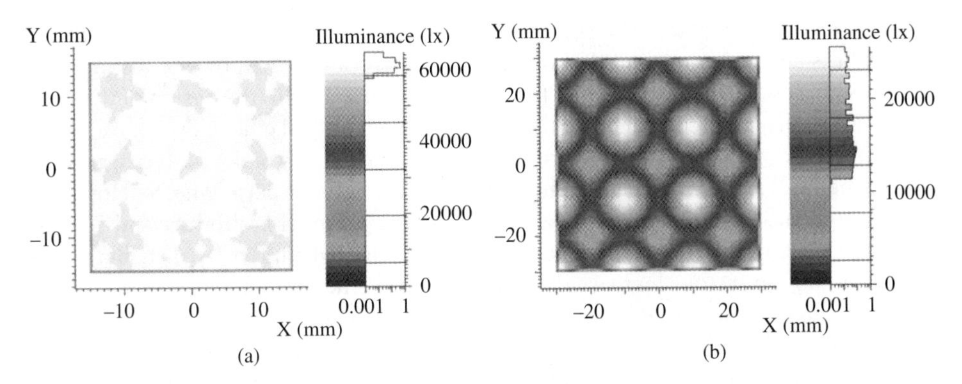

Figure 7.6 Lighting performance of an SMD LED module array on the plate 10 mm away when the distance between the two adjacent LED modules is (a) 10 mm, uniformity = 0.902; and (b) 20 mm, uniformity = 0.446.

optical efficiency is at a high level of more than 90%. Therefore, many companies adopt this design concept for direct-lit LED backlighting products. However, the number of LED modules used in a large LED TV is huge when the TV is thin, which causes high cost and assembly inconvenience. For example, about 4500 LED modules will be needed for a 40-in. LED TV to achieve uniform illumination when the thickness is 10 mm. As shown in Figure 7.6, if we increase the distance between two adjacent LED modules when the TV is thin, equal to increasing the distance–height ratio (DHR), uniformity of illuminance on the plate will degrade significantly, resulting in lighting performance unacceptable for LED TVs. Thus, if we want to reduce the number of LED modules, the thickness should be increased (e.g., ~20 mm) to provide enough light-mixing space to achieve uniform illumination. Therefore, the high cost and thickness of direct-lit LED TVs are two major barriers to extended marketing. A new approach is needed to achieve uniform illumination using fewer LED modules when the thickness is less than 10 mm in direct-lit LED BLUs.

Besides the thickness and number of LED modules, the co-design with BEF is another important issue in the design of direct-lit LED BLUs, as shown in Figure 7.7. The

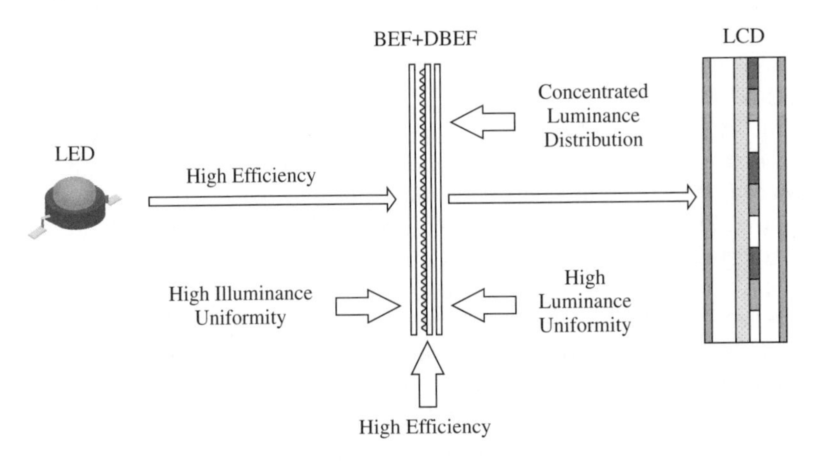

Figure 7.7 Schematic of a co-design concept in a direct-lit LED BLU.

system optical efficiency (considering the efficiencies of diffuse plate, BEF, and DBEF) of direct-lit is always lower than that of edge-lit, which causes the power consumption and cost of direct-lit to be higher than that of edge-lit. In edge-lit, the design of the LGP and BEF is usually coupled to achieve the highest system optical efficiency. For example, BEF is integrated with the LGP. The structure of BEF, the density of dot patterns, and the shape of the LGP are all co-designed to achieve better lighting performance. However, little research has been performed on the co-design of an LED module and BEF in direct-lit. Therefore, there is still space to optimize the light intensity distribution of the LED module to further increase the system optical efficiency of direct-lit.

According to this analysis, it is clear that two problems need to be solved for the development of direct-lit LED BLUs in the future: one is how to achieve uniform illumination using fewer LED modules when the thickness is as thin as 10 mm (i.e., a large DHR), and the other is how to couple the design of light intensity distribution of LED modules and BEF to achieve the highest system optical efficiency. If both of these two issues could be overcome through this research, it will provide a new way to realize high-efficiency LED BLUs for ultra-large LED TVs (>40 in.) that are thinner and have fewer LED modules, lower power consumption, and lower cost. Performing this research is meaningful to promote the development of ultra-large LED TVs in the future. In the remainder of this chapter, we will present how to design freeform optics to address these two issues.

7.3 Freeform Optics for Uniform Illumination with a Large DHR

As mentioned in Section 7.1, LED backlighting, especially its application in large-scale LED TVs, became the main driver of LED marketing growth in 2010. According to the proposed design method in Section 3.6, new LED modules integrated with freeform lenses for slim direct-lit backlighting will be designed as examples in this section. Actually, this new design method is not only suitable for LED backlighting applications but also available for other LED illumination applications, such as LED interior lighting, commercial lighting, office lighting, warehouse lighting, and so on. Since the design processes of these applications are quite similar, we take only the optical design of new, slim, direct-lit LED backlighting as an example in the following discussion.[3]

Figure 7.8a shows an optical model of a traditional LED backlighting module. Power consumption of this LED module is 0.068 W when driven by 20 mA, and its total luminous flux is 6.5 lm. The size of the LED chip is 280 × 280 μm. As shown in Figure 7.8b, the light intensity distribution curve (LIDC) of this LED module is a Lambertian type of $I(\theta) = I_0 cos\theta$, where I_0 is the light intensity when the emitting angle $\theta = 0$. The distance $z0$ between a square LED modules array and the receiving plane is 10 mm, as shown in Section 3.6. Considering the periodical distribution of an LED modules array, an array area with a size of 6 × 6 LED modules on the receiving plane is set as the testing area, which is able to reflect the whole lighting performance on the receiving plane. In addition, since a part of the lights emitted by the outside LED modules will be reflected by the reflection box in direct-lit backlighting, only the uniformity of illuminance U (E_{min}/E_{max}) of the central area of an array of 4 × 4 LED modules is considered in this design.

Ray-tracing simulation results of a traditional LED module array with 7 million rays are shown in Figure 7.9. The value of the coefficient of variation of root mean square

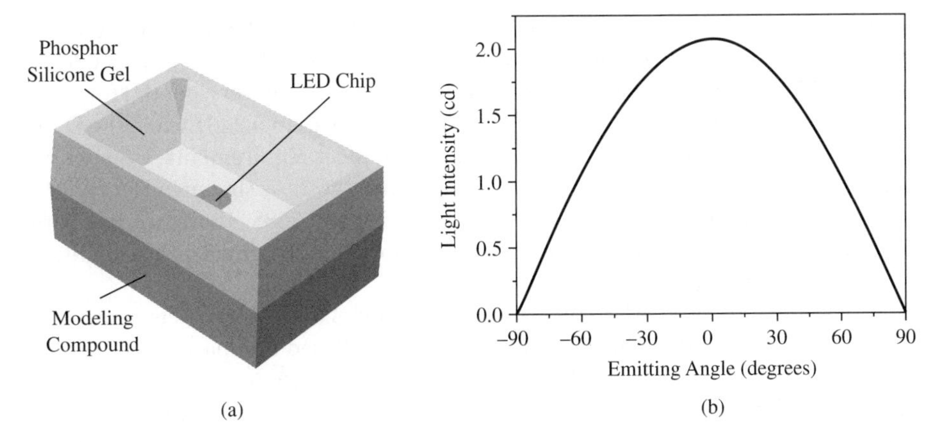

Figure 7.8 (a) Optical model of a traditional 0.068 W LED module for direct-lit backlighting and (b) its LIDC.

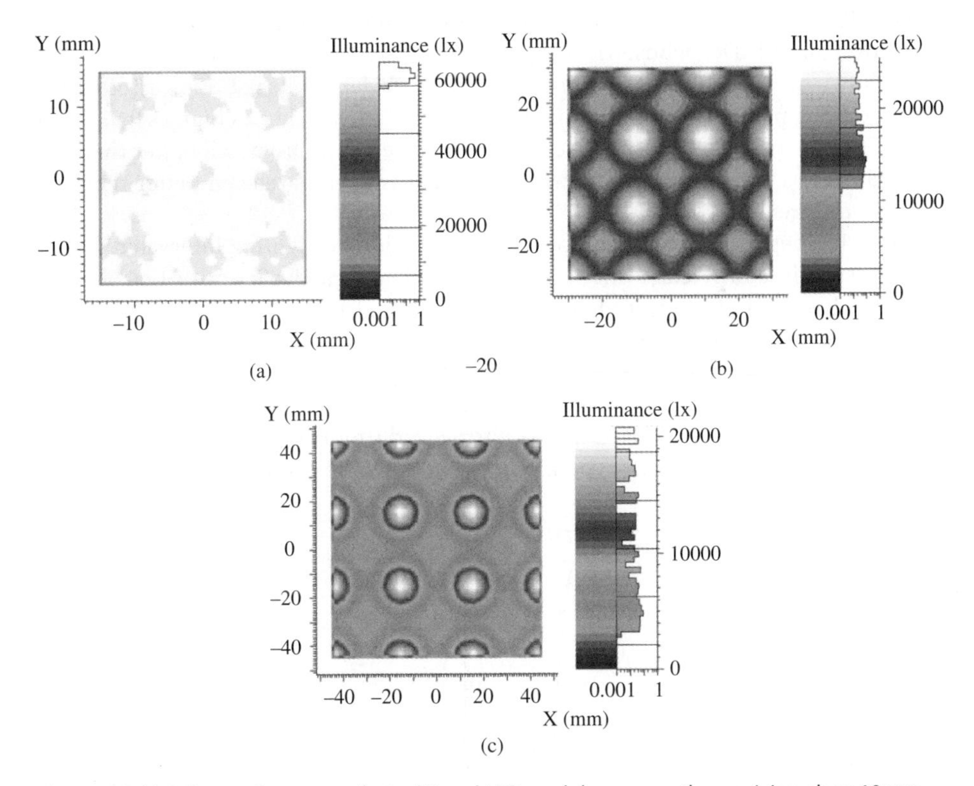

Figure 7.9 Lighting performance of a traditional LED module array on the receiving plane 10 mm away when DHR increases from 1 to 3: (a) DHR = 1, $U = 0.902$, CV(RMSE) = 0.0167; (b) DHR = 2, $U = 0.446$, CV(RMSE) = 0.2201; and (c) DHR = 3, $U = 0.155$, CV(RMSE) = 0.5688.

error, or CV(RMSE), is calculated through 23 × 23 = 529 sampled points. We can find that the uniformity U is at a high level of 0.902, and CV(RMSE) is as low as 0.0167 when the DHR is 1 and $d = 10$ mm. Its lighting performance is very good. However, the uniformity U decreases to 0.446 and CV(RMSE) increases to 0.2201 when the DHR increases to 2 (as shown in Figure 7.9b), which cannot meet requirements. Even worse lighting performance is obtained when the DHR increases to 3. As shown in Figure 7.9c, the uniformity U is only 0.155, and CV(RMSE) is as high as 0.5688. Therefore, it is hard for the traditional LED module to achieve good performance in direct-lit backlighting when the DHR is larger than 1.

According to the design method mentioned in Section 3.6, the LIDC is optimized to achieve good lighting performance when the DHR is 2. Based on Sparrow's criterion, $a_0 = a_0$, $a_1 = 6.994809a_0 - 4.261381a_2 - 4.099808a_3 - 1.781556a_4$, $a_2 = a_2$, $a_3 = a_3$, and $a_4 = a_4$ are obtained first. Then, according to the second criterion, four verification points of $R(d/2, d/2, z0)$, $R(7/6d, 7/6d, z0)$, $R(3/2d, 3/2d, z0)$, and $R(3/2d, d, z0)$ are introduced to evaluate the uniformity of the whole testing area. Finally, $a_0 = 1.0$, $a_1 = 0.878827$, $a_2 = 1.0$, $a_3 = 0.8$, and $a_4 = -0.8$ are obtained when $R(d/2, \mathrm{d}/2, z0) = 1.01$, $R(7/6d, 7/6d, z0) = 0.967$, $R(3/2d, 3/2d, z0) = 0.946$, and $R(3/2d, d, z0) = 0.956$. The optimized normalized LIDC, whose view angle is much larger than that of the Lambertian type, is shown in Figure 7.10a. In addition, Figure 7.10b depicts its lighting performance on the receiving plane. We can find that the uniformity U reaches as high as 0.936 and CV(RMSE) is lower than 0.01, which can meet requirements, are much better than the performance of a Lambertian LIDC when the DHR is 2, and are even better than the performance when the DHR = 1.

Figure 7.11a depicts a new LED module for direct-lit backlighting integrated with a special silicone freeform lens, which is designed with a refractive index of 1.54 according to the algorithm of freeform lenses for the required LIDC. The *optical efficiency*, defined as the ratio of light energy of incident light to that of emergent light of one optical component, of this lens is 94.5% when considering material absorption and Fresnel loss. The silicone freeform lens is easy to manufacture by a molding process. The height of the freeform lens is set as 1.8 mm. The normalized simulated LIDC of the freeform lens irradiated by a Lambertian-type LED chip is shown in Figure 7.11b. Normalized

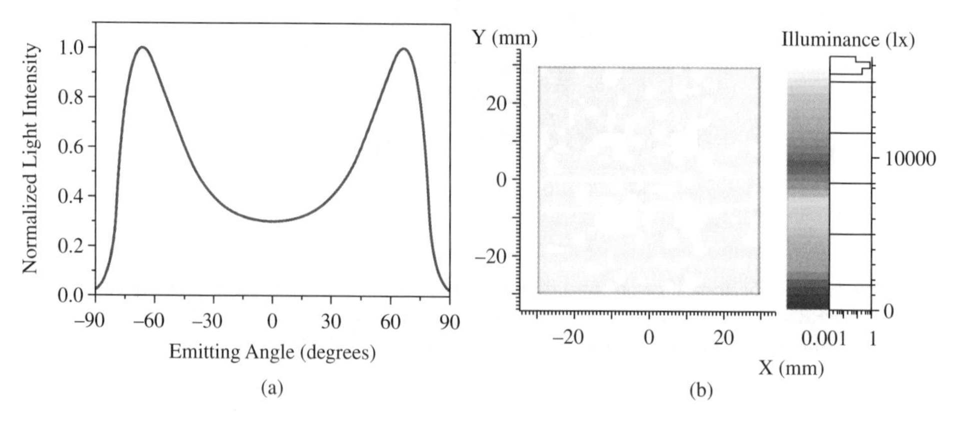

Figure 7.10 (a) Optimized LIDC and (b) its lighting performance on the receiving plane when DHR = 2: $U = 0.936$, CV(RMSE) = 0.0097.

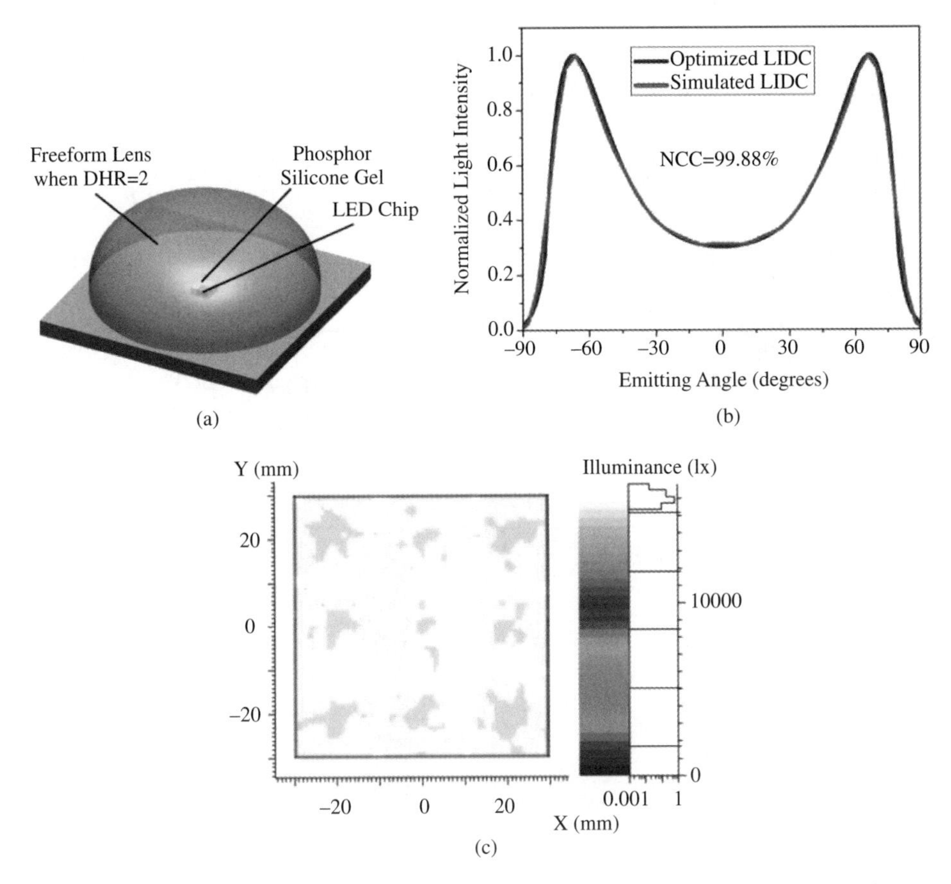

Figure 7.11 (a) A new LED module integrated with a special silicone freeform lens for direct-lit backlighting when DHR = 2; (b) a comparison of LIDCs between the simulated and the optimized results; and (c) lighting performance of these new LED module arrays, $U = 0.915$, CV(RMSE) = 0.0128.

cross-correlation (NCC)[4] is used to quantify the similarity between the simulated LIDC and the optimized LIDC. We can find from Figure 7.11b that simulated LIDC is quite similar to optimized LIDC and the NCC reaches as high as 99.88%, which demonstrates that the new algorithm for a freeform lens is effective. Note that the point source assumption could not be acceptable when the NCC was lower than 97% during the freeform lens design. The lighting performance of these new LED module arrays on the receiving plane is also shown in Figure 7.11c with results of $U = 0.915$ and CV(RMSE) = 0.0128. This lighting performance is similar to that of the optimized LIDC and is also quite good for backlighting.

By using this reversing design method, LIDC is optimized and a special freeform lens is also designed to achieve good lighting performance in terms of high illuminance uniformity when the DHR is 3. The new LED backlighting module integrated with the freeform lens and its lighting performance are shown in Figure 7.12. We can find that the uniformity of illuminance on the receiving plane is much better than that of the Lambertian LIDC when the DHR is 3, achieving $U = 0.887$ and CV(RMSE) = 0.0224. Although these results are a little worse than those of an optimized LIDC when the

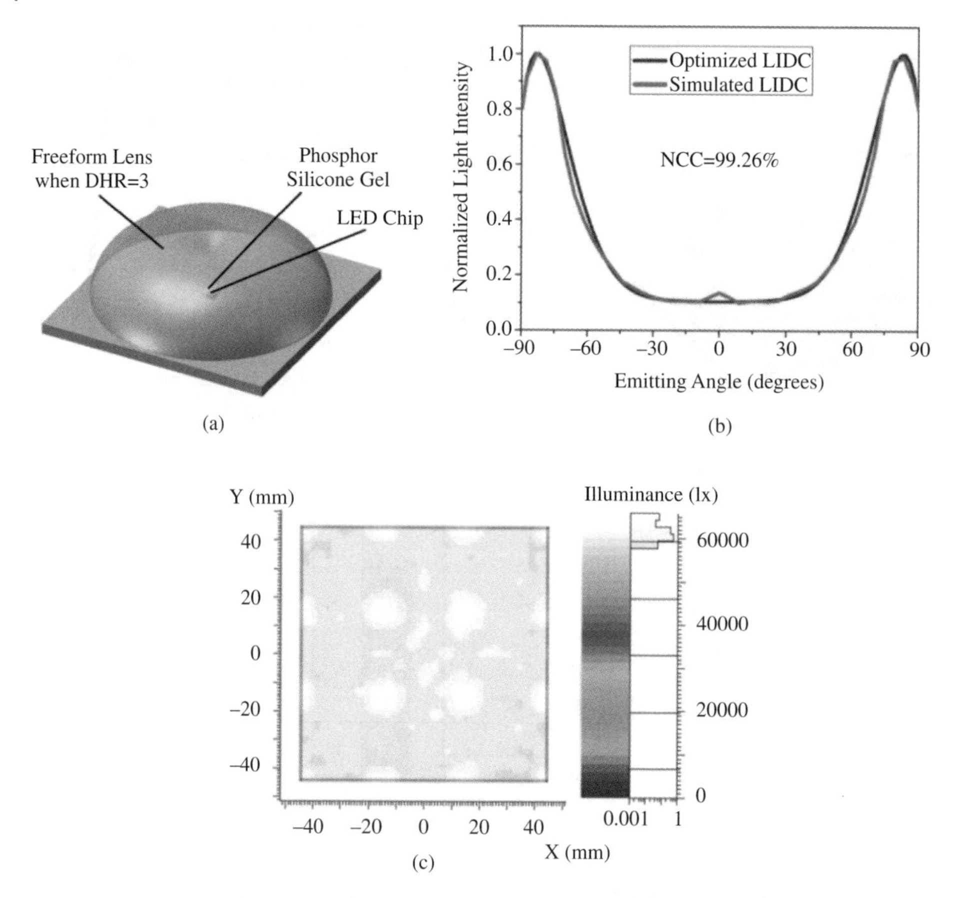

Figure 7.12 (a) A new LED module integrated with a special silicone freeform lens for direct-lit backlighting when DHR = 3; (b) comparison of LIDCs between the simulated and the optimized results; and (c) lighting performance of these new LED module arrays, $U = 0.887$, CV(RMSE) = 0.0224.

DHR is 2, they can meet requirements very well. Therefore, uniform illumination can be achieved easily by this new method when the DHR is much larger than 1. Therefore, in this design, the number of new LED modules is only one-ninth of the traditional LED modules when achieving similar uniform illumination in backlighting. Consequently, this reversing design method provides an effective way to dramatically decrease the number of LED modules and also is able to achieve good lighting performance in slim LED backlighting with a thickness of 10 mm.

As shown in Figure 7.13, a PMMA DHR = 3 freeform lens is manufactured according to the design method mentioned in this chapter. A direct-lit BLU with a DHR = 3 at the vertical direction and DHR = 2 at the horizontal direction is set to evaluate the illumination performance of the freeform lens. From Figure 7.13, we can find that uniform illumination on the receiver plane is obtained when using the freeform lens, which is much more uniform than the situation without the freeform lens.

In this section, slim direct-lit LED backlighting with thickness of 10 mm is designed as an example. According to this new design method, two new LED backlighting modules integrated with freeform lenses are successfully designed, generating optimized LIDCs

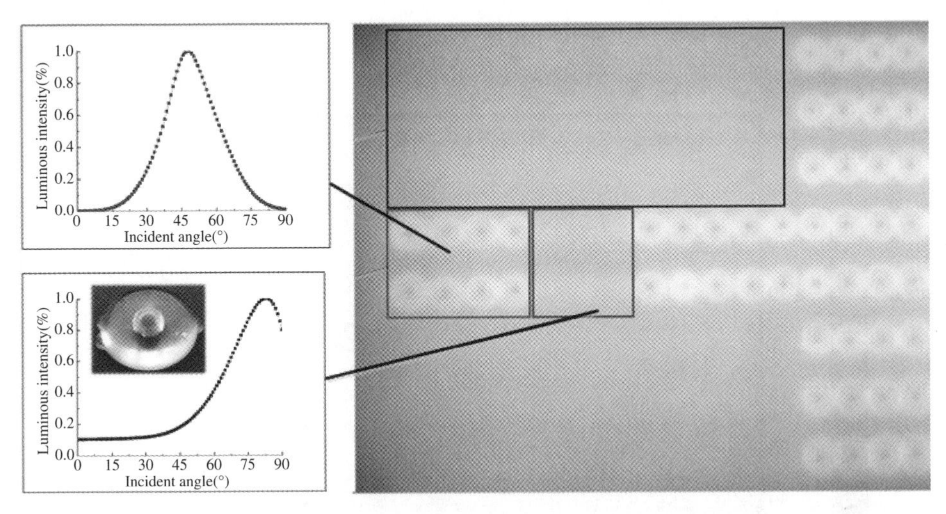

Figure 7.13 Experimental comparisons of illumination performance of the traditional LED module and the new LED module integrated with a freeform lens of DHR = 3.

to achieve uniform illumination. Uniformities of illuminance increase from 0.446 to 0.915 and from 0.155 to 0.887 when DHRs are 2 and 3, respectively. The range of the DHR to achieve uniform illumination is much larger than the maximum range reported in a previous reversing design. This reversing-design method provides an effective way to dramatically decrease the number of LED modules while also achieving good lighting performance in slim direct-lit LED backlighting. This new method provides a practical and simple way for the optical design of LED uniform illumination when the DHR is much larger than 1.

7.4 Freeform Optics for Uniform Illumination with an Extended Source

With the increase of the size of LED backlighting, the power of each LED module also increases, which means a larger LED chip will be needed for the so-called middle-power LED packaging module. For example, a 0.5 W LED module with a chip size of 600 × 600 μm is widely used in the market for larger scale BLUs, which is much larger than the size mentioned in Section 7.3. Therefore, a modified design method for an LED extended source to achieve uniform illumination with a large DHR is needed for the fast development of the LED BLU market.

As shown in Figure 7.14, high DHR (DHR ≥ 2) and uniform illuminance at the target plane can be achieved for a point source, but for an extended source, the uniformity at the target plane will decrease. Uniformity U represents the variance between the minimum and maximum illuminance E and could be expressed as follows:

$$U = E_{min}/E_{max} \tag{7.1}$$

Besides U, we also use *CV(RMSE)* to measure the uniformity of the target plane, which is equal to the RMSE divided by the mean value of the illuminance E as in Eq. 7.2.

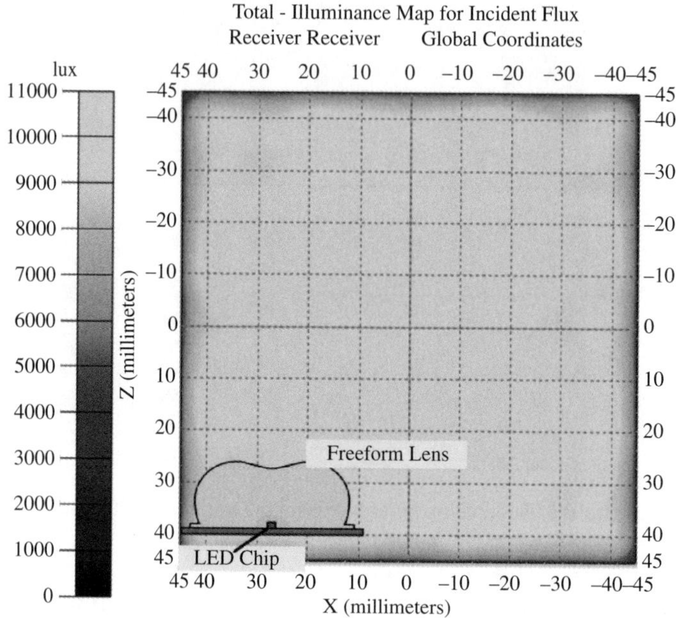

(a)

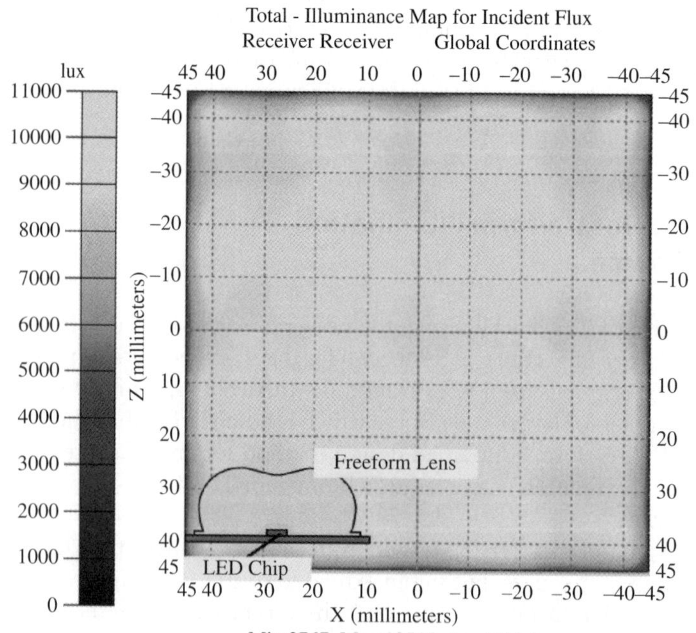

(b)

Figure 7.14 When DHR = 3, the average uniformity decreases with the light source size increasing. The size of the light source is: (a) 0.28 × 0.28 mm, $U = 0.89$, CV(RMSE) = 0.02; (b) 0.4 × 0.4 mm, $U = 0.74$, CV(RMSE) = 0.05; and (c) 1 × 1 mm, $U = 0.54$, CV(RMSE) = 0.15.

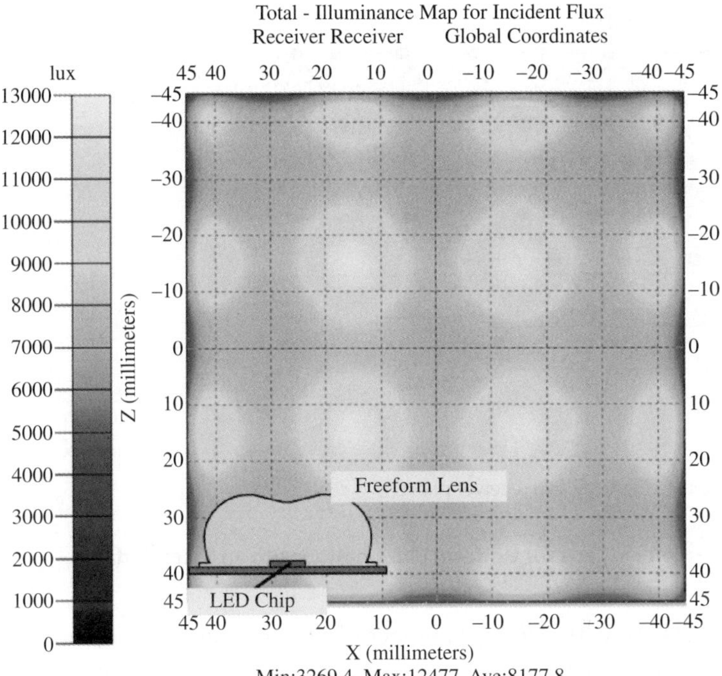

(c)

Figure 7.14 (*Continued*)

Comparing these two standards,[5] *U* measures the maximum illuminance variation, while *CV(RMSE)* can evaluate small illuminance variation across the target plane:

$$CV(RMSE) = RMSE/\overline{x} \tag{7.2}$$

The insert schematic pictures show that each light source is composed of an LED chip and a designed freeform lens. In Figure 7.14, only the size of the LED chip changes, and this leads to deterioration of illuminance uniformity. When the DHR = 3 and the size of the LED chip increases from 0.28 × 0.28 mm to 1 × 1 mm, *U* drops from 0.89 to 0.54. *CV(RMSE)* increases from 0.02 to 0.15 with the increase of the size of the light source. In our design, if the size of the light source, specifically the LED chip size, will be larger than or equal to one-fifth of the height of the lens, the light source belongs to the extended source category.

The design target is to achieve uniform illumination of a square LED extended source array with a high DHR value. In this section, a new feedback reversing design method for uniform illumination with an LED extended source is proposed, including the calculation of feedback optimization ratios (FORs) to achieve uniform illumination for an extended source and the generation of freeform lenses with the required light intensity distribution. This method is practical and simple. High uniform illumination with an extended source is achieved successfully when the DHR = 3. The uniformity of an LED extended source array enhances significantly by adopting this new design method.

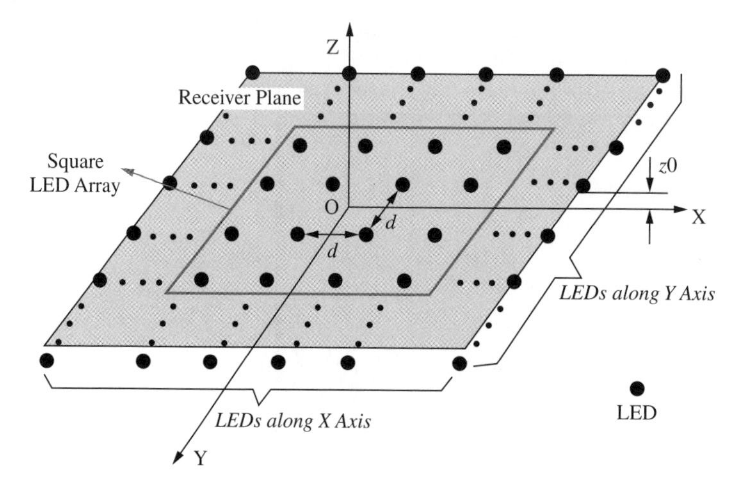

Figure 7.15 Schematic of the LED array and receiver plane.

7.4.1 Algorithm of a Freeform Lens for Uniform Illumination with an Extended Source

We extract a square LED array from a large number of LEDs as shown in Figure 7.15. Each LED light source is defined as an LED package with a freeform lens. The distance between the centers of each two adjacent LED light sources is d. We include a receiver plane to collect the energy emitting from the LED light source and observe the light pattern. The distance between the receiver plane and the light source is set as $z0$. The commonly adopted DHR = $d/z0$ denotes the ratio of adjacent light sources' distance and the distance between the target plane and the light source. We will present a new feedback reversing design method to realize highly uniform lighting at the receiver plane with a large DHR and high optical efficiency for an extended source.

The flow chart for our feedback optimization design method of uniform illuminance with an extended source is shown in Figure 7.16. Firstly, we use a design algorithm[3] to generate a freeform lens for point source uniform illumination at the target plane. Secondly, we replace the point source with an extended source and calculate the FOR. In the following, LI_{Point_input} and $LI_{Extended\_input}$ represent the light intensity (LI) incident into the lens from the point source and extended source, respectively. LI_{Point_output} and $LI_{Extended\_output}$ represent the LI emitting from the lens with a point source and extended source. LI_{Cal} represents the optimized LIDC to achieve uniform illumination. As in Eq. 7.3, FOR is defined as the ratio of light intensity that emits from the lens and the optimized light intensity.

$$k = LI_{Extended\_output}/LI_{Cal} \tag{7.3}$$

In the next step, the FORs are applied to energy grids of the light source, an energy mapping relationship is rebuilt, and a freeform lens for the extended source is constructed. After that, our design is validated and optimized through ray-tracing simulation until it meets our requirement.

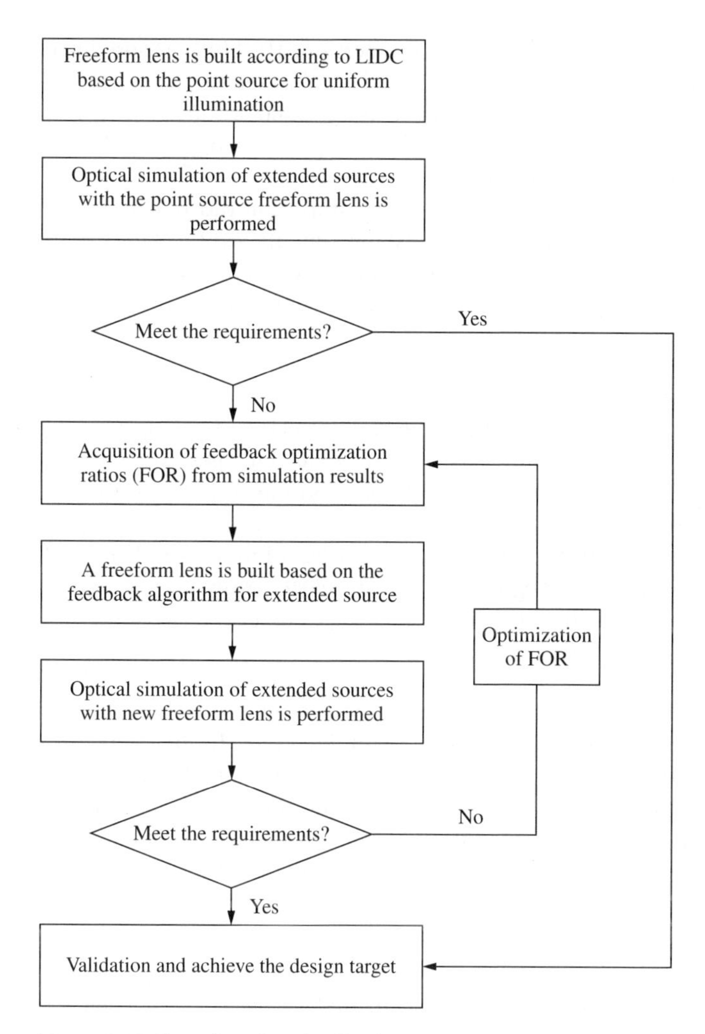

Figure 7.16 Flow chart for a feedback optimization design method for an extended source.

7.4.2 Design Method of a Freeform Lens for Extended Source Uniform Illumination

First of all, we optimize an LIDC to meet the two uniform requirements on the target plane. Then, we construct a freeform lens to generate the calculated LIDC. Finally, we validate the whole design and realize uniform illumination for the point source. For a detailed design method, refer to Section 3.6.

When we replace the point source with an extended source, both the efficiency and the uniformity at the target plane are deteriorated. A new design method for extended source uniform illumination will be proposed.

Previously, for the point source, we divide both the light source and the output energy into M grids with equal energy. An energy mapping relationship is built between a couple of grids. However, this equal distribution of energy is no longer suitable for an extended source. We redistribute the energy for an extended source by applying FORs on each energy grid and maintain the total energy conservation. The FORs are obtained by comparing the simulation results of $LI_{Extended\_output}$ and LI_{Cal}. Based on the optimized grids division, we rebuild an energy relationship between the extended source and output light to reconstruct a freeform lens. The simulation is performed, and FORs are optimized until the results meet our requirement.

In this design method, three key steps are feedback optimization ratios acquisition, redivision of energy grids, and energy relationship establishment, and the other steps are similar to those of the point source design method discussed in Section 3.6. In the following, we specifically describe the extended source freeform lens construction procedures.

7.4.2.1 Step 1. Calculation of FORs

Since light source intensity distribution and the shape of the lens for a single LED package are all circular-symmetrical, from Figure 7.17 we know that θ changes from 0 to $\pi/2$ with $\gamma = 0$ are able to reflect the whole light intensity distribution. When the division number of grids is large enough, we can use $LI_{Point\_output\_i}$, $LI_{Point\_input\_i}$, $LI_{Extended\_output\_i}$, and $LI_{Extended\_input\_i}$ to represent both the input and output light intensity of the point source and extended source at a given degree θ_i. FORs for each energy grid at a given degree θ_i are set as k_i, which equals the ratio of $LI_{Extended\_output\_i}$ and LI_{Cal_i} as in Eq. 7.4.

$$k_i = LI_{Extended\_output\_i}/LI_{Cal\_i} \tag{7.4}$$

We can obtain only limited LI values from the simulation result. These LI values, as a function of θ, can be provided with a limited number of θ degrees. In the simulation result, these degrees are restricted by the software but are not designed and selected according to the feedback optimization algorithm. As a result, these degrees are probably not in accordance with light source and output light intensity energy grid division degrees. In order to get k_i at each desired θ_i, we use a 10th-order algebraic polynomial to fit k_i, as $k_i = f(\theta_i)$ specifically in Eq. 7.5.

$$k_i = b_0 + b_1x + b_2x^2 + \ldots + b_{10}x^{10} \tag{7.5}$$

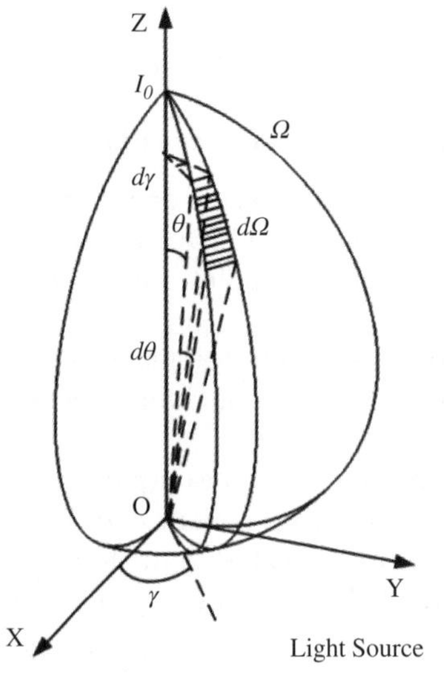

Figure 7.17 Schematic for the light intensity distribution of a light source.

7.4.2.2 Step 2. Energy Grids Division for an Extended Source

As shown in Figure 7.17, we set the total light energy of light source Φ_{input} which is then divided into M parts. Each part of the energy is Φ_{input_i} $i = 1, 2, \ldots M$. In the point source algorithm, $\Phi_{Point\_input\_i} = \Phi_0 = \Phi_{input}/M$. According to spatial distribution of point source light intensity, the unit conical object Φ_0 can be expressed with field angle $d\gamma$ in the latitudinal direction and $d\theta$ in the longitudinal direction. With Eq. 7.6, we can calculate the Φ_0. Since the LIDC of the light source is circular-symmetrical, Eq. 7.6 can be simplified as Eq. 7.7, and the total luminous flux of the light source can be described as Eq. 7.8. In this case, we can calculate the nth energy map by Eq. 7.9.

$$\Phi_0 = \int I_{input}(\theta)\ d\omega = \int_{\gamma_1}^{\gamma_2} d\gamma \int_{\theta_1}^{\theta_2} I_{input}(\theta) \sin\theta\ d\theta \tag{7.6}$$

$$\Phi_0 = \int_0^{2\pi} d\gamma \int_{\theta_1}^{\theta_2} I_{input}(\theta) d\theta = 2\pi \int_{\theta_1}^{\theta_2} I(\theta)\ \sin\theta d\theta \tag{7.7}$$

$$\begin{aligned}\Phi_{total} &= 2\pi \int_0^{\theta_1} I(\theta)\ \sin\theta d\theta + 2\pi \int_{\theta_1}^{\theta_2} I(\theta)\ \sin\theta d\theta + \ldots \\ &+ 2\pi \int_{\theta_i}^{\theta_{i+1}} I(\theta)\ \sin\theta d\theta + \ldots + 2\pi \int_{\theta_{M-1}}^{\theta_M} I(\theta)\ \sin\theta d\theta \\ &= \Phi_0 + \Phi_0 + \ldots + \Phi_0 = M \times \Phi_0\end{aligned} \tag{7.8}$$

$$N \times \Phi_0 = 2\pi \int_0^{\theta_N} I(\theta)\ \sin\theta d\theta \tag{7.9}$$

For an extended source, each grid energy $\Phi_{Extended\_input\_i}$ is equal to $k_i\Phi_0$. The directions of rays, which define the boundary of one subsource $\Phi_{Extended\_input\_i}$, also have been calculated as θ_{input_i} and θ_{input_i+1}. We divide the output light energy into M parts with equal energy. Based on Eq. 7.9, we can transfer the energy mapping relationship to the LI mapping as shown in Figure 7.18. Correspondingly, by Snell's law, we know each grid output light energy $\Phi_{Extended\_output\_i}$. Rebuild the energy mapping relationship, and design a freeform lens for an extended source.

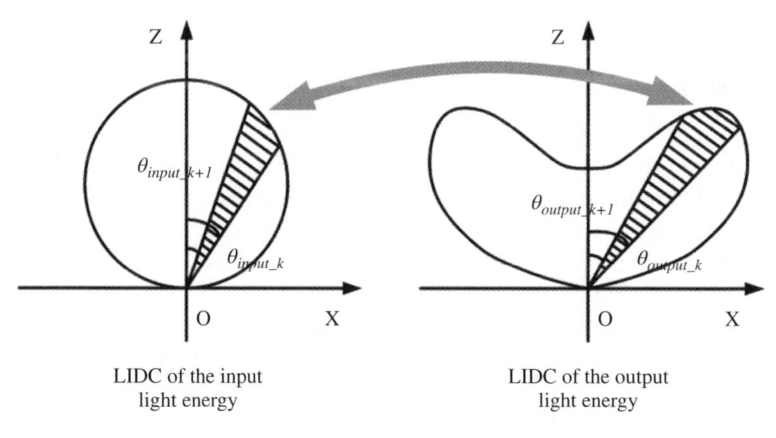

Figure 7.18 Energy mapping relationship between input and output light.

In this method, after the first-time optimization of the grids' division, the ith grid luminous flux of the light source is changed into $\Phi_{Extended\_input\_i}$. However, during the second-time optimization, FORs possess different values from the first time. Meanwhile, each grid of the light source is optimized based on the first-time optimized result instead of the original point source grids' even division. To easily illustrate and understand, we introduce subscript j to represent the times of optimization. Light source grids and corresponding FORs are expressed as follows: $\Phi_{Extended\_input\_ji}$ and k_{ji}. For example, we use k_{23} to represent the changing ratios of the second-time optimization applied on the third grid of the light source.

According to the light source energy division method, we can obtain emitting angles θ_{input_ji} of an optimized light source grid with Eq. 7.10.

$$\begin{aligned} 2\pi \int_0^{\theta_{input\_ji}} I(\theta)\sin\theta d\theta &= \sum_{i=0}^{N} \Phi_{Point\_input\_ji} \\ &= \sum_{i=0}^{N} k_{ji}\Phi_{Extended\_input\_j-1i} \end{aligned} \tag{7.10}$$

Among them, Φ_{input_0i} means "before optimization of the luminous flux of each point source grid" (i.e., Φ_0); θ_{input_j0} represents the first emitting light ray direction that is vertically pointed, and the value is 0. In this way, we can build an energy mapping relationship between the newly divided light source and the target plane.

7.4.2.3 Step 3. Construction of a Freeform Lens for an Extended Source

Since we have obtained the emitting directions of input rays and the corresponding exiting directions of output rays, it is easy to design a freeform lens to meet this mapping relationship according to the edge ray principle, Snell's law, and the surface lofting method.

7.4.2.4 Step 4. Ray-Tracing Simulation and Circulation Feedback Optimization

We analyze the light pattern illuminance of an extended source with the new designed freeform lens. If the evaluation results satisfy our preset standards, we accept the design; otherwise, we circulate more times to optimize the design until the results satisfy the illuminance requirements.

7.4.3 Freeform Lenses for Direct-Lit BLUs with an Extended Source

There are two types of LED backlighting, edge-lit and direct-lit. With limitation by the light transport distance in the light guide plate (LGP), edge-lit is not suitable for large or ultra-large LED backlighting. Direct-lit has advantages of size scalability and local dimming that could enhance the contrast ratio and decrease power consumption for LED TVs. Therefore, direct-lit probably will become a development trend for ultra-large LED backlighting.

Figure 7.19a shows an optical model of a LED package that includes a substrate, an LED chip, and a freeform lens. The height of our freeform lens generated by our feedback design method is 1.8 mm, and the refractive index is 1.54. This lens can be easily manufactured by a molding process. The size of the LED chip varies from 0.28 × 0.28 mm to 1 × 1 mm, which is larger than one-fifth of the height of a freeform lens and therefore

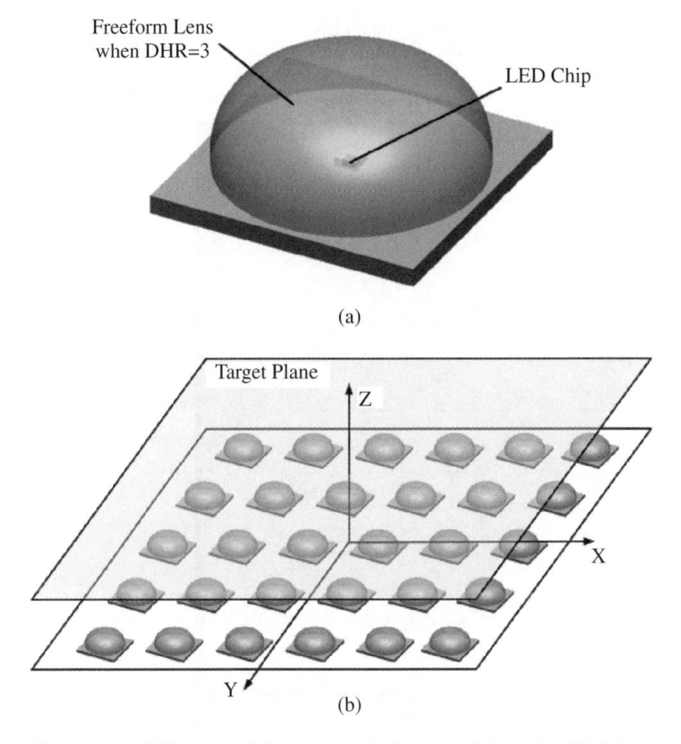

Figure 7.19 LED array with an extended source: (a) single LED light source, and (b) LED array optical model.

is considered as an extended source. As shown in Figure 7.19b, the distance between a square LED module array and the receiving plane is 10 mm, and the distance between each adjacent LED light source is 30 mm. Considering the periodical distribution of an LED module array, an area with the size of a 6 × 6 array of LEDs on the receiving plane is set as the testing area, which is able to reflect the whole lighting performance on the target plane. In addition, since a part of the lights emitted by the outside LED modules will be reflected by the reflection box in the side walls, only the uniformity of illuminance U of the central area of a 4 × 4 LED module array is considered in this design.

First of all, we calculated the uniform illumination LIDC according to the two criteria, and the expression is $LIDC = 0.032 - 0.1\theta^2 + 0.33\theta^4 - 0.142\theta^6 + 0.0099\theta^8$. A freeform lens for a point source is generated. The simulation results in Figure 7.20 show that when the DHR = 3, the point source algorithm is able to produce uniform illumination at the target plane by adopting a point source with a size of 0.28 × 0.28 mm. The values of U and $CV(RMSE)$ are calculated through 102 × 102 = 10,404 sampled points. The results show that $U_{Point} = 0.89$ and $CV(RMSE)_{Point} = 0.02$. For most lighting applications, if U is 0.85–1.15, the design satisfies the requirements and we accept the design. This step guarantees that uniform illumination is obtained for the point source. The *optical efficiency* is defined as the optical energy emitted from the freeform lens divided by that incident into the lens. As for the 0.28 × 0.28 mm point source, simulation results show the optical efficiency is 91%.

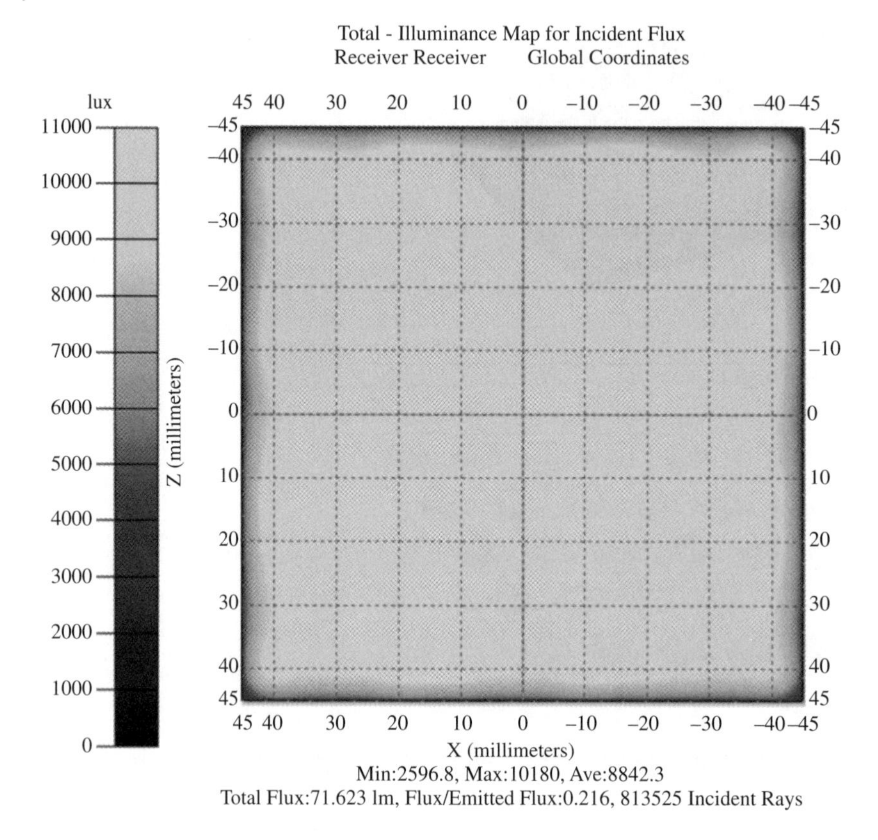

Figure 7.20 Illuminance distribution of a 0.28 × 0.28 mm point source LED array at the receiver plane.

Secondly, we replace the point source with a 1 × 1 mm extended source and rerun the simulation. From Figure 7.21, we can find the $U_{Extended}$ decreases to 0.54 and $CV(RMSE)_{Extended}$ increases to 0.15. This uniform deterioration is caused simply by the size of the LED chip. Figure 7.21 shows an obvious light spot that is the LED light source location. The optical efficiency is 86%.

Thirdly, we calculate the FORs. FORs at various degrees are obtained by the ratios of the normalized values of $LI_{Extended}$ to LI_{Point}. Then FORs are fit by a 10th-order algebraic polynomial $FOR = 90 + 40\theta + 340\theta^2 - 7460\theta^3 + 89970\theta^4 - 372060\theta^5 + 744410\theta^6 - 822530\theta^7 + 515820\theta^8 - 172550\theta^9 + 23970\theta^{10}$. Therefore, at each given θ, we can easily obtain a corresponding FOR value. We apply these FORs to build a new freeform lens for an extended source and run a simulation to test the result. In Figure 7.22, we can find that the uniformity is improved greatly. $U_{Extended}$ increases to 0.85, and $CV(RMSE)_{Extended}$ decreases to 0.04. Besides, the optical efficiency increases to 91%.

The light intensity distribution of a single LED light source before and after the feedback reversing design method is shown in Figure 7.23. The energy emitting from the light source is redistributed. Optical energy from −30° to 30° is reduced by the feedback

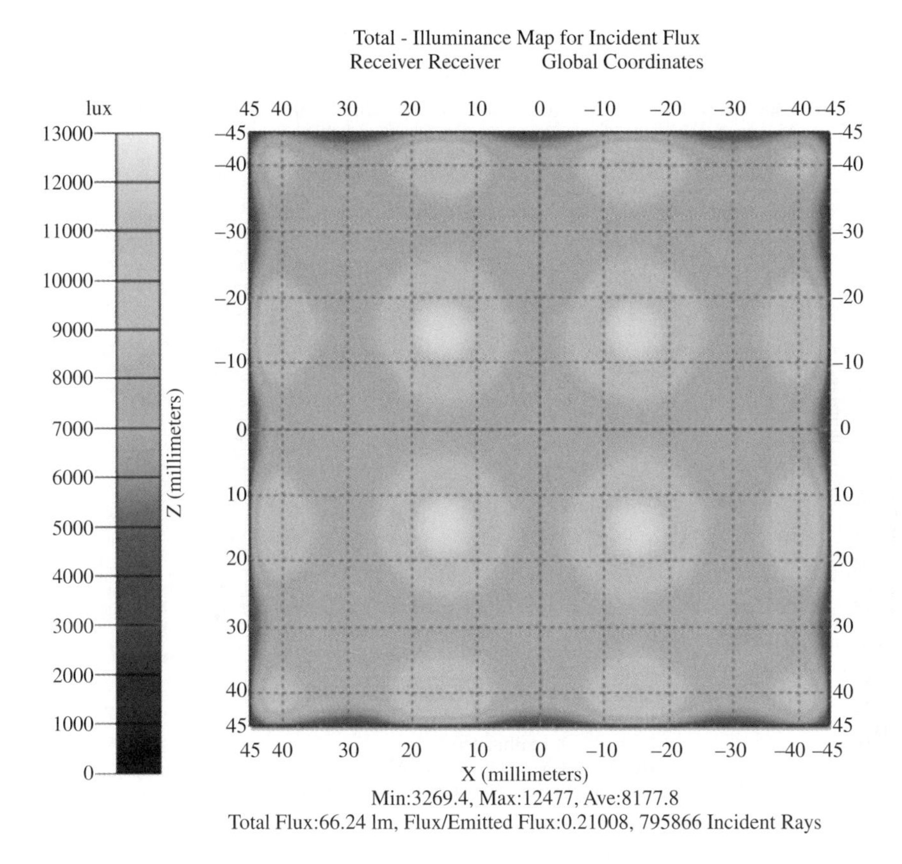

Figure 7.21 Illuminance distribution of an LED array consisting of 1 × 1 mm extended chip size LEDs at the receiver plane.

reversing method, which leads to uniform illuminance distribution at the receiver plane.

In this section, a feedback reversing design method for uniform illumination in LED backlighting with extended source has been verified through simulation. Good lighting performance in terms of high illuminance uniformity is achieved when DHR = 3. We can find that the uniformity deterioration of illuminance caused by an extended source on the receiving plane is greatly improved, achieving $U_{Extended} = 0.85$ and $CV(RMSE)_{Extended} = 0.04$ from $U_{Extended} = 0.54$ and $CV(RMSE)_{Extended} = 0.15$. Therefore, uniform illumination can be achieved easily by this new method when the DHR is much larger than 1. Consequently, this reversing design method provides an effective way to overcome the extended source problem and to achieve high uniform illumination in LED backlighting with an extended source. Based on this method, integrated with the application-specific LED packaging (ASLP) design concept mentioned in Chapter 4, a new kind of ASLP with a modified large DHR freeform lens could be designed for the larger or even ultra-large-scale LED BLUs in the near future.

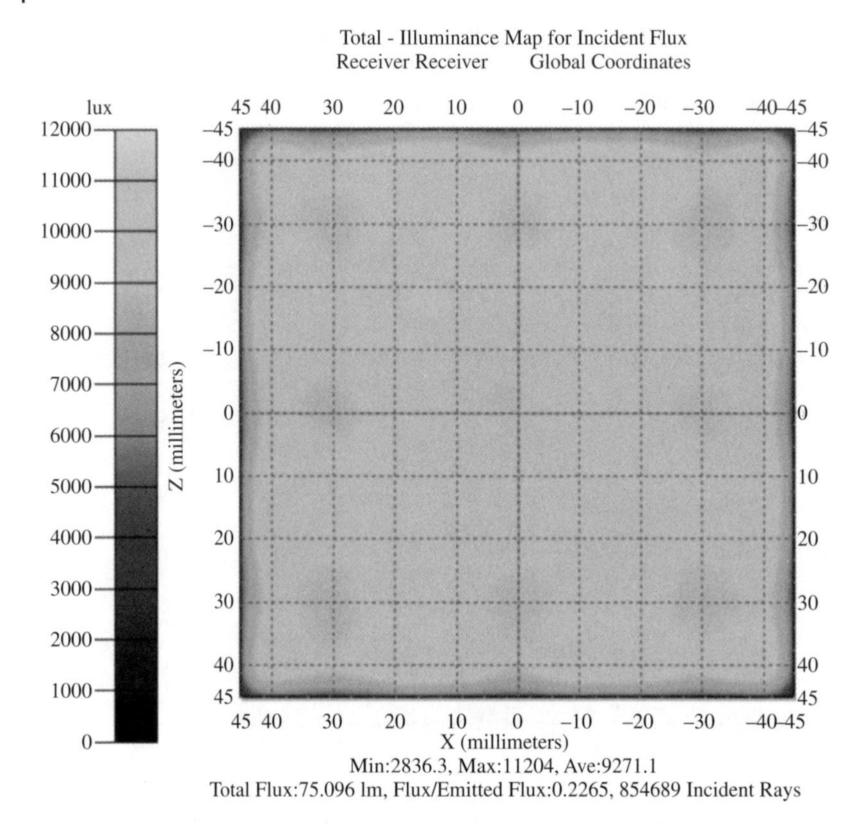

Figure 7.22 Illuminance distribution of an LED array consisting of 1 × 1 mm extended chip size LEDs with a feedback-optimized freeform lens at the receiver plane.

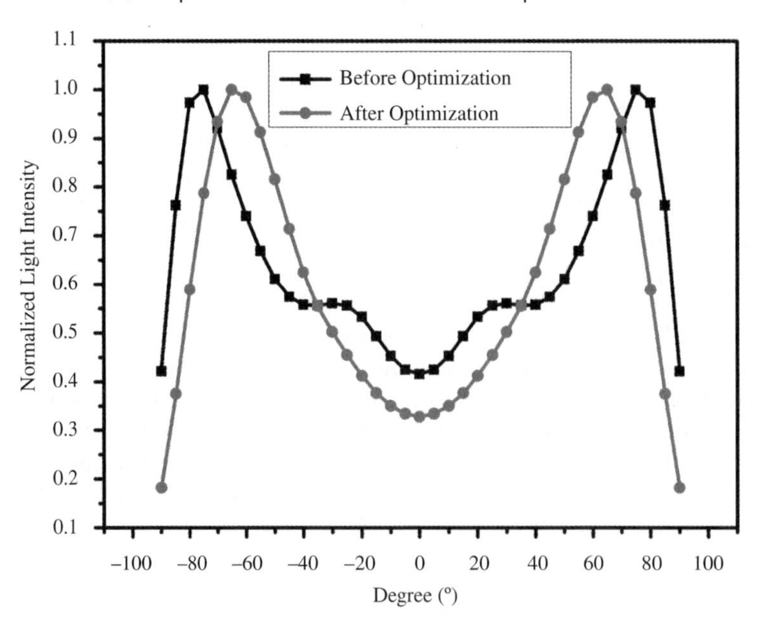

Figure 7.23 Light intensity distributions before and after application of the feedback optimization method.

7.5 Petal-Shaped Freeform Optics for High-System-Efficiency LED BLUs

7.5.1 Optical Co-design from the System Level of BLUs

Among solutions for LED backlighting, direct-lit structure in which an LED array is placed facing directly toward the LCD panel is a popular one that can provide feasibility of ultra-large modules and local dimming. However, optical films (including diffusers, prism sheets, and reflective polarizers) that are necessary to modify the illuminance and luminance distribution to satisfy the display requirements can cause significant light loss.[6–10]

In traditional usage of films, two plates of diffusers, as upper diffuser and down diffuser, are usually used on both sides of the prism sheets, which converge the light to obtain an on-axis luminance gain, spread light energy, and eliminate visible patterns. For direct-lit LED backlighting, uniform illumination distribution can be obtained by increasing the packaging density of an LED array compared with CCFL tubes, so the upper diffuser is always used alone to enhance efficiency and reduce cost. In this case, the LED array also provides light with Lambertian luminous intensity distribution to the BEF, as the down diffuser does originally. As mentioned in the simulation analysis in Section 7.6.2.2, light with such an angular energy distribution is not highly efficient for passing through the BEFs. In this section, we will focus on the relation of transmittance of BEFs on the incident angle of the light propagating onto them. Meanwhile, an optimized system configuration comprising a freeform lens that can control the angle range of light emitting from an LED is proposed to provide enhanced efficiency. With the proposed configuration, LEDs can have greater effects on energy saving with displays.

7.5.2 Optimization of a High-Efficiency LIDC for BEFs

Among various kinds of optical films, BEF is an essential element and is widely adopted; it is designed to collimate the incident light and improve the axial brightness. Since a single prism sheet can confine light in only one direction, BEFs usually consist of two orthogonal prism sheets. As shown in Figure 7.24a and 7.24b, the typical BEF is actually a prism sheet, and its optical characteristics are shown in Figure 7.24c.[12] In Figure 7.24c, when a light ray is incident upon the interface of prism sheet and air, about 36.8% of the incident light can transmit from the BEF, and about 46.3% is reflected back. Although the light reflected back can be recycled by being bounced back from the reflective sheet and being diffused by other elements, the efficiency of the BEF is still at a low level (~50%)[12] and leads to a *low output efficiency* of BLUs, which is defined as the percentage of light energy exiting from BLUs compared to the light energy emitting from LED sources.

An optical model is constructed for simulation. A Lambertian LED with a 1×1 mm emitting area is placed at the bottom of a $1000 \times 1000 \times 20$ mm cavity that has four side walls and an open top. Two typical 90/50 BEFs with an apex angle of 90°, a prism pitch of 50 μm, and a material refractive index of 1.517, as shown in Figure 7.24, are orthogonally crossed over the cavity. The transmittance of BEFs is calculated as the ratio of luminous flux exiting from prism sheets to that emitting from the LED. As the transmittance depends on the reflectivity property of backlight cavity walls that contain mirror

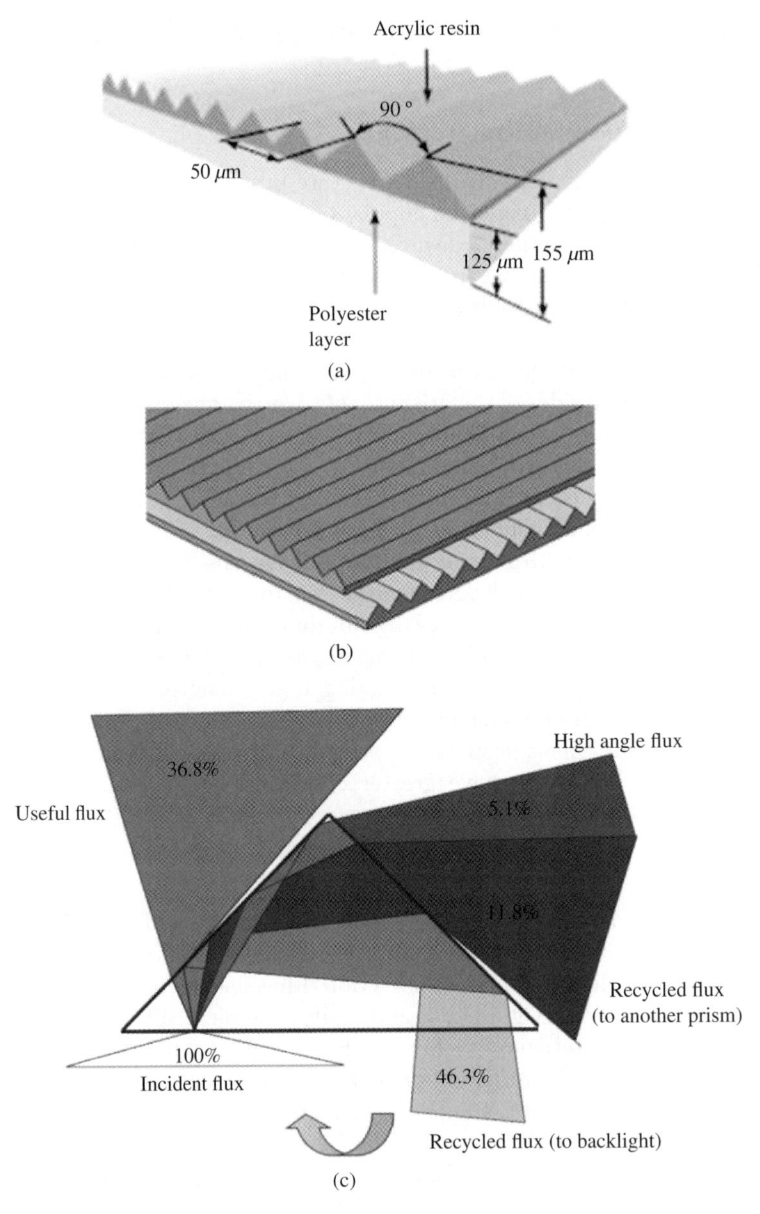

Figure 7.24 Schematic of BEF: (a) optical, (b) structures, and (c) function.[11]

reflections and diffuse the reflection of backlight cavity walls, two situations of specular and completely diffuse reflection cavity walls (both with a reflectance of 95%) are considered to investigate the role of the incident angle.

As shown in Figure 7.25, the Lambertian LED source is divided into different parts every 5° and 10° in the latitudinal angle θ and longitudinal angle φ, respectively, to verify the transmittance efficiency. The LED is divided into 18 parts with 0~5°, 5~10°, 10~15°, … 85~90° in the latitudinal angle θ direction and 36 parts with 0~10°, 10~20°, 30~40°,

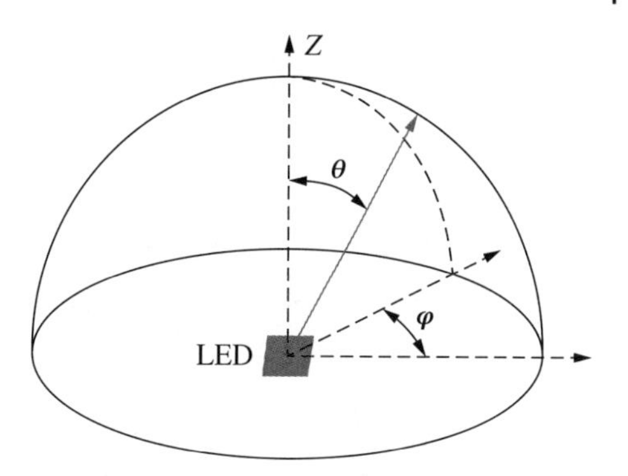

Figure 7.25 Division of a Lambertian LED in the latitudinal θ direction and longitudinal φ direction.

... 350~360° in the longitudinal φ angle direction. Every segmental source is simulated in the model in sequence instead of a complete Lambertian source.

Under the situation of specular reflection walls, where the transmittance of prism sheets is 52.8% for a single Lambertian LED, the complete Lambertian source is divided into 18 segmental ones in the latitudinal direction with θ = 0~5°, 5~10°, ... 85~90°. Simulated transmittances produced by each latitudinal segmental source are shown in Figure 7.26a, from which we can see that light with θ = 25~75° is efficient to pass through the prism sheets. Furthermore, segmental Lambertian LED emitting light within 25~75° in the latitudinal direction is divided into 36 segments with φ = 0~10°, 10~20°, ... 350~360° in the longitudinal direction. Simulated transmittances corresponding to these longitudinal angles are shown in a relative plot in Figure 7.26b. We can find that light with φ = 30~60°, 120~150°, 210~240°, and 300~330° is efficient to pass through the BEFs.

In addition, the situation of diffuse reflection walls is also conducted. The relations of transmittance of BEFs on latitudinal and longitudinal incident angles corresponding

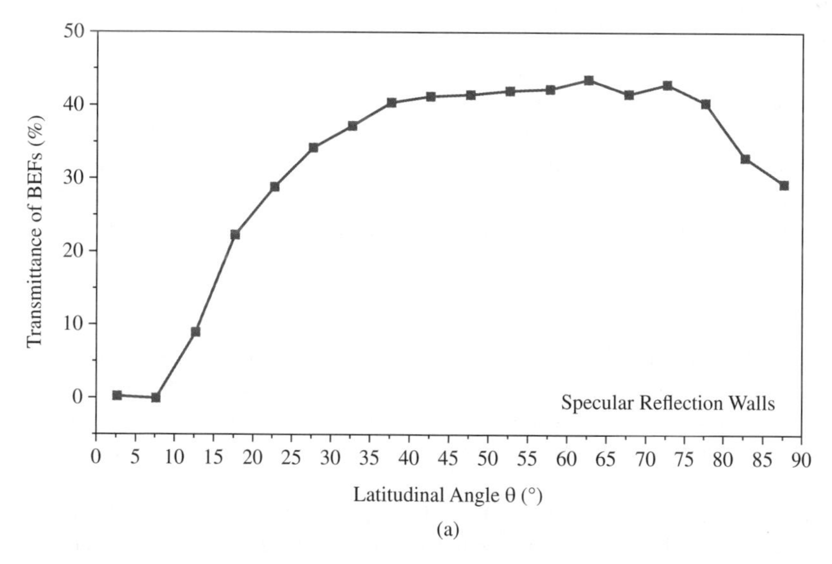

Figure 7.26 Relations of transmittance of BEFs on (a) latitudinal and (b) longitudinal incident angles corresponding to specular reflection walls.

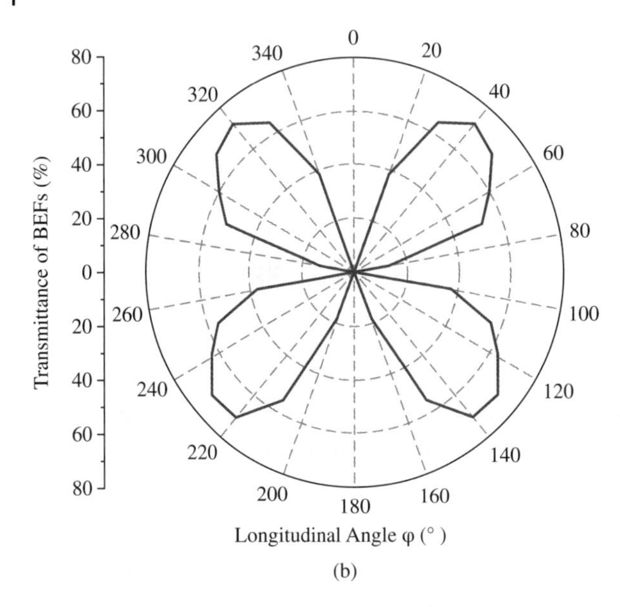

Figure 7.26 (*Continued*)

to diffuse reflection walls are shown in Figure 7.27a and 7.27b. Due to the effects of diffusion, the emitting windows with high light output efficiencies are changed to $\theta = 30\sim75°$ and $\varphi = 25\sim65°$, $115\sim155°$, $205\sim245°$, and $295\sim335°$.

7.5.3 Petal-Shaped Freeform Lenses, and ASLPs for High-Efficiency BLUs

In this section, we will design a freeform lens whose light intensity distribution matches the incident angle with high transmittance as described in Section 7.5.2. Since the design processes of freeform lenses for specular reflection and diffuse reflection are quite similar, we will take the case of specular reflection as an example in this section.

Taking into consideration both Figure 7.26a and 7.26b, which demonstrate that the transmittance is strongly dependent on the incident angle, the output light intensity distribution is generated only by light with "useful" incident angles. In this case, with the aid of the algorithm mentioned in Chapter 3, a freeform lens is designed to confine light in a desired angle range, as 25~75° in the latitude direction and 30~60°, 120~150°, 210~240°, and 300~330° in the longitude direction. The profile of the lens and produced intensity distribution are shown in Figure 7.28. It is important that output efficiency of the lens is 95.7% (mainly caused by Fresnel loss), which will not add a new loss factor in the system. Since both the shape and the light intensity distribution of the freeform lens look like a petal, we call this novel lens a *petal-shaped freeform lens*. Moreover, based on this petal freeform lens and the design concept mentioned in Chapter 4, a series of ASLP modules are also designed for high-efficiency direct-lit LED BLUs, as shown in Figure 7.29.

In a 20-mm-thick 16-inch direct-lit BLU, a LED array integrated with the petal freeform lenses with a pitch of 18 mm is placed in a cavity with 95% specular reflection walls, and the BLU with BEFs has an overall output efficiency of 51.9% originally. An output efficiency of 86.5%, which leads to an on-axis luminance gain of 2.96, is achieved

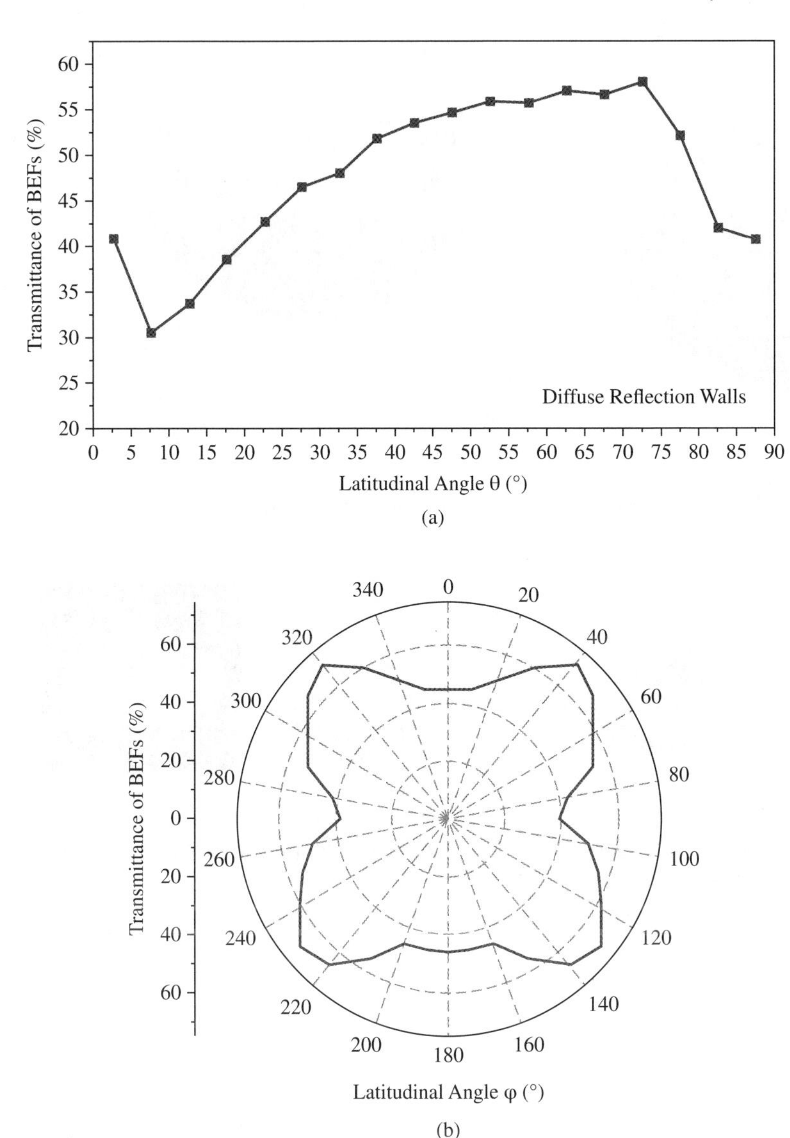

Figure 7.27 Relations of transmittance of BEFs on (a) latitudinal and (b) longitudinal incident angles corresponding to diffuse reflection walls.

from simulation. A comparison of angular luminance distribution of the central point on the output plane between LED arrays with and without the petal freeform lenses is shown in Figure 7.30a. The illuminance distribution with uniformity of 88.0% (the ratio of the minimal illuminance to the maximal one among nine sampled points) is also shown in Figure 7.30b. The designed freeform lens has enhanced the efficiency significantly.

With the significant effect on the situation of specular reflection walls, this approach is also applied to the direct-lit system with completely diffuse reflection walls. Analogously, relations of the transmittance of prism sheets on latitudinal and

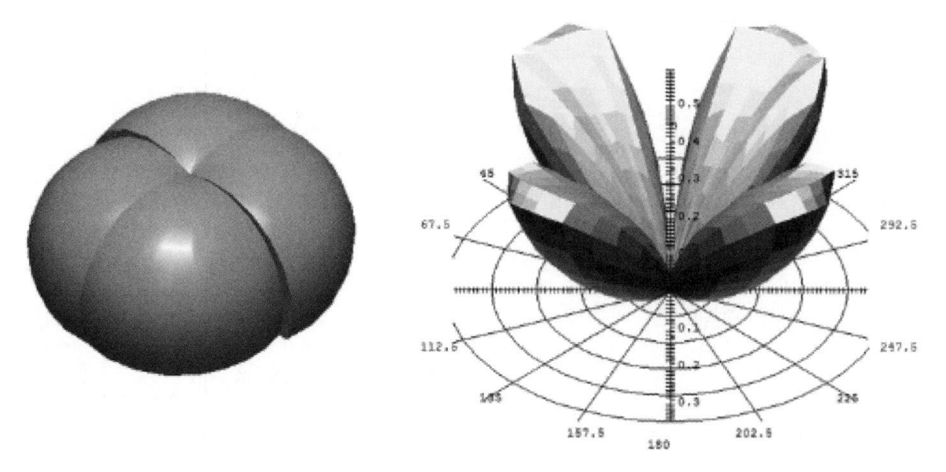

Figure 7.28 A petal-shaped freeform lens and its light intensity spatial distribution.

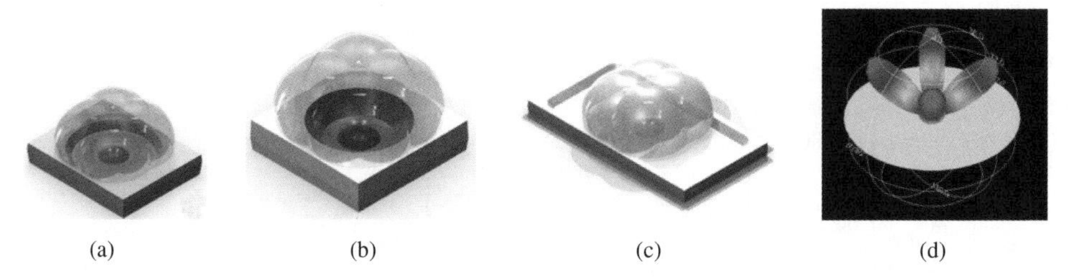

Figure 7.29 ASLP integrated with a petal freeform lens for a high system optical efficiency direct-lit BLU: (a) 2835, (b) 5050, (c) 5730, and (d) its light intensity spatial distribution.

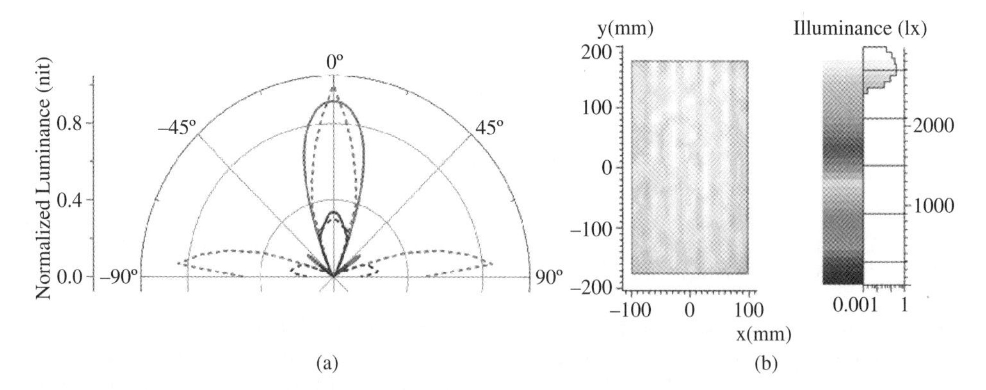

Figure 7.30 Under the situation of specular reflection walls. (a) Normalized luminance distributions of the central point produced by an LED array. Upper or down solid line denotes an array with or without lenses corresponding to 0° in the longitudinal direction; upper or down dotted line denotes array with or without lenses corresponding to 90° in the longitudinal direction. (b) Illuminance distribution produced by an LED array with freeform lenses.

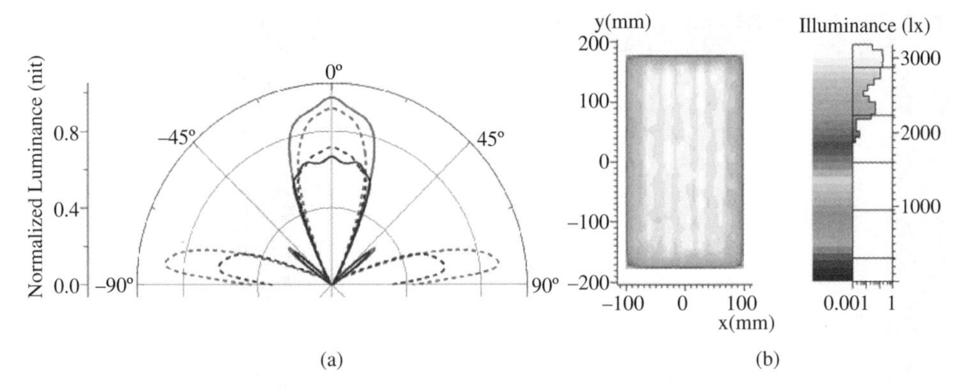

Figure 7.31 Under the situation of completely diffuse walls. (a) Normalized luminance distributions of the central point produced by an LED array. Upper or down solid line denotes an array with or without lenses corresponding to 0° in the longitudinal direction; upper or down dotted line denotes an array with or without lenses corresponding to 90° in the longitudinal direction. (b) Illuminance distribution produced by an LED array with freeform lenses.

longitudinal incident angles are investigated with the angle-by-angle method used before. A petal freeform lens is designed to confine light in the range of 30~75° latitudinally and 25~65°, 115~155°, 205~245°, and 295~335° longitudinally. When adding the petal freeform lenses to the 16-in. module used before (except the walls are changed to completely diffuse reflection), the overall output efficiency is enhanced from 82.4% to 88.6%, resulting in an on-axis luminance gain of 1.36. The comparisons of angular luminance distribution and illuminance distribution with a uniformity of 87.5% are shown in Figure 7.31a and 7.31b, respectively. The enhancement of efficiency and luminance gain is smaller than that corresponding to specular reflection walls, because diffuse reflection weakens the dependency of transmittance on incident angle.

The petal freeform lens is fabricated by a molding process as shown in Figure 7.32. A 19-in. BLU is made to verify the luminance gain of the freeform lens, as shown in Figure 7.33. The cavity of the BLU is made of aluminum in order to obtain efficient heat dissipation and because the reflection property of aluminum is very good. The left part is fixed with the petal freeform lenses, and the right part is not. Then, add diffuser film

Figure 7.32 A PMMA petal freeform lens for a high system optical efficiency direct-lit BLU.

Figure 7.33 Brightness comparisons of BLU with the lens (left) and without the lens (right).

and BEFs on the BLU, and the luminance gain effect is very clear by vision observation as shown in Figure 7.33. We can obviously find that the left part is brighter than the right part because light emitted from the LED is accurately controlled by the petal freeform lens, which is able to increase the luminance gain even after adding BEFs.

7.6 BEF-Adaptive Freeform Optics for High-System-Efficiency LED BLUs

7.6.1 Design Concept and Method

The petal freeform lens mentioned in Section 7.5 has advantages of high transmittance of BEFs so as to increase the BLU system optical efficiency as well as luminance gain factor significantly. However, the petal freeform lens will be limited to a small DHR (<1.5) if we just use the optical design method mentioned in Section 7.5. Therefore, an improved design method that can balance the enhancement of system efficiency and luminance gain as well as a large DHR should be proposed. In this section, we will introduce a BEF-adaptive freeform optics design method to solve this problem.

From Figure 7.24c, we may notice that the incident angle is the key parameter to determine how many rays can be transmitted and reflected. It is predicted from Snell's law that the larger the incident angle is, the more rays can be transmitted. To design a BEF-adaptive lens, the first problem is to find out the best incident angle range in which the more incident rays fall in, the more rays will be transmitted through the BEFs. The second problem is to design a freeform lens that can adapt to the BEF's characteristics and redistribute more rays from an LED light source into the best incident angle range, so as to enhance the lighting performance of BLUs.

The design flowchart is shown in Figure 7.34. Firstly, the best incident angle range for high BEF transmittance is investigated. Secondly, a BEF-adaptive lens is designed to match this incident angle range. Then, the lighting performance of BLUs on the target plane is conducted by Monte Carlo ray-tracing simulations. If the simulation result

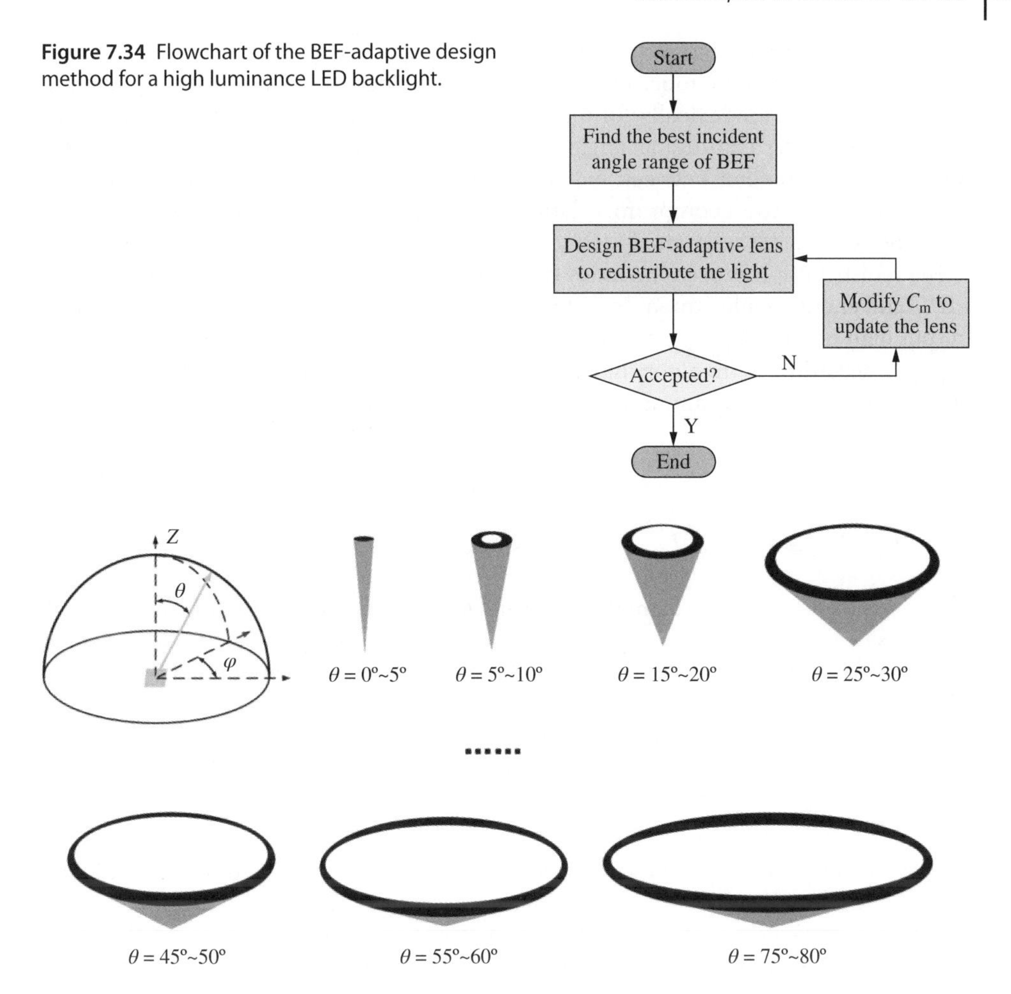

Figure 7.34 Flowchart of the BEF-adaptive design method for a high luminance LED backlight.

Figure 7.35 Segmental light sources within a different latitudinal incident angle θ.

cannot significantly improve the BEF efficiency and axial luminance simultaneously, we optimize the lens by modifying a constriction factor until its performance is accepted.

7.6.1.1 Step 1. Finding Out the Best Incident Angle Range

As shown in Figure 7.35, the light source is divided into 18 segments in the latitudinal direction with $\theta = 0{\sim}5°, \ldots 85{\sim}90°$, respectively. Then, BEF transmittance of each segmental source is investigated by Monte Carlo ray tracing. By doing so, a best incident angle range $\theta = \theta_1 \sim \theta_2$ is selected to generate the original output LIDC.

The LIDC can be denoted as the function of the emitting angle θ, $I_{in}(\theta)$, where θ is the angle between the ray and the z-axis. In the same way, the original output LIDC mentioned before can be expressed as follows:

$$I_{ou\_origin}(\theta) = \begin{cases} 0 & \theta < \theta_1 ς\ \theta > \theta_2 \\ I_{in}(\theta) & \theta_1 \leq \theta \leq \theta_2 \end{cases} \tag{7.11}$$

7.6.1.2 Step 2. Redistribution of Original Output LIDC

Now we can redistribute more light into the best incident angle range to improve the output efficiency. Paradoxically, for the sake of high axial luminance, light energy should be concentrated on a latitudinal angle range that deserves high axial luminance. Therefore, there is a tradeoff between enhancement of output efficiency and gain of axial luminance. We chose a compromise latitudinal incident angle θ_m to conduct the redistribution process of the original LIDC.

The redistribution process includes two steps: energy division of the original output LIDC and energy readjustment by compressing the original output LIDC. The original output light energy Φ_{ou_origin} could be regarded as composed of N parts of energy unit ϕ_{ou_origin}, according to the algorithm of freeform lens design.[3] The luminous flux of each energy unit and total luminous flux of output energy can be expressed as follows:

$$\phi_{ou\_origin} = \int I_{ou\_origin}(\theta) d\omega = \int_{\varphi_1}^{\varphi_2} d\varphi \int_{\theta_1}^{\theta_2} I_{ou\_origin}(\theta) \sin\theta d\theta \tag{7.12}$$

$$\Phi_{ou\_origin} = \int_0^{2\pi} d\varphi \int_0^{2\pi} I_{ou\_origin}(\theta) \sin\theta d\theta \tag{7.13}$$

where φ is the longitudinal azimuth angle; and θ is the latitudinal azimuth angle of light energy. Since total luminous flux of output energy is divided into N parts, the field angle $\Delta\theta_{ou\_i}$ of each energy unit along the latitudinal direction can be obtained by iterative calculation as follows:

$$2\pi \int_{\theta_{ou\_i}}^{\theta_{ou\_i+1}} I_{ou\_origin}(\theta) \sin\theta d\theta = \frac{\Phi_{ou\_origin}}{N} \quad (\mathrm{i} = 1,2 \ldots N,\ \theta_1 = 0) \tag{7.14}$$

$$\Delta\theta_{ou\_i} = \theta_{ou\_i+1} - \theta_{ou\_i} \ (\mathrm{i} = 1,2, \ldots \mathrm{N}) \tag{7.15}$$

Thus, the original output LIDC has been divided into N parts with equal luminous flux, and the boundary of each energy unit has been also calculated as $\Delta\theta_{ou\_i}$.

Then we conduct the energy division of the optimized output LIDC in the same method. In order to redistribute the original output LIDC, a constriction factor C is introduced to compress the original output LIDC into a more compact pattern; since θ_m has been determined, the constriction factor C can be expressed as follows:

$$C_i = C_1 \cdot q_1^{\,i-1} \ (\mathrm{i} = 1,2 \ldots \mathrm{m}, 0 < q_1 < 1) \tag{7.16}$$

$$C_i = C_m \cdot q_2^{\,i-m} \ (\mathrm{i} = \mathrm{m}+1, \mathrm{m}+2 \ldots \mathrm{N}, q_2 > 1) \tag{7.17}$$

where C consists of two geometric progressions, of which the first one is in decreasing sequence with a common ratio of q_1, and the second one is in increasing sequence with a common ratio of q_2. After integration with the constriction factor, the output energy division (Eq. 7.14 and Eq. 7.15) can be expressed as follows:

$$2\pi \int_{\psi_i}^{\psi_{i+1}} I_{ou\_origin}(\theta) \sin\psi d\psi = C_i \cdot \frac{\Phi_{ou\_origin}}{N} \quad (\mathrm{i} = 1,2 \ldots N,\ \psi_1 = 0) \tag{7.18}$$

$$2\pi \int_{\psi_i}^{\psi_{i+1}} I_{ou\_origin}(\psi) \sin\psi d\psi = C_i \cdot \frac{2\pi}{N} \int_{\theta_i}^{\theta_{i+1}} I_{ou\_origin}(\theta) \sin\theta d\theta$$

$$(\mathrm{i} = 1,2 \ldots N,\ \psi_1 = \theta_1 = 0) \tag{7.19}$$

$$\Delta\psi_{ou\_i} = \psi_{ou\_i+1} - \psi_{ou\_i} \ (\mathrm{i} = 1,2, \ldots \mathrm{N}) \tag{7.20}$$

In this process, the light energy within the range of $\theta_{\text{ou\_i}}$ to $\theta_{\text{ou\_i+1}}$ is redistributed to the range of $\psi_{\text{ou\_i}}$ to $\psi_{\text{ou\_i+1}}$. Due to the addition of the constriction factor, the output light energy on both sides of latitudinal angle θ_{m} is converged on θ_{m}. According to the law of energy conservation, the relationship between C_1, q_1 and $C_{\text{m+1}}$, q_2 can be expressed as follows:

$$\sum_{i=1}^{m} C_i = C_1 \cdot \frac{1 - {q_1}^{\text{m}}}{1 - q_1} = m \tag{7.21}$$

$$\sum_{\text{i=m+1}}^{\text{N}} C_{\text{i}} = C_{\text{m+1}} \cdot \frac{1 - {q_2}^{\text{(N-m)}}}{1 - q_2} = N - m \tag{7.22}$$

To avoid a saltation of luminous intensity between both sides of θ_{m}, C_{m} and $C_{\text{m+1}}$ should be approximately equal; this principle provides us with a reference to select and modify the common ratio q_1 and q_2. Once C_{m} is selected, $C_{\text{m+1}}$, q_1, and q_2 are selected as well. Therefore, we can take C_{m} as a criterion to measure the compression degree of the original output LIDC.

7.6.1.3 Step 3. Construction of a BEF-Adaptive Lens

To establish the light energy mapping relationship between the light source and the optimized output LIDC, the light source energy is divided into N parts with equal luminous flux. The boundary of each energy unit can be expressed as follows:

$$2\pi \int_{\theta_{\text{in\_i}}}^{\theta_{\text{in\_i+1}}} I_{\text{in}}(\theta) \sin\theta d\theta = \frac{2\pi}{N} \int_0^{\pi/2} I_{\text{in}}(\theta) \sin\theta d\theta \tag{7.23}$$

$$\Delta\theta_{\text{in\_i+1}} = \theta_{\text{in\_i+1}} - \theta_{\text{in\_i}} \tag{7.24}$$

As the mapping relationship between an emitting angle of input rays and exiting angle of output rays has been established, the method to design a freeform lens to meet this relationship is convenient, as described in Chapter 3.

7.6.2 BEF-Adaptive Lens Design Case

We will take the direct-lit LED backlight as an example to validate this design method. The optical components in a typical LCD monitor include a back reflector, LED light sources, a first diffuser, BEFs, a second diffuser, and a liquid crystal panel. A back reflector is usually a material with diffuse reflection or specular reflection, which recycles the reflected light rays. Diffuser sheets like first and second diffusers are used to uniformize spacing and angular distribution of output light energy.[13] Therefore, here we mainly discuss the optical properties of a typical direct-lit LED BLU with a back reflector, LED light sources, and optical stacks (as mentioned above), and then optimize the output LIDC by a BEF-adaptive lens to improve output efficiency and axial luminance simultaneously in the whole radiation angle.

7.6.2.1 Basic Setup of a BLU

As shown in Figure 7.36, an optical model of a conventional direct-lit LED BLU is established. A backlight cavity functions as the back reflector, by which the internal surface is covered with a diffuse reflection material. A typical 2835 SMD LED is used as a light

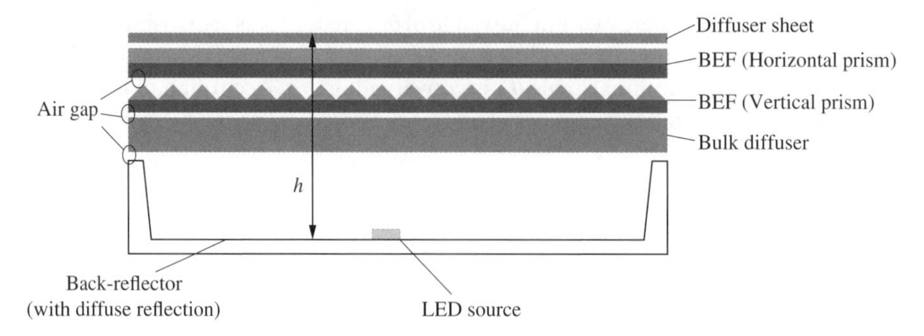

Figure 7.36 Optical model of a traditional direct-lit BLU.

Table 7.1 Detailed configurations of optical stacks in the BLU.

Optical component	Dimensions	Optical properties
Back reflector	420 × 270 × 19 mm	90% diffuse reflection
Bulk diffuser	420 × 270 × 2 mm	94.5% transmittance, 98.2% haze
BEF	420 × 270 × 0.155 mm	Vikuiti 90/50 BEF
Diffuser sheet	420 × 270 × 0.1 mm	98.9% transmittance, 96.8% haze

source. Power consumption of this LED module is 0.1 W when driven by 30 mA, and its total luminous flux is 9 lm. The size of the LED module is 3.5 × 2.8 mm and its LIDC is Lambertian type of $I_{in}(\theta) = I_0\, cos\theta$, where I_0 is the light intensity when the emitting angle $\theta = 0°$. The wavelength of this light source is set as 550 nm.[14] In front of the LED light source, there are a bulk diffuser, two orthogonal BEFs, and a diffuser sheet. Detailed configurations of these optical components are listed in Table 7.1. Then, a receiving plane with an angular luminance meter is set above the diffuser sheet. The size of the back reflector is 410 × 260 mm, and the distance h between the LED light source and the receiver is 18 mm. Considering the real situation, an air gap is inserted between every two optical stacks.

7.6.2.2 Design Results and Optical Validation

Next, simulation of BEF output efficiency under a different latitudinal angle range is conducted, and the results are shown in Figure 7.37a. Simulated analysis shows that emitted light with $\theta = 20\sim80°$ is efficient to pass through the double-layer BEFs. Based on this, we constructed a BEF-adaptive lens.

As the efficient incident angle range is $\theta = 20\sim80°$, the original output LIDC can be expressed as follows:

$$I_{ou\_origin}(\theta) = \begin{cases} 0 & \theta < 20°\text{ç } \theta > 80° \\ I_0 \cos\theta & 20° \leq \theta \leq 80° \end{cases} \tag{7.25}$$

According to Figure 7.37, to improve the output efficiency of the backlight module, light energy should be concentrated to incident angle $\theta = 60°$. However, for the purpose of improving the axial luminance, light energy should be compressed to an angle range

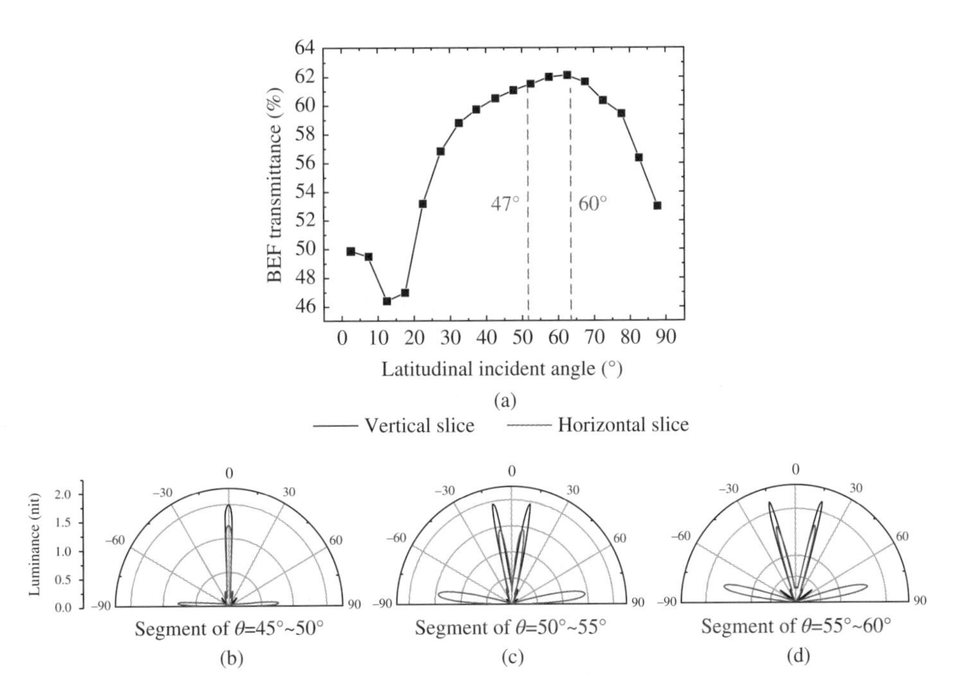

Figure 7.37 (a) Simulated BEF transmittance produced by each latitudinal segmental source; (b) luminance distribution produced by a segmental source of $\theta = 45 \sim 50°$; (c) luminance distribution produced by a segmental source of $\theta = 50 \sim 55°$; and (d) luminance distribution produced by a segmental source of $\theta = 55 \sim 60°$.

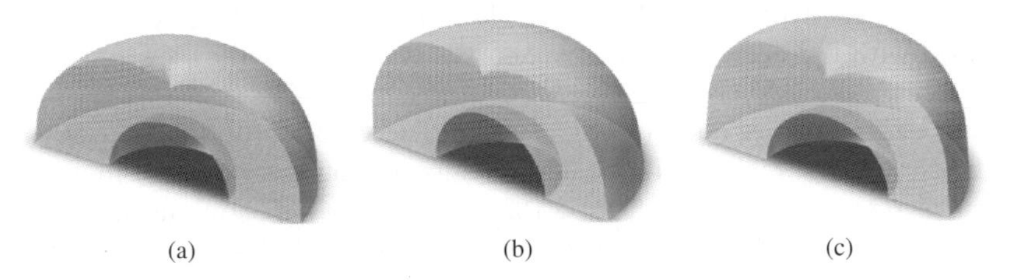

Figure 7.38 Models of a BEF-adaptive lens under a different constriction factor: (a) $C_m = 1$; (b) $C_m = 0.6$; and (c) $C_m = 0.1$.

of $\theta = 45 \sim 50°$, which will not cause divarication of luminance distribution. Therefore, to achieve both an improvement of output efficiency and axial luminance, the compromise incident angle is chosen as $\theta_m = 47°$.

Based on the algorithm of freeform lens design, a series of BEF-adaptive lenses under different constriction factors C_m are designed based on PMMA material, as is shown in Figure 7.38. The height of the BEF-adaptive lens is 5 mm, and the radius of the inner hemispherical surface is 3 mm.

Then, Monte Carlo ray-tracing simulations by 5 million rays are conducted. As shown in Figure 7.39, the results show that axial luminance and output efficiency vary with the constriction factor C_m. For the BLU without a BEF-adaptive lens, the output efficiency is 50.16% and axial luminance is 35.3 nit. For the BLU with a BEF-adaptive lens of $C_m = 1$,

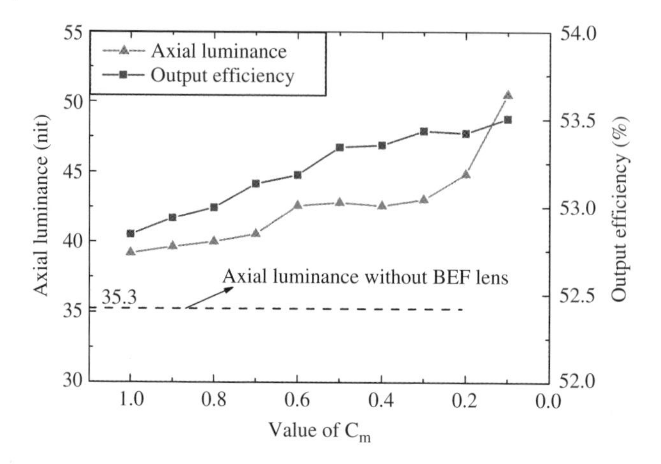

Figure 7.39 Output efficiency and axial luminance of BLUs under different C_m.

the output efficiency is improved to 52.84%, and the axial luminance improved to 39.2 nit. Moreover, axial luminance and output efficiency have similar trends with the variation of C_m, which illustrates that the compression process makes the output light energy more adaptive to BEFs.

For BLU, the viewing angle is another critical evaluation index. In the simulation, a viewing angle with different C_m is investigated as well, and the results are shown in Figure 7.40 and Figure 7.41. Figure 7.40 shows the horizontal and vertical luminance distribution under different C_m. In Figure 7.41, the horizontal axis is the C_m varying from 1.0 to 0.1, and the left vertical axis and right vertical axis are the viewing angle and output efficiency, respectively. From the figures, both horizontal and vertical viewing angles decreased with the decrease of C_m. When C_m is below 0.3, the viewing angle decreased sharply.

Taking both axial luminance and viewing angle into account, an optimal C_m of 0.6 is obtained. Under this optimal C_m, the axial luminance and output efficiency increase from 35.3 nit to 42.6 nit and from 50.16% to 53.18%, respectively, while the viewing angle only decreases slightly by 2°. Therefore, coupled with an optimal BEF-adaptive lens, the overall output efficiency and the axial luminance can be enhanced by 6.02% and 20.67%, respectively, without causing a large decrease of the viewing angle.

Furthermore, realistic objectives for an LED BLU are high efficiency, high axial luminance within a specific viewing angle, and high uniformity of spatial luminance. Since comparison of output efficiency, axial luminance, and the viewing angle of the BLU with and without a BEF-adaptive lens is illustrated as in this chapter, it is essential to validate the spatial luminance uniformity of the BLU integrated with the optimal BEF-adaptive lens.

Based on the direct-lit LED BLU established in this chapter, a typical 2835-LED array with square distribution is placed in the center of a backlight cavity with 90% diffuse reflection walls. Considering periodical distribution of the LED modules array, an area with the size of a 6 × 6 LED array is set as the testing area, which is able to reflect the whole lighting performance on the receiving plane, and the horizontal and vertical

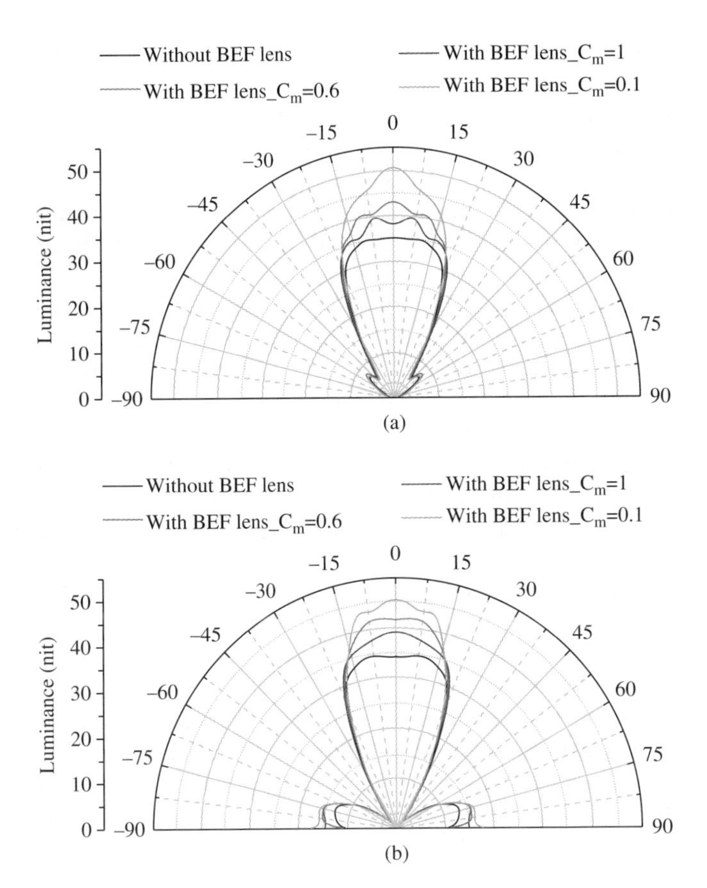

Figure 7.40 (a) Horizontal axial luminance distribution and (b) vertical axial luminance distribution under different C_m.

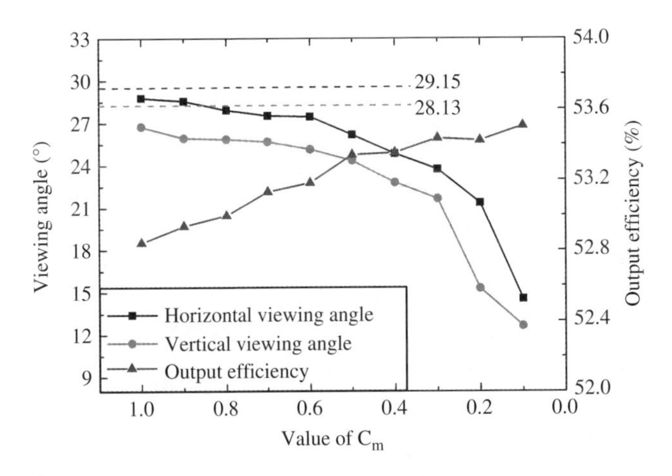

Figure 7.41 Horizontal viewing angle and vertical viewing angle of BLUs under different C_m.

distance of every two light sources is 18 mm. A spatial luminance meter is set above the receiving plane with the longitudinal angle and latitudinal angle at 0, which indicates that the spatial luminance meter is facing directly toward the backlighting module. Distance from the spatial luminance meter to the top of optical stacks is 100 mm, and the half cone angle of the meter is set as 30°. In addition, since a part of the lights emitted by the outside LED modules will be reflected by the backlight cavity, only the luminance uniformity U (L_{min} / L_{ave}) of the central area of a 4 × 4 LED module array is considered in this validation.

Then Monte Carlo ray-tracing simulations by 7 million rays are performed, and the spatial luminance distributions of the backlight module without and with the optimal BEF-adaptive lens are obtained, as shown in Figure 7.42a and 7.42b. For a traditional BLU without a BEF-adaptive lens, the spatial luminance uniformity is at a high level of 0.87 (5872 nit/6775 nit), and when integrated with the optimal BEF-adaptive lens, the spatial luminance uniformity is 0.86 (6031 nit/7049 nit); this indicates that spatial luminance uniformity is approximately constant in these two cases. Therefore, a BLU with the optimal BEF-adaptive lens can achieve good performance in spatial luminance uniformity.

In this section, a BEF-adaptive design method is proposed to achieve high output efficiency and high axial luminance in LED BLUs. An optimization is performed to obtain the best C_m. Under the optimal BEF-adaptive lens of C_m = 0.6, the output efficiency increases from 50.16% to 53.18% and axial luminance increases from 35.3 nit to 42.6 nit, while the spatial luminance uniformity is approximately constant at a high level of more than 0.8, and the viewing angle only decreases slightly by 2°. Therefore, this design method provides a flexible design freedom to achieve high axial luminance and overall output efficiency. Since a different backlight system has different requirements toward brightness and viewing angle, contributions of the BEF-adaptive lens depend on the specific demand, which means the brightness enhancement by an optimal BEF-adaptive lens is not confined to 20.67%. For a BLU that requires higher efficiency and brightness, this BEF-adaptive design method is effective to achieve high overall performance with more design freedom.

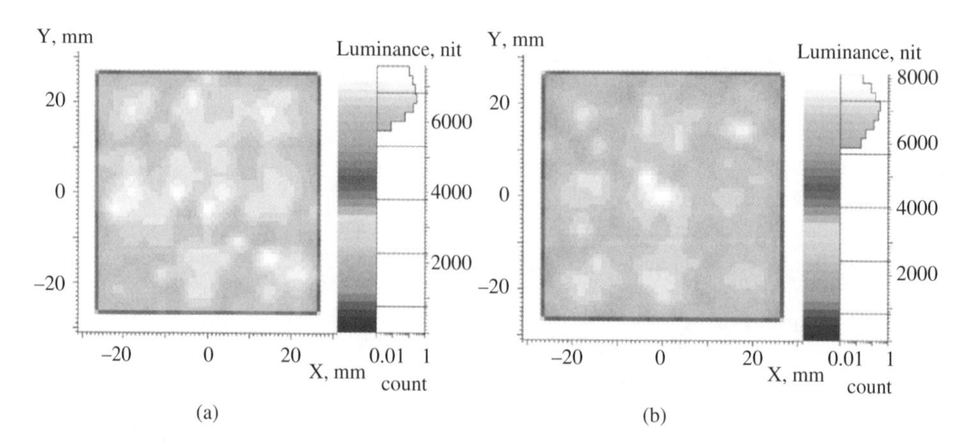

Figure 7.42 (a) Spatial luminance distribution of a BLU without a BEF-adaptive lens and (b) spatial luminance distribution of a BLU integrated with the optimal BEF-adaptive lenses.

7.7 Freeform Optics for Uniform Illumination with Large DHR, Extended Source and Near Field

Although freeform optics design methods for uniform illumination with large DHR or extended source have been introduced in the above sections, to obtain high uniformity in the harsh condition of large DHR, extended source and near field at the same time is a key as well as challenging issue. As shown in Figure 7.43, there are four types of 4 × 4 square array LED illumination modules. The distance between light source and target plane increases from 10 mm to 1 m, while the size of LED chip changes from 0.2 mm × 0.2 mm to 1 mm × 1 mm. The same lens with height of 3.5 mm is applied in the simulation and the DHR is constant as 4. In Figure 7.43a, high DHR and uniform illuminance at the target plane can be achieved with $U = 0.89$ and CV(RMSE) = 0.02, when the target is far field and the light source is point source. When the height or diameter of lens will be larger 1/10 of the distance of target, the lighting system is defined as near field illumination. As shown in Figure 7.43b, the lighting system becomes near field. As a result, the uniformity of LED array is decreased, and U is 0.67 and CV(RMSE) is 0.09. While the diameter of LED chip will be larger 1/5 of the height of lens, the light

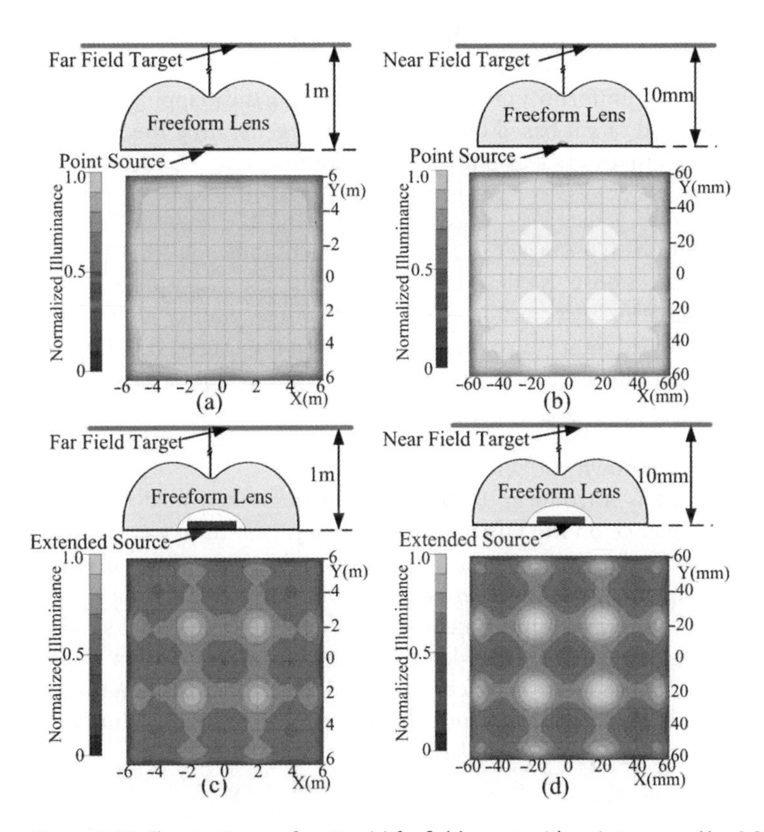

Figure 7.43 Illumination uniformity: (a) far field target with point source, $U = 0.89$, CV(RMSE) = 0.02, (b) near field target with point source, $U = 0.67$, CV(RMSE) = 0.09, (c) far field target with extended source, $U = 0.55$, CV(RMSE) = 0.15, and (d) near field target with extended source, $U = 0.45$, CV(RMSE) = 0.24.

source belongs to extended source. As shown in Figure 7.43c, the extended source result in the uniformity deterioration, while U is 0.55 and CV(RMSE) is 0.15. As shown in Figure 7.43d, the size of LED chip is 1 mm × 1 mm and the distance between light source and target plane is 10 mm which is close to the practical application condition. The optical performances with $U = 0.45$ and CV(RMSE) = 0.24 deteriorate seriously with near field and extended source.

In this chapter a new design method is presented to achieve uniform illumination with high DHR when light source is extended source and the target plane is near field. The general principle of algorithm is illuminance distribution instead of traditional light intensity distribution which is related to light source's position and size. This method can obtain the illuminance distribution function (IDF) of single LED extended source fitting the data measured by light ray tracing software. Meanwhile, IDF of freeform lens can be achieved according to the Sparrow's criterion and selected point uniformity requirement. Finally, based on the relationship between extended source and the target plane, novel freeform lens is generated and optical simulation is conducted to verify the design method.

7.7.1 Design Method

The design flow for our illumination fitting method includes four main steps: (1) obtaining the IDF of single LED extended source fitting reference data; (2) getting IDF of freeform lens according to two uniformity criteria; (3) establishing the mapping relationship between light source and target based on their IDFs and generating the freeform lens; (4) verifying the design method's feasibility by simulation.

7.7.1.1 IDF of Single Extended Source

Since illumination distribution on the target and the shape of lens for single LED package are circular symmetrical, we choose the illuminance $E(r)$ along the radius direction to represent the whole illumination distribution of extended source, while r is the space between the projection point of light source and test point on the target. IDF $E(r)$ can be expressed with 10th order algebraic polynomial as in equation (7.26)

$$E_r = a_0 + a_1 r + a_2 r^2 + \cdots + a_{10} r^{10} \tag{7.26}$$

where $a_0, a_1, a_2 \cdots a_{10}$ are unknown variables to be solved.

Light ray tracing software is introduced to simulate the single extended source to obtain the reference illuminance on the target. The simulation condition based on the practical chip size and specified target height will not bring in the deviation result from point source assumption. After simulation, there are a set of M discrete reference point r_i on target and the corresponding illuminance E_i can be obtained. The least squares method is used for curve fitting, and we consider $E(r)$ as the objective function, and E_i as the fitting reference point. Based on this concept, we used the least squares method to fit the known illuminance of the target plane into an illuminance distribution function. In the least squares method, the square error ϕ is given by

$$\phi = \sum_{i=1}^{M} (E(r_i) - E_i)^2 \tag{7.27}$$

where r_i is the coordinates of the ith reference point, E_i is illuminance obtained by simulation. When minimizing $\phi$and substituting $E(r_i)$ with coefficient polynomial of matrix $[a_0, a_1, a_2 \cdots a_{10}]$, a continuous and complete IDF of single extended source can be obtained.

Normalized Cross Correlation (NCC) is introduced to evaluate the accuracy of the fitting, which is expressed as

$$NCC = \frac{\sum_i (E(r_i) - \overline{E}(r_i))(E_i - \overline{E}_i)}{\left[\sum_i (E(r_i) - \overline{E}(r_i))^2 \sum_i (E_i - \overline{E}_i)^2\right]^{1/2}} \tag{7.28}$$

when NCC ≥95%, the fitting is satisfied. As shown in Figure 7.44, the fitting IDF *E(r)* and E_i can be obtained and NCC is as high as 97.8%.

7.7.1.2 IDF of Freeform Lens

Firstly, we define the IDF $E'(r)$ of freeform lens referring to the IDF of extended source as follows:

$$E'_r = b_0 + b_1 r + b_2 r^2 + \cdots + b_{10} r^{10} \tag{7.29}$$

For LED array illumination, the illuminance values at various positions on the target plane are superimposed with the neighboring multi-LEDs. The total illumination distribution *E'(x,y)* can be expressed as

$$E'(x, y) = \sum_{i=1}^{N} E'(r_i) \tag{7.30}$$

where

$$r_i = \sqrt{(x - x_i)^2 + (y - y_i)^2} \tag{7.31}$$

Here, (x,y) are the coordinates of the test point and (x_i, y_i) are the coordinates of the *i*th LED in the array, N is the number of calculated LED.

Secondly, in order to obtain illuminance distribution satisfying uniformity requirement, we introduce two uniform illumination criterions in this study. First criterion is

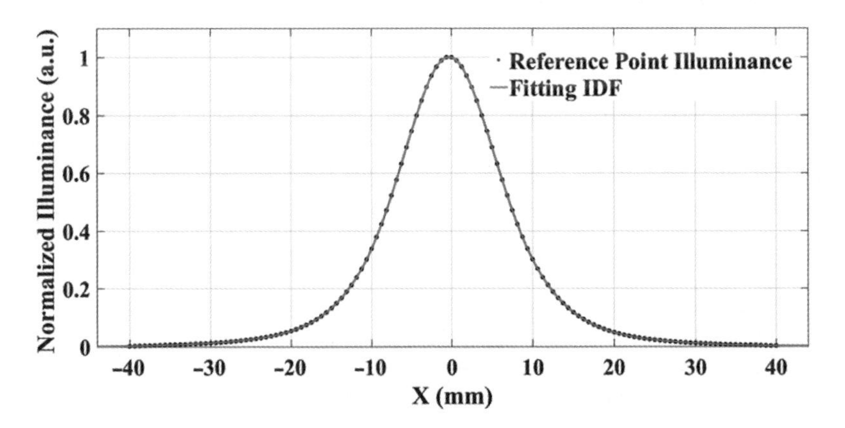

Figure 7.44 Schematic of fitting IDF and reference point illuminance.

the Sparrow's criterion which can be expressed as

$$\left.\frac{\partial E'(x,y)}{\partial x}\right|_{x=0,y=0} = 0 \tag{7.32}$$

Combining the equations (6) (7) (9), the equation $f(b_0, b_1, b_2, \cdots, b_{10}) = 0$ can be obtained. By solving the equation, the coefficient of $E'(r)$ can be determined partly.

Another criterion $R(x_i, y_i)$ is introduced to get all coefficients of $E'(r)$:

$$R(x_i, y_i) = \frac{E'(x_i, y_i)}{E'(0,0)} \tag{7.33}$$

where $E'(x_i, y_i)$ is the illuminance of selected special point on the target plane. Ratio $R(x_i, y_i)$ reflects the illuminance variance between side point and the central point. To obtain high uniformity, the range of $R(x_i, y_i)$ is set as between 0.85 and 1.15, which further confines the ranges of variations of solved coefficient. In this design, selected point is regarded as verification point and proper selection of these points is a key issue. For example, if all selected points are around the central point, illuminance uniformity of side points cannot be guaranteed. Therefore, suitable verification points which can reflect the whole illuminance distribution across the target plane should be selected as references. Selecting sufficient verification points is essential for solving the IDF of freeform lens. As shown in Figure 7.45, the IDF of freeform lens can be obtained based on meeting above criterions. Compared with the two IDFs, the center of light emitting from source is transformed to edges by freeform lens to realize uniform illumination.

7.7.1.3 Construction of Freeform Lens

In this design, light emitted from the extended source is regarded as the input light, and light exited from the freeform lens is regarded as the output light. Since we have obtained IDF of input rays and the corresponding IDF of output rays, next step is to divide energy grids and establish the mapping relationship between light source and target. Total light

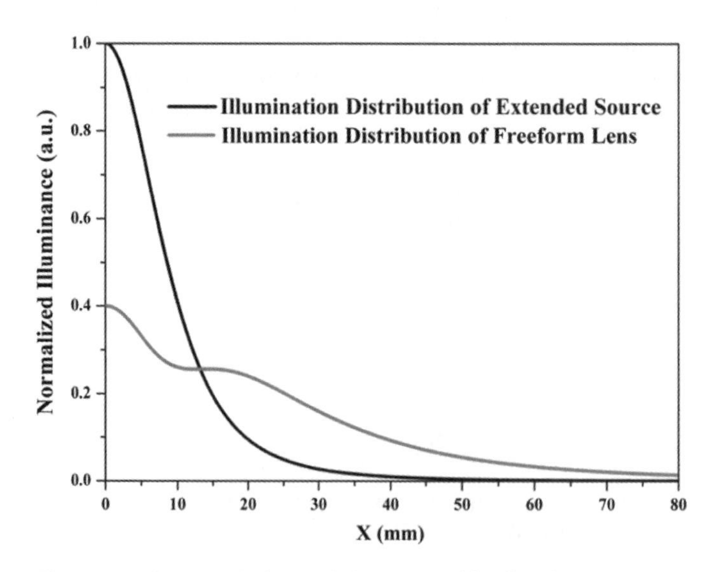

Figure 7.45 Fitting IDF of extended source and freeform lens.

energy of light source is divided into M × N grids equally along X axis direction and Y axis direction. The luminous flux φ_{input_0} of each grid can be expressed as follows:

$$\varphi_{input\_0} = \int_{x_1}^{x_2} \int_{y_1}^{y_2} E(x, y)\ dxdy \tag{7.34}$$

where $E(x, y)$ is the IDF of extended source. The total luminous flux of the light source can be expressed as follows:

$$\begin{aligned}\varphi_{input\_total} &= \int_{0}^{x_1} \int_{0}^{y_1} E(x, y)\ dxdy + \int_{x_1}^{x_2} \int_{y_1}^{y_2} E(x, y)\ dxdy + \cdots \\ &+ \int_{x_{i-1}}^{x_i} \int_{y_{j-1}}^{y_j} E(x, y)\ dxdy + \cdots + \int_{x_{M-1}}^{x_M} \int_{y_{N-1}}^{y_N} E(x, y)\ dxdy \\ &= \text{M} \times \text{N} \times \varphi_{input\_0}\end{aligned} \tag{7.35}$$

Since luminous flux of each grid is constant, then the coordinate *(x, y)* of each input ray emitting on the target can be obtained by iterative calculation with above equation. Therefore, using the same calculation method, the coordinate *(x;' y')* of each output ray striking on the target can be also obtained. According to the edge ray principle, input rays from the edge of the source should strike the edge of the target. Since the emitting directions of input rays and the corresponding exiting directions of output rays have been obtained, a freeform lens meeting mapping relationship can be constructed according to Snell's law and surface lofting method.

7.7.1.4 Ray Tracing Simulation and Verification

We analyze the light performance of extended source and near field target with the new designed freeform lens. To verify the new method is valid, we also generate the lens based on traditional method for comparison.

7.7.2 Design Example

In order to evaluate the performance of this new design method, simulation is performed with new lens to compare it with the traditional design method based on point source and far field. We choose DHR = 4 as the design condition which has been applied in existing LED TV. A higher DHR will realize a thinner thickness of backlighting module and need less number of LEDs. According to the size of LED chip and freeform lens, there are two types of receiving target plane, near field and far field, and two types of light sources, point source and extended source. The lens material adopts the PMMA with refractive index of 1.49. The reversing design method is used to generate the traditional lens based on point source assumption. The profiles of the new lens and the traditional lens are as shown in Figure 7.46. Ray tracing simulation with 6 × 6 LEDs square array is performed. We conduct four groups of simulation for different conditions as follow: (1) the light source is point source and target plane is far field, (2) the light source is point source and target plane is near field, (3) the light source is extended source and target plane is far field, and (4) the light source is extended source and target plane is near field.

As shown in Figure 7.47, the distance between the receiving surface and the LED array plane is 1000 mm, so it can be considered as the far-field condition. The distance

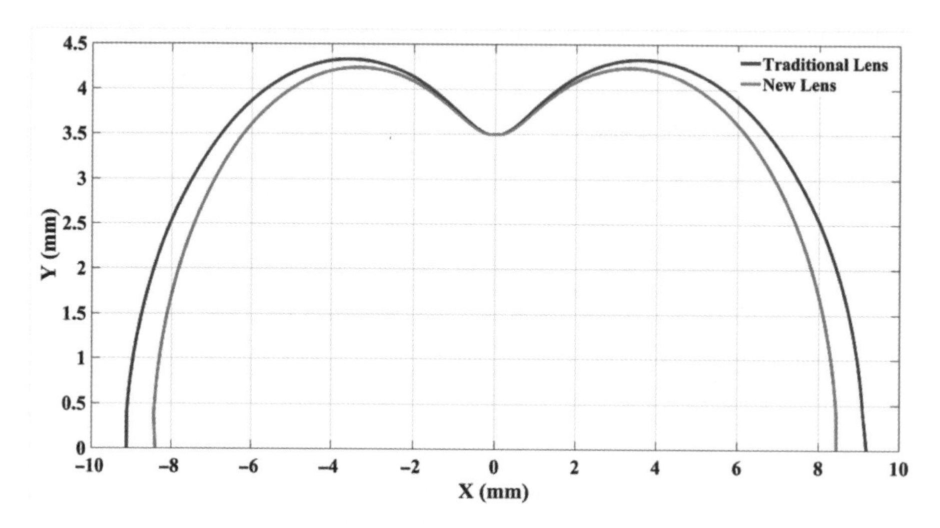

Figure 7.46 Comparison of profiles of the traditional freeform lens and the new freeform lens.

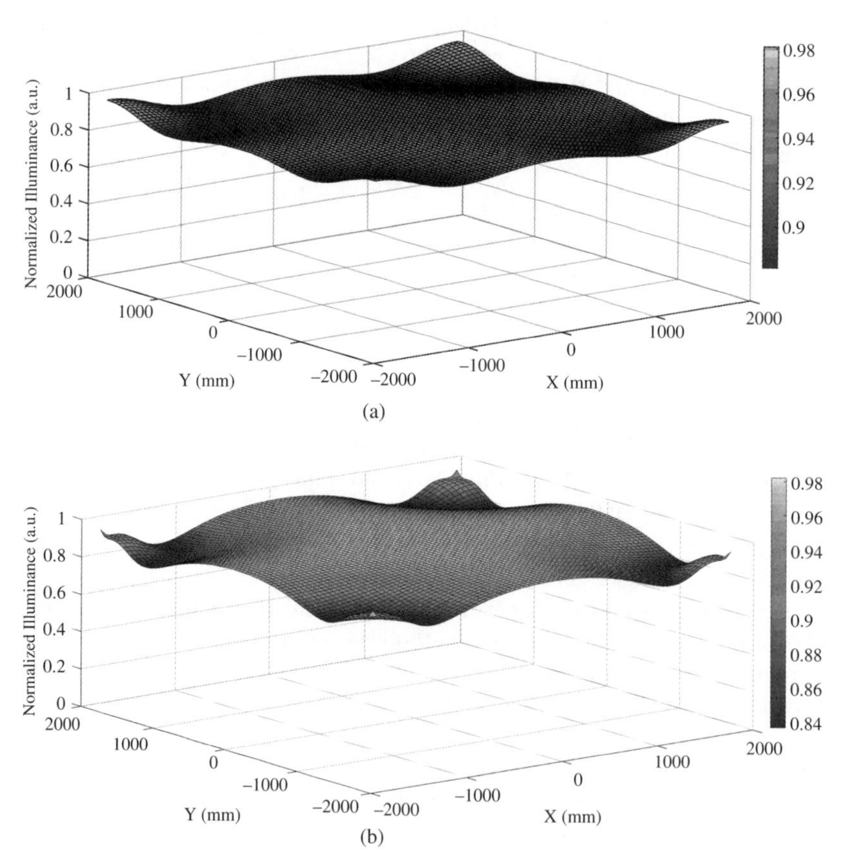

Figure 7.47 Illuminance distribution of 0.2 mm × 0.2 mm point source at the 1000 mm far field receiver plane, (a) LED array with traditional lens, and (b) LED array with new lens.

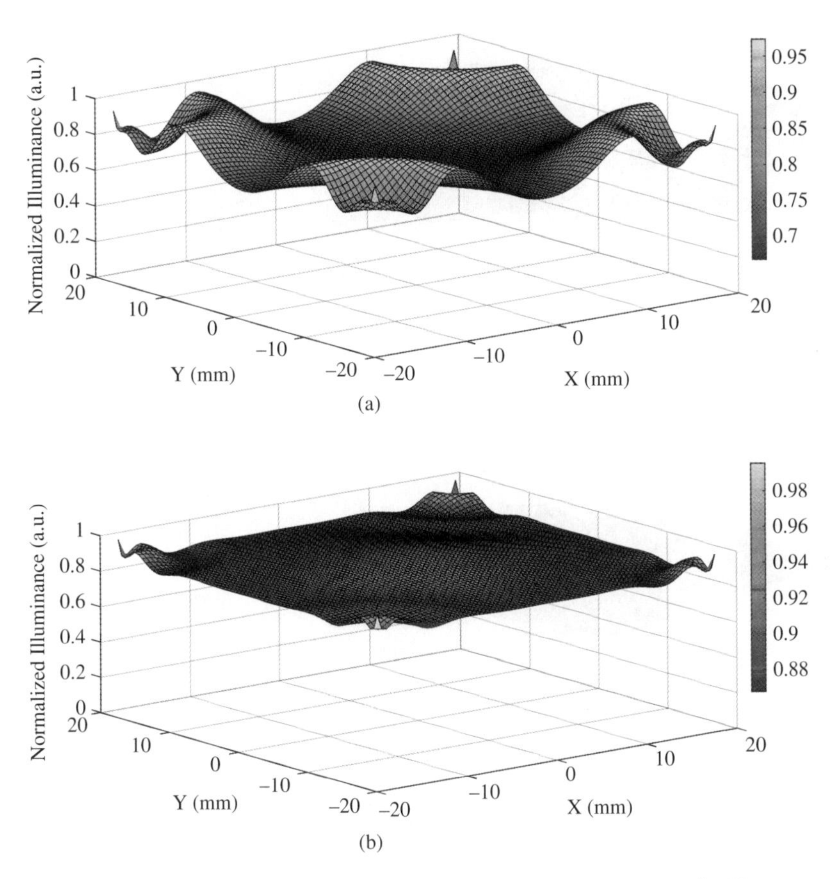

Figure 7.48 Illuminance distribution of 0.2 mm × 0.2 mm point source at the 10 mm near field receiver plane, (a) LED array with traditional lens, and (b) LED array with new lens.

between two adjacent LEDs is D = 4000 mm. The size of LED chip is 0.2 mm × 0.2 mm, which can be considered as point source compared with the height of lens. LED array are arranged in the axis of symmetry distribution, so the square receiving surface from (-D/2, -D/2) to (D/2, D/2) unit area is chosen as the evaluation area to reflect the whole illumination target. Figure 7.47a shows that the uniformity, $U_{traditional}$ of the traditional lens is 0.89 with $CV(RMSE)_{traditional} = 0.02$, and in Figure 7.47b the uniformity U_{new} of the new lens simulation is 0.84. In this case, the uniformity of traditional lens is a little better, indicating that traditional method is feasible when the simulation conditions satisfy the point source assumption.

In order to analyze the effect of target location, we only change the distance between target plane and light source to 10 mm, while other conditions remain constant. Compared with far field condition, Figure 7.48 shows $U_{traditional}$ decreases to 0.67 and $CV(RMSE)_{traditional}$ increases to 0.09. The height of lens is about 4 mm which is approximately half of the height of target plane. Considering the whole LED emitting light module as point source to design lens will results in serious performance deterioration. But with the new lens, the uniformity keep a high level of 0.87 and

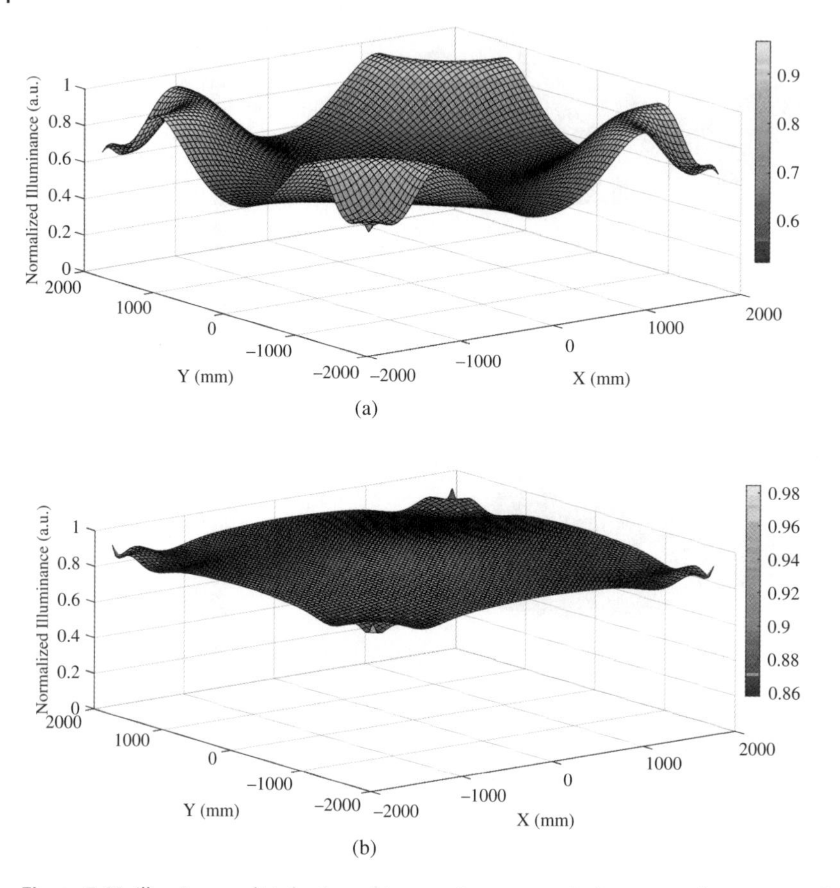

Figure 7.49 Illuminance distribution of 1 mm × 1 mm extended source at the 1000 mm far field receiver plane, (a) LED array with traditional lens, and (b) LED array with new lens.

$CV(RMSE)_{new} = 0.03$, because target plane energy is based on illuminance division rather than light intensity in new method.

Then, it is verified that the size of LED chip will also affect the uniformity of array illumination. The light source is replaced with 1 mm × 1 mm extended source and we rerun simulation. Figure 7.49 shows that the uniformity decreases more than near field condition for traditional lens design method. $U_{traditional}$ is 0.55 and $CV(RMSE)_{traditional}$ increase to 0.15, while the uniformity of new lens is 0.86 and $CV(RMSE)_{new}$ is 0.03. The light performances of new method remain excellent which proves the fitting method of extended source is reliable and accurate.

Finally, the two factors (i.e. small target location and large size of chip) are together integrated in the simulation model. In Figure 7.50, simulation results reveal the uniformity of new lens remain at 0.84 and $CV(RMSE)_{new}$ is 0.03, while the optical performances of array backlighting module with traditional lens are deteriorated seriously with $U_{traditional} = 0.45$ and $CV(RMSE)_{traditional} = 0.24$. Therefore, it is indicated the new method

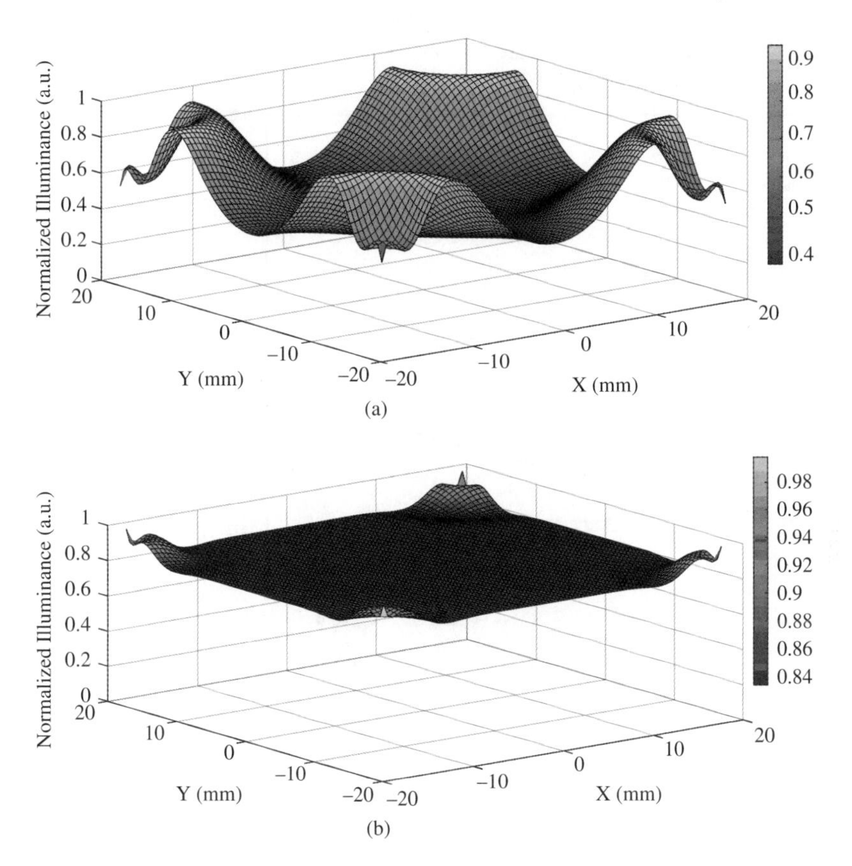

Figure 7.50 Illuminance distribution of 1 mm × 1 mm extended source at the 10 mm near field receiver plane, (a) LED array with traditional lens, and (b) LED array with new lens.

can meet the requirement of backlighting with severe design condition. As shown in Figure 7.51, the uniformity variation trends of above four cases demonstrate our design method can overcome the negative effect of the small target location and the large size of chip.

Therefore, a new reversing freeform lens design algorithm, based on the IDF instead of traditional light intensity distribution, has been proposed successfully for LED uniform illumination in the very harsh condition. IDF of freeform lens can be obtained by the proposed mathematical method, considering the effects of large DHR, extended source and near field target at the same time. Slim direct-lit LED backlighting with DHR of 4 is designed. Compared with traditional lens, illuminance uniformity of LED backlighting with the new lens increases significantly from 0.45 to 0.84, and CV(RMSE) decreases dramatically from 0.24 to 0.03 in the harsh condition. This new method provides a practical and effective way to solve the problem of large DHR, extended source and near field for LED array illumination.

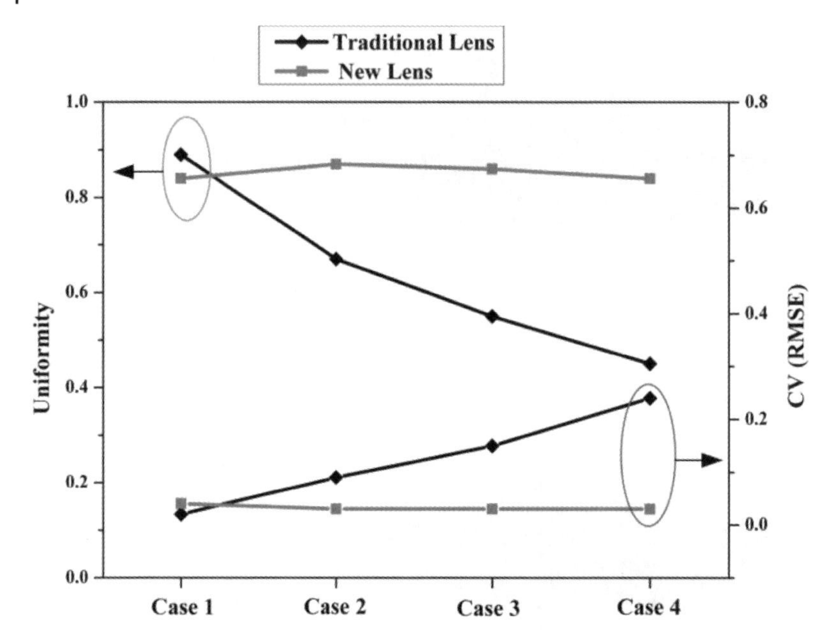

Figure 7.51 Comparison of uniformity and CV(RMSE) between traditional lens and new lens for four different simulation condition.

References

1 DisplaySearch. www.displaysearch.com
2 West, R.S., Konijn, H., Smitt, W. S., Kuppens, S., Pfeffer, N., Martynov, Y., and Takaaki Y. High brightness direct LED backlight for LCD-TV. *SID 03 Digest 43.4* (2003).
3 Wang, K., Wu, D., Qin, Z., Chen, F., Luo, X., and Liu, S. New reversing design method for LED uniform illumination. *Opt Expr.* **19** Suppl. 4, A830–A840 (2011).
4 Sun, C.-C., Lee, T.-X., Ma, S.-H., Lee, Y.-L., and Huang, S.-M. Precise optical modeling for LED lighting verified by cross correlation in the midfield region. *Opt. Lett.* **31**, 2193–2195 (2006).
5 Moreno, I. Illumination uniformity assessment based on human vision. *Opt. Lett.* **35**, 4030–4032 (2010).
6 Qin, Z., Wang, K., Wang, S., and Liu, S. Energy-saving bottom-lit LED backlight with angle-control freeform lens. *Renew. Ener. Envir.* (2011). doi:10.1364/SOLED.2011.SDThB6
7 Sun, C.-C., Chien, W.-T., Moreno, I., Hsieh, C.-T., Lin, M.-C., Hsiao, S.-L., and Lee, X.-H. Calculating model of light transmission efficiency of diffusers attached to a lighting cavity. *Opt. Expr.* **18**, 6137–6148 (2010).
8 Wang, M.-W., and Tseng, C.-C. Analysis and fabrication of a prism film withroll-to-roll fabrication process. *Opt. Expr.* **17**, 4718–4725 (2009).
9 Park, G.-J., Kim, Y.-G., Yi, J.-H., Kwon, J.-H., Park, J.-H., Kim, S.-H., Kim, B.-K., Shin, J.-K., and Soh, H.-S. Enhancement of the optical performance by optimization of

optical sheets in direct-illumination LCD backlight. *J. Opt. Soc. Korea* **13**, 152–157 (2009).

10 Lee, W. G., Jeong, J. H., Lee, J.-Y., Nahm, K.-B., Ko, J.-H., and Kim, J. H. Light output characteristics of rounded prism films in the backlight unit for liquid crystal display. *J. Info. Disp.* **7**(4), 1–4 (2006).

11 Kobayashi, S., Mikoshiba, S., and Lim, S. *LCD Backlights*. Chichester: John Wiley & Sons (2009).

12 Pan, J.-W., and Fan, C.-W. High luminance hybrid light guide plate for backlight module application. *Opt. Expr.* **19**, 20079–20087 (2011).

13 Chang, J. G., and Fang, Y. B. Dot-pattern design of a light guide in an edge-lit backlight using a regional partition approach. *Opt. Eng.* **46**(4), 043002 (2007).

14 Qin, Z., Wang, K., Chen, F., Luo, X., and Liu, S. Analysis of condition for uniform lighting generated by array of light emitting diodes with large view angle. *Opt Expr.* **18**, 17460–17476 (2010).

8

Freeform Optics for LED Automotive Headlamps

8.1 Introduction

Light-emitting diodes (LEDs) have been widely adopted in automotive lighting for many years, such as LED turning lights, high mount stop lights, taillights, backlights of dashboards, and others. However, for the most important lighting, the automotive headlamp, halogen lamps and high-intensity-discharge (HID) xenon lamp have always dominated the market, and LEDs were not considered for adoption until recent years. One of the major reasons is that the luminance or the luminous flux per a certain emitting area of a single high-power LED source increased significantly in these years, and now is able to reach around 70~100 Mnit, which larger than a halogen lamp and is comparable with HID xenon. In addition, another major reason is the dramatic drop in cost of high-power LED modules, whose price is close to or even lower than that of HID xenon lamps with the same luminance or luminous flux. High-power LEDs have recently broke through the critical point of automotive headlamp applications in the market and will rise in popularity in the near future. Moreover, besides the basic lighting function, LEDs have more design flexibilities for intelligent control, advanced front-lighting system (AFS) function, and even visible light communication function for intelligent traffic in the future. Therefore, although it has not been a long time since high-power LEDs were adopted as a light source for automotive headlamps, with the highly cost-effective LED module products and active promotion from famous car companies (e.g., Audi), LED headlamps will quickly become an automotive trend in the near future.

In this chapter, first we will introduce the optical regulations of low-beam and high-beam headlamps. Then, the specific requirements of LED packaging modules for headlamp application as well as related application-specific LED packaging (ASLP) will be introduced. Based on basic information on LED headlamps, different kinds of optical design methods based on freeform optics will be introduced, including a Fresnel freeform lens, a freeform optics integrated poly-ellipsoid system (PES), freeform reflectors, and low-beam–high-beam integrated double-beam freeform optics.

8.2 Optical Regulations of Low-Beam and High-Beam Light

8.2.1 Low-Beam

Usually the headlamp consists of low-beam light, high-beam light, and fog light. There are many regulations, such as Economic Commission for Europe (ECE) regulations in

Freeform Optics for LED Packages and Applications, First Edition. Kai Wang, Sheng Liu, Xiaobing Luo and Dan Wu.

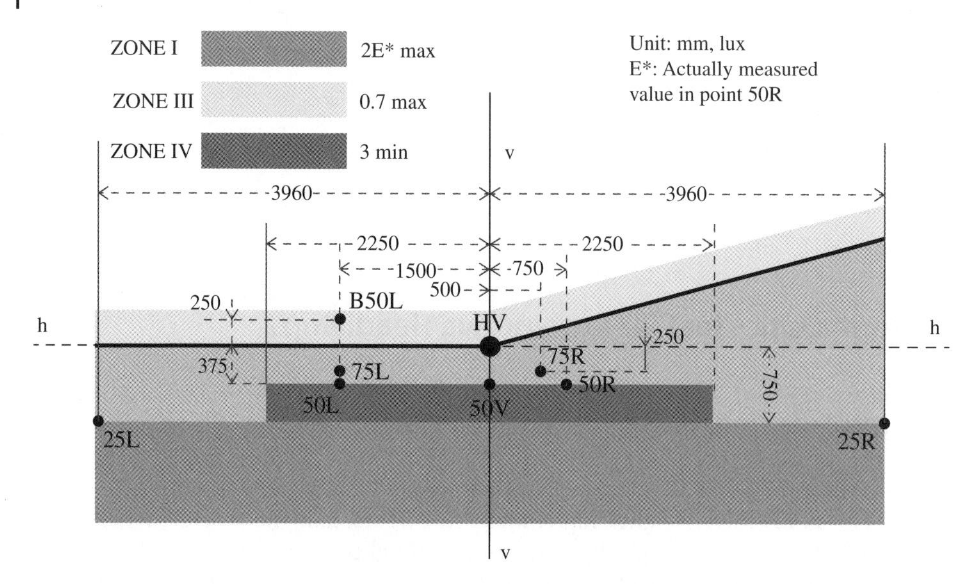

Figure 8.1 Low-beam pattern on the measuring screen provided by ECE R112 Regulation (reproduced from Ref. [1]).

Europe, SAE International (Society of Automotive Engineers) regulations in the United States, and Guobiao (GB, or National Standards) regulations in China, to specify the optical performance of headlamps. As low-beam light should give adequate illumination within a safe braking distance and cannot cause glare for drivers in the opposing lane, there are many restrictions for low-beam light in various regulations. That is why the optical design of low-beam light is the most difficult part in design of the whole lamp. Figure 8.1 shows the low-beam pattern provided by ECE R112 regulation,[1] which specifies the range of illuminance on specific points or segment areas and defines the minimum gradient of the cutoff line.

In order to avoid glare, the minimum gradient of the horizontal cutoff line is claimed in the regulation. In ECE R112 regulation, the gradient is normally defined at 2.5° to the left of the v–v line by searching the maximum value G obtained by scanning the illuminance at different vertical angles β as shown in Figure 8.2, which is expressed in Eq. 8.1. The legal minimum value of G is 0.13. Moreover, the major specifications of low-beam are shown in Table 8.1, which is an important standard to evaluate optical designs that will be introduced in this chapter.

$$G = \log E_{\beta} - \log E_{(\beta+0.1^{\circ})} \tag{8.1}$$

8.2.2 High-Beam

The optical regulation of high-beam is relatively simpler than that of low-beam, as shown in Table 8.2.

8.2.3 Color Range

Besides the shape and intensity of light emitted from a headlamp, the color performance is another important issue to regulate the headlamp's illumination effects.

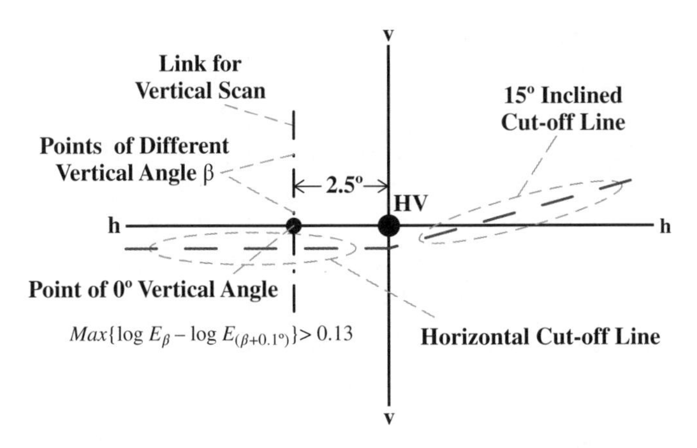

Figure 8.2 Sketch diagram of a gradient of the horizontal cutoff line.

Table 8.1 Value of ECE R112 (Class B) for low-beam light.

Point on measuring screen	Required illumination in lx
Point B50L	≤0.4
Point 75R	≥12
Point 75L	≤12
Point 50L	≤15
Point 50R	≥12
Point 50V	≥6
Point 25L	≥2
Point 25R	≥2
Any point in zone I	$\leq 2 \cdot E^*$
Any point in zone III	≤0.7
Any point in zone IV	≥3

Note: E^* is the actually measured value in point 50R.

Table 8.2 Value of ECE R112 (Class B) for high-beam light.

Point on measuring screen	Required illumination in lx	Simulated illumination in lx
E_{max}	≥48 and ≤240	55.1
Point HV	$\geq 0.8\ E_{max}$	51.2
Point HV to 1125L and R	≥24	≥33.3
Point HV to 2250L and R	≥6	≥28.7

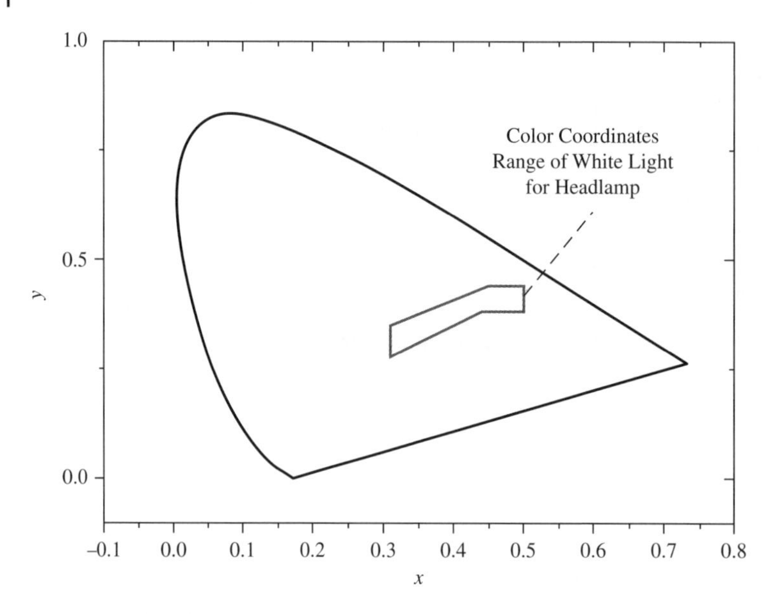

Figure 8.3 White-light color coordinate range for an LED automotive headlamp.

The regulations have also provided the color coordinate range for headlamp application. The white xy color coordinates should follow the range in Eq. 8.2:

$$\begin{aligned} & x \geq 0.310 \\ & x \leq 0.500 \\ & y \leq 0.150 + 0.640x \\ & y \leq 0.440 \\ & y \geq 0.050 + 0.750x \\ & y \geq 0.382 \end{aligned} \tag{8.2}$$

And the color coordinate range is also shown in Figure 8.3.

8.3 Application-Specific LED Packaging for Headlamps

Many high-power LED packaging types exist in the market for LED general lighting, including types based on lead frame, ceramic board, metal board, and chip-on-board (CoB). However, most of them are unqualified for LED headlamp application due to low luminance. In this section, we will firstly analyze the requirements in both optical and thermal fields in theory. With the results of theoretical analysis, we will design an ASLP module with optimized structure parameters and appropriate packaging materials for headlamp applications. This ASLP LED module will be used as a light source for many optical designs in this chapter.

8.3.1 Small Étendue

From nonimaging optics, we know that the Étendue (E) of light cannot be reduced without loss in an optical system.[2,3] In other words, it means that the Étendue of a light source (E_{Source}) should not be more than the Étendue of an optic system (E_{System}) to ensure high light transmittance efficiency. This relationship can be described by Eq. 8.3.

$$E_{Source} \leq E_{System} \tag{8.3}$$

Étendue E is defined to describe the extension of lights in both area and angle, and it is expressed as follows:

$$E = n^2 \iint d\Omega \cdot dA \tag{8.4}$$

where n is the refraction index of the material the light source is embedded in (for traditional packaged LED, this material is silicone); Ω is the solid angle occupied by lights; and A is the emitting area.

As most LED modules are Lambertian emitters, E_{Source} can be expressed as:

$$E = n^2 \cdot \pi \cdot \sin^2\alpha \cdot A \tag{8.5}$$

where α is the half angle of the cone the light is in (if an LED emits light in a half space, α is 90°). In automotive headlamps application, the divergence angle of lights in the vertical direction is restricted strictly to avoid glare to drivers in the opposing lane. Additionally, the size of the lamp aperture is limited by vehicle design styling. Therefore, the Étendue of a lamp's optic system is always limited. From Eq. 8.3, we can conclude that LED packaging with a small Étendue is needed for high optical efficiency in headlamp application. Equation 8.5 shows that the Étendue of LED packaging can be decreased by using encapsulating material with a small refraction index n and decreasing the emitting area.

8.3.2 High Luminance

For low-beam light, the light intensity varies from a range of several hundreds to 20,000 cd. At a certain direction, the light intensity is very large, which is called a *hotspot* (e.g., a 75R point), while only about 300 cd in zone III in Figure 8.1. From Ref. [2], we can get the light intensity of a hotspot as Eq. 8.6:

$$I_{hotspot} = \frac{1}{\Omega_{hotspot}} L_s A_s \Omega_s = A_{lamp} L_s \tag{8.6}$$

where $\Omega_{hotspot}$ and Ω_s are the solid angular spreads of the hotspot and light source; L_s is luminance of the LED; and A_s and A_{lamp} are the apertures of the LED and lamp. Therefore, in order to achieve a large light intensity at the hotspot, the aperture of the lamp should be large enough. However, the aperture of the lamp is always limited by the vehicle styling design. Hereby, the luminance of the LED should be increased greatly. For Lambertian emitters, L_s can be expressed as follows:

$$L_s = \frac{I_0}{A_s} = \frac{\phi_s}{\pi A_s} \tag{8.7}$$

In order to increase the luminance, LED packaging should emit more light flux Φ_s and have a small emitting area A_s. For LED packaging, the efficient way to decrease the emitting area is conformal coating technology.

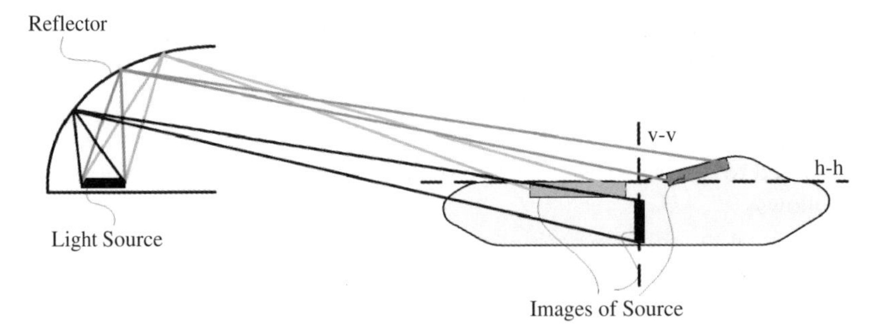

Figure 8.4 Schematic diagram of overlapping images of a light source on a test screen.

8.3.3 Strip Shape Emitter with a Sharp Cutoff

This method, in common use for optical design, overlaps the images of a light source on a test screen to form a regulation-compliant beam pattern. As shown in Figure 8.4, since the beam patterns specified by regulations have a larger spread in the horizontal direction than in the vertical direction, the light source should be a strip-shaped emitter. Additionally, as low-beam light should produce a sharp horizontal cutoff line on the test screen, the emitting area of the light source should also provide a sharp cutoff in luminance to form images with sharp cutoff too. In conventional LED packaging, there is a phosphor layer with larger size than the LED chip, which increases the emitting area and blurs the cutoff.[4] That is why LED of traditional packaging is not suitable for headlamps.

8.3.4 Small Thermal Resistance of Packaging

Besides the requirements in optics, there are also many requirements on the thermal performance of LED packaging. As a headlamp is usually installed in the engine compartment, the ambient temperature of the headlamp can reach 85 °C.[5] The optical performance of LEDs may degenerate greatly when working at a high ambient temperature. Therefore, LED packaging should have a low thermal resistance to keep a low junction temperature for the LED chip. In automotive headlamp application, as one of the requirements in optics, an LED chip array should be arranged densely, which increases the difficulty of thermal management greatly. Therefore, some packaging structures with little thermal resistance are preferred in this ASLP, such as CoB technology.

8.3.5 ASLP Design Case

In order to comply with the requirement in optics, we design a new LED packaging with small Étendue and high luminance. Additionally, the emitting area of this LED packaging is a rectangular area with a very sharp cutoff. Figure 8.5 shows the structure of this LED packaging.

In this ASLP, 10 conventional high-power blue LED chips are arranged in a 2 × 5 array and covered by a phosphor layer with conformal coating. Gold wires connect the 10 chips into two series circuits (five chips for each branch) and are also packaged in the phosphor layer. As the phosphor layer is exposed in the air directly, the refraction index

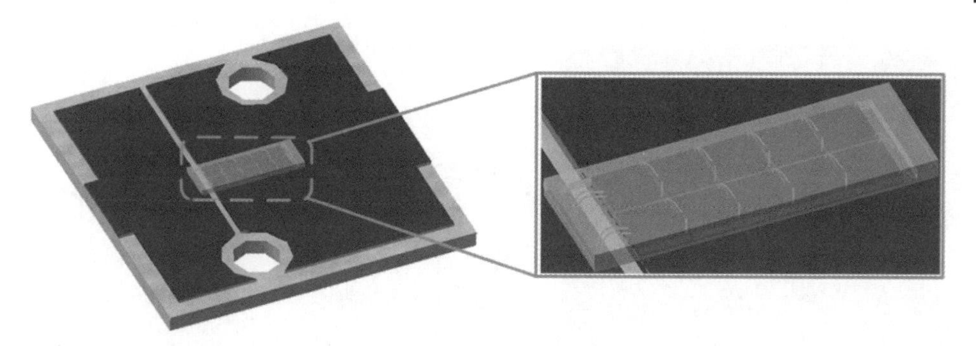

Figure 8.5 Schematic diagram of LED packaging.

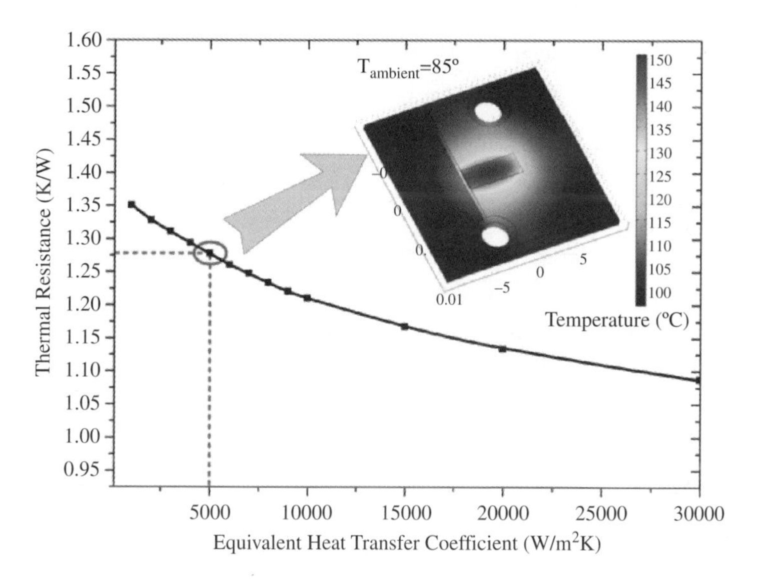

Figure 8.6 Simulation results on thermal resistance and heat distribution of LED packaging.

n is 1. Furthermore, the sizes of the emitting area for both blue light and yellow light are as small as the size of the LED chips array (~2.1 × 5.3 mm); therefore, this packaging has a small Étendue. Also, this packaging can provide a rectangular emitting area with a sharp cutoff.

The board used in the packaging is direct bond copper (DBC) board, and LED chips are bonded on the board directly. The thicknesses of both the ceramic and copper are optimized to obtain minimum thermal resistance. We simulate the thermal resistances of LED packaging at different equivalent heat transfer coefficients (EHTCs) loaded at the bottom surface of the packaging. We get the heat distribution on the LED packaging when the EHTC is 5000 W/(m^2·K) and the ambient temperature is 85 °C. From the simulation results, the thermal resistance can be less than 1 K/W when the EHTC of the cooling system is large enough. The simulation result is shown in Figure 8.6.

We also have made some packaging samples, shown in Figure 8.7. The test results of our samples are listed in Table 8.3. In the table, xy is the chromaticity coordinates,

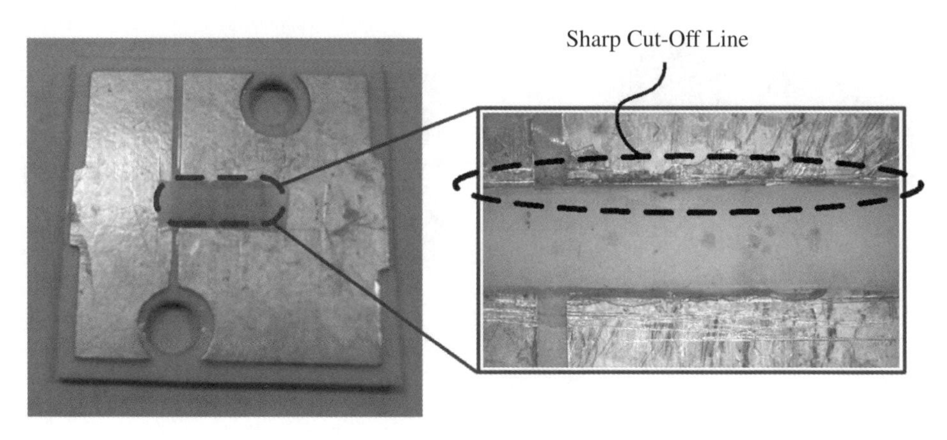

Figure 8.7 Picture of an ASLP sample with a sharp cutoff line.

Table 8.3 Test results of the ASLP sample.

x	*y*	T_C (K)	Φ(lm)	R
0.3389	0.3549	5244	1062	13.4%

T_C is CCT, Φ is luminous flux, and R is the percentage of red light that is required to be greater than 5% according to ECE regulation.

8.3.6 Types of LED Packaging Modules for Headlamps

In recent years, some large LED manufacturers have released different special LED packaging types for headlamps. There are two main types of LED packaging for headlamp application. The first one is LED packaging with a single LED die. In this packaging, as the emitting area is quite smaller than the optical system, the light source can be treated as a point light source approximately. Another type is packaging with an LED die array. In this type of LED packaging, multiple LED dies are arranged in a rectangular area, and this type of LED can be treated as a line source in optical design. A sharp cutoff edge of the emitting area is needed to easily form a light pattern with a horizontal cutoff line by the optical system. In order to decrease the Étendue and increase luminance of the light source, no injection silicone and conformal coating technology are used in both types of LED packaging. Figure 8.8 shows three different LED packaging modules for headlamp application in the market, wherein Figure 8.8a refers to LED packaging of the point source type with a single LED die, and Figure 8.8b and 8.8c refer to LED packaging of the line source type with four LED dies.

According to optical principles, LED packaging of these two types is very suitable for headlamps. However, as there are many differences between these special LED packaging and conventional light sources of headlamps (halogen bulb or HID) in light viewing angles and in sizes of emitting area, the optical efficiency will be very low even if it does not work when using an existing optical system with special LED packaging directly. Therefore, novel optics or improved optical systems from existing ones should

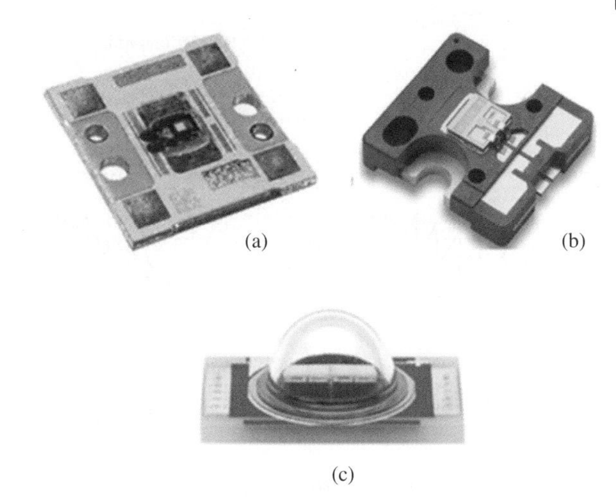

Figure 8.8 Three different LED packaging modules for headlamp application in the market. Pictures from (a) OSRAM, (b) Lumileds, and (c) Nichia.

be developed for different types of LED packaging. Freeform optical design methods with the LED point source type will be introduced in Section 8.4, and methods with the LED line source type will be presented in Sections 8.5 to 8.7.

8.4 Freeform Lens for High-Efficiency LED Headlamps

8.4.1 Introduction

In LED headlamps, if the optical efficiency of the optical device is low, more LEDs are needed to comply with the regulation, which will raise the electrical power consumption of the headlamp greatly. As a small part of the electrical power is converted to light, the headlamp will bring great heat. Usually a headlamp is installed near the engine; therefore, the typical ambient temperature can be about 80 °C. If the headlamp generates a large amount of heat, it will be a heavy load to the cooling system. In addition, if more LEDs are used, the size and price will also increase. Therefore, it is quite necessary to enhance the optical efficiency of the optical device.

In recent years, with the development of nonimaging optics, some LED headlamp optical design methods based on nonimaging optics have been suggested. Oliver Dross et al. have invented an excellent approach to designing the optical device of an LED headlamp based on the simultaneous multiple surfaces (SMS) method.[6] In their design, the optical efficiencies of both low-beam and high-beam lamps are more than 75%, which are very high values so far. In addition, Fresnel lens and micro-lens arrays have been attempted in low-beam lens designs.[7] In this section, we will present a new method of design of both low-beam and high-beam freeform lenses, which can fully comply with the ECE regulation without any other lens or reflector. Through the numerical simulation, both lenses have optical efficiencies of more than 88%.

8.4.2 Freeform Lens Design Methods[8]

In our design, the lens is separated into two parts: collection optics and refraction optics (shown in Figure 8.9). The collection optics is used to collimate the lights radiating from

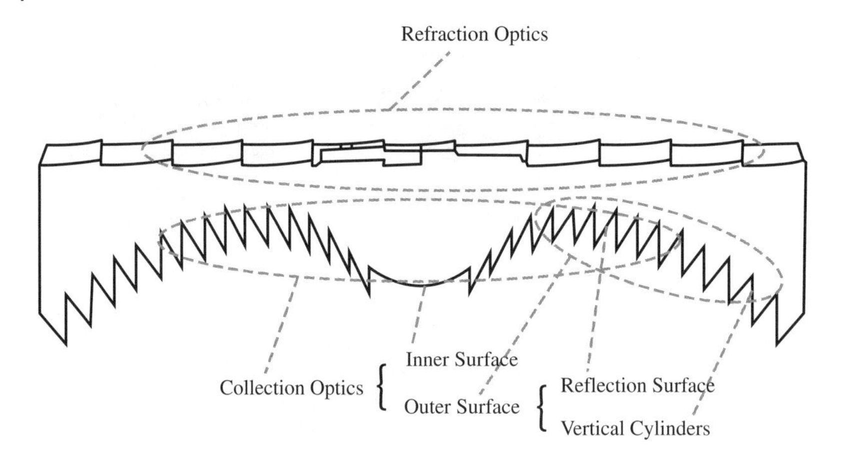

Figure 8.9 Sketch diagram of the lens' structure.

the LED, and the refraction optics is used to redirect the collimated rays to obtain a regulation met beam pattern.

8.4.2.1 Design of Collection Optics

Collection optics is used to collimate lights radiated from LEDs. There are two main methods to collimate rays: a total internal reflection (TIR) lens and total internal reflection–refraction (TIR-R) lens, also called a Fresnel lens. The TIR lens works well when the light source can be regarded as a point source, but if the source has a finite size, rays cannot be collimated accurately. However, the Fresnel lens structure can meet the properties of an extended light source like a 1 ×1 mm LED chip.[9] The Fresnel lens consists of two parts, an inner surface and outer surface, which are shown in Figure 8.4.

The Cartesian oval method is always used to design lenses for concentrating or collimating lights.[10,11] In this design, we use the Cartesian oval method to calculate the inner surface, which is constructed in the following steps. First of all, as shown in Figure 8.10, we fix a point as the vertex of the surface's curve that is the first point (P_i^1) on the inner surface, and the normal vector of this point is also determined to be vertical up. The second point on the curve can be determined by the intersection of the incident ray and the tangent plane of the previous point. Secondly, as the exit ray is vertical up, we calculate the present point's normal vector by incident ray and exit ray using the inverse procedure of Snell's law. Finally, we can get all the points and their normal vectors on the inner surface's curve in this chain of calculation. Then we fit these points to form the curve and rotate this curve to get the inner surface.

The outer surface uses the rays' TIR to collimate the incident rays. Incident rays are refracted first by the vertical cylinder surface and then reflected vertically by TIR on the reflection surface, which is shown in Figure 8.9. For the outer surface, we just need to calculate the points on the curve of the reflection surface. The calculation procedure is specified as follows. Firstly, we fix a point as the first point P_T^1 of the curve, and then calculate the normal vector of this point according to the vertical exit ray and the inverse procedure of Snell's law. Secondly, the second point on the curve can be determined by the intersection of the incident ray, which is refracted by the cylinder surface, and the tangent plane of the previous point. Thirdly, we use the inverse procedure of Snell's law

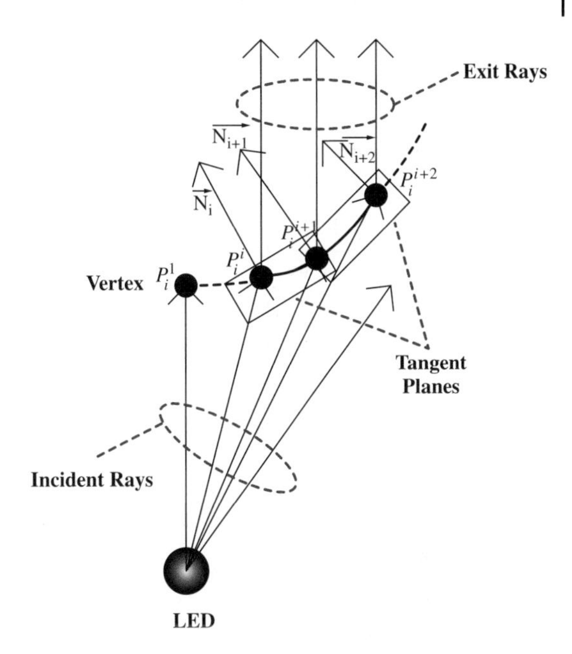

Figure 8.10 Calculation of the points on the inner surface's curve.

again to calculate the normal vector of the second point. Therefore, we can obtain all the points on the curve in the same manner, and then the reflection surface is obtained. The calculation procedure is denoted in Figure 8.11.

We design the collection optics using the method mentioned here. The collection optics collected the lights with a range of 0° to 80°, lights that take up about 97% of the whole lights radiated by LEDs.

8.4.2.2 Design of Refraction Optics

Freeform micro lenses are sometimes used as the refraction optics in the illumination optical design.[7,12] In Ref. [7], a microlens array is also used to design the low-beam lens, but the lens cannot fully comply with the ECE regulation and the optical efficiency is not quite high (70%). The main disadvantage of freeform microlens arrays is that high-precision machining is needed to fabricate the micro surfaces. In this section, we use several freeform surfaces to construct the refraction optics. The procedure of designing the freeform surface consists of meshing the target plane and calculating the points on the surface.

Step 1. Meshing of the target plane. As the beam pattern of the headlamp can be regarded as a nonuniform rectangular pattern or be resolved into several nonuniform rectangular patterns such as a low-beam light pattern, the freeform surface also has a rectangular appearance. In this chapter, we can obtain a nonuniform beam pattern by making each incident ray intersect at different points on the target plane, which is shown in Figure 8.12. Assume there are $M \times N$ rays refracted by a freeform surface.

The intersection points are determined by meshing the target plane. A meshing method of unequal area grids on the target plane is adopted in this design. According to the regulation, illuminance in the center area is always higher than that in the outer area on a measuring screen, so that the grids in the center of the target plane are small.

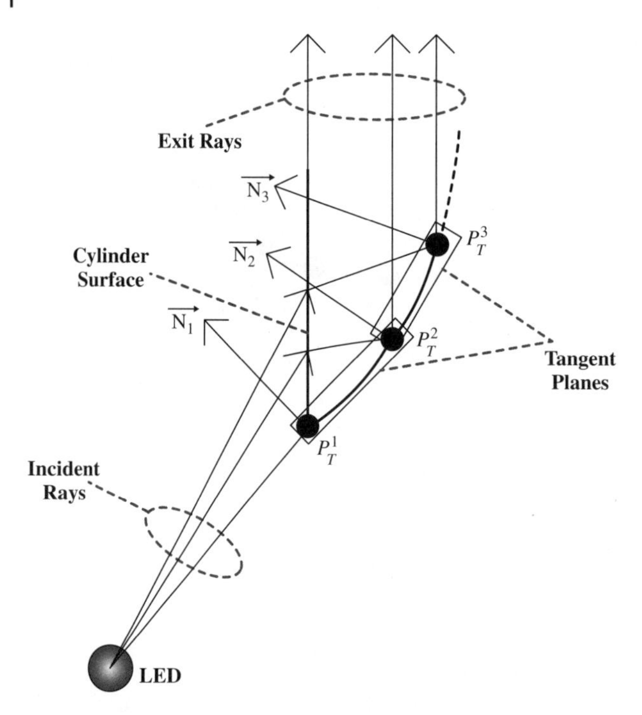

Figure 8.11 Calculation of the points on the TIR surface's curve.

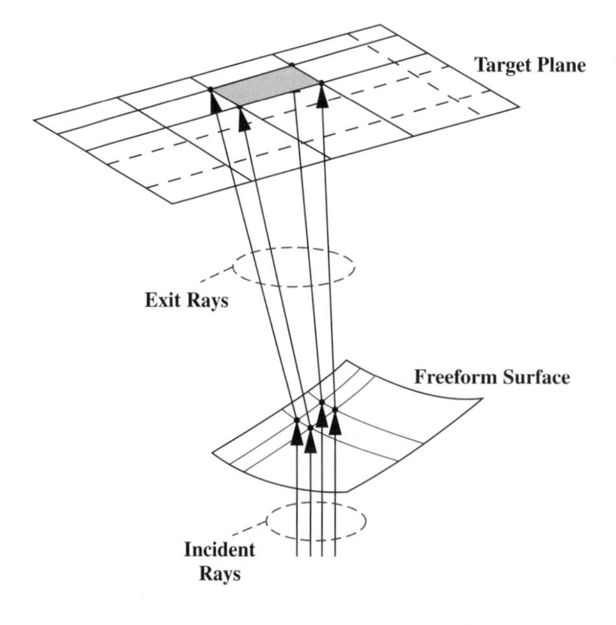

Figure 8.12 Control of the incident collimating rays.

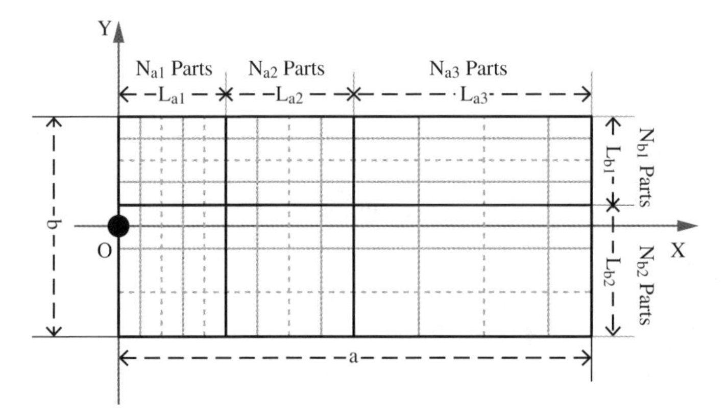

Figure 8.13 Sketch diagram of meshing the target plane.

Now we use several parameters to mesh the target plane nonuniformly. Half of the target plane is divided into three parts in length and two parts in width. The method of meshing the target plane is shown in Figure 8.13.

In Figure 8.13, a is the half length of the target plane; b is the width of the target plane; L_{a1}, L_{a2}, and L_{a3} are the lengths of the subregions; and L_{b1} and L_{a2} are the widths of the subregions on the target plane. L_{a1} is divided into N_{a1} equal copies, and L_{a2}, L_{a3}, L_{b1}, and L_{a2} are also divided into different parts equally, which is shown in Figure 8.13. Therefore, we can obtain the length and width of each grid by Eq. 8.8 and Eq. 8.9.

$$\begin{cases} a_1 \times N_{a1} = L_{a1} \\ a_2 \times N_{a2} = L_{a2} \\ a_3 \times N_{a3} = L_{a3} \end{cases} \tag{8.8}$$

$$\begin{cases} b_1 \times N_{b1} = L_{b1} \\ b_2 \times N_{b2} = L_{b2} \end{cases} \tag{8.9}$$

In the same way, we can mesh the other half of the target plane. Through optimizing these parameters, we can design the rectangular beam pattern that is needed for headlamps.

Step 2. Calculation of a freeform surface. The freeform surface is constructed by lofting with a large number of curves on the surface. We used the seed curve iteration method to design a freeform LED packaging lens in Chapter 3. In this section, this method is used again to calculate the freeform surface. The procedure of calculation is shown in Figure 8.14.

8.4.3 Design Case of a Freeform Lens for Low-Beam and High-Beam

In this section, we have designed a low-beam lens and a high-beam lens using the method introduced in this chapter. To validate our design, an OSRAM OSTAR headlamp LED is used in the numerical simulation. According to the data sheet of this LED, its flux can reach 224 lm when the LED is driven at a specified current of 700 mA in 2010.[13] We assume the flux of each LED is 200 lm in the simulation. The material used

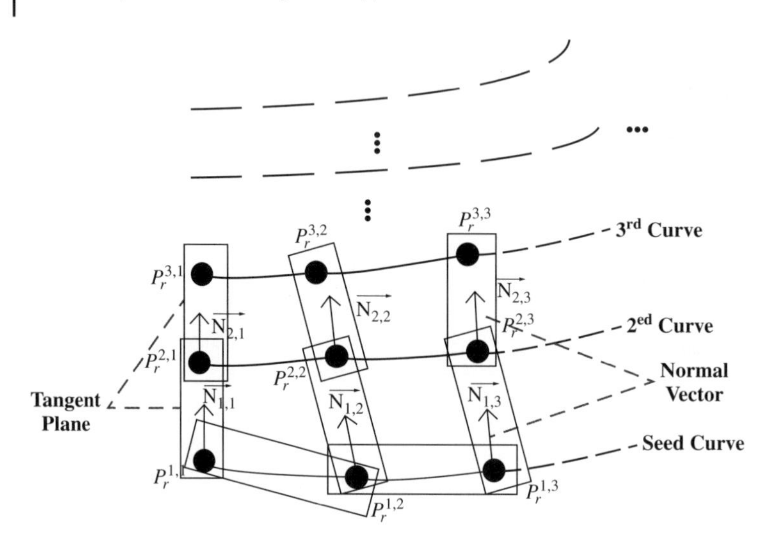

Figure 8.14 Calculation of the freeform surface.

in the design of both the low-beam and high-beam lens is PMMA, the refractive index of which is 1.493.

8.4.3.1 Design of a Low-Beam Lens

As a 15° inclined cutoff line is needed for a low-beam light pattern in regulation, the refraction optics is divided into two parts. The first part generates the horizontal beam pattern, and the second part is used to generate a 15° inclined beam pattern; therefore, a regulation met beam pattern can be obtained by overlapping the two parts' beam patterns. In our design, the first part consists of a large freeform surface, and the second part is made up of six same freeform surfaces. The included angle of the two parts is 15°. Although the second part is excised in the center area and border, the beam pattern of this part can still be accepted as the overlap effect of the six freeform surfaces. The diameter of the low-beam lens is 78.9 mm, and the thickness is 26.52 mm. Figure 8.15 shows the low-beam lens.

Through the Monte Carlo ray-tracing method based numerical simulation, we can obtain the illuminance distribution on the measuring screen. Figure 8.16 shows the simulated beam pattern on the measuring screen and the vertical sectional curve.

In the simulation, flux of the light source is 400 lm. Therefore, two LEDs and two low-beam lenses are used for low-beam light, which are shown in Figure 8.17. The simulation result shows the optical efficiency of the low-beam is 88% (Fresnel losses included), and the maximum value of G is 0.35. The optical efficiency in this study is defined as the ratio of flux into the 20 × 10 m area on a measuring screen to the flux of the LED.

The comparison of simulated illumination and required illumination in lx for ECE R112 regulation is shown in Table 8.4, which indicates the optical performance of the low-beam lens can fully comply with the ECE regulation.

As the regulation for low-beam light has strict restrictions, we have considered the splitting of lights at the air–PMMA interface by using the Fresnel formulae during the ray trace of the above simulation. Compared with the simulation, in which the splitting

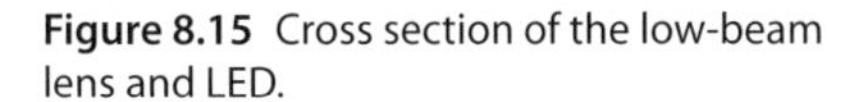

Figure 8.15 Cross section of the low-beam lens and LED.

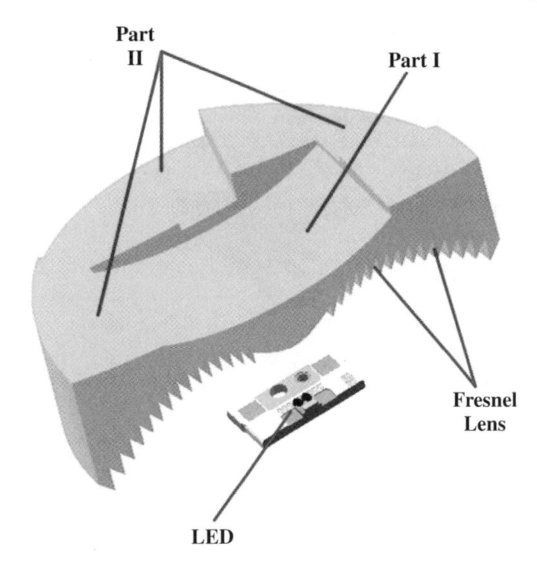

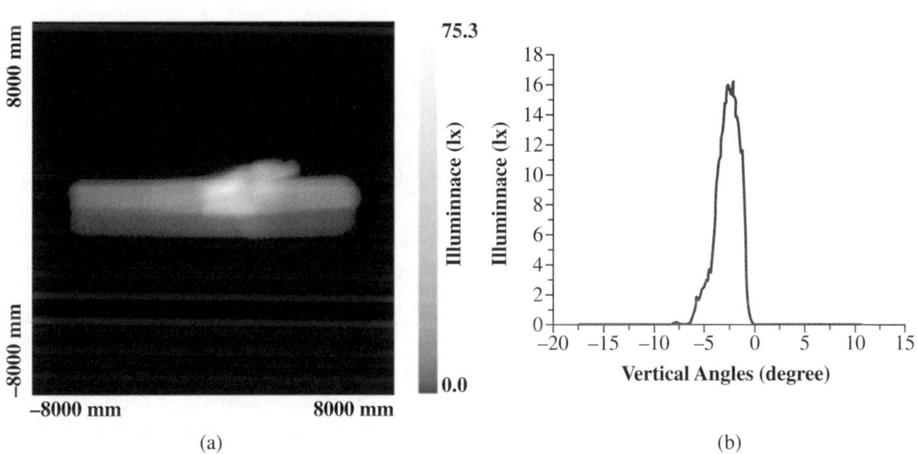

Figure 8.16 (a) Simulated illuminance distribution on the measuring screen, and (b) vertical sectional curve.

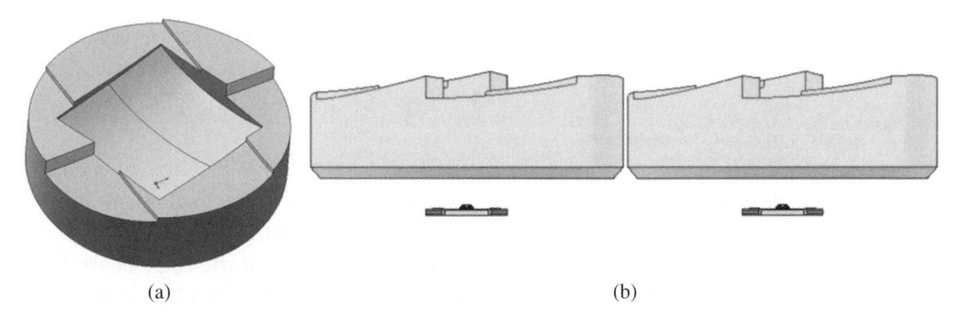

Figure 8.17 Schematic view of (a) low-beam lens, and (b) low-beam light.

Table 8.4 Simulated illumination result compared with the corresponding value of ECE R112 (Class B) for low-beam light (sharp edges).

Point on measuring screen	Required illumination in lx	Simulated illumination in lx
Point B50L	≤0.4	0.02
Point 75R	≥12	26.11
Point 75L	≤12	1.64
Point 50L	≤15	6.18
Point 50R	≥12	49.72
Point 50V	≥6	27.27
Point 25L	≥2	7.48
Point 25R	≥2	14.75
Any point in zone I	≤2·E*	√
Any point in zone III	≤0.7	√
Any point in zone IV	≥3	√

Note: E* is the actually measured value in point 50R.

of lights is not considered, the optical efficiency decreases from 94% to 88%. The cutoff line is blurred, and illuminance at point 75R decreases from 35.23 lx to 26.11 lx when splitting of lights is considered.

Another issue we must consider is the effect of the rounded edges of the lens facet on the optical performance of the lens. Therefore, we have also simulated the low-beam with rounded edges. In this simulation, the radii of the edges or corners on the refraction optics are 0.3 mm, the tooth radii of the Fresnel lens are set as 0.05 mm, and also the splitting of lights is considered during the ray tracing. Figure 8.18 shows rounded edges of the low-beam lens. The simulated beam pattern of a low-beam lens with rounded edges is shown in Figure 8.19.

Figure 8.19 shows that disordered lights appear on the measuring screen, and the max illuminance is decreased by 7.4%. The comparison of simulated illumination of a low-beam lens with rounded edges considered and required illumination in lx for ECE R112 regulation is shown in Table 8.5. From the simulation, a low-beam lens with rounded edges has an optical efficiency of 79%, and the illuminance on the measuring screen is lower than the case in which the edges of the low-beam lenses are not rounded. However, a low-beam lens with rounded edges shown in Figure 8.19 can still comply with the regulation.

8.4.3.2 Design of a High-Beam Lens

As restraints for high-beams are looser in the ECE R112 regulation, the size of the lens can be smaller than that of a low-beam lens. In this design, the refraction optics is also divided into two parts. The center area of the refraction optics is a freeform surface, and the outer part is a flat surface. The diameter and thickness of the high-beam lens are, respectively, 25 mm and 6.9 mm. Figure 8.20 shows the high-beam lens and simulated beam pattern on a measuring screen correspondingly. The optical efficiency of the high-beam lens is as high as 92% (Fresnel losses included).

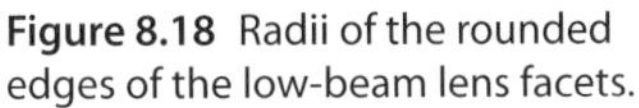

Figure 8.18 Radii of the rounded edges of the low-beam lens facets.

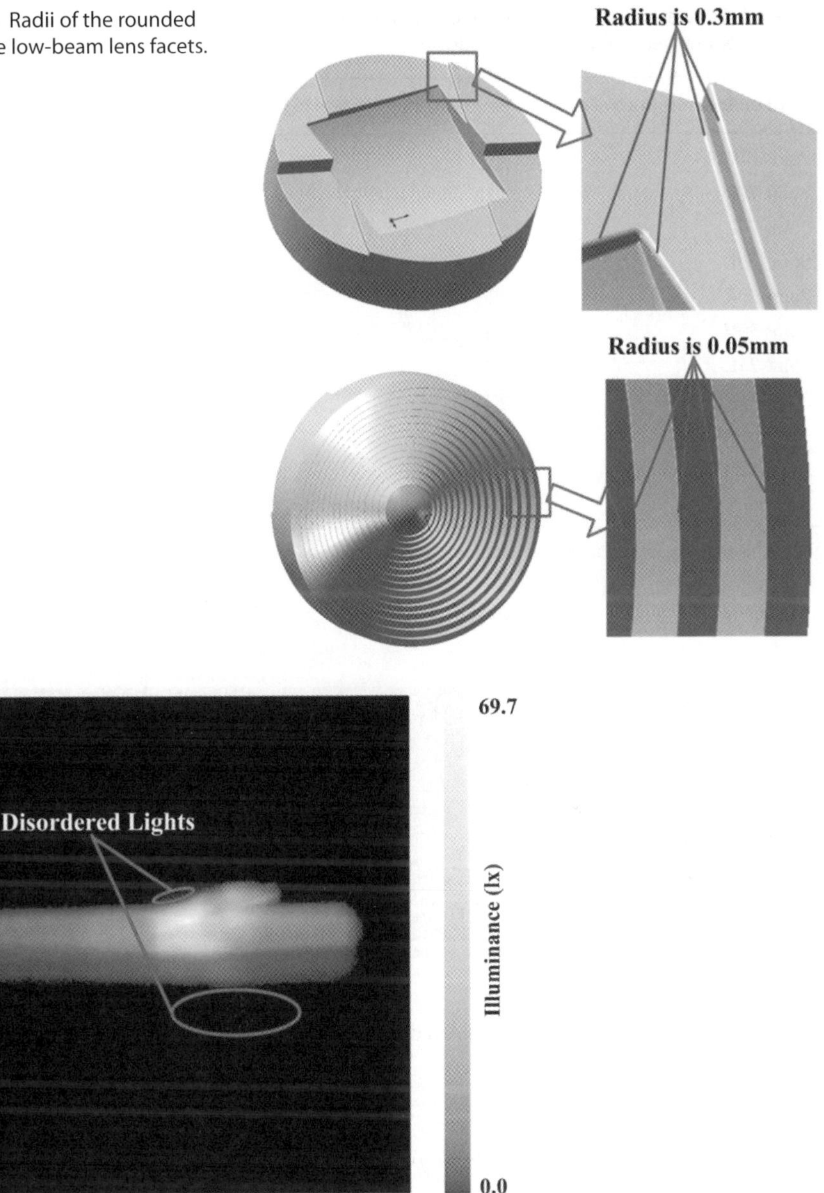

Figure 8.19 Simulated beam pattern of a low-beam lens with rounded edges.

The simulated illuminance on a measuring screen is listed in Table 8.6, and the flux of the light source is 600 lm. Therefore, three LEDs are needed in high-beam light. The simulation result indicates the optical property of the high-beam lens can also fully comply with the regulation.

In this section, a high-efficiency LED headlamp freeform lens design method is presented. There are abundant parameters to optimize the illuminance distributed on the

Table 8.5 Simulated illumination result compared with the corresponding value of ECE R112 (Class B) for low-beam light (rounded edges).

Point on measuring screen	Required illumination in lx	Simulated illumination in lx
Point B50L	≤0.4	0.04
Point 75R	≥12	23.24
Point 75L	≤12	1.37
Point 50L	≤15	5.49
Point 50R	≥12	45.10
Point 50V	≥6	24.30
Point 25L	≥2	12.20
Point 25R	≥2	13.08
Any point in zone I	≤2·E*	√
Any point in zone III	≤0.7	√
Any point in zone IV	≥3	√

Note: E* is the actually measured value in point 50R.

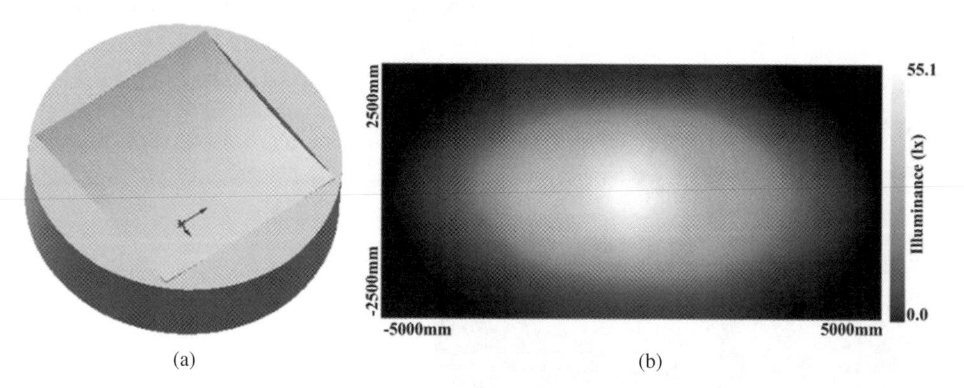

Figure 8.20 (a) High-beam lens, (b) simulated beam pattern of a high-beam lens on a measuring screen.

Table 8.6 Simulated illumination result compared with the corresponding value of ECE R112 (Class B) for high-beam light.

Point on measuring screen	Required illumination in lx	Simulated illumination in lx
E_{max}	≥48 and ≤240	55.1
Point HV	≥0.8 E_{max}	51.2
Point HV to 1125L and R	≥24	≥33.3
Point HV to 2250L and R	≥6	≥28.7

measuring screen. Using this method, the optical lens of both low-beam and high-beam light can be designed. Through numerical simulation, the headlamp with these lenses can fully comply with the ECE regulation without any other lens or reflector. And simulation results show the two lenses have optical efficiencies more than 88% in theory. However, it will take a relatively long time to optimize all the parameters; we hope to solve this problem in future work by letting the optimization performed by computers automatically shorten the design cycle.

8.4.4 Design Case of a Freeform Lens for a Low-Beam Headlamp Module

In order to reduce the difficulty of fabrication of freeform lenses, we can separate the freeform surfaces that can generate a horizontal rectangular light pattern and the 15° inclined rectangular light pattern. In this way, the top surface of the freeform lens will be a continuous smooth surface. This will not only reduce the fabrication difficulty but also improve the aesthetics. In order to coordinate with the Fresnel lens, the four corners of the freeform surface are cut, which does not affect the cutoff line and the illuminance distribution in the center area. Figure 8.21a and 8.21b show, respectively, the freeform lens that can produce a horizontal rectangular light pattern and a 15° inclined rectangular light pattern. With a certain combination method, an illumination light pattern can be generated that complies with the regulations. Figure 8.21c shows the use of three

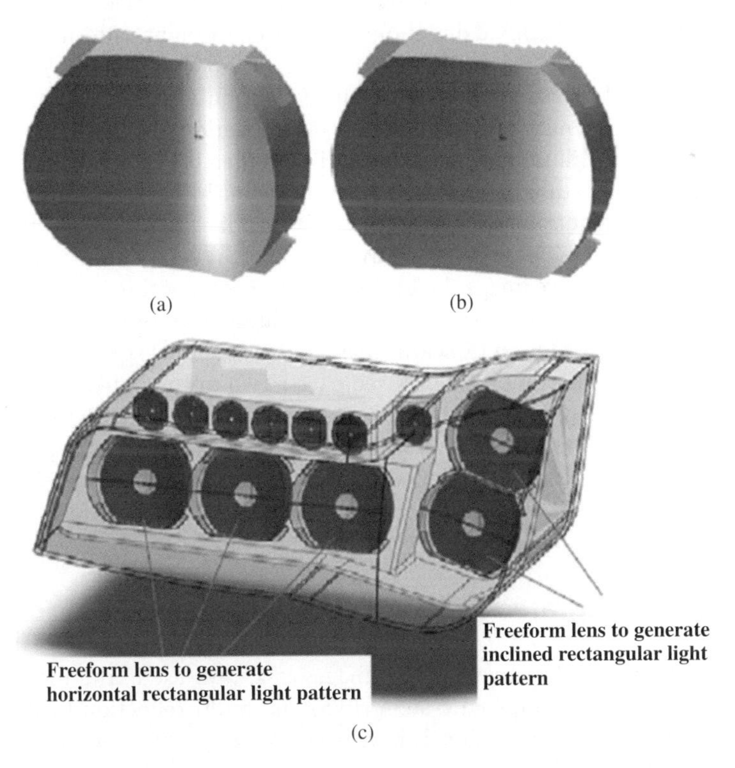

Figure 8.21 (a) Freeform lens that generates a horizontal rectangular light pattern, (b) freeform lens that generates a 15° inclined rectangular light pattern, and (c) schematic of a low-beam headlamp module with freeform lenses.

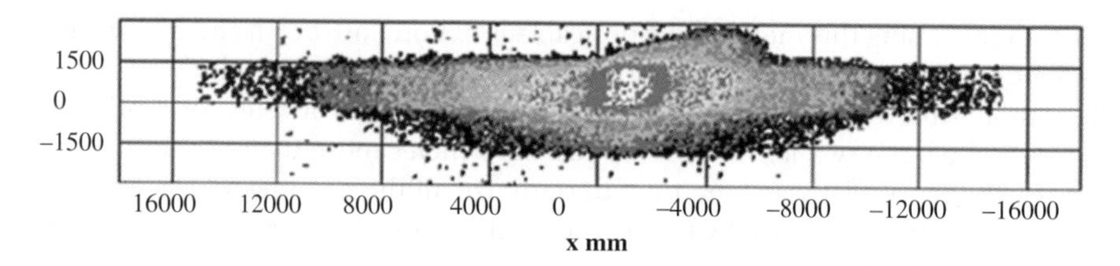

Figure 8.22 Illuminance of a smooth top surface freeform lens.

freeform lenses with three LED modules to generate a horizontal light pattern with a clear cutoff line, and the other two to generate a 15° inclined light pattern. The generated inclined rectangular light pattern is inclined 15° with respect to the lens, which produces a horizontal light. Due to using a large number of LEDs in the optical design, low luminous flux LED modules will meet the design requirement (relative to the two LED light intensity distribution designs discussed here). Therefore, a small LED current drive is enough to reduce the difficulty of thermal management. Meanwhile, separated LED and optical system methods will facilitate the outer shape design flexibility of LED lights and AFS characteristics.

The system optical efficiency of the module reaches as high as 85%. The illumination performance on the test screen of the low-beam headlamp module with freeform lenses is shown in Figure 8.22. Simulation results demonstrate that there exist very clear cutoff lines at both the horizontal and 15° inclined directions. Moreover, the light pattern also has a higher width.

However, in the design of an LED low-beam light intensity distribution curve, illuminance of the 75R point usually is considered as an important evaluation of light pattern quality. In order to guarantee the illuminance in zone III is less than 0.7 lx, and due to the 75R point near the cutoff line, the illuminance of the 75R point has great difficulty reaching the high illuminance requirement or even the 12 lx standard requirement. In order to improve illuminance at the 75R point, we introduce the imaging method in the freeform lens design. That is to say, we image the LED chip at the point of the 75R position, and therefore this method can effectively improve the illuminance of point 75R. Figure 8.23 shows the freeform lens for an LED low-beam with a spherical surface. The top surface of this freeform lens is composed of three parts. Two of them, like those mentioned, will generate the horizontal and inclined rectangular light pattern. Besides, there is a spherical lens in the center of the freeform lens, and its function is imaging the LED chip to the targeted 75R position. In this way, the illuminance value at the 75R point can reach as high as 32.66 lx.

8.5 Freeform Optics Integrated PES for an LED Headlamp

For a line light source, the method of overlapping images is always used. The most common optical system is the poly-ellipsoid system (PES) and multi-reflection (MR) basing on this method. These two methods already have been applied to traditional light sources for many years. However, the viewing angle of the LED and traditional light source is quite different. Therefore, the PES and MR optical system should be improved to apply to LED packaging of the line light source type. We will introduce the improved

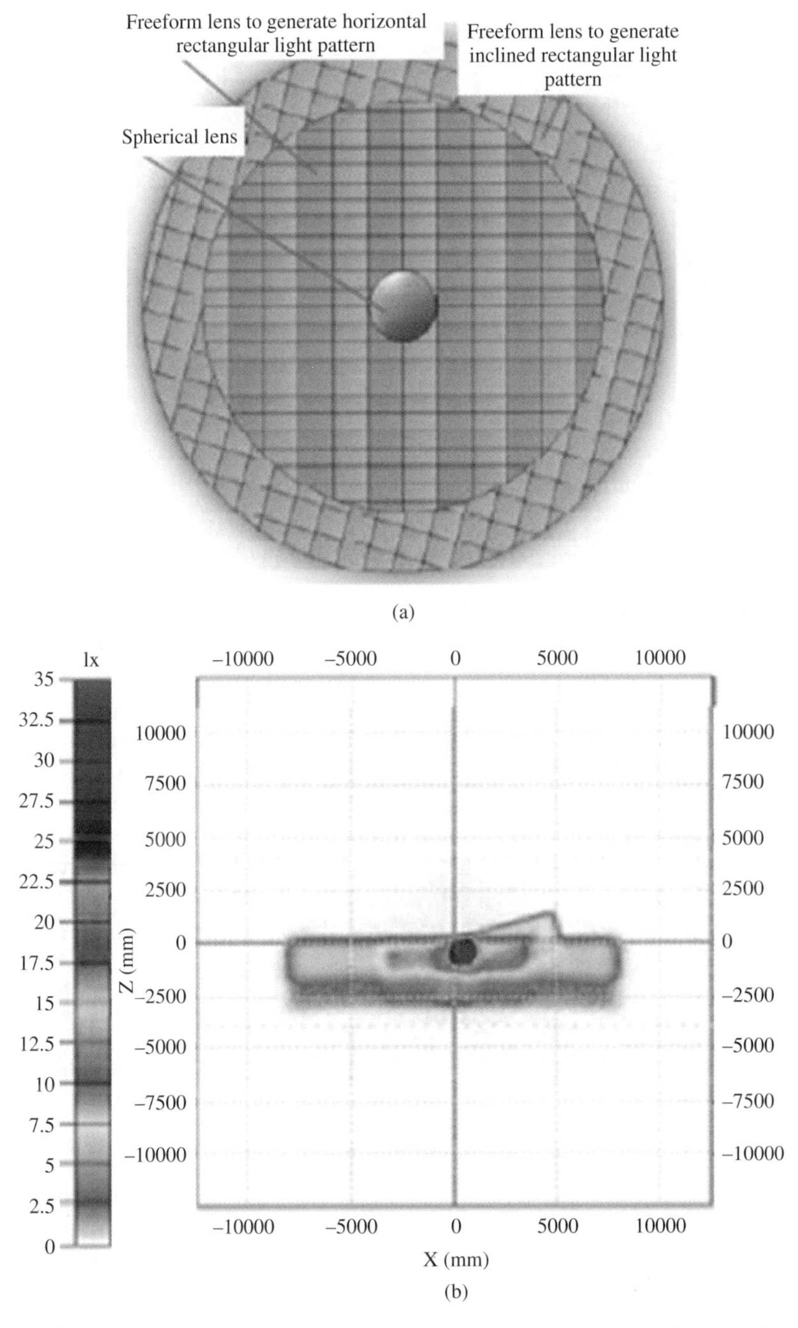

Figure 8.23 (a) LED low-beam freeform lens integrated with a spherical lens, and (b) light pattern of this freeform lens.

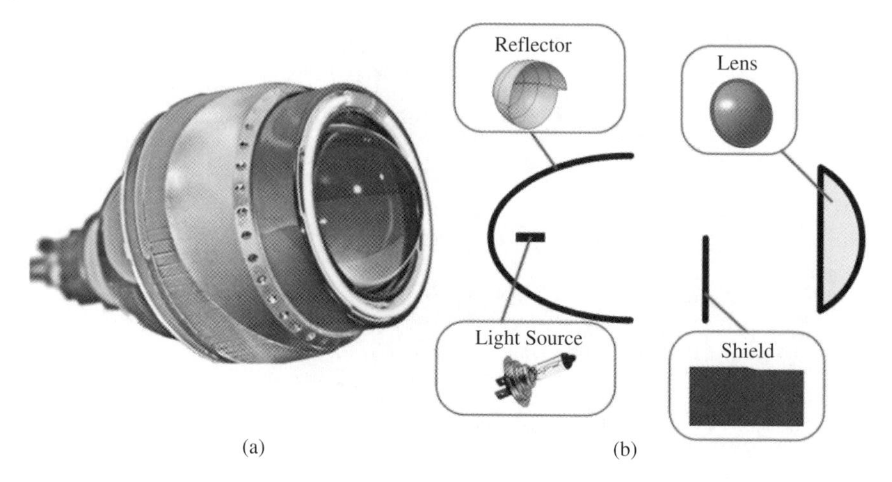

Figure 8.24 (a) PES-based headlamp module, and (b) schematic of a PES system.

optics based on PES and MR for LED packaging of the line light source type in this section and Section 8.6, respectively.

The PES system is mainly built up by a light source, a reflector, a shield, and a projection lens. By changing the shape of the shield, the PES system can achieve different beam patterns required by different regulations, such as ECE and SAE. Figure 8.24 shows a PES-based headlamp module and the schematic diagram of a PES system.

The reflective mirror is the critical part of the PES optical system. In this section, we use a new method for the reflective mirror. The structure is similar with the ellipsoid surface. We also use the ellipsoid surface as the fundamental surface for the reflective mirror. Only the top surface of the LED emits light, and therefore we only need half of the ellipse. We divide the half fundamental ellipse into several regions and obtain subsurfaces. The surface parameters of each subsurface are adjusted separately, and we check the light pattern at the visor simultaneously until the optimized light pattern is obtained. We use Eq. 8.10 to describe the quadratic curve equation.

$$y^2 - 2Rx + (k + 1)x^2 = 0 \tag{8.10}$$

where R is the radius of the curvature of the curve; and k is the quadratic curves constant and can be expressed as Eq. 8.11.

$$k = -e^2 \tag{8.11}$$

where e is the eccentricity of quadratic curves. When k is equal to 0, Eq. 8.10 can be used to describe circle curve. When k is equal to -1, the curve described by the equation is a parabola. When $-1 < k < 0$, the curve is an ellipse. When $k < -1$, the curve is a hyperbola. We choose $k = -0.5$ and $R = 25$ mm as the base ellipse surface parameter, and the half width and height is 32 mm. Figure 8.25 shows the front view of the base ellipse, and the base ellipse surface is divided into eight subsurfaces. And the final designed reflector for an LED PES is shown in Figure 8.26 consisting of eight subsurfaces.

In order to satisfy the ECE low-beam requirement of a light lamp, the baffle size is fixed as shown in Figure 8.27a. However, in order to counteract the impact of the distortion of the projection lens and guarantee the flatness of the low-beam pattern, the baffle is

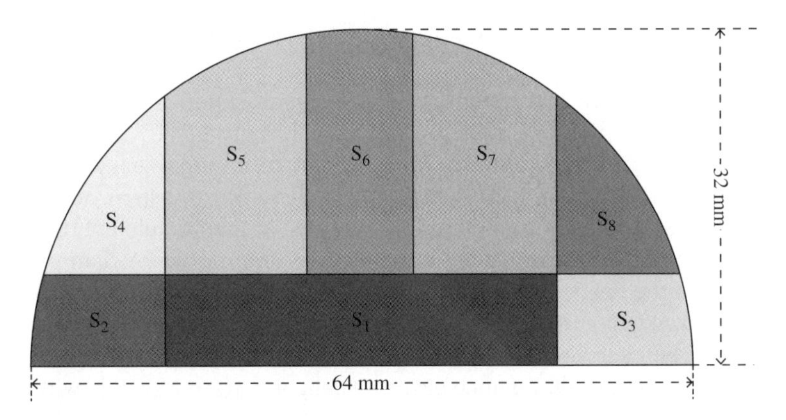

Figure 8.25 Base ellipsoid cross section size and segmentation method.

Figure 8.26 Designed reflector for an LED PES.

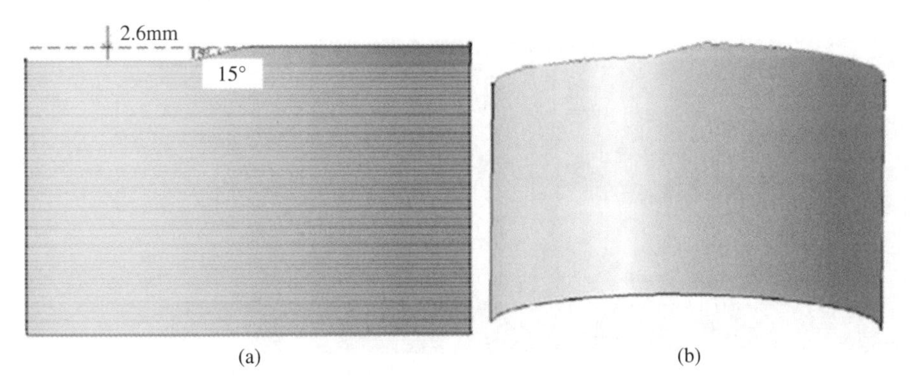

Figure 8.27 Baffle of the PES optical system.

designed into a cylinder instead of the flat plane as shown in Figure 8.27b. The radius of the cylinder is 39.7 mm.

The projection lens in the PES optical system is a plano-convex spherical lens made of borosilicate glass with a refractive index of 1.47 and a focal length of 41.7 mm, and the pore size of the lens is 76 mm. The complete LED PES low-beam module is shown in Figure 8.28. This module includes an LED light source, a PES optical system, active cooling, a heat dissipation sink, and a driving module. The size of the module is 110 × 120 × 150 mm, which can be easily integrated in various types of automotive headlamps.

The simulation results of the beam pattern of the LED PES low-beam module on the receiver 25 m away are shown in Figure 8.29. And the simulation result shows the optical efficiency is 35% (the total efficiency of the reflector's reflectivity and lens' transmission is assumed to be 70%). The highest illuminance of the near-light pattern is 25.5 lx with

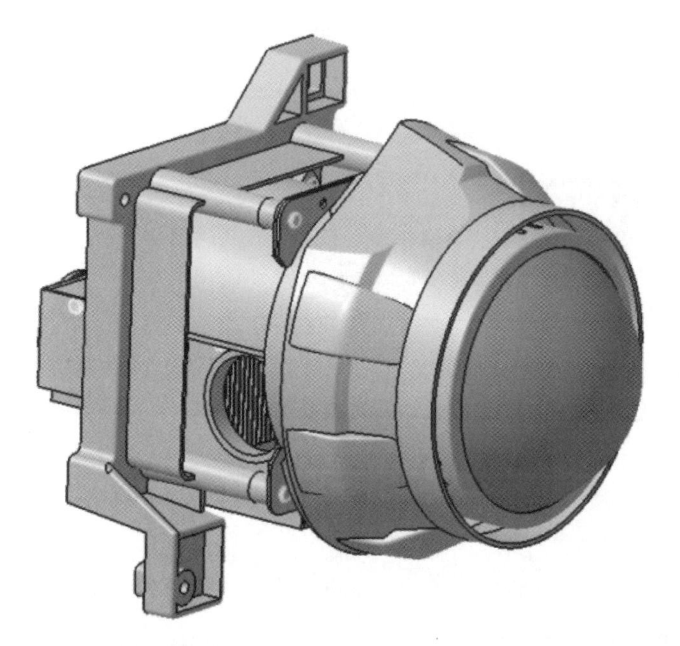

Figure 8.28 LED PES low-beam module with multiple functions.

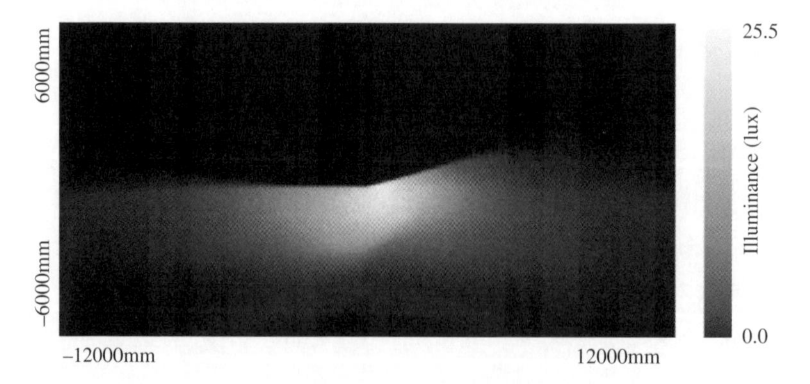

Figure 8.29 Simulated beam pattern for low-beam light with an improved PES system.

Table 8.7 Simulated illumination result compared with the corresponding value of ECE R112 (Class B) for low-beam light with improved PES system.

Point on measuring screen	Required illumination in lx	Simulated illumination in lx
Point B50L	≤0.4	0.0
Point 75R	≥12	24.9
Point 75L	≤12	4.8
Point 50L	≤15	12.8
Point 50R	≥12	24.0
Point 50V	≥6	20.1
Point 25L	≥2	6.0
Point 25R	≥2	6.1
Any point in zone I	≤2·E*	√
Any point in zone III	≤0.7	√
Any point in zone IV	≥3	√

Note: E* is the actually measured value in point 50R.

a clear cutoff line, and the width from the left to the right is ±10 m. The illuminance values are shown in Table 8.7. These values demonstrate that our design has satisfied the GB regulations, and the maximum illuminance closely approaches the 75R and 50R positions. The design has left enough space for fabrication error.

8.6 Freeform Optics Integrated MR for an LED Headlamp

The method of optical design overlaps the images of light source by a paraboloid-based multi-reflector on the measuring screen to form a regulation compliant beam pattern. As the beam patterns specified by regulations have a larger spread in the horizontal direction than the vertical direction and LED emits light in half space, the LED should be a strip-shaped emitter and be placed toward the horizontal direction. Figure 8.30 shows the LED installation and the MR optics.

The MR optics consists of two parts: part I can generate a horizontal light distribution, and part II can generate a 15° inclined light distribution on the measuring screen. The simulated beam pattern on the measuring screen is shown in Figure 8.31, and simulated illuminance on the measuring screen is listed in Table 8.8. The optical efficiency of this MR optics can reach 65% (mirror reflectivity is assumed to be 80%).

Besides low-beam light, improved MR optics can also be used in the high-beam light with LED packaging of the line light source type. Figure 8.32 shows the LED a far-light lamp module with an MR reflective lens. This module includes the LED light source package module, an MR reflective lens, a heat sink, and the cooling fan. In order to satisfy the modeling requirement, the reflective surface of the MR is trimmed accordingly but has little effect on the light pattern. The simulated beam pattern and illuminance values on the testing screen of the LED high-beam module are shown in Figure 8.33 and Table 8.9, respectively. The test results show that the LED MR low-beam and high-beam modules can fully comply with ECE regulation and provide a high optical efficiency.

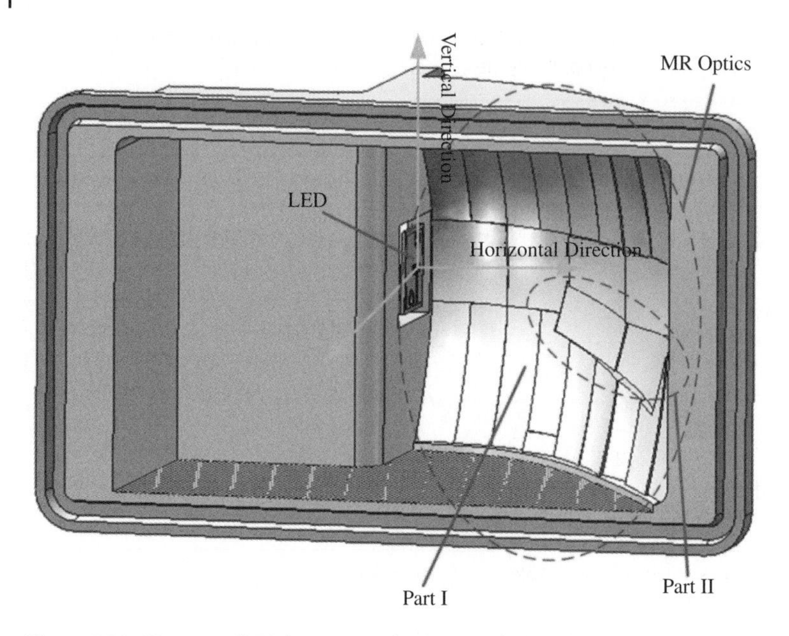

Figure 8.30 Diagram of LED location and MR optics for low-beam light.

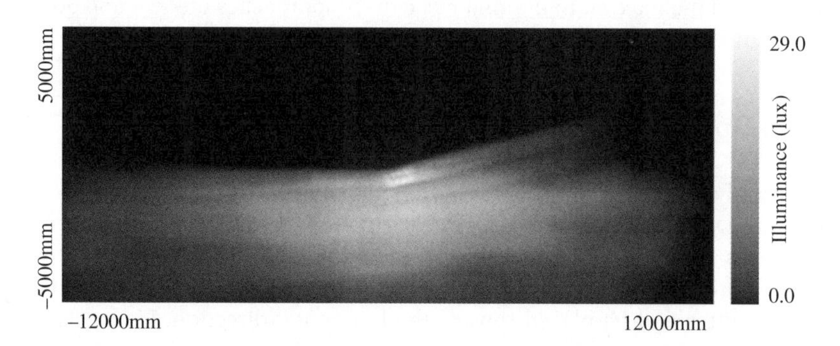

Figure 8.31 Simulated beam pattern on a measuring screen with MR optics for low-beam light.

Based on the above-mentioned method, one LED headlamp module integrated with improved MR low-beam and high-beam modules is designed, as shown in Figure 8.34. According to the design, we also fabricate the real reflector as shown in Figure 8.35. The fabrication process is machining rapid prototyping, and therefore the parts accuracy and surface smoothness are inferior to what is produced with the mold injection method, which may lead to deviation of the light type.

The lighting performance of an MR low-beam and high-beam is shown in Figure 8.36. We can find that for the MR low-beam lamp, the cutoff line becomes vague, but the width and uniformity are satisfactory. The highest illuminance can reach 150 lx and above.

Table 8.8 Simulated illumination result compared with the corresponding value of ECE R112 (Class B) for a low-beam light with an improved MR system.

Point on measuring screen	Required illumination in lx	Simulated illumination in lx
Point B50L	≤0.4	0.17
Point 75R	≥12	21.43
Point 75L	≤12	6.06
Point 50L	≤15	8.73
Point 50R	≥12	24.30
Point 50V	≥6	19.45
Point 25L	≥2	8.81
Point 25R	≥2	10.50
Any point in zone I	≤2·E*	√
Any point in zone III	≤0.7	√
Any point in zone IV	≥3	√

Note: E* is the actually measured value in point 50R.

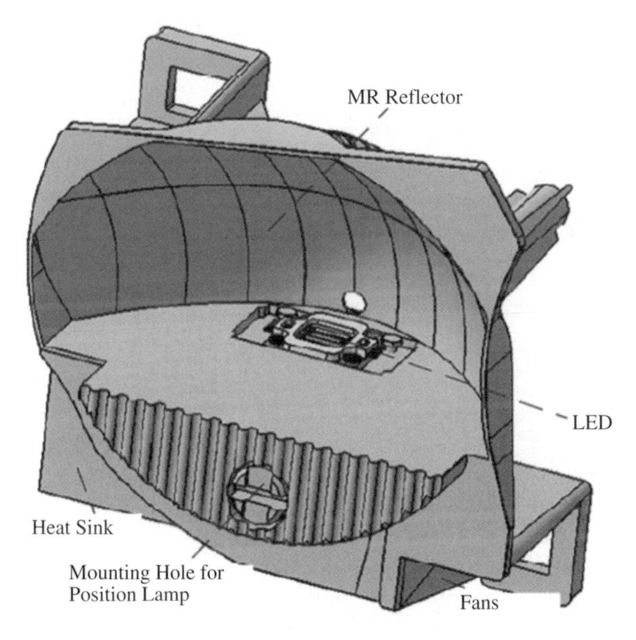

Figure 8.32 MR optics for an LED high-beam light.

Besides the optical design, we also have performed some work on thermal management. In our design, a heat sink with optimized sizes and structures and a fan are used in the active cooling system shown in Figure 8.37. The simulated heat distribution on the heat sink is shown in Figure 8.38. The ambient temperature is 85 °C, which is close to the temperature near the headlamp in the engine compartment. The simulation results

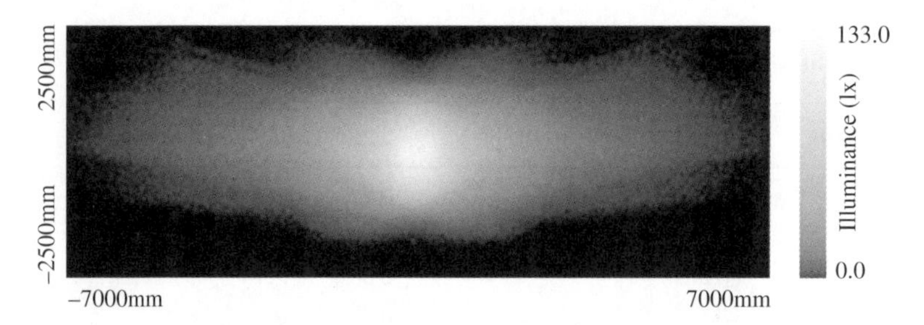

Figure 8.33 Simulated beam pattern on a measuring screen with MR optics for high-beam light.

Table 8.9 Simulated illumination result compared with the corresponding value of ECE R112 (Class B) for a high-beam light with an improved MR system.

Point on measuring screen	Required illumination in lx	Simulated illumination in lx
E_{max}	≥48 and ≤240	125.0
Point HV	≥0.8 E_{max}	0.96 E_{max}
Point HV to 1125L and R	≥24	53.2
Point HV to 2250L and R	≥6	22.5

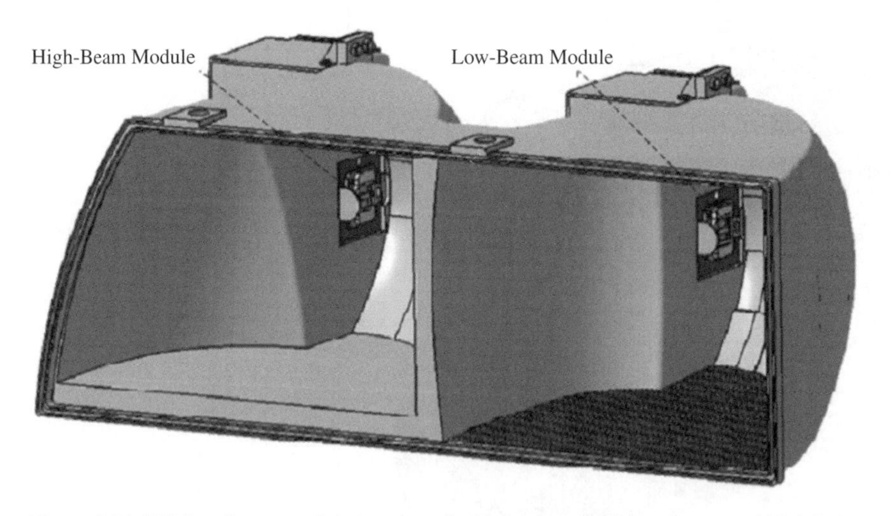

Figure 8.34 LED headlamp module integrated with improved MR low-beam and high-beam modules.

show that the temperature of the surface on which the ASLP is attached is about 110 °C; therefore, the junction temperature would be less than 150 °C.

8.7 LED Headlamps Based on Both PES and MR Reflectors

Based on the freeform optics improved PES and MR reflector optical system mentioned in this chapter, we design a LED headlamp integrated with these optical elements,

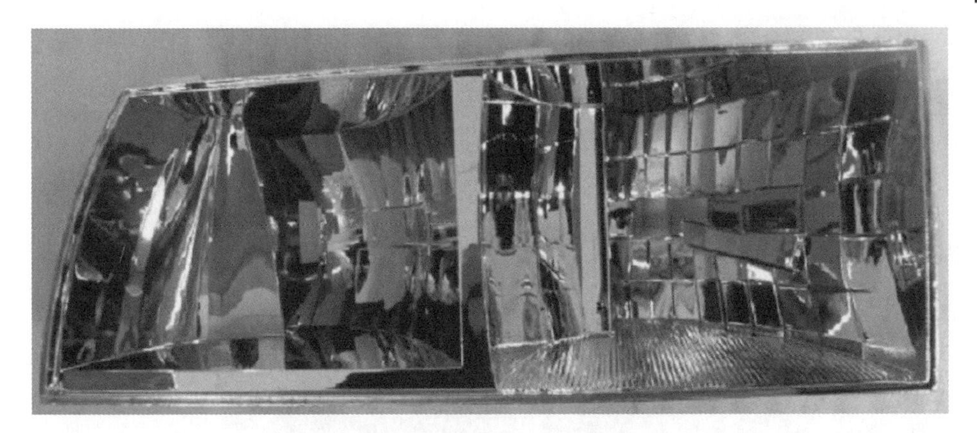

Figure 8.35 Fabricated MR reflector module for an LED headlamp.

(a)

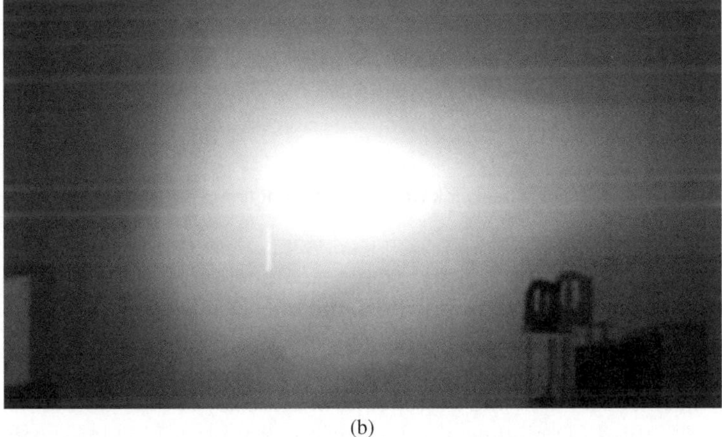

(b)

Figure 8.36 MR LED (a) low-beam and (b) high-beam light pattern.

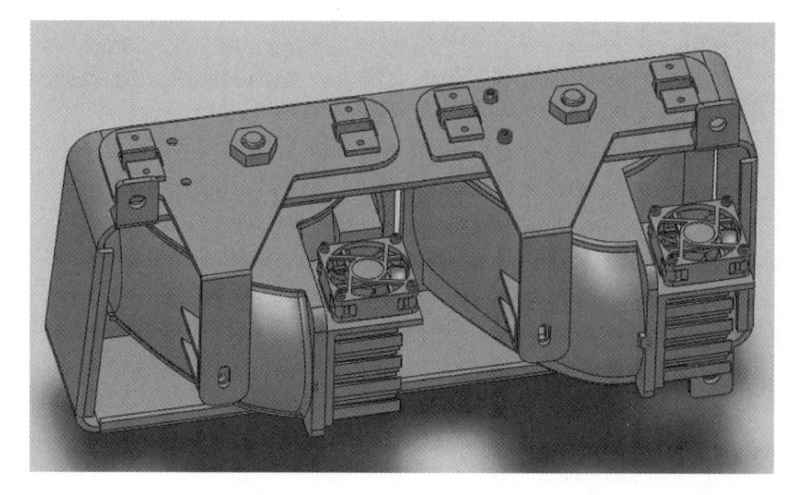

Figure 8.37 Schematic diagram of an active cooling system in an LED headlamp.

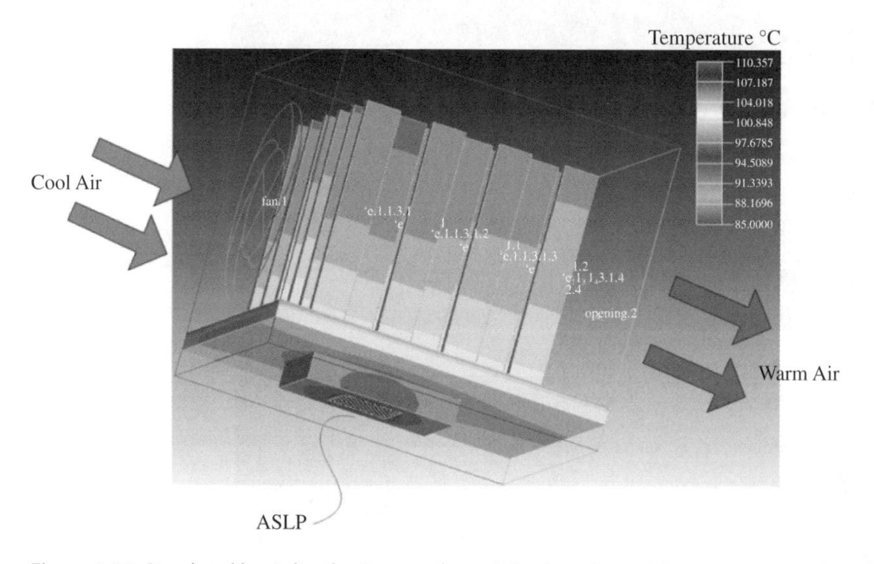

Figure 8.38 Simulated heat distribution on a heat sink when the ambient temperature is 85 °C.

where low-beam function is realized by PES and high-beam function is realized by MR reflector. The design model and practical prototype of the LED headlamp are shown in Figure 8.39 and Figure 8.40, respectively.

In this LED headlamp, the LED low-beam module integrated with functions of freeform optics, thermal management, and power control is the key of the whole lamp, as shown in Figure 8.41. This LED module also has the concept of an LED low-beam light engine for a headlamp, and many different kinds of LED headlamps could be designed by using the same LED low-beam engine. This kind of LED module has the potential to become one of the standard lighting modules for LED headlamps.

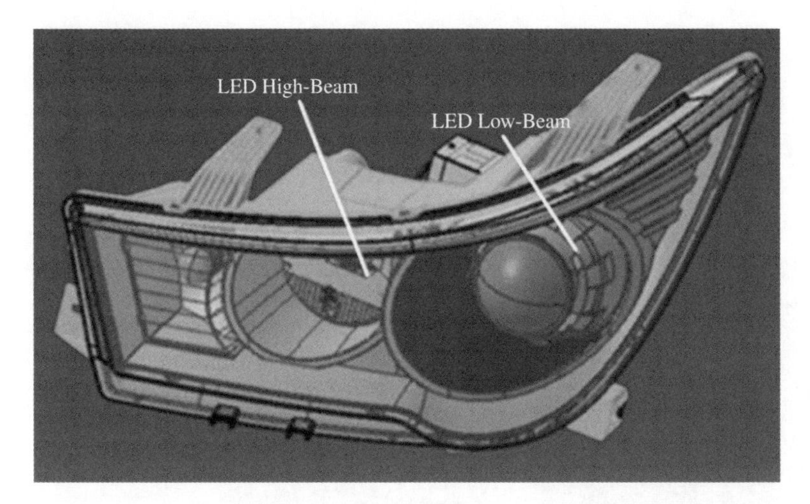

Figure 8.39 Design model of the LED headlamp.

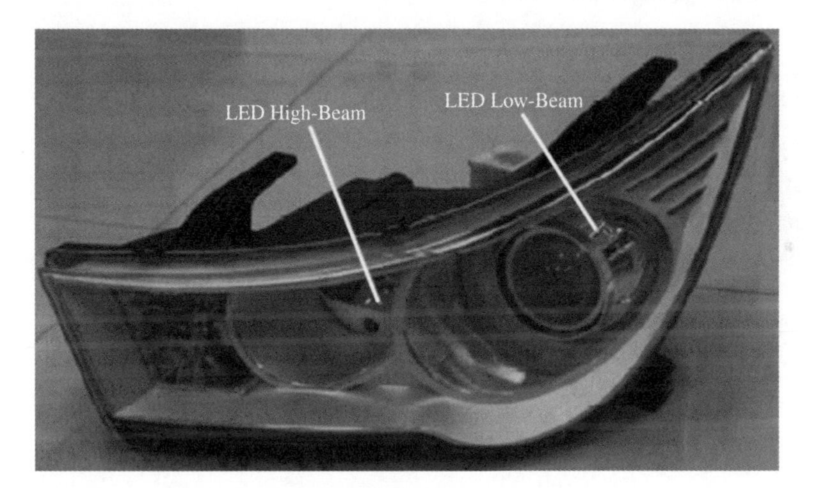

Figure 8.40 Practical prototype of the LED headlamp.

The experimental low-beam lighting performance on the test wall of the LED headlamp is shown in Figure 8.42a and 8.42b. We can find that a clear cutoff line appears at the horizontal and inclined 15° directions. To observe the practical illumination performance on the road, we test this LED headlamp at the parking place. From Figure 8.43, we can find that the range of illumination on the road is wide and the illuminance is distributed uniformly. The experimental high-beam lighting performance on the test wall of the LED headlamp is shown in Figure 8.44a and 8.44b.

We use the LMT GO-H 1300 luminosity distribution meter to precisely measure the light intensity and illuminance distributions of LED low-beam and high-beam. The testing results and their comparison with the GB 25991 regulation are shown in Table 8.10.

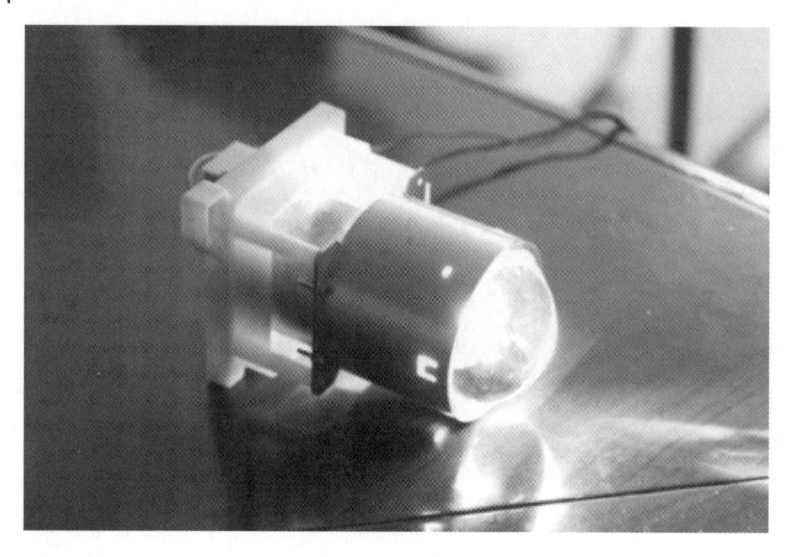

Figure 8.41 LED low-beam light engine for headlamps.

Figure 8.42 (a) LED low-beam illuminated performance and (b) LED low-beam light pattern on the vertical wall.

We can find that the illumination performance of the LED headlamp is able to meet the GB 25991 requirements very well.

8.8 LED Module Integrated with Low-Beam and High-Beam

Due to the LED light source only emitting light in a half-spherical space, the reflective lens in the optical system only uses the half ellipse as the base curve surface. In real PES design, the projection lens in the optical system only needs a half ellipse as the base curve surface. Figure 8.45 shows the LED PES optical system ray-tracing schematic. The light ray-tracing results show that the rays, after passing the beam-shaping shielding plate, are projected at the receiving surface by the lower half of the projection lens system, and

Figure 8.43 LED headlamp low-beam lighting performance.

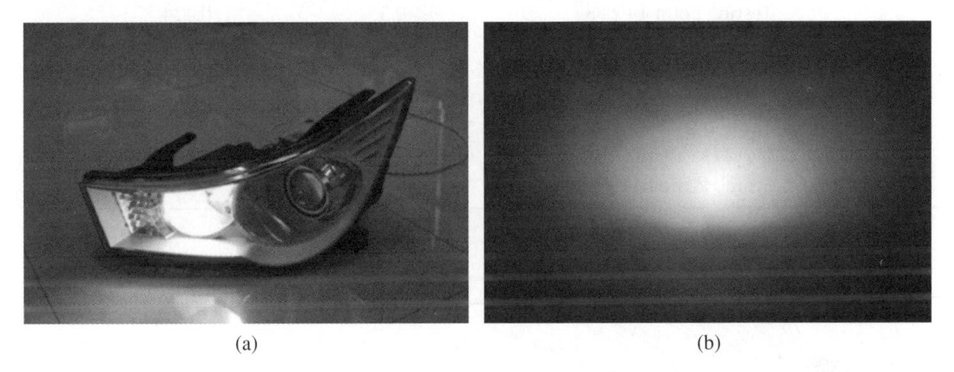

(a) (b)

Figure 8.44 (a) LED high-beam illuminated performance and (b) LED high-beam light pattern on the vertical wall.

therefore the upper half of the projection lens is not in use. We remove the redundancy lens part and use the space to design LED high-beam optics, in order to achieve one LED light module integrated with low-beam and high-beam functions.

The high-beam function in the integrated LED headlamp module can be realized by three CREE XP-G LED packaging modules and three TIR collimating lenses. When the high-beam lamp needs to be lit, the three low-beam LEDs and the other three high-beam LEDs are lit simultaneously. The design model of the integrated LED module is shown in Figure 8.46, and the simulated beam patterns of low-beam and high-beam of this module are shown in Figure 8.47a and 8.47b, respectively.

According to the above design, we fabricate the integrated LED headlamp module. This prototype module includes an LED source for low-beam and high-beam, light intensity redistribution optical elements, and the heat dissipation system (heat sink and cooling fan), as shown in Figure 8.48. Compared with the LED PES module mentioned

Table 8.10 Comparisons of test results of an LED headlamp with the GB requirements.

Type	Testing points or region	Illuminance requirement (lx)	Tested illuminance (lx)
Low-beam	B50L	≤0.4	0.259
	75R	≥12	16.228
	75L	≤12	2.010
	50L	≤15	9.158
	50R	≥6	17.271
	50V	≥2	15.135
	25L	≥2	4.989
	25R	≥2	6.155
	Any points in region I	$\leq 2\times E_{50R}$*	≤17.659
	Any points in region III	≤0.7	0.581
	Any points in region IV	≥3	7.040
	Testing point 7	≥0.1 and ≤0.7	0.229
	Testing point 8	≥0.2 and ≤0.7	0.356
	Testing point 1+2+3	≥0.3	0.434
	Testing point 4+5+6	≥0.6	0.753
High-beam	E_{max}	≥48 and ≤240	96.1
	HV point	$\geq 0.8\times E_{max}$	95.8
	From HV point to 1125L and R	≥24	≥55.9
	From HV point to 2250L and R	≥6	≥24.7

* E_{50R} denotes the illuminance value at 50R.

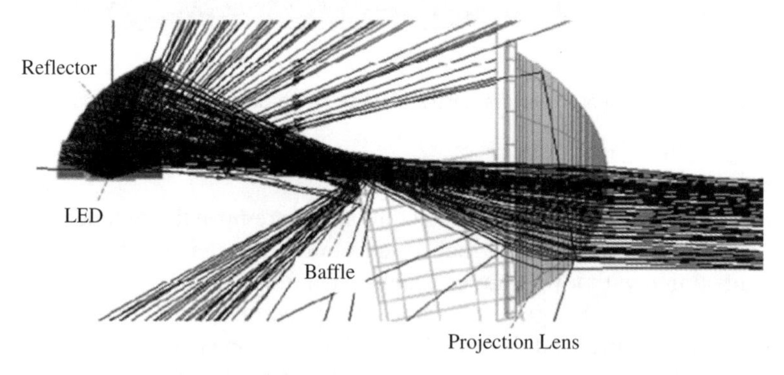

Figure 8.45 Schematic of ray tracing of the LED PES optical system.

in Section 8.5, the new LED headlamp module has only a small change in size but is integrated with high-beam function.

The low-beam and high-beam lighting performance of the integrated LED headlamp module is shown in Figure 8.49a and 8.49b, respectively. The illuminance at the center testing point of the high-beam reaches 96 lx.

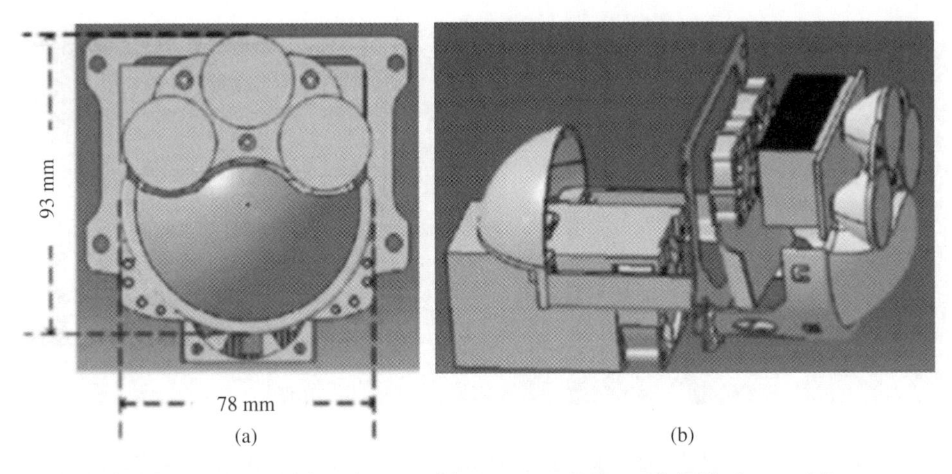

Figure 8.46 (a) Front view and (b) side view of the integrated LED module (design model).

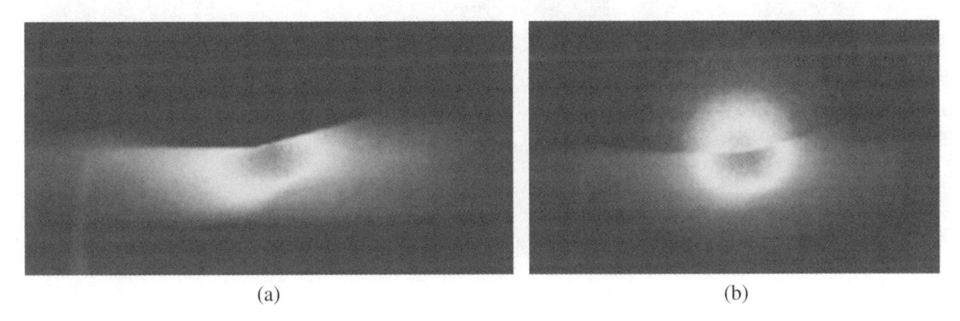

Figure 8.47 Simulated (a) low-beam and (b) high-beam beam pattern of the integrated LED module where the highest illuminance is at the center area of the light pattern.

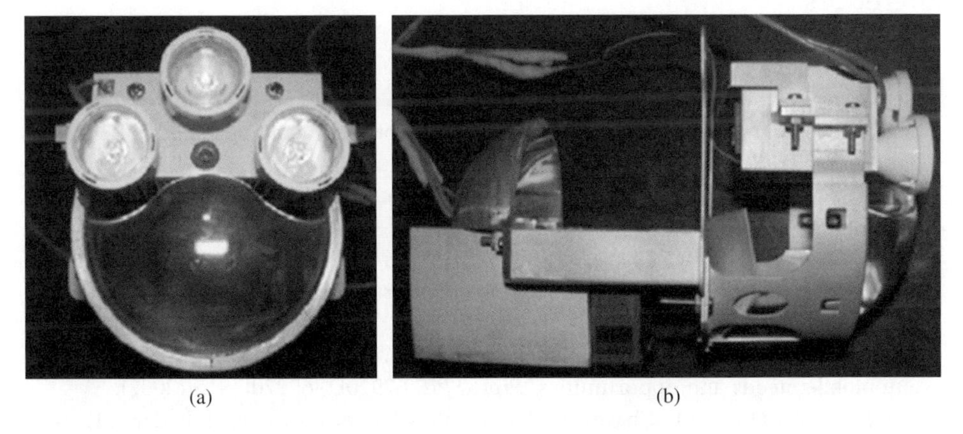

Figure 8.48 (a) Front view and (b) side view of the integrated LED module (prototype).

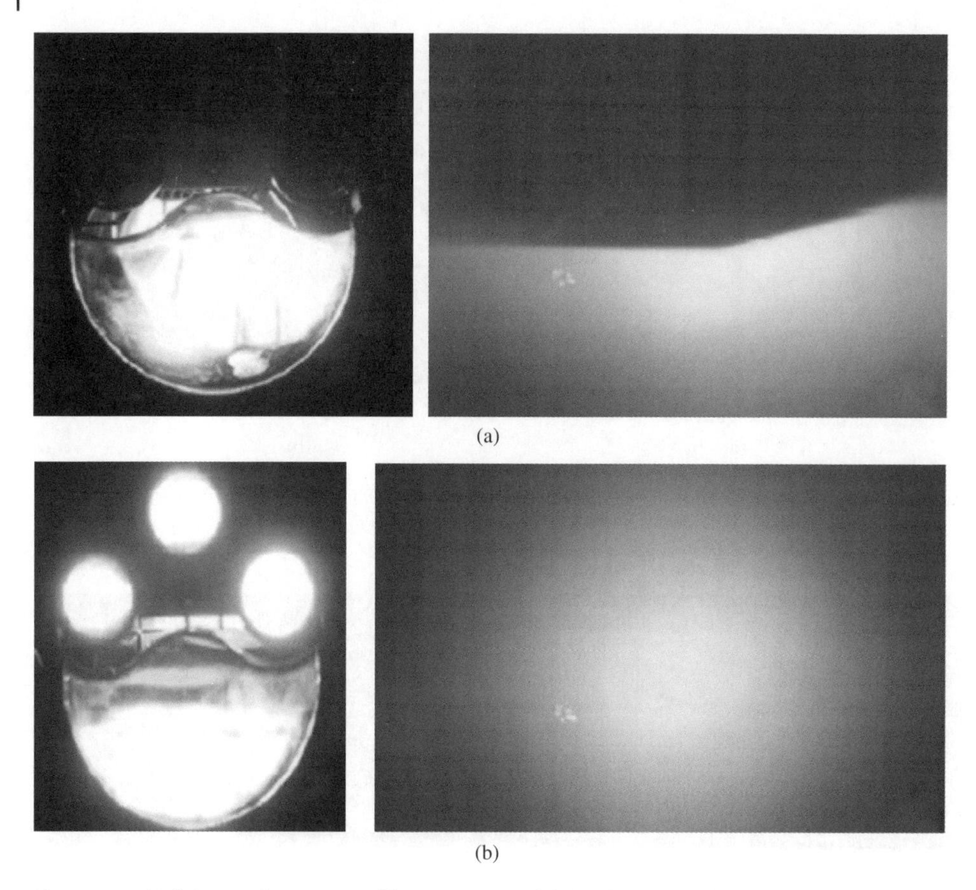

Figure 8.49 Lighting performance and beam pattern of the (a) low-beam and (b) high-beam of the integrated LED headlamp module.

References

1 United Nations Economic Commission for Europe. Vehicle regulations, Reg. 112, Rev. 2.

2 Jiao, J., and Wang, B. Etendue concerns for automotive headlamps using white LEDs. *Proc. SPIE* **5187**, 234–242 (2004).

3 Jiao, J., and Wang, B. High-efficiency reflector optics for LED automotive forward lighting. *Proc. SPIE* 66700M–66700M–10 (2007).

4 Dross, O., Cvetkovic, A., Chaves, J., Benitez, P., and Minano, J.C. LED headlight architecture that creates a high quality beam pattern independent of LED shortcomings. *Proc. SPIE* 59420D–59420D–10 (2005).

5 Pohlmann, W., Vieregge, T., and Rode, M. High performance LED lamps for the automobile: needs and opportunities. *Proc. SPIE* 67970D–67970D–15 (2007).

6 Cvetkovic, A., Dross, O., Chaves, J., Benitez, P., Minano, J. C. and Mohedano, R. Étendue-preserving mixing and projection optics for high-luminance LEDs, applied to automotive headlamps. *Optics Expr.* **14**(26), 113014 (2006).

7 Domhardt, A., Rohlfing, U., and Weingaertner, S. New design tools for LED headlamps. *Proc. SPIE* **7003** (2008).
8 Chen, F., Wang, K., Qin, Z., Wu, D., Luo, X., and Liu, S. Design method of high-efficient LED headlamp lens. *Opt Expr.* **18**, 20926–20938 (2010).
9 Domhardt, A., Weingaertner, S., Rohlfing, U., and Lemmer, U. TIR optics for non-rotationally symmetric illumination design. *Proc. SPIE* **7103** (2008).
10 Alvarez, J.L., Hernandez, M., Benitez, P., and Minano, J.C. TIR-R concentrator: a new compact high-gain SMS design. *Proc. SPIE* **4446** (2002).
11 Munoz, F., Benitez, P., Dross, O., Minano, J.C., and Parkyn, B. Simultaneous multiple surface design of compact air-gap collimators for light emitting diodes. *Opt. Engin.* **43**(7), 1522–1530 (2004).
12 Sun, L.W., Jin, S.Z., and Cen, S.Y. Free-form microlens for illumination applications. *Appl. Opt.* **48**(29), 5520–5527 (2009).
13 OSRAM OSTAR. Headlamp LED. Data sheet. http://www.osram-os.com

9

Freeform Optics for Emerging LED Applications

9.1 Introduction

From the above chapters, we can find that due to the significant advantages of precise light energy distribution control, high design freedom, compact size, and high light output efficiency, freeform optics have been widely recognized and are playing more and more important roles in the optical design of light-emitting diode (LED) packaging and applications. In several chapters of this book, we introduced a lot of freeform optics design methods and many major LED packaging and application design cases, including LED indoor lighting, LED road lighting, LED backlighting for large-scale liquid displays, and LED automotive headlamps. Actually, with the rapid development of LED technologies (e.g., efficiency and cost), some emerging LED applications have appeared in recent years in which freeform optics also will have opportunities to develop. In this chapter, we will introduce some freeform optics designs based on the algorithms mentioned in Chapter 3 for several emerging LED applications, including an LED pico-projector, LED stage light, LED taxiway light, and LED search light. We hope readers will broaden their vision regarding freeform optics for LED applications.

9.2 Total Internal Reflection (TIR)-Freeform Lens for an LED Pico-Projector

9.2.1 Introduction

Projection displaying technology has been widely used in daily life and scientific research. However, the weight and volume of the traditional projectors will limit the mass acceptance of them to some extent. Supposing that a projector can be miniature and portable, it will further satisfy user demands. A pico-projector is a newly emerging technology that aims at making projectors portable and available to be integrated with other devices. The core element of a projector, how to reduce the size of the optical engine, is the most critical step to get a miniature projector. The major components constituting an optical engine include: the light source, illumination system, micropixelized panel, and projection system.[1]

As shown in Figure 9.1, the working principle of the optical engine is summarized as: (1) an illumination system collects light energy of the light source onto the surface of a micropixelized panel, (2) light transmissivity of the individual pixels of the micro

Freeform Optics for LED Packages and Applications, First Edition. Kai Wang, Sheng Liu, Xiaobing Luo and Dan Wu.

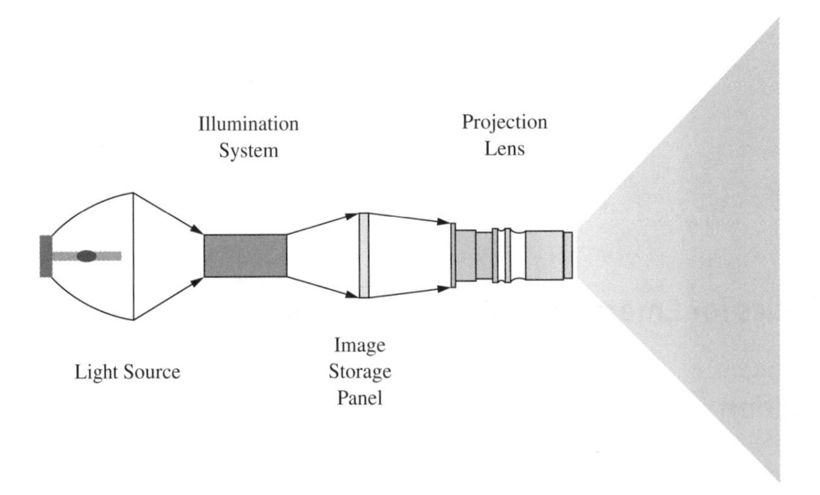

Figure 9.1 Schematic of construction components of an optical engine.

panel is modulated to form the grayscale image, and (3) the image on the micro panel is magnified and projected onto the screen by the projection system.

Compared with a traditional projector, a pico-projector has limitations of size and weight of its inner devices. Thus, light sources like a metal halogen lamp and super-high-pressure mercury lamp, which are commonly used for traditional projectors, are not suitable for the pico-projector. Due to its advantages of low power consumption, high conversion efficiency, and especially small packaging size, LED is the most ideal light source for a pico-projector.

According to the irradiation characteristics of LEDs, the light pattern in front of it is circular. However, the contour of the micro panel is rectangular. To increase the light energy utilization efficiency, the light pattern should be reshaped by an illumination system to match the contour of the micro panel. The luminous flux ratio of the micro panel and light source is determined as the collection efficiency of the illumination system.

On the premise that the illumination distribution of a micro panel before modulating is uniform, the grayscale image can be formed by modulating the light transmissivity of each pixel. Thus, *illumination uniformity*, defined as the minimum illumination to the average illumination of the micro panel, is another parameter that indicates the light performance of an illumination system.

As shown in Figure 9.2, light pipe and microlens arrays are two commonly used illumination system design methods for projectors. Generally, to collect light rays into a light pipe as much as possible, an ellipsoidal reflector is added with a light source locating at its primary focal point and the entrance surface of the light pipe locating at its secondary focal point. Illumination uniformity on the exit surface of the light pipe is related to the reflection times, which depend on the length of the light pipe.[2] Thus, the light pipe should be long enough to obtain high illumination uniformity at the exit surface.

In a microlens array illumination system, light rays are collimated at first. Then, the collimated beam is sliced by a micro lens into small beams. Finally, each of the small beams is magnified by condenser lenses to cover the whole micro panel.[3] Obviously, the two illumination systems described here are complicated in structure, and that will result in complexity of the optical engine.

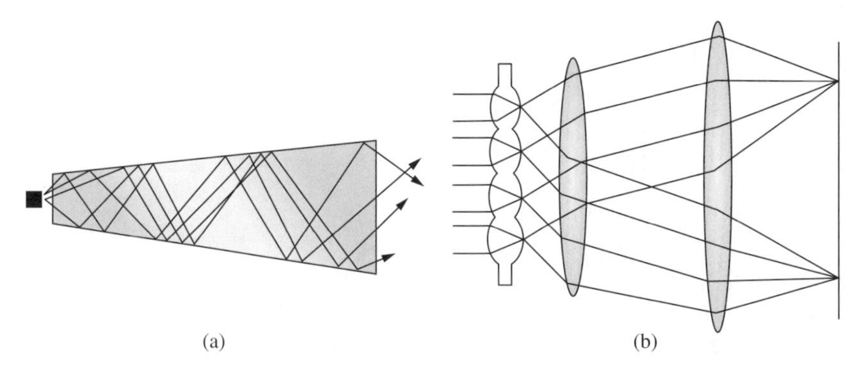

Figure 9.2 (a) Light pipe illumination system and (b) microlens array illumination system.

With its advantages of design freedom and precise light irradiation control, freeform optical components are considered as the best technique to get the desired illumination.[4–7] In this section, a novel integral illumination lens design method is proposed that is based on a freeform surface and responds to the requirement of a more compact illumination system.

9.2.2 Problem Statement

9.2.2.1 Defect of a Refracting Freeform Surface for Illumination with a Small Output Angle

The optical engine of a projector is a system that contains both nonimaging and imaging components. The light performance of an illumination system plays a role in the determination of the technical parameters of a projection system. For example, an F-number of projection lenses, which is defined as the ratio of the lens' focal length to the diameter of the entrance pupil, equals the reciprocal of the maximum light angle of the illumination system. To reduce the aberration and simplify the projection system, the light angle of an illumination system should be as small as possible. In this design, we set the maximum light angle of an illumination system as 15°. However, defects exist when concentrating light rays into a small angle with a single refracting freeform surface.

As shown in Figure 9.3, a freeform lens with a single refracting surface is designed to realize a central symmetrical light distribution, of which the maximum output angle equals 15°. The lens material is set to be poly(methyl methacrylate) (PMMA) for simulation, and the ray-tracing result shows that some light rays whose emitting angles from the light source exceed a certain value are out of control.

As we know, the refraction index of materials such as plastic or glass used for the manufacturing of optical lens is greater than that of air. TIR exits when light rays transmit from an optically denser medium to a thinner medium. Supposing that θ_1 and θ_2 are the incident and output angles, respectively, the angle between the incident light ray and output light ray can be expressed as:

$$\beta = \frac{\pi}{2} + \theta_1 + \left(\frac{\pi}{2} - \theta_2\right) = \pi + \theta_1 - \theta_2 \tag{9.1}$$

Considering the critical circumstance, $\theta_1 = \arcsin(1/n)$ and $\theta_2 = \pi/2$. Therefore, the minimum value of β is $\pi/2 + \arcsin(1/n)$. Substitute n with the refraction index of PMMA (1.49), and the value of the minimum angle equals 132°.

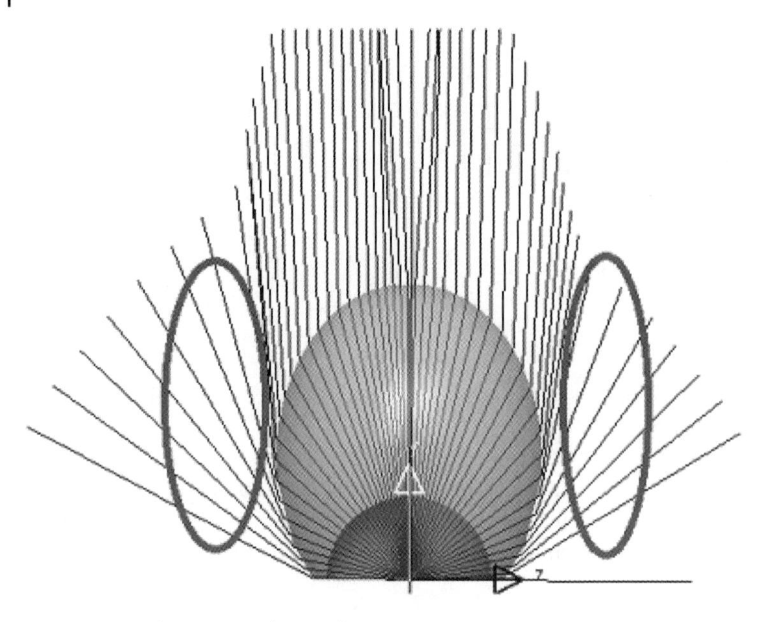

Figure 9.3 Light rays out of control.

9.2.2.2 Problem of an Extended Light Source

Another problem that exists in all current optical design is extended light sources.[8,9] As shown in Figure 9.4, a refracting surface is designed to shape a prescribed light pattern based on point source *S*. Since the normal vector of any point on the surface is fixed, every incident light ray emitting from *S* is refracted to a certain direction.

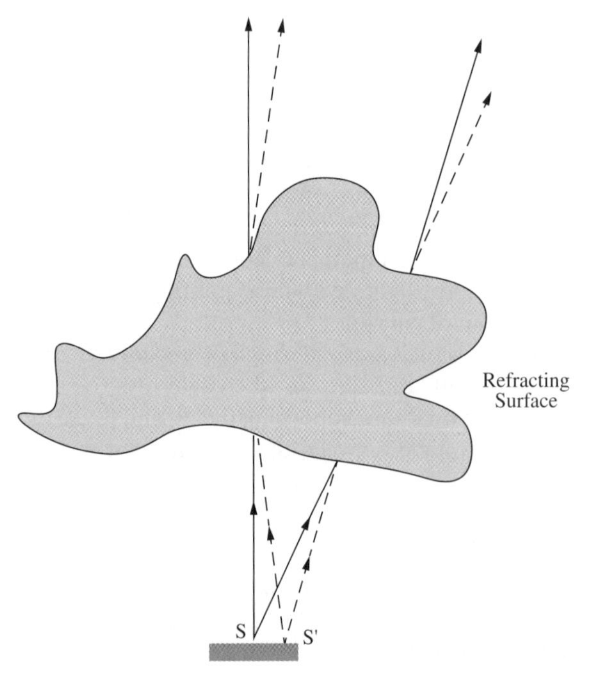

Figure 9.4 Effect of an extended light source.

Considering another point source S' on the emitting surface, the incident directions of light rays coming from it are different from those of S, which leads to the difference of the output directions. Obviously, the light pattern of S' cannot overlap with the prescribed light pattern. Major effects of extended light source on the performance of optical lens include the decreasing of light efficiency and illumination uniformity. Thus, it is a big challenge for the illumination system design of a pico-projector.

9.2.3 Integral Freeform Illumination Lens Design Based on an LED's Light Source

The emitting angle of LEDs reaches as high as 90°; as a result of TIR, the single refracting surface cannot satisfy the illumination design requirement of a pico-projector. TIR lenses are currently used widely in illumination design.[10,11] In this section, the TIR lens is employed to collimate light rays. An improving design method for the top surface of the TIR will be proposed.

9.2.3.1 Freeform TIR Lens Design

Actually, the reflecting surface of TIR takes advantage of TIR to collimate light rays with a big emitting angle. Because that reflecting surface is unenclosed, light rays with a small emitting angle are projected into the air without being collimated by it. Therefore, a TIR lens combines the refracting surface as well.

The first step for the design of a freeform TIR lens is partition of light rays. In other words, the critical light angle should be determined at first; light rays whose emitting angles are smaller than the critical angle will be collimated by the refracting surface (the bottom surface of the TIR lens) or the reflecting surface (the side surface of the TIR lens). To guarantee that each light ray is controllable, the critical light angle should be smaller than the incident angle where TIR occurs for a refracting surface. PMMA is often used to manufacture TIR lenses; the minimum angle between the incident direction and output direction of the light ray is 132°. Since all of the light rays are collimated, their output directions are vertical up. According to the geometry of the triangle, the maximum incident angle for the refracting surface equals 48°. Thus, the critical light angle should be set to be smaller than 48°.

In Figure 9.5, h stands for the height of the initial point P_1, which refracts the central light ray and should be determined at first. In this application, h is set to be 4 mm to keep a suitable size for the TIR lens. A normal vector of P_1 can be calculated according to Snell's law:

$$[1 + n^2 - 2n(\vec{O}\cdot\vec{I})]^{1/2}\vec{N} = \vec{O} - n\vec{I} \tag{9.2}$$

At the intersection of incident light ray I_2 with a tangent plane of P_1, the intersection is the location of the second refracting point P_2. Calculate the normal vector of P_2 according to Snell's law, then obtain the location of P_3 by intersecting I_3 with a tangent plane of P_2. Continue the procedure until all of the refracting points are obtained.

Calculation procedures for the reflecting points are similar to those for the refracting points; however, calculation of the normal vector is based on reflection law instead of Snell's law.

9.2.3.2 Top Surface Design of the TIR Lens

A rectangular light pattern with uniform illumination is required. As mentioned in this chapter, TIR with a plane top surface just collimates all the light rays without doing

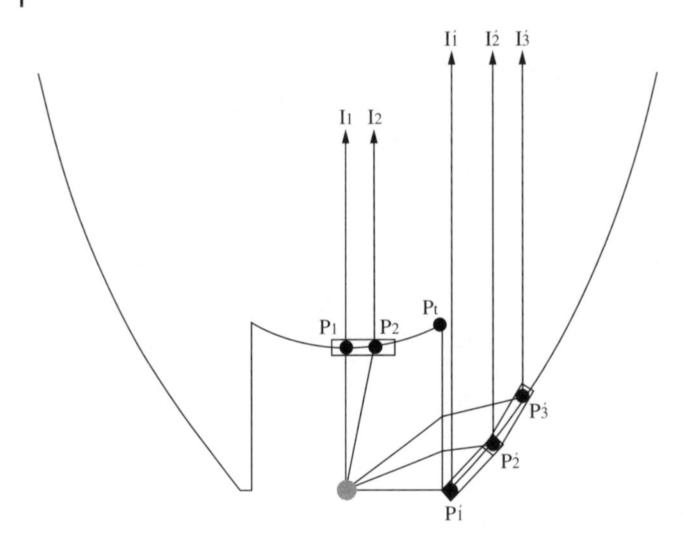

Figure 9.5 Cross section of a freeform TIR lens.

energy redistribution. Due to its advantages of unique design freedom, small size, and accurate light irradiation controlling, a freeform surface is adopted for the design of the top surface of TIR.

Step 1. Energy mapping for the design of the top surface of TIR. In a single freeform surface design, energy mapping is done between the light source and target plan. However, even though all the light rays are collimated by TIR, light energy has been transferred onto the top of the TIR, and it forms an energy distribution plane that corresponds to the energy distribution space of the light source. Thus, for the design of the top surface of TIR, energy mapping is built between the energy distribution plane and target plane.

The two important steps of energy mapping are energy slicing and target plane slicing, which result in grids with equal luminous flux and equal area, respectively. Slicing of the energy distribution plane should be on the basis of the slicing of the light source. As shown in Figure 9.6, the light source is divided into M and N parts along the latitudinal and longitudinal directions.[12] Supposing that the luminous flux within M parts is equal to each other and that the luminous flux within N parts is too, luminous flux in each grid formed by a slicing light source along two directions is equal.

The radiation characteristic of most commercially available LEDs adheres to Lambert's cosine law, which is expressed as $I = I_0 \cos\theta$, in which I_0 is the light intensity of the central light ray, and θ is the angle between the light ray and the central light ray. According to photometry theory, the relationship between luminous flux and light intensity is expressed as:

$$I = \frac{d\phi}{d\omega} \tag{9.3}$$

Where $d\omega$ is the solid angle of an energy grid, it can be written as:

$$d\omega = \sin(\theta) d\theta d\gamma \tag{9.4}$$

Figure 9.6 Slicing of the light source.

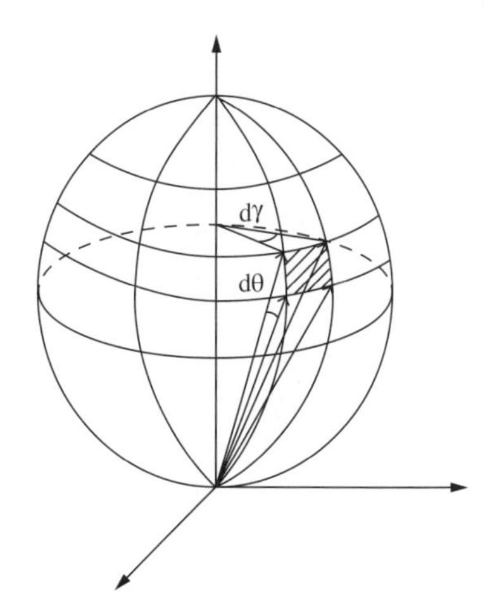

Substituting Eq. 9.4 into Eq. 9.3, the luminous flux within an energy grid is expressed as:

$$d\phi = I(\theta)\sin(\theta)d\theta d\gamma \tag{9.5}$$

Total luminous flux of the light source is the integration of $d\phi$ in the whole radiation space:

$$\phi_0 = \int_0^{2\pi}\int_0^{\pi/2} I(\theta)\sin(\theta)d\theta d\gamma \tag{9.6}$$

In the latitudinal direction, light sources are divided into M parts with equal luminous flux. The value of the edge angle θ_i $(i = 1, 2, \ldots M)$ should obey Eq. 9.7:

$$2\pi\int_0^{\theta_i} I(\theta)\sin(\theta)d\theta = \frac{i\phi_0}{M} \tag{9.7}$$

As a result of the central symmetry radiation property of LEDs, the value of the edge angle $\gamma_j(j = 1, 2, \ldots N)$ in the longitudinal direction obeys Eq. 9.8:

$$\gamma_j = \frac{j}{N}2\pi \tag{9.8}$$

Direction angles θ_i and γ_j determine the edge ray $I(\theta_i, \gamma_j)$. As mentioned in this chapter, a number of light rays in the cross section of the TIR lens should be selected to calculate the discrete points of the seed curve. We chose a cross section that contains a longitude line, and edge rays $I(\theta_i, \gamma_j)$ $(i = 1, 2, \ldots M, j)$ are picked to do the calculation.

All of the edge rays $I(\theta_i, \gamma_j)$ $(i = 1, 2, \ldots M, j = 1, 2, \ldots N)$ are collimated by TIR and then intersect with the top of it. As shown in Figure 9.7, these intersections are edge points that form the grids of an energy distribution plane within which luminous flux is equal.

The second step of energy mapping is slicing of the target plane, which should correspond to energy slicing as well.

As shown in Figure 9.8a, the size of the target plane is $a \times b$. Rectangular frames whose areas equal $(i/\mathrm{M})\, a \times b\,(i = 1, 2, \ldots M)$, and whose diagonals overlap each other, slice the target plane into M parts with equal areas. In Figure 9.8b, two diagonals of the target plane divide it into four triangles. Then, straight lines radiate from the center of the

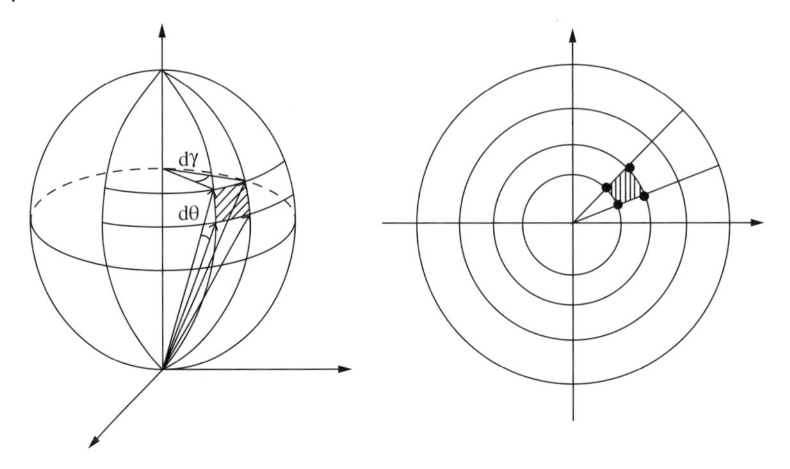

Figure 9.7 Intersections of the edge rays with the top surface of TIR.

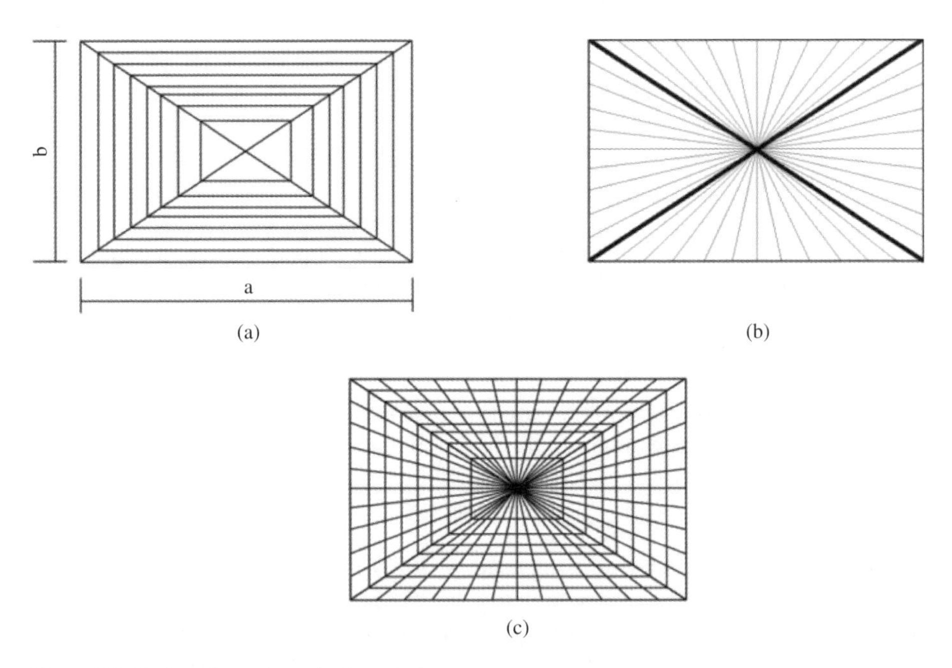

Figure 9.8 Slicing of a rectangular target plane.

target plane and further slice them into small triangles. In the above and below triangles, radiation lines intersect with the top and bottom boundaries of the target plane, and the distance between the adjacent intersections equals $4a/N$. Similarly, in the left and right triangles, the distance between the adjacent intersections made by radiation lines and the boundary of the target plane equals $4b/N$. According to the slicing method, the area of each small triangle is ab/N. Overlapping the rectangular frames and radiation lines shown in Figure 9.8a and Figure 9.8b, the target plane is sliced into grids with an area equaling ab/MN, as shown in Figure 9.8c.

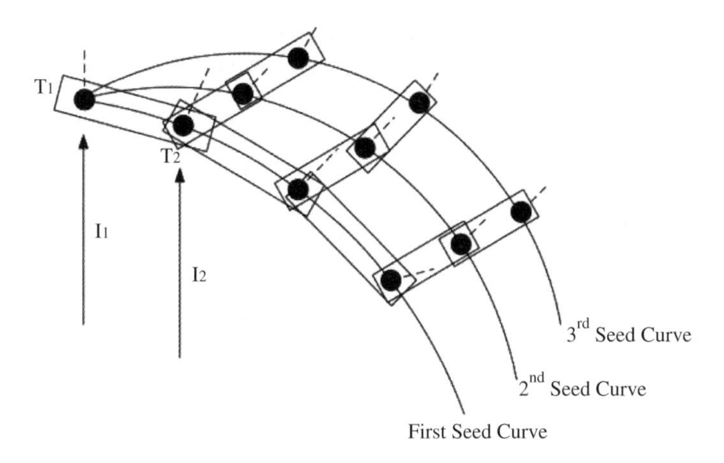

Figure 9.9 Construction of the seed curves of a freeform top surface.

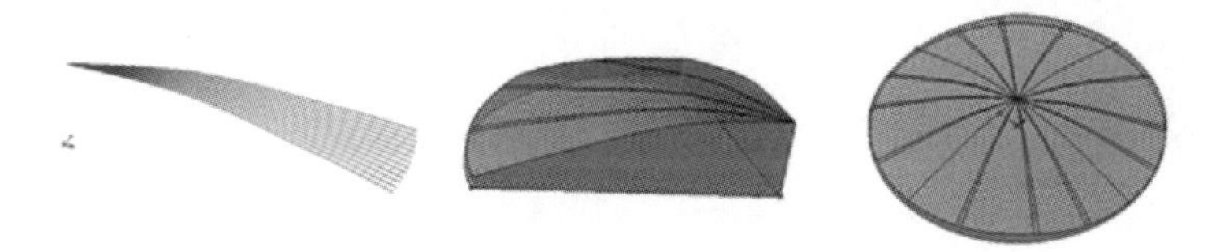

Figure 9.10 Validation of the freeform top surface.

Edge rays and edge points are obtained by slicing the light source and target plane, respectively. Discrete refracting points need to be calculated to realize the mapping between them.

Step 2. Calculation and construction of a freeform top surface. Mapping between $M \times N$ pairs of edge rays and edge points should be realized; thus, $M \times N$ discrete refracting points will be calculated. However, since both the light source and target plane are central-symmetrical, we only need to calculate one-quarter of the discrete refracting points, and the other points can be obtained by mirroring them twice.

As shown in Figure 9.9, height of the public initial point T_1 should be determined before calculation. Like the calculation steps described in Chapter 3, M discrete points forming the first seed curve are obtained. Construction of the rest of the seed curves is different from that of the first one. According to the surface construction principle, discrete points of the rest of the seed curves are intersected by incident rays and the tangent plane of the previous seed curve.

As shown in Figure 9.10, after finishing the calculation of seed curves, one-quarter of the freeform top surface can be constructed by connecting these curves using the lofting method,[13] and the whole freeform top surface can be validated by mirroring it twice. To validate the freeform surface to be a freeform lens, material is filled at the bottom of the surface with 2 mm thickness.

Step 3. Simulation results of the integral freeform illumination lens based on a point light source and an extended light source. A cross section of the integral freeform illumination lens with traced light rays is shown in Figure 9.11. The distance between the light source and

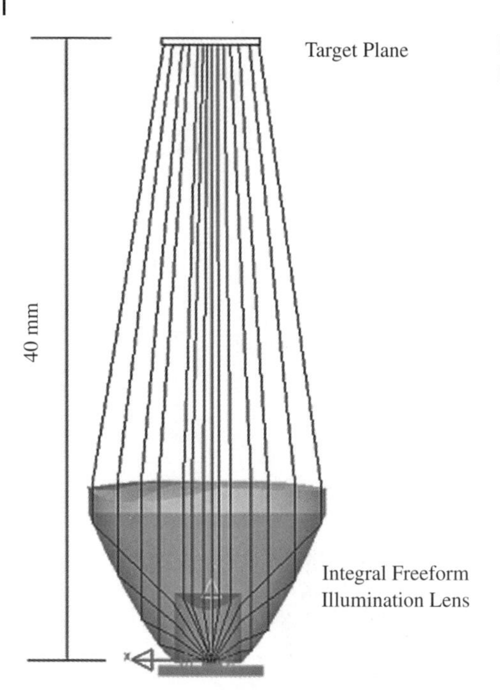

Figure 9.11 Cross section of an integral freeform illumination lens with light rays traced.

target plane is 40 mm. Compared with the light pipe and microlens arrays introduced in this chapter, it is much more compact in structure. The height and diameter of the integral freeform illumination lens are 10 mm and 16 mm, respectively.

To test light performance of the illumination lens, a model including a light source, lens, and target plane is set up. The size of the target plane accords with a real liquid crystal on silicon (LCoS) panel with a ratio of 16:9 and a receiving area of 10×5.6 mm. Figure 9.12 shows ray-tracing results of the illumination distribution on the target plane using a point light source. It is found that light efficiency and illumination uniformity achieve 98% and 0.97 under the point light source.

However, light performance deteriorates a lot under an extended light source, as shown in Figure 9.13. An emitting area of the extended light source is 1.6×1 mm. We learned from ray-tracing results provided by the simulation software that light

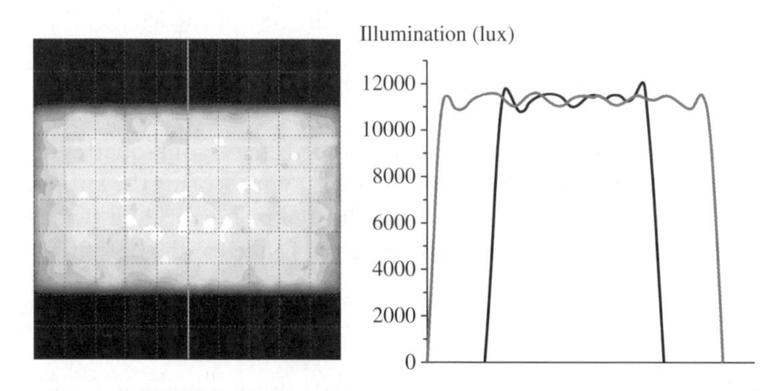

Figure 9.12 Illumination distribution on the target plane based on a point light source.

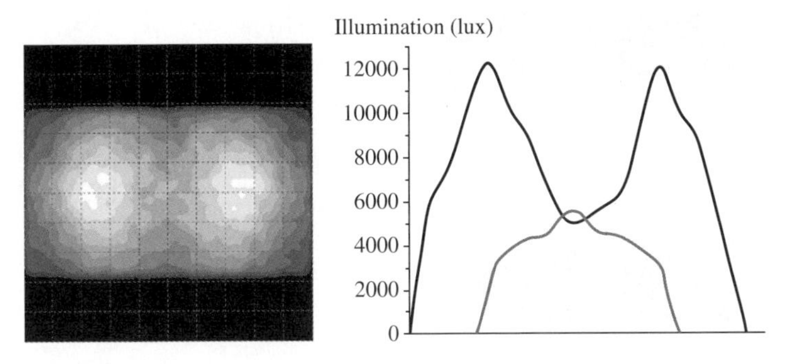

Figure 9.13 Illumination distribution on the target plane based on an extended light source.

efficiency and illumination uniformity decrease to 62% and 0.5, respectively. Therefore, to make this available in real applications, the lens should be optimized.

9.2.4 Optimization of the Integral Freeform Illumination Lens

Actually, both TIR and the top surface are affected by an extended light source. However, merely optimizing the top surface is more straightforward than optimizing them simultaneously to redistribute the illumination on the target plane.

As shown in Figure 9.14, we match the area grids of the target plane with illumination distribution. Light energy in the grids with higher illumination should be transformed partially to the grids with lower illumination. Generally, adjusting slicing of the target plane is an effective method for energy transforming.[14] The area of those grids with higher illumination is increased, and the area of the grids adjacent is decreased. Because the light energy of each grid is designed to be equal, according to photometry theory, illumination in a grid decreases with the increasing of its area that can realize energy transforming.

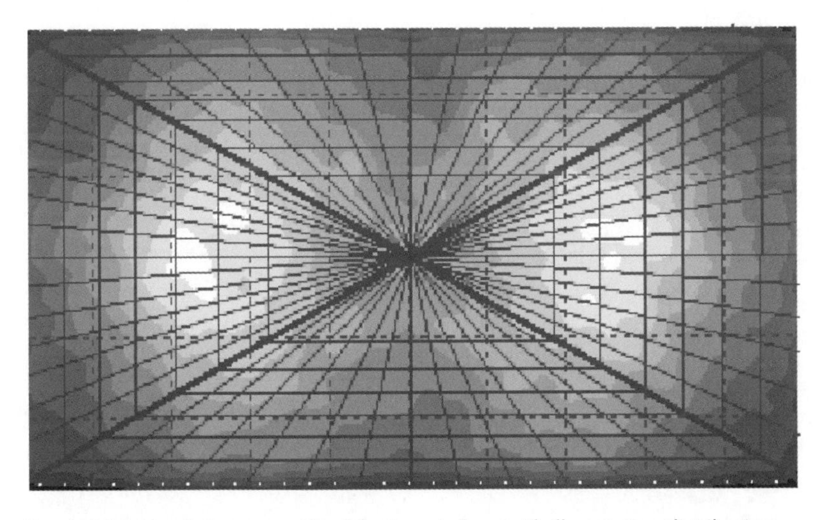

Figure 9.14 Match the area grids of the target plane with illumination distribution.

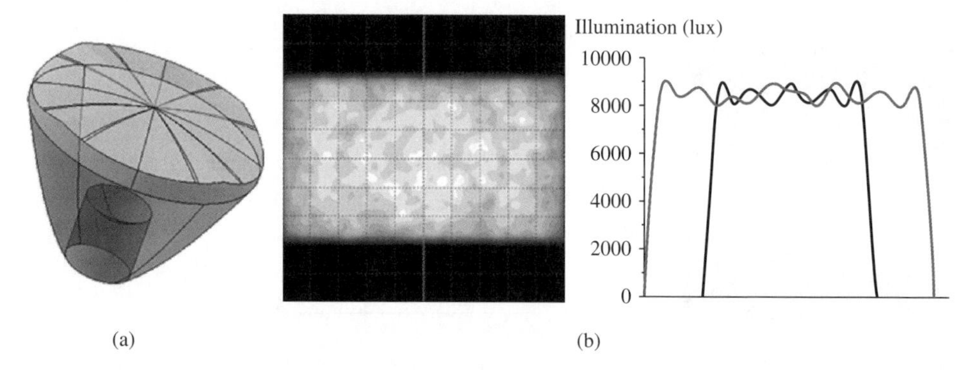

Figure 9.15 (a) The final optimized illumination lens and (b) its illumination distribution.

Cartesian coordinates (x_i, y_j, z) specify the edge points on the target plane being sliced equally. Coefficients a_i and b_j are employed to adjust the area of the grids, and optimized coordinates are expressed as:

$$x_i' = a_i x_i \ (i = 0, 1 \ldots M) \tag{9.9}$$

$$y_j' = b_j y_j \ (j = 0, 1 \ldots N) \tag{9.10}$$

For those grids whose area should be increased, coefficients a_i and b_j of their edge points are greater than 1. Oppositely, a_i and b_j should be less than 1. Values of a_i and b_j can be determined roughly by the ratio of real illumination of the grid and average illumination on the whole target plane. A new top surface will be obtained with the adjusted area grids. If its ray-tracing result cannot match the requirement of the projector, slicing of the target plane should be optimized for another time.

The final optimized illumination lens and simulation result with an extended light source is shown in Figure 9.15a and 9.15b. Compared with the result before optimization, light efficiency and illumination uniformity are increased to 77% and 0.92, respectively.

9.2.5 Tolerance analysis

Tolerance analysis is important because installation errors cannot be avoided in real application. As shown in Figure 9.16, the installation errors can be summarized as

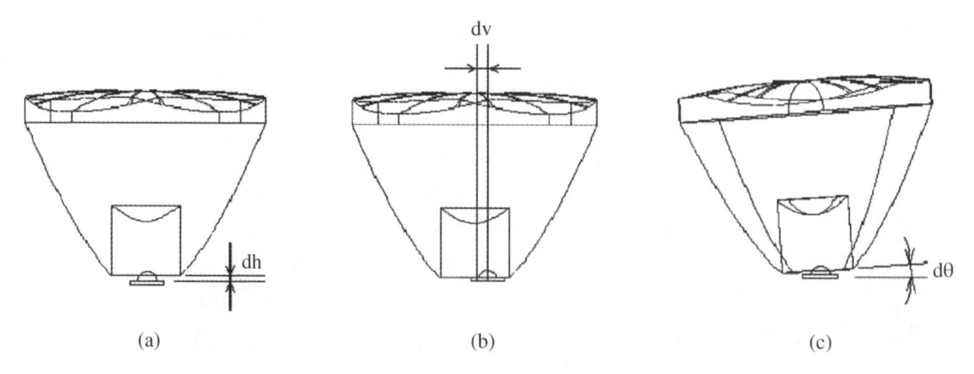

Figure 9.16 (a) Vertical deviation, (b) horizontal deviation, and (c) rotational error.

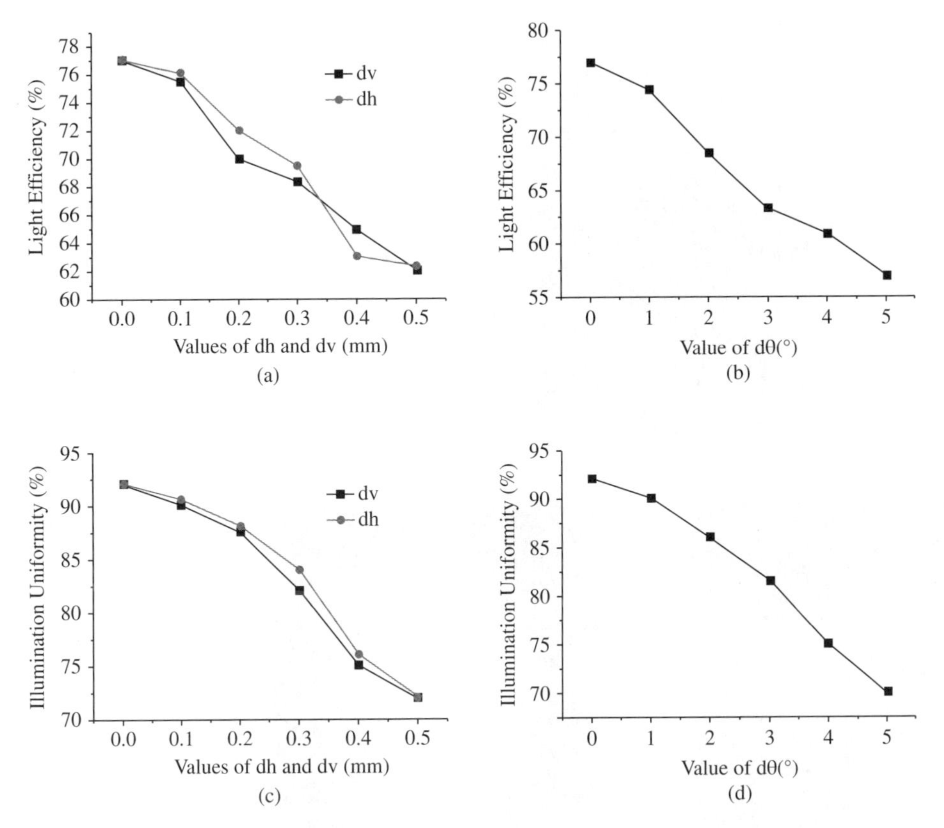

Figure 9.17 (a) Curves of light efficiency versus *dh* and *dv*; (b) curve of light efficiency versus $d\theta$; (c) curves of illumination uniformity versus *dh* and *dv*; and (d) curve of illumination uniformity versus $d\theta$.

vertical deviation, horizontal deviation, and rotational error represented as *dh*, *dv*, and $d\theta$, respectively.

For this analysis, the value range of *dh* and *dv* is set to be (0, 0.5 mm), and the value range of $d\theta$ is set to be (0, 5°). The effects of the three types of the installation error on light efficiency and illumination uniformity are given by graphs shown in Figure 9.17a and 9.17b.

It is found from Figure 9.17 that both light efficiency and illumination uniformity are sensitive to all three types of installation error. Light performance will deteriorate rapidly when *dh* or *dv* exceeds 0.1 mm or $d\theta$ exceeds 1°. Thus, high installation accuracy is required for the integral illumination lens.

9.2.6 LED Pico-Projector Based on the Designed Freeform Lens

In order to verify the actual optical performance of the freeform lens, we select a commercial LED pico-projector on the market to perform comparative analysis. The internal structure of the projector optical engine is shown in Figure 9.18. The image display chip uses a transmission-type liquid crystal display (LCD) with size of 50 × 42 mm. A light beam is reflected onto the LCD panel for modulation by reflecting mirror 1, and then

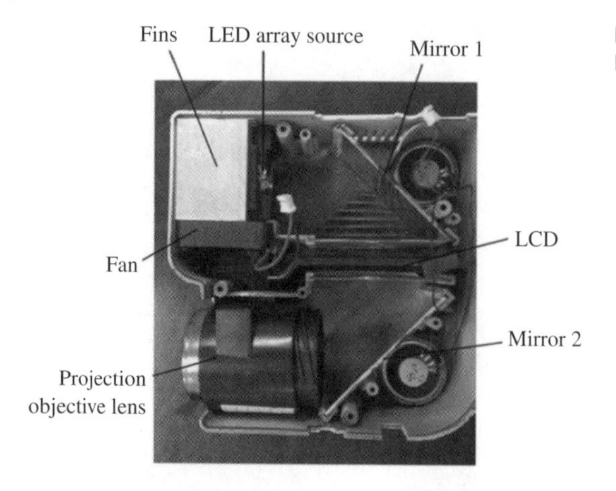

Figure 9.18 Optical engine inside the LED pico-projector.

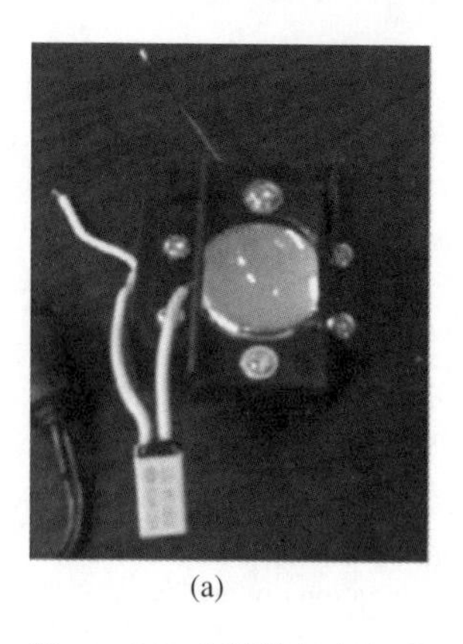

(a)

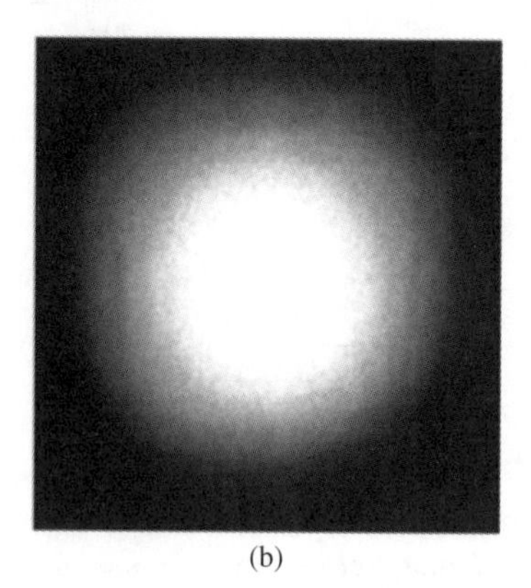

(b)

Figure 9.19 (a) LED lighting module of the pico-projector and (b) illumination distribution on the target plane with a 70 mm distance.

reflected onto the surface of the projection lens for amplification imaging by reflecting mirror 2.

Figure 9.19a shows the lighting module, which comprises a chip-on-board LED light source and an illumination lens. The optical distance between the light-emitting surface and the LCD panel is 70 mm, and the illumination distribution with 70 mm in front of the lighting module is shown in Figure 9.19b, which is also the actual illumination distribution on the LCD panel. Obviously, the output light pattern of the projector is circular, and it cannot match the outline of the LCD panel that is rectangular, resulting in most light energy being wasted.

According to the actual size of the LCD, we designed a PMMA freeform lens. Because the face of the freeform type is complex and there is a requirement for high processing precision, we adopt the injection-molding process. Compared to mechanical processing, the freeform lens processed by injection molding can avoid edge-cutting machining errors. Figure 9.20 shows the freeform lens. The support columns 1 and 2 at the bottom are used for lens installation. The largest diameter of the freeform lens is 22 mm, and the height is 19 mm.

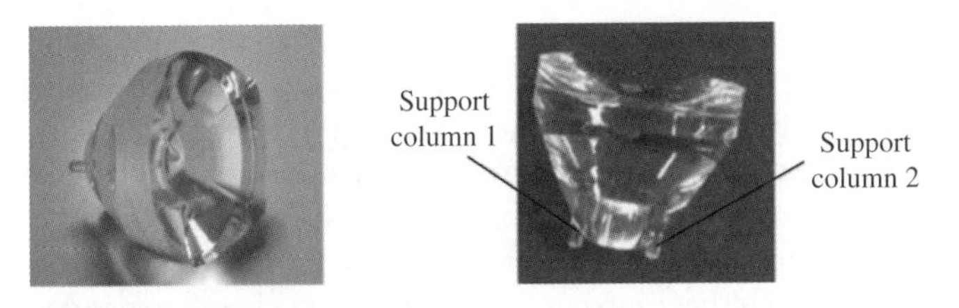

Figure 9.20 PMMA freeform lens for a high-efficiency LED pico-projector.

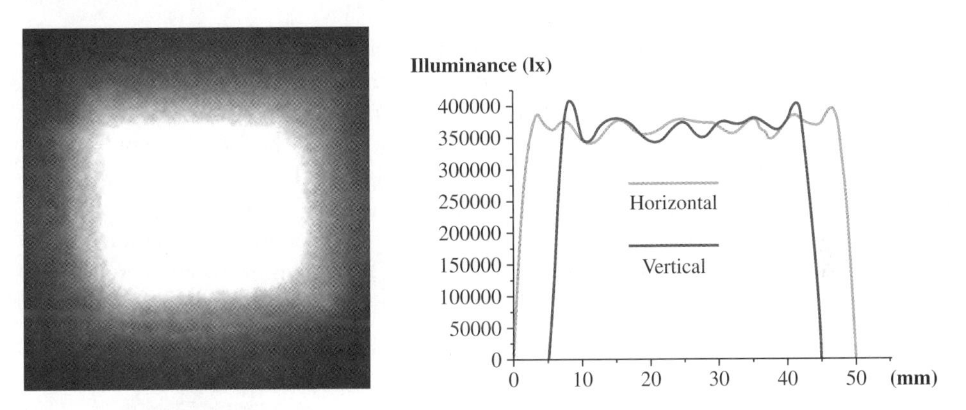

Figure 9.21 Illumination distribution at the target plane, 70 mm from the LED illumination module with freeform lines.

The LED light source used with the freeform lens is CREE XL-M. The packaging size of the LED module is 5 mm^2, the full viewing angle is 125°, and the luminous flux can reach 1040 lm under a driving current of 3A. This means that the advantage of this LED source applied in the pico-projector lies in the small chip light-emitting area and compact packaging size, so that the Étendue is small, whereas the output lumen flux is large. A light pattern 70 mm in front of the LED lighting module with the freeform lens is measured, as shown in Figure 9.21. The light pattern is rectangular, and its size is in accordance with the LCD panel contour. By calculation, the light energy utilization ratios on the LCD panel of these LED lighting modules with a hemisphere lens and freeform lens are 54% and 77%, respectively. The enhancement of the light energy utilization ratio reaches as high as 42.6% when adopting the freeform lens.

9.3 Freeform Lens Array Optical System for an LED Stage Light

In this section, an optical system based on a freeform array configuration is proposed to realize the light distribution control of a large-distance LED light source array. Taking LED stage lighting as an example, an adjustable light optical system that consists of a high-power colored LED light source array, a freeform collimating lens array board, and a freeform array beam-expanding plate is designed. The system can realize different lighting requirements by replacing the beam-expanding plate. The replacement process is more convenient compared with the traditional lens installation process. Taking the LED stage light multiplex control system as an example, we design an eight-color stage-lighting control system based on the DMX512 protocol. Compared with traditional stage light, the system can achieve a wider range of lighting in the chromaticity diagram and produce more colors. At the same time, the operation is simple and strongly compatible. Figure 9.22 shows the schematic of controlling the light pattern of a large gap colored LED array by a freeform lens array. In this section, we will introduce the collimating board, the one-dimensional beam expander, the rectangle beam expander algorithm, surface construction, and their practical applications in LED stage light, respectively.

According to the TIR collimating lens construction method mentioned in Chapter 3, a freeform collimating lens for this chapter can be constructed. Specific steps are more or less the same and therefore will not be narrated. We put 40 monochromatic LED chips of eight colors on the printed circuit board (PCB) substrate for the formation of the LED light source array. As shown in Figure 9.23, when designing the collimation system, we adopt a hexagon arrangement to make the lens array as close as possible. Compared to the square matrix, this configuration can save about 14% area, make the parallel beam more intensive, take up less space, and also enable LEDs to be more evenly mixed at the same time.

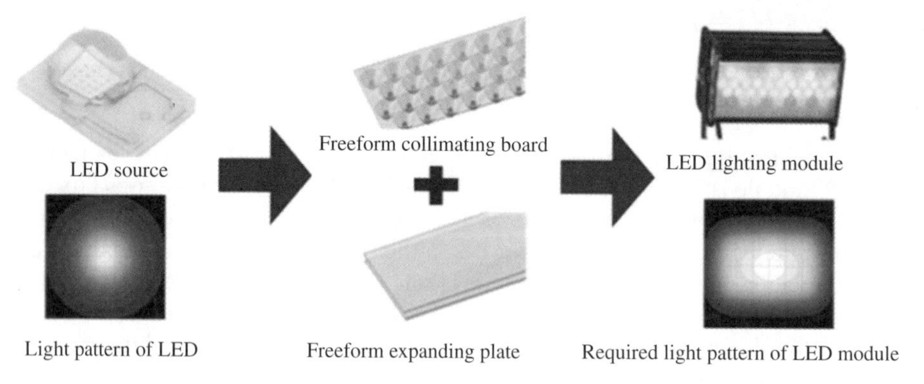

Figure 9.22 Schematic of light pattern control of a large-distance color LED array by an LED freeform lens.

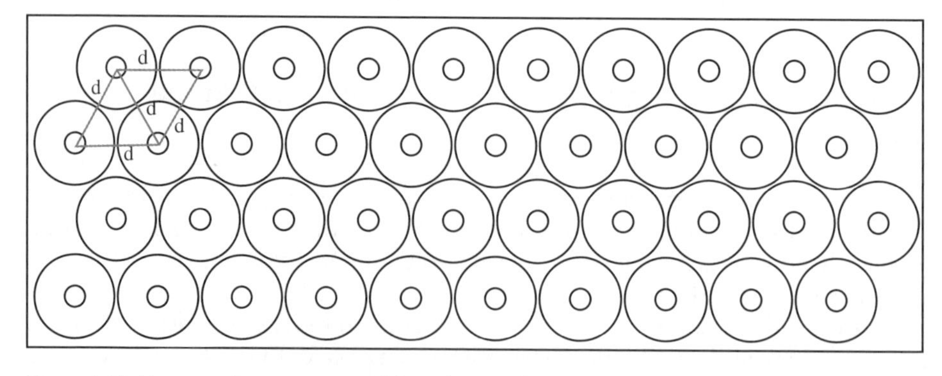

Figure 9.23 Hexagonal arrangement of the collimator lens array.

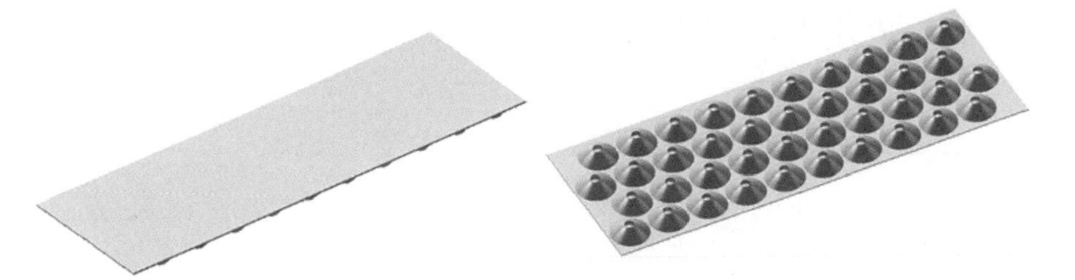

Figure 9.24 Collimating lens array board of front view (left) and back view (right).

As shown in Figure 9.24, because the large number of independent collimating lenses is not easy for installation and integration, we decide to design the collimating lens on one plate.

9.3.1 Design of a One-Dimensional Beam Expander Based on a Freeform Lens Array

For the design of a small spacing LED array light source, the LED light source adopts polar coordinates (γ, θ, ρ), and the lens and the target plane use a rectangular coordinate system (x, y, z). In this section, we will deeply study the situation when incident light is parallel, and construct a light control system for a complete set of large LED light source arrays by the planar microfreeform lens array.

9.3.1.1 Part 1. Gridding of the One-Dimensional Target Plane

When discussing the beam expanding of a single direction, we only need to consider realizing the one-dimensional light energy distribution in a freeform microstructure array. Therefore, when designing the freeform lens, we just extrude the relative connection and structure inside the plane to get the microstructure array.

The receiver plane is set 10 m in front of the freeform. In a one-dimensional beam-expanded light source, we just need to divide the receiver plane along the x-axis direction by equal energy, and the y-axis direction needs no consideration. As shown in Figure 9.25, the receiver plane is divided along the x-axis into M subareas parallel with the y-axis, and regional boundaries of each division and x-axis have an intersection point, which can be regarded as energy projecting points on the target plane. So the light energy projecting relationship from the freeform surface to the receiver plane is established.

9.3.1.2 Part 2. Algorithm of a One-Dimensional Freeform Microstructure

1. As shown in Figure 9.25, we design the bottom surface of the microstructure as a plane surface; in this way, when parallel light is incident to the bottom surface, it will not bend as it is a vertical incident parallel beam. At the same time, vertical incidence can reduce specular reflectance and Fresnel loss to their minimum values. We successively set the location of the incident point as (0, 0, 0), (x_2, 0, 0), (x_3, 0, 0). The expressions of the incoming light corresponding to these points are: $\begin{cases} x = 0 \\ z = 0 \end{cases}, \begin{cases} x = x_2 \\ z = 0 \end{cases}, \begin{cases} x = x_3 \\ z = 0 \end{cases}$ .... First in the two-dimensional

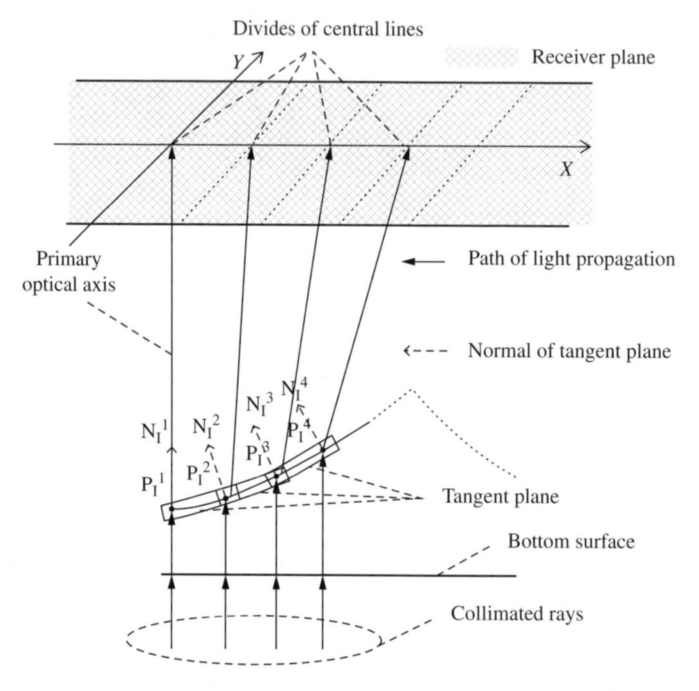

Figure 9.25 Schematic of a one-dimensional freeform microstructure beam-extending algorithm.

plane, the coordinates of the first point P_I^1 on the freeform surface of the microstructure are $(0, y_1, 0)$. Its location lies right above the main optical axis. The height of point P_I^1 determines the height of the lowest point of the microstructures Z_0. The first ray and the optical axis should coincide, and the normal vector N_I^1 of point P_I^1 is parallel to the optical axis direction and vertical.

2. Calculating the second point P_I^2 based on the first point P_I^1 in the free curve. P_I^1's normal direction vector N_I^1 is set as the normal vector, a tangent plane is made out of point P_I^1 to intersect with the second light at P_I^2 $(x_2, y_2, 0)$, and $\begin{cases} x = x_2 \\ z = 0 \end{cases}$ is used as the equation for the second parallel incident light. By combining Snell's law with the incident light vector I_I^2 and outgoing light vector O_I^2, we can get the normal vector of the point P_I^2 as N_I^2. The direction vector of the incident light is I, which can be normalized as $(0, 1, 0)$, and the vector of output light O is determined by the coordinates $(x_2, 10000, 0)$ of the corresponding point Q_2 on the target plane and the location of the point on the lens. Based on this method, we can get the coordinates $(x_{i+1}, y_{i+1}, 0)$, I_I^{i+1}, O_I^{i+1}, and N_I^{i+1} of the next point P_I^{i+1} by using the coordinates $(x_i, y_i, 0)$, I_I^i, O_I^i, and N_I^i of the point P_I^i. By repeatedly using this method, we can construct the whole free curve.

Compared to the Lambertian light emitted by a point light source, the incident position of the parallel beam is more flexible. The adjacent two incoming light beams can be equally spaced or not; the specific spacing value can be determined according to the size of the designed lens microstructure.

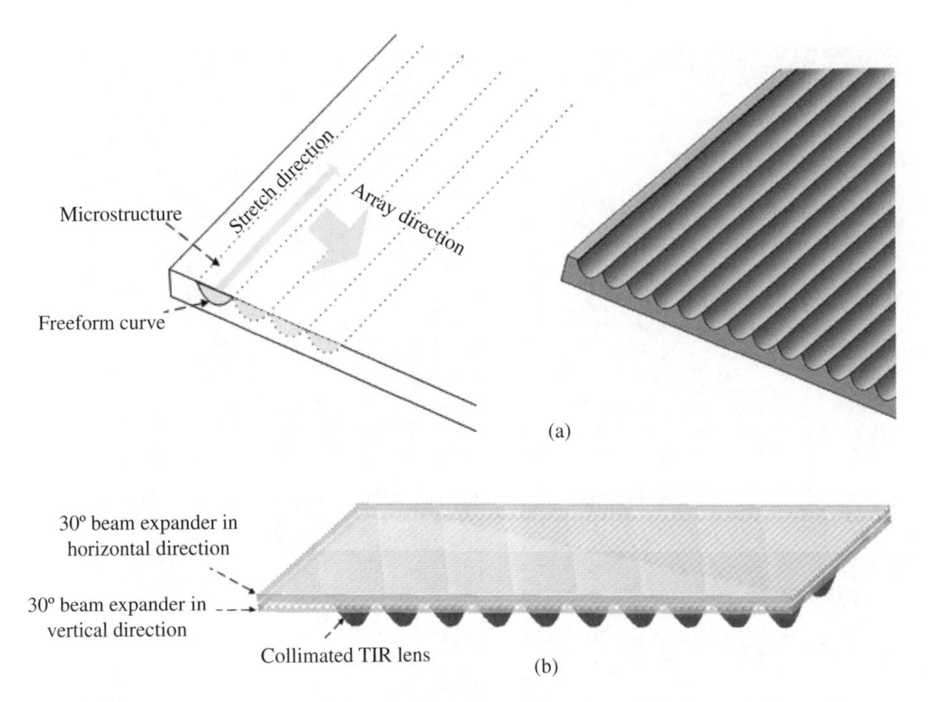

Figure 9.26 Schematic of (a) a freeform microstructure groove, and (b) the optical system consisting of a one-dimensional freeform microstructure array beam expander and collimating TIR lens board.

9.3.1.3 Part 3. Optical Simulation Results of the Optical System

As shown in Figure 9.26, a horizontal beam expander is added in the upper part of the vertical beam expander, and the stretching directions of the microstructure groove on the two surfaces are perpendicular to each other. The Monte Carlo ray-tracing method is used to observe the light pattern change when a beam expander is used alone or two beam expanders are used in conjunction with each other.

As showed in Figure 9.27a, when the beam expander is not used, the calibrated parallel beams illuminate on the receiver plane, which is 10 m away, and form a concentrated light spot. In the case of ignoring the absorption of dielectric material and Fresnel losses, the efficiency of the optical system is 90.41%. The installation of the beam expander in the vertical or horizontal directions can stretch the light patterns in the corresponding direction. It can even be obviously observed in Figure 9.27b and 9.27c that the beam expander is able to stretch the spot light pattern uniformly on the receiving screen, showing an outstanding performance of beam expanding. Meanwhile, the light efficiency is 82.03% and 82.05%, respectively. In Figure 9.27d, when combining use of the vertical and horizontal beam expander with a $\pm 30°$ inclined angle, a quasi-square spot with uniform illuminance distribution appeared on the receiver plane, and the efficiency of the optical system decreased to 72.95%.

9.3.2 Design of a Rectangular Beam Expander Based on a Freeform Lens Array

In order to eliminate the "pillow"-type distortion caused by the twice refraction, we will use one refractive to obtain a giant distortion-free light pattern. Different from a

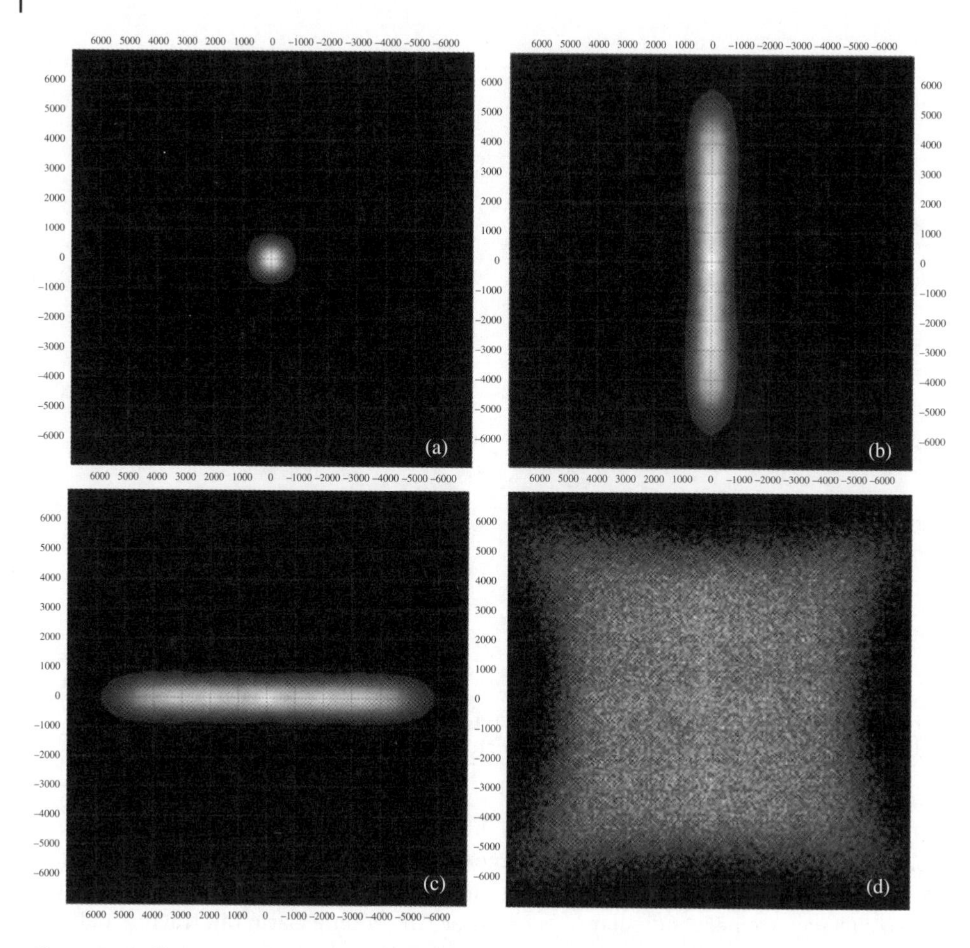

Figure 9.27 Illuminance distribution of a freeform array optical system on the receiver plane: (a) without beam expander, (b) expanding the beam in the vertical direction, (c) expanding the beam in the horizontal direction, (d) and expanding the beam in both the vertical and horizontal directions.

rotation symmetric freeform surface, the forming of a microfreeform surface with a two-dimensional rectangular beam requires implementation of multiple freeform surfaces lofting to obtain the microlens surface.

As shown in Figure 9.28, the receiver plane is set in front of the microstructure surface at a distance of 10 m. We need to implement the equal energy dispersing on the receiver plane along the x-axis and y-axis direction simultaneously. As an example, the square receiving screen is divided into M number of equal subregions along the x-axis and y-axis directions. Moreover, each divided vertical subregion has an intersection point on the boundary. These points would be the energy mapping point of the target plane, for the purpose of establishing the mapping between the incident ray energy.

9.3.2.1 Part 1. Algorithm of the Rectangular Freeform Structure

1. As shown in Figure 9.28, in order to reduce the influence on the optical path, the bottom surface of the beam expander is still a planar plate. Expressions of the incident

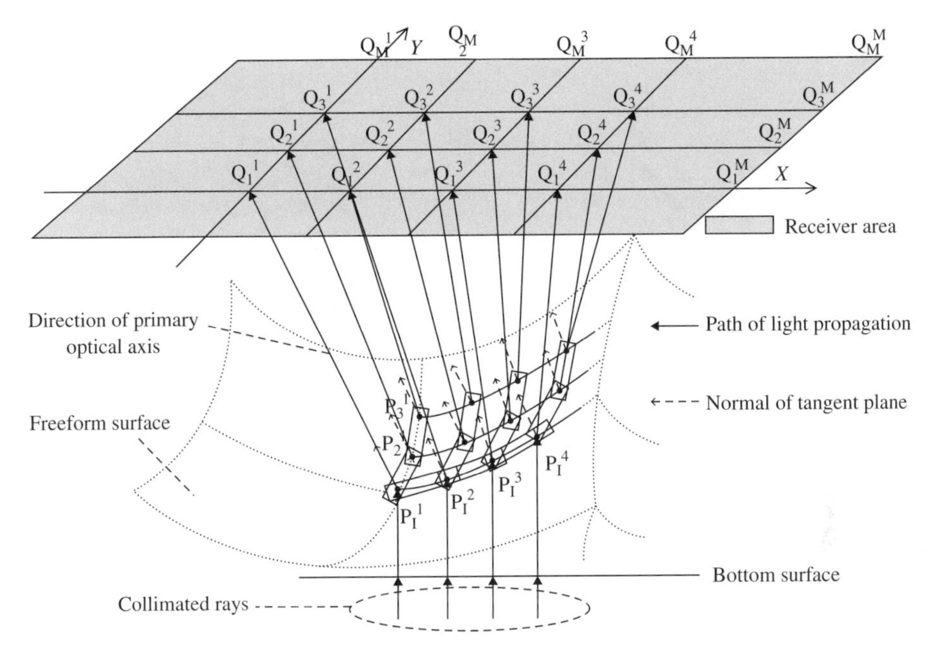

Figure 9.28 Schematic of a freeform lens algorithm for a rectangular target plane.

ray are sequentially set as:

$$\begin{cases} x = 0 \\ z = 0 \end{cases}, \begin{cases} x = x_2 \\ z = 0 \end{cases}, \begin{cases} x = x_3 \\ z = 0 \end{cases} \cdots \text{(the incident ray of the first free curve)}$$

$$\begin{cases} x = 0 \\ z = z_2 \end{cases}, \begin{cases} x = x_2 \\ z = z_2 \end{cases}, \begin{cases} x = x_3 \\ z = z_2 \end{cases} \cdots \text{(the incident ray of the second free curve)}$$

Firstly, based on the requirements, we determine the initial point of the first freeform curve on the microstructure as P_I^1 $(0, y_1, 0)$ in three-dimensional space. Since both point P_I^1 and the mapping point of the receiving screen are on the optical axis, the first ray should be parallel with the optical axis. The normal vector of P_I^1 point, which is denoted as N_I^1, should be parallel to the optical axis and sticking straight up.

2. The second point P_I^2 is calculated based on the first point of the first freeform curve. We set the normal vector of point P_I^1 as the reference normal vector and obtain a tangent plane through point P_I^1. The tangent plane has an intersection point P_I^2 $(x_2, y_2, 0)$ with the second ray. The equation of the second incident ray can be expressed as: $\begin{cases} x = x_2 \\ z = 0 \end{cases}$. Then, we can cross-link the incident ray vectors I_I^2 to the exit ray vector O_I^2 based on Snell's law, and the normal vector N_I^2 of point P_I^2 can be obtained. The expression of the direction vector of the incident ray is normalized as I $(0, 1, 0)$. The emitted ray vector O_I^2 ought to be determined by the position of corresponding point Q_1^2 $(x_2, 10000, 0)$ on the target plane and a relative point on the optical lens. Based on this method, we can calculate the coordinates of the followed point P_I^{i+1} $(x_{i+1}, y_{i+1}, 0)$, I_I^{i+1}, O_I^{i+1}, and N_I^{i+1} after acquiring the position of P_I^{i+1} $(x_{i+1}, y_{i+1}, 0)$, I_I^i, O_I^i, and N_I^i. By repeating this way, the entire free curves can be constructed.

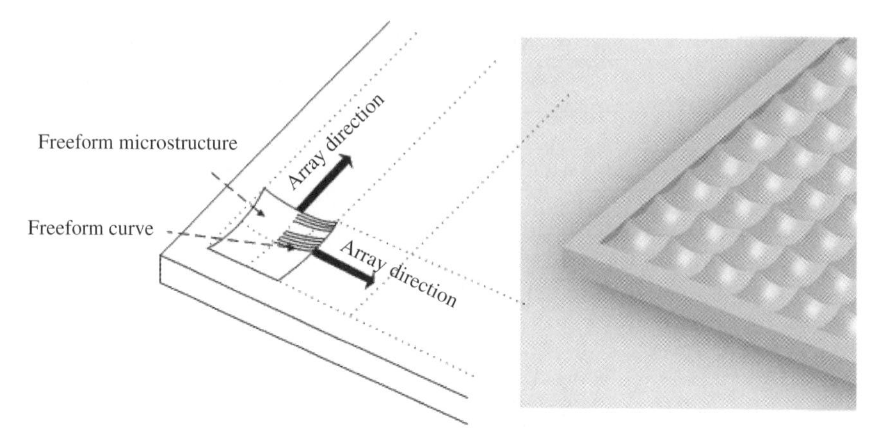

Figure 9.29 Schematic of a rectangular beam expander based on a rectangular freeform microstructure array.

3. Based on the coordinates of points on the first free curve, coordinates of points on the second curve, third curve, and so on can be calculated in sequence. A tangent plane is generated by setting the first point of the first free curve as a tangent point and the vector $N_1^{\,1}$ as a normal vector. As shown in Figure 9.28, where this tangent plane meets with the first incident ray $I_2^{\,1}$ of the second free curve is the position of the first points $P_2^{\,1}$ on the second free curve. Then, the normal vector of point $P_2^{\,1}$ can be determined by the corresponding position of point $Q_2^{\,1}$ on the receiving screen and the direction vector of the incident ray $I_2^{\,1}$, emergent ray $O_2^{\,1}$. In this way, we can calculate the coordinates of the followed point $P_I^{\,i+1}$ $(x_{i+1}, y_{i+1}, 0)$, $I_I^{\,i+1}$, $O_I^{\,i+1}$, and $N_I^{\,i+1}$ after acquiring the position of $P_I^{\,i+1}$ $(x_{i+1}, y_{i+1}, 0)$, $I_I^{\,i}$, $O_I^{\,i}$, and $N_I^{\,i}$. By repeating the steps mentioned above, the entire free curves can be constructed.

As shown in Figure 9.29, based on the point calculated in advance, the separation-point-lofting method is conducted on the rectangular plate with microstructure to obtain independent freeform curves. Then, these freeform curves are lofted to obtain a complete resection plane. Next, the microstructure with rectangular freeform can be formed by implementing the stretching and cutting operation on the board. Finally, with the help of an array operation, we can obtain a surface with matrix microstructure.

9.3.2.2 Part 2. Optical Simulation Results of the Optical System

From the simulation results as shown in Figure 9.30, we can find that the expanding of a uniform rectangular spot can be realized with only one rectangular beam expander, which is much more convenient than the combing method mentioned in the last section.

9.4 Freeform Optics for a LED Airport Taxiway Light

9.4.1 Introduction

The data from the National Transportation Safety Board (NTSB) has shown that the aviation accident rate during aircraft take-off and landing is more than 85%, especially

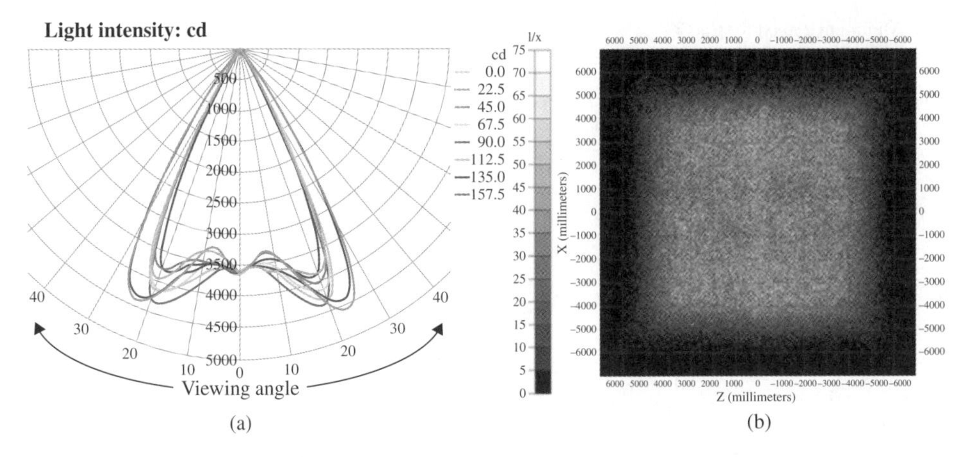

Figure 9.30 (a) Light intensity distribution and (b) illuminance distribution of the optical system with the rectangular beam expander.

at night.[15] Therefore, the luminosity quality of an aeronautical ground light system is essential for aviation safety. As a fourth-generation light source, LED has many advantages over the conventional light sources, including energy and space savings, high reliability, and long life.[16] In traditional optical design, system efficiency is a key issue for the aeronautical ground light when LEDs are adopted as the light source. In addition, there is a conundrum in the optical system design to obtain any wanted luminous intensity distribution from an extended source. In this section, we introduce a design method for an optical system with a TIR freeform lens and prism for LED chip array packaging (LCAP). Normal luminous intensity distribution (NLID) is achieved by an initial ray control and optimization process. As an example, the optical system of a runway centerline light (exit taxiway centerline light) is designed that can fully comply with the International Civil Aviation Organization (ICAO) regulation. Through the numerical simulation, the optical system has optical efficiencies of more than 85% in theory.

9.4.2 Requirement Statement

Usually, many different kinds of aeronautical ground lights are arranged in a normal airport. As the main part of a visual landing and docking guidance system, there are many optical restrictions for the centerline light, especially in the distribution and intensity of luminosity.[17] Firstly, centerline lights should give adequate illumination within a safe flight direction and cannot make glare. Secondly, these lights should give accurate and clear indications for aircraft landing and docking. Therefore, the optical design of the centerline light is one of the most difficult parts of aeronautical ground light design. Figure 9.31 shows the runway centerline beam pattern provided by ICAO regulation,[17] which specifies the luminance intensity on specific areas (I, II, and III) or their borderlines.

9.4.3 Design Method of an Optical System

In our design, normal distribution is adopted as the prospective luminous intensity distribution. With the increasing of emergence angle, light intensity is monotonically and

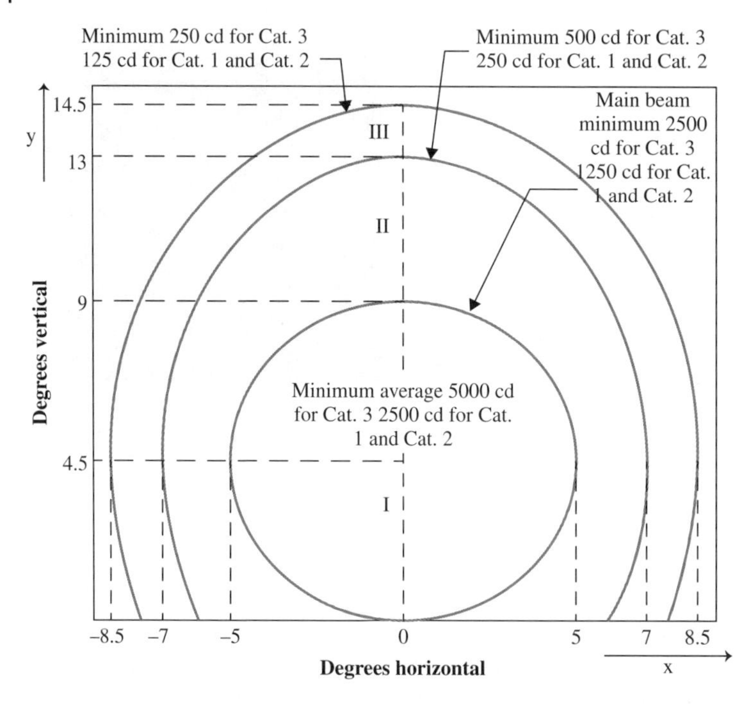

Figure 9.31 Isocandela diagram for a runway centerline light with 15 m longitudinal spacing and a rapid exit taxiway indicator light.[17]

smoothly decreasing.[18] This kind of distribution will give a clear visual guidance to the pilot, and it cannot make glare at the center line. A freeform optical system is separated into three parts: the TIR collimating optics, freeform surface, and prism. The TIR collimating optics is used to collect and collimate the lights radiating from the light source, which consists of two parts, the inner surface and outer surface of the TIR lens, and they collimate rays by refraction and TIR, respectively.[19] We use the Cartesian oval method to calculate the inner surface, which is constructed in the following steps.[20,21]

In order to obtain NLID, a freeform surface is used as the refraction optics on the TIR lens. Firstly, as shown in Figure 9.32b, NLID from 0° to 13° was divided into 500 equal luminous flux annular zones in polar coordinates. Equal luminous flux degrees of each zone in the coordinated plane can be obtained. Secondly, we can determine the direction vector of the exit rays of TIR collimating optics, and the element of the freeform surface can be calculated by Snell's law, which is shown in Figure 9.32a. Finally, rotate the three elements to get the integrated TIR collimating optics with the freeform surface.

To validate our design, as shown in Figure 9.33a, an OSRAM OSTAR SAT LE UW S2W-NZPZ-FRKV white LED with satisfactory color coordinates 0.31(Cx)/0.32(Cy) is used in the numerical simulation.[22] According to the data sheet of this LED, its flux can reach 710 lm when the LED is driven at a specified current of 700 mA.[21] We assume the flux of each LED is 320 lm in the simulation. As Figure 9.33b shows, a prism with a silver plating reflector is used to reflect rays to the ultima direction with 4.5° to the horizon. The material of the TIR lens and prism are PMMA and borosilicate glass, respectively, the refractive indexes of which are 1.49 and 1.47. The dimension of the LED optical system is 94.8 (H) × 55 (L) × 50 (W) mm.

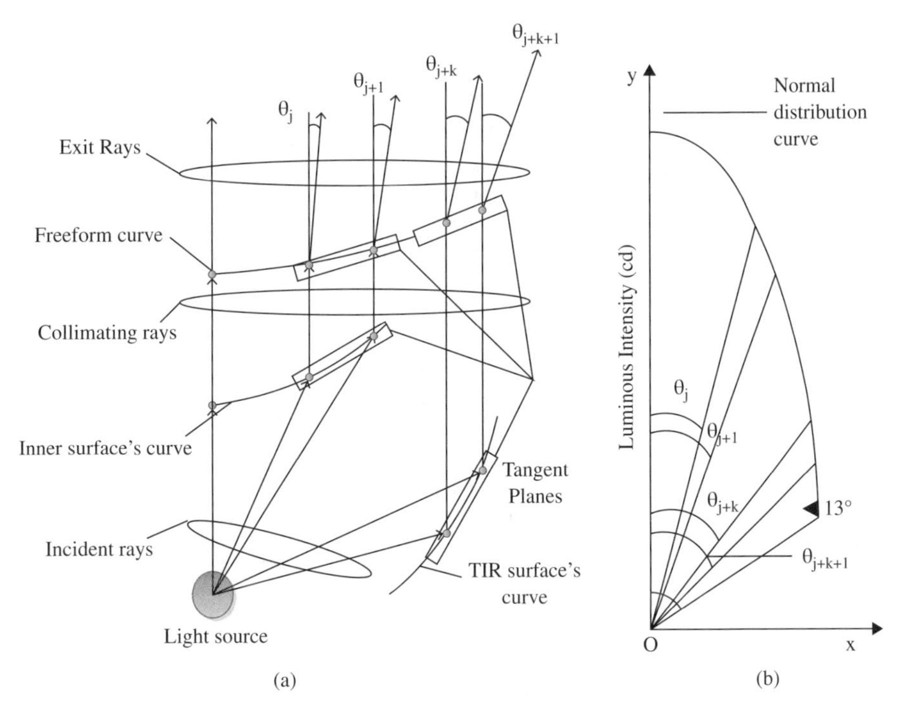

Figure 9.32 Schematic diagrams of collimation and control of rays.

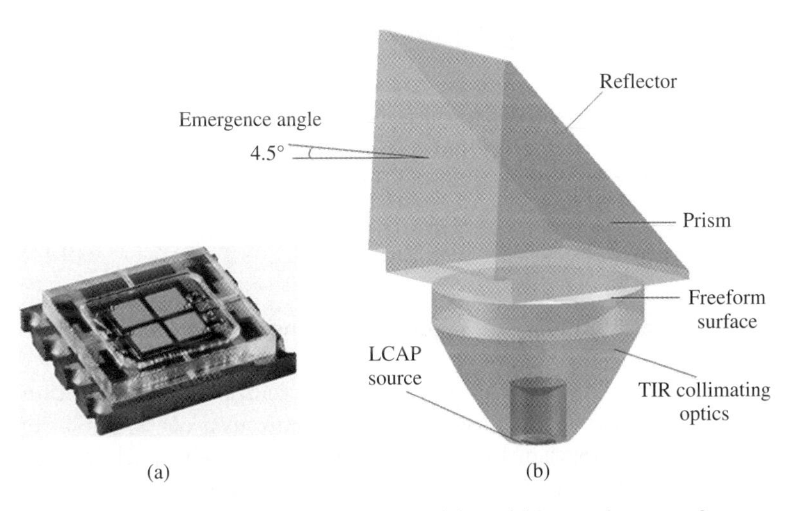

Figure 9.33 Schematic of (a) LCAP source module and (b) optical system of a runway centerline light.

9.4.4 Simulation and Optimization

As shown in Figure 9.34a, through the Monte Carlo ray-tracing method based numerical simulation, results demonstrate that the luminous intensity distribution can fully comply with the ICAO regulation. However, the schematic shows the differences between the initial simulation result and normal distribution, which are coursed by the extended

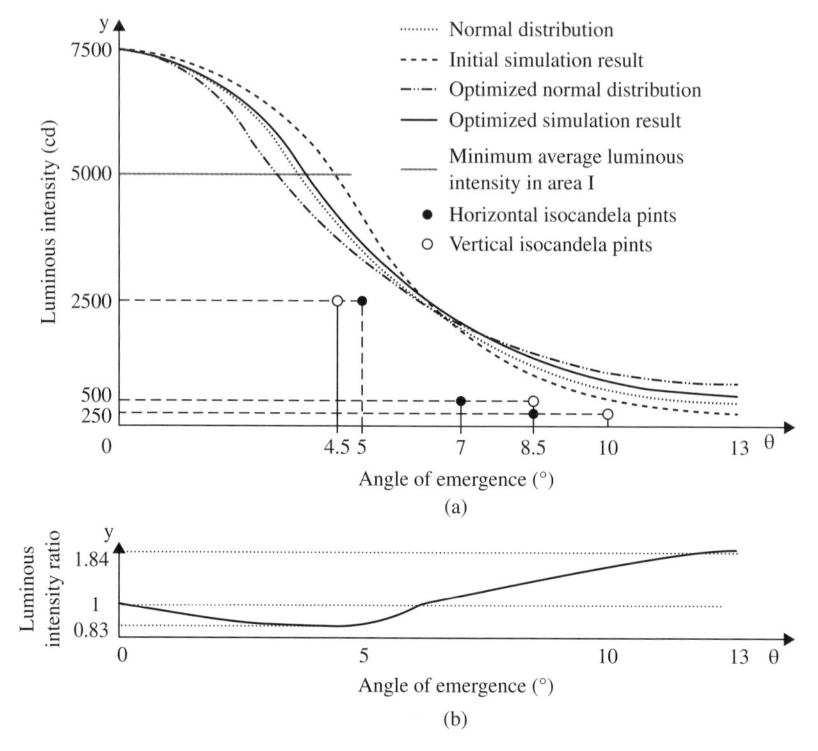

Figure 9.34 (a) Simulation luminous intensity distribution result and (b) primary optimization curve $\chi_1(\theta)$ in a two-dimensional plane.

source. Figure 9.34b shows that the primary optimization curve $\chi_1(\theta)$, which is the ratio variation of the luminous intensity distribution to normal distribution, is adopted to form the optimization normal distribution function (Eq. 9.11):

$$f(\theta) = \frac{7500}{\sqrt{2\pi}\sigma} \exp\left(\frac{\theta - \mu}{2\sigma}\right)^2 \bullet \chi_1(\theta) \quad \theta \in (0^\circ, 13^\circ) \tag{9.11}$$

where the expected value μ and standard deviation σ are defined as 0 and 0.08. respectively.

Through primary optimizing, NLID is achieved, and the normalized cross-correlation (NCC)[23] increases from 83.5% to 93.2%. Moreover, by a repetitious process, similarity can be further improved. The comparison of simulated luminous and required luminous for ICAO regulation is shown in Table 9.1. Considering Fresnel losses, the optical efficiency of the lenses system is as high as 85%. Based on this optical system, a rapid exit taxiway indicator light and taxiway centerline light can be obtained only by substituting the white LCAP source for the yellow one and setting two prisms.

9.4.5 Tolerance Analysis

In practice products, other issues we must consider are a rounded edge in the TIR lens and installation error. A rounded edge is one of the most common manufacturing errors.

Table 9.1 Simulated illumination result compared with the corresponding value of ICAO for a runway centerline light.

Measuring range on screen		Required luminous in cd for Cat. 3	Simulated luminous in cd
I	BLI*	≥2500	4436.2 (5°) ~ 4909.4 (4.5°)
	ALI**	≥5000	6270.7
II	BLI	≥500	1024.1 (8.5°) ~ 2541 (7°)
	ALI	–	3904.3
III	BLI	≥250	350.8 (10°) ~ 1024.1 (8.5°)
	ALI	–	2241.6

* Borderline luminous intensity.

** Average luminous intensity.

By comparing Figure 9.35a with Figure 9.35b, since the rays in the rounded edge cannot be collimated correctly, v the average luminous intensity in area I linearly decreases as the radius of the rounded edge increases, as shown in Figure 9.35c. When vertical deviation *d*, as shown in Figure 9.35d, is increasing from −1 mm to 2 mm, the average luminous intensity in area I appears at a maximum value of 8054.4 cd at 0.5 mm. This phenomenon is caused by the focus of collimated rays that are reflected by the TIR surface, as shown in Figure 9.35f. With the hoisting of the source position, the focus rays diverge again and the luminous intensity value gradually declines.

With the exception of average intensity, these two kinds of error also have negative effects on luminous intensity distribution. In this design, the limits on the radius of the rounded edge and vertical deviation are 0.3 mm and −0.1~0.2 mm, respectively, whose luminous intensity variations are less than 15%, which are acceptable for mass production. This method may provide an effective way to design other kinds of aeronautical ground lights with an LED chip array source.

9.4.6 Design of an LED Taxiway Centerline Lamp

One LED airport taxiway centerline lamp is designed based on the optical system mentioned in the last section. As shown in Figure 9.36a, due to the reason that the lamp will be under huge pressure when aircrafts fly over the lamp, the top plate needs to be relatively thick to strengthen its pressure-bearing capacity. Consequently, the space for emergent light is small and the emergent angle is low. This leads to the high requirement for optical design. As shown in Figure 9.36b, the optical system of the lamp consists of a freeform TIR lens and total reflection prism. The whole system is designed to be bidirectional emergent, and the TIR lens with double freeform surfaces is adopted to meet the requirement of light intensity distribution. Meanwhile, the LED light source module, adjustable stand, and fins for heat dissipation are installed at the bottom of the TIR lens. Compared with the airport runway centerline lamp, this kind of taxiway centerline lamp can be adjusted to accommodate more requirements of ground navigation.

As shown in Figure 9.37, by adjusting the included angle of the optical system, this LED airport taxiway centerline lamp can be adapted to a variety of airport navigation conditions. This lamp achieves the goal of using one taxiway centerline lamp for different

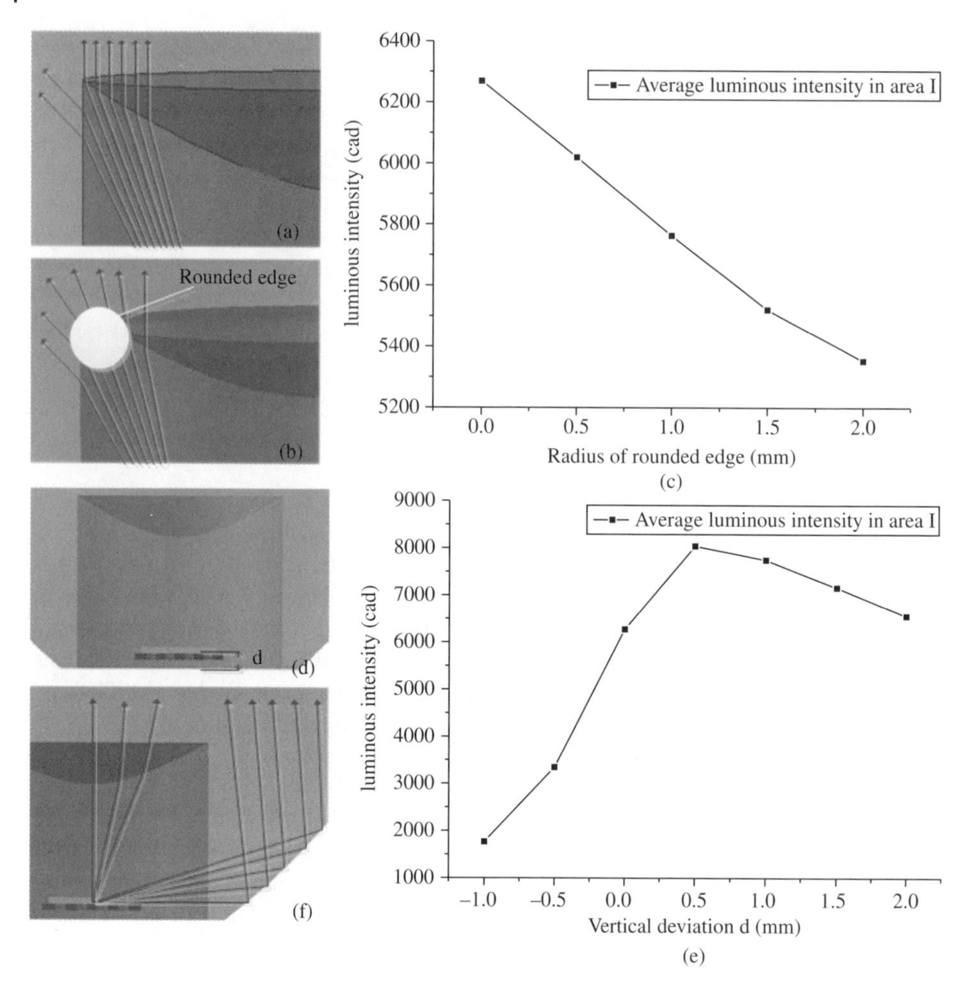

Figure 9.35 Effects of rounded edges and installation errors on lighting performance of the LED runway centerline light. (a) Perfect edge; (b) rounded edge; (d) vertical deviation *d*; (f) focus of rays; and (c, e) average luminous intensity variation in area I.

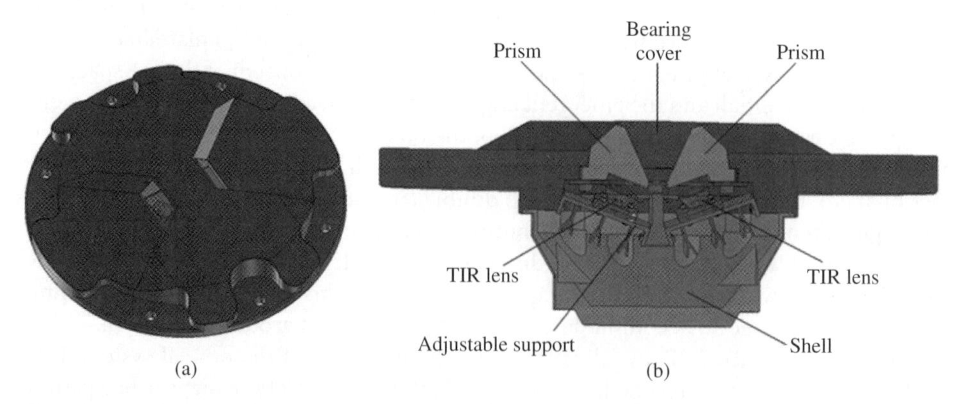

Figure 9.36 Schematic of an LED taxiway centerline lamp: (a) top view and (b) cross-sectional view.

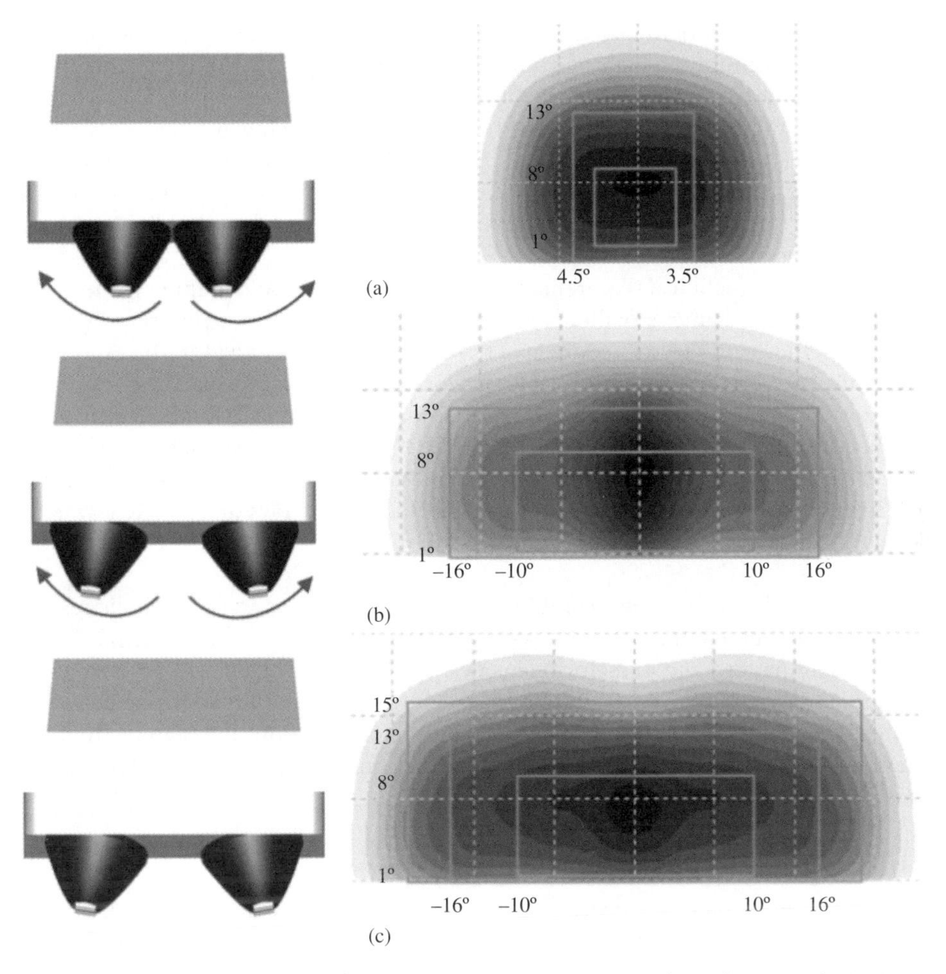

Figure 9.37 LED taxiway centerline lamp with adjusting angles of (a) 0°, (b) 7°, and (c) 9°, which can satisfy different illumination requirements.

purposes and greatly saves the manufacturing and maintenance costs. Compared with lamps that use traditional light sources, the LED navigation lamp shows a good performance on energy saving. Ignoring the absorption and Fresnel losses, the efficiency of the optical system is above 80%, while the life and reliability of the lamp have been improved greatly. Meanwhile, due to its prominent monochromatic performance, the LED light source is more suited to the diverse needs of airport illumination.

9.5 Freeform Optics for LED Searchlights

9.5.1 Introduction

A searchlight installed on a boat plays an important role in coast guarding, searching on lake or sea, and other important offshore operations. Since it should offer good illumination for emergencies, high performance of the light is very crucial. At present, the luminous efficiency of LED lighting has increased rapidly, and due to its advantages of low power consumption, high reliability, long lifetime, and environment protection, it is becoming one of the new-generation light sources. Nevertheless, the light distribution of LED chips, which is considered as Lambertian with the radiance property given by $I = I_0 \cos\theta$, is quite different from that of traditional light sources like incandescent lamps, halogen lamps, fluorescent lamps, and so on. Therefore, it is necessary to design a lens to make the LED light fit the illumination requirement of a searchlight, and we call this *LED secondary optical design*.

According to the technical requirement of a searchlight, it should illuminate the place 250 m away, and the diameter of the spot is 11 m with an average flux no less than 50 lx. In other words, the beam divergence angle of the searchlight is $arctan(11/250) = 2.5°$, and the luminous flux on the target area is at least 4750 lm. Therefore, it would be considered as a highly collimated problem, with which methods such as the TIR mechanism[10,11] and reflecting mirrors[24] are usually employed to deal. As to the two methods, the TIR lens will cause problems of bulkiness and heaviness, and it is sensitive to the position of the light source.[24] Reflecting mirrors inevitably lose some rays that lead to low light efficiency. However, an available optical design should balance weight, size, and energy efficiency.

To achieve the balance, a novel combination of a freeform lens and a parabolic reflector is proposed. In the remainder of this chapter, the design steps of a freeform lens for a circular light pattern and a freeform lens design of the searchlight will be presented. Then, the studies about the critical angle of a freeform lens for collimation and its limits in light efficiency are discussed. This research aims to offer an optical design method for all high-brightness, far-spot LED lighting. Thus, the construction of the novel combination is also given in detail.

9.5.2 Freeform Lens Design of a Small Divergence Angle

As shown in Figure 9.38, considering the LED as a point source, it irradiates light rays with a full angle of 180°. The freeform lens should concentrate these light rays into a beam with a full angle of 2.5°.

An effective way is to collimate all the light rays emitting from the LED. Now, the norm vector $\vec{O}$ is vertical up, and the discrete points of the seed curve can be calculated by the two steps mentioned in Chapter 3. Note that the seed curve is more accurate as the discrete points are calculated more, and 100 is appropriate.

A PMMA freeform lens is designed according to the method mentioned above with a refractive index of 1.49. Figure 9.39 shows the designed freeform lens, and a set of ray fans emitting from a point source at the cross section is added to it. It is found that part of the light rays propagating upward within the divergence angle of 2.5° can reach the

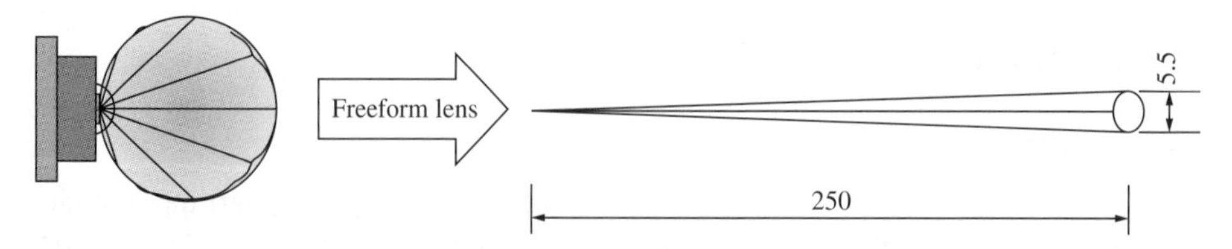

Figure 9.38 Light distribution of an LED and the searching light.

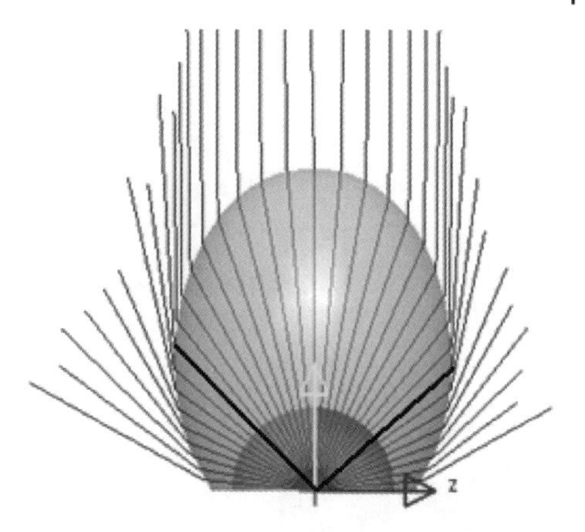

Figure 9.39 Ray trace of the freeform lens at the cross section.

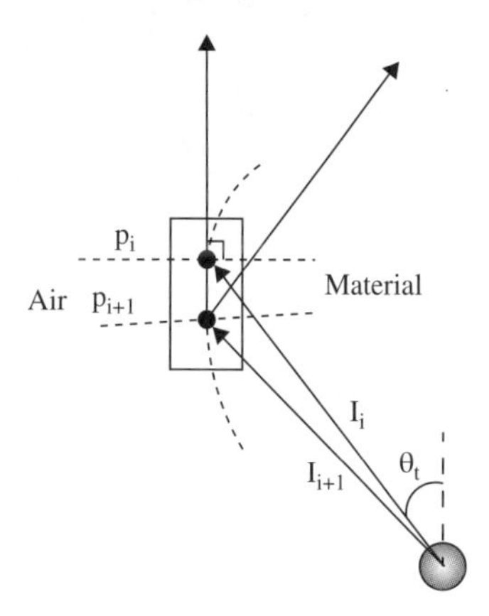

Figure 9.40 The critical circumstance of a freeform lens for collimation.

target plane, while those with a big divergence angle are wasted. We mark the boundary rays with bold lines, and the angle they enclose is the critical angle of the freeform lens for collimation.

In fact, the result can be explained by theoretical analysis, and the critical angle of the freeform lens for collimation can be calculated. As shown in Figure 9.40, the incident ray I_i is refracted upward to air by p_i, and the tangent plane of p_i is vertical up. According to the design step of the freeform lens, the location of the next point p_{i+1} is right below p_i. It is visible that due to the effect of total internal reflection, the reflected light ray deviates from the upward direction, as it is designed to. Referring to the contour of the freeform lens, points before p_i can refract incident rays upward, while points after p_i cannot. Figure 9.40 presents the critical circumstance.

See from Figure 9.40: as to p_i, the incident angle is $(\pi/2) - \theta_t$, and the output angle is $\pi/2$. Snell's law should be employed to calculate the critical angle, and what we obtain is $\theta_t = (\pi/2) - arcsin(1/n)$. Obviously, it is related to the refractive index of the material. As to PMMA, its critical angle equals 48°. This value coincides with the ray=-tracing result in Figure 9.39.

Theoretically, luminous flux of the light source is:

$$\int_0^{2\pi} \int_0^{0.5\pi} I_0 \cos\theta \sin\theta d\theta d\varphi \tag{9.12}$$

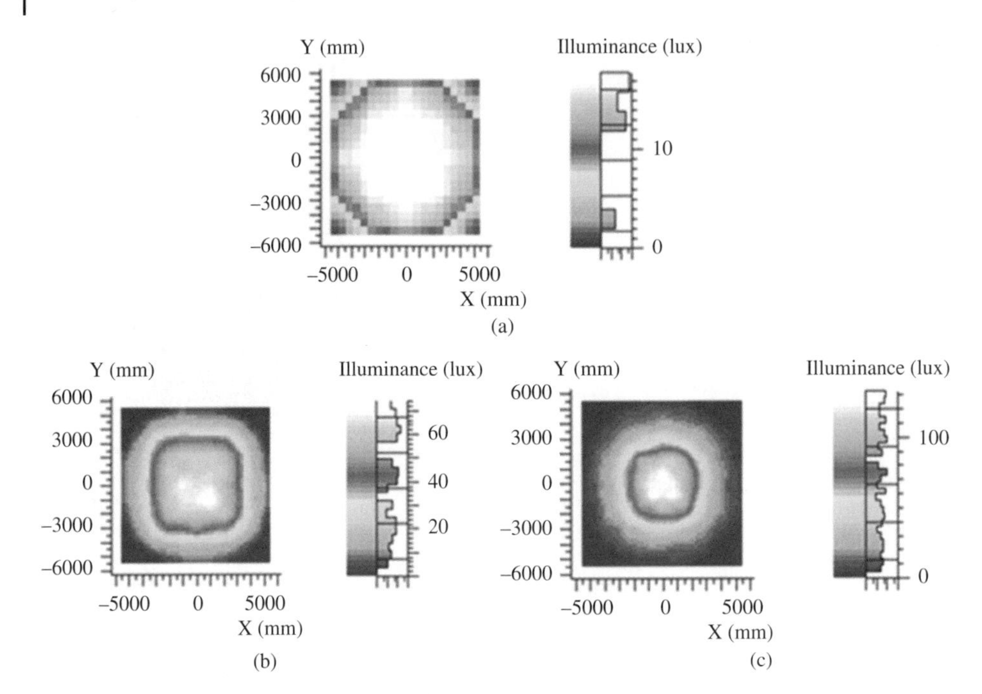

Figure 9.41 Illumination distribution on the target plane of three size freeform lenses with (a) h = 10 mm, (b) h = 15 mm, and (c) h = 20 mm.

and flux in the solid angle of 48° is:

$$\int_{0}^{2\pi}\int_{0}^{0.27\pi} I_0 \cos\theta \sin\theta d\theta d\varphi \tag{9.13}$$

Thus, the limit light efficiency of the freeform lens is the ratio of Eq. 9.13 to Eq. 9.12, and the value calculated is 55%.

Based on the Monte Carlo ray-tracing method, the illumination distribution on the target plane of three different-size freeform lenses is obtained and shown in Figure 9.41. In the simulation, heights of the freeform lenses are 10, 20, and 30 mm; the size of the light source is 1 × 1 mm; and, to simulate the illumination of LED arrays, total flux of the light source is set to be 9000 lm. From Figure 9.41, it is found that the flux on the target plane increases with the increment of the freeform lenses' size as a result of an extended source. That is because the light source is considered a point during freeform lens design; however, the real chip is a rectangular surface, and the smaller the lens' size is, the more the chip cannot be considered as a point.

In addition, based on the simulation results of more different-size freeform lenses, the effect of the size of the freeform lenses on light efficiency, which is defined as the ratio of luminous flux on the target plane to luminous flux of the light source, is studied as shown in Figure 9.42. The emitting surface of the light source is still set to be 1×1 mm^2, which is the real size of the LED chip. The trend of the curve indicates that the maximum value of light efficiency is 55%, in agreement with the theoretical calculation; and, from Figure 9.42, the effect of the extended source will disappear when h is

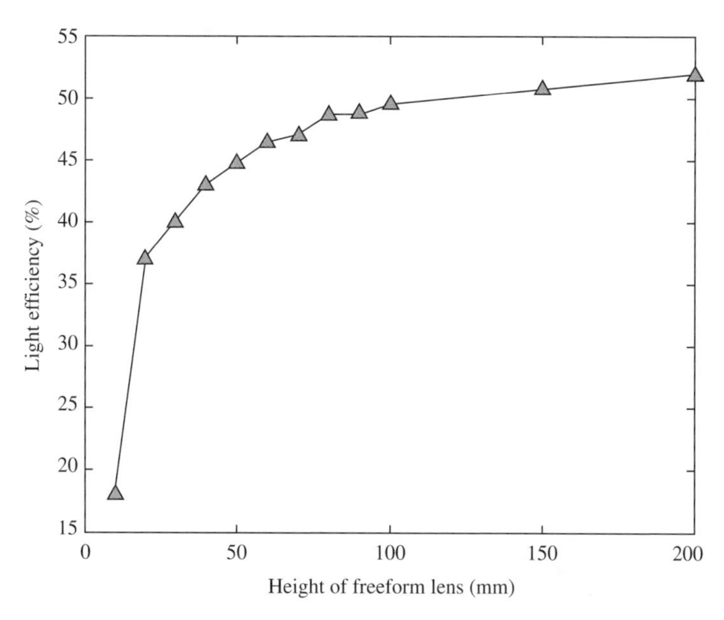

Figure 9.42 The curve of light efficiency to the height of a freeform lens.

bigger than 200 mm. Besides, it is found that light efficiency is enhanced about 19% when h increases from 10 to 20 mm, Afterwards, the light efficiency increases comparatively slowly. Especially when $h > 100$ mm, the improvement is lower than 1 percentage point as h increases by 50 mm. Thus, it needs to balance between the light efficiency improvement and the lens' volume.

9.5.3 Improving Methods and Tolerance Analysis

Since a part of the light rays emitting from the LED cannot be collimated by a freeform lens, the big emitting angle rays should be reflected to improve the light efficiency, and there are many ways to achieve this. In this chapter, we adopt a parabolic reflector and will introduce the design of a combination of a freeform lens and parabolic reflector.

9.5.3.1 The Design of a Freeform Lens and Parabolic Reflector

In Figure 9.43, section planes of two parabolic reflectors that share the same focus are shown with different line shapes. And P (twice the distance from focus to vertex) represented by the broken line is bigger than that by the continuous line. According to the geometric properties of the parabolic reflector, light rays coming from its focus and projecting on the reflector surface are collimated forward, while light that misses the reflector does not get collimated.

From Figure 9.43, we find that when P is bigger, the dimension of the parabolic reflector is larger to collimate the same quantities of light rays. Therefore, a basic rule of this design is to choose the parabolic reflector of small P. When leaving room for the LED package module in the bottom of the reflector, $P = 10$ is suitable. Then, the diameter of the reflector should be determined. In Figure 9.43, we also find that to a parabolic reflector, it collimates more light rays when its diameter is bigger. However, since the

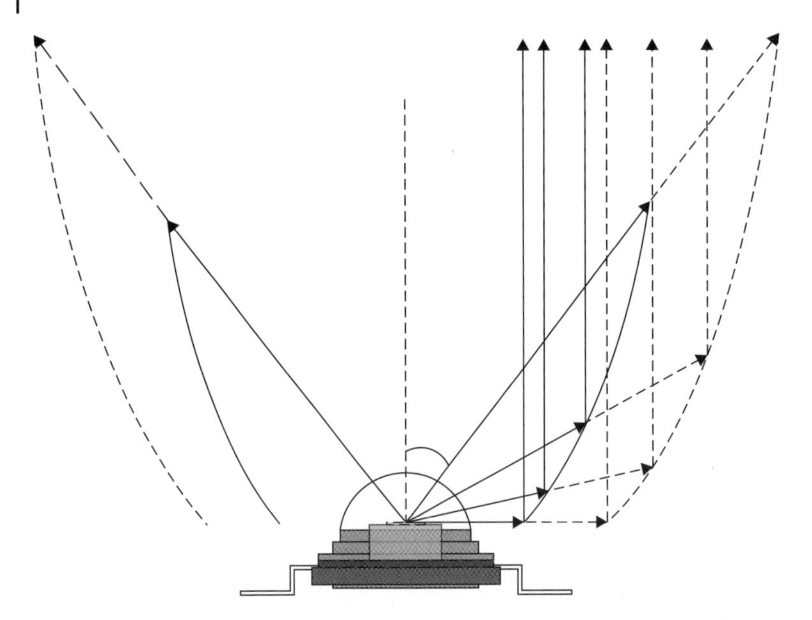

Figure 9.43 Schematic of parabolic reflectors with the LED locating on the focus.

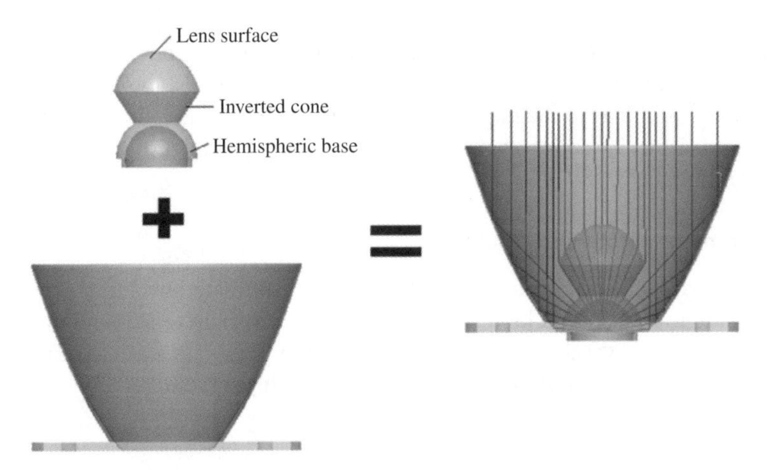

Figure 9.44 Optical design of a freeform lens and parabolic reflector.

dimension of the reflector decides the whole dimension of the searching light, we set the diameter of the reflector at 60 mm, which is appropriate in this project. According to parabolic function, the height of the reflector is 40 mm; thus, it collimates light rays with an emitting angle larger than 37°. Therefore, the freeform lens surface is responsible for collimating light rays in the solid angle of 37°, which is smaller than the critical angle.

The combination of the freeform lens and parabolic reflector is demonstrated in Figure 9.44. To support the freeform lens segment, a special sustaining structure is designed. The light source locates at the center of the hollow hemispheric base

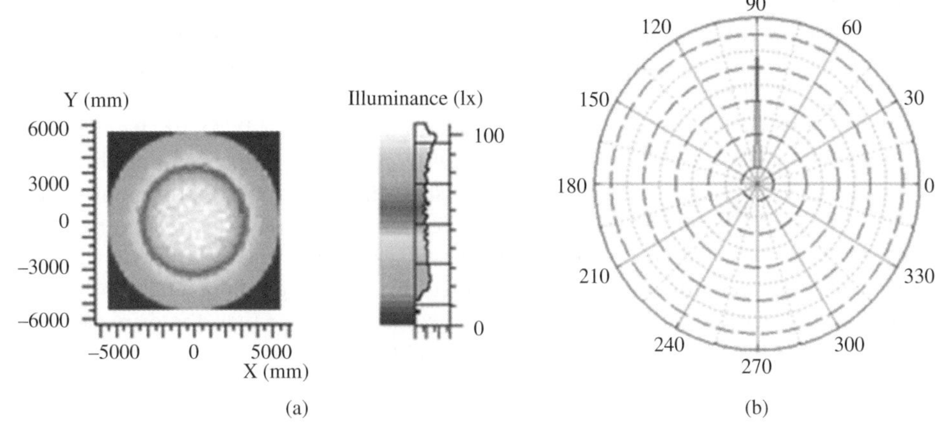

Figure 9.45 (a) Illumination distribution on the target plane and (b) light distribution curve of the combined optical design.

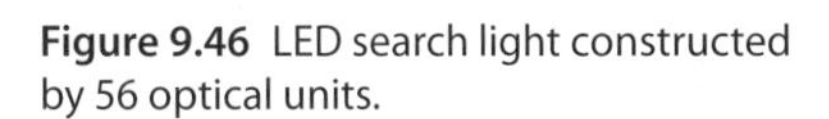

Figure 9.46 LED search light constructed by 56 optical units.

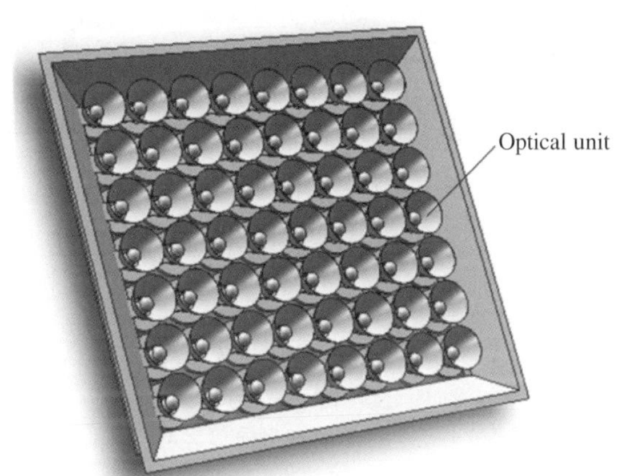

through which rays transmit without refraction, nor does the inverted cone change the propagating direction of rays. The ray-tracing result shows that the combination works well in collimating all the rays emitting from a point light source.

Illumination distributions on the target plane and light distribution curve are displayed in Figure 9.45a and 9.45b. In this simulation, the size of the light source is 1×1 mm, and the total flux of the light source is 7000 lm, of which 4900 lm is incident on the target plane. That means the average flux of the spot is 51.58 lx, which exceeds the required 50 lx, and light efficiency has been improved greatly to 70%. As shown in Figure 9.45b, the light rays from the LED have been concentrated into a small divergence beam.

In this design, a CREE Xlamp LED is employed due to its high luminous efficiency, which is 125 lm/W. The total number of LEDs we need is 56. Thus, the optical system of the lamp is constructed by 7×8 units of the combination module, as shown in Figure 9.46. The size of the optical part of the searchlight is about 42×48 mm, and the weight is about 15 kg.

9.5.3.2 Tolerance Analysis

The tolerance of a freeform lens is an important issue in the freeform optical component design. The installation errors, including horizontal deviation (*dh*), vertical deviation (*dv*), and rotational error ($d\theta$), are shown in Figure 9.47.

Their effects on average illumination are presented in Figure 9.48. Figure 9.48a shows the light efficiency curves when *dh* or *dv* changes from 0 to 0.5 mm. It is found that light efficiency decreases rapidly when one of the displacement errors exceeds 0.2 mm, and the deterioration of light efficiency is more serious under the effect of *dh*. Figure 9.48b shows the light efficiency curve of $d\theta$ when it increases from 0° to 10°. Different from the displacement errors, light efficiency only reduces by 5.5% when $d\theta$ reaches 10°. Obviously, the combination is more tolerant to the rotational error than the displacement errors, and *dh* is the most important influence on light efficiency.

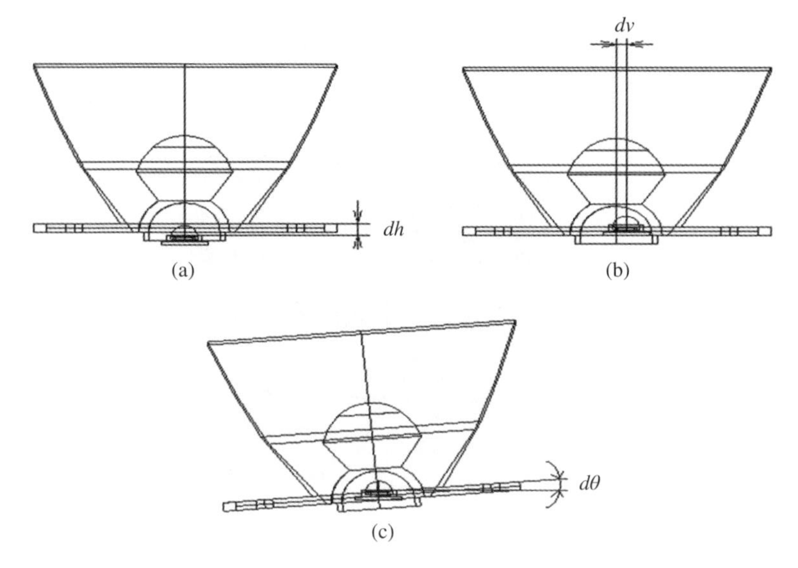

Figure 9.47 Installation errors of (a) horizontal deviation, (b) vertical deviation, and (c) rotational error.

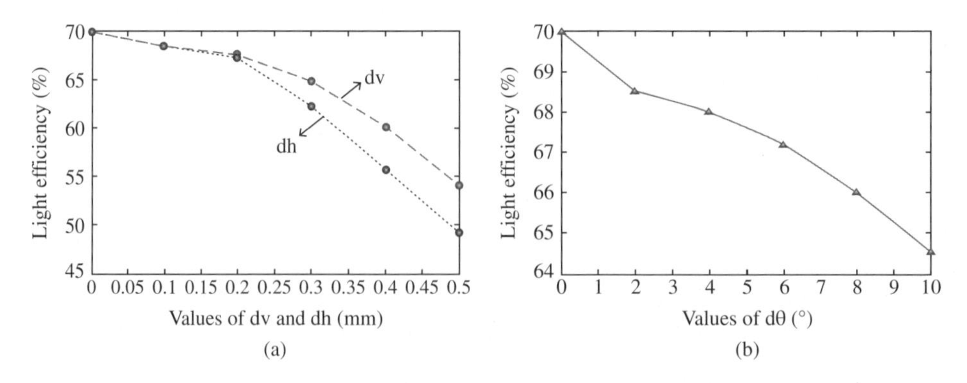

Figure 9.48 (a) Curves of light efficiency to *dv* and *dh* when they range from 0 to 0.5 mm; (b) curve of light efficiency to $d\theta$ as it changes from 0° to 10°.

References

1 Yu, X.J., Ho, Y.L., Tan, L., Huang, H.-C., and Kwok, H.-S. LED-based projection systems. *U. Disp. Tech.* **3**, 295–303 (2007).
2 Jacobson, B.A., Gengelbach, R.D., and Ferri, J.M. Beam shape transforming devices in high-efficiency projection systems. *Proc. SPIE* 3139, 141–150 (1997).
3 Pan, J.-W., Wang, C.-M., Lan, H.-C., Sun, W.-S., and Chang, J.-Y. Homogenized LED-illumination using microlens arrays for a pocket-sized projector. *Opt. Expr.* **15**, 10483–10491 (2007).
4 Shatz, N.E., Bortz, J.C., Kirkpatrick, D., and Dubinovsky, M. Optimal design of a nonimaging projection lens for use with an RF-powered source and a rectangular target. *Proc. SPIE* 4446, 171–184 (2001).
5 Wang, L., Qian, K., and Luo, Y. Discontinuous free-form lens design for prescribed irradiance. *Appl. Opt.* **46**, 3716–3723 (2007).
6 Ding, Y., Liu, X., Zheng, Z., and Gu, P. Freeform LED lens for uniform illumination. *Opt. Expr.* **16**, 12958–12966 (2008).
7 Zhang, W., Li, X., Liu, Q., and Yu, F. Compact LED based LCOS optical engine for mobile projection. Proc. SPIE 7506, 75061Y–75061Y–8 (2009).
8 Goldstein, P. Radially symmetric freeform lens design for extended sources. *Proc. SPIE* 8487, 84870C (2012).
9 Fournier, F.R., Cassarly, W.J., and Rolland, J.P. Designing freeform reflectors for extended source. *Proc. SPIE* 7423, 742302 (2009).
10 Parkyn, W.A., and Pelka, D.G. New TIR lens applications for light-emitting diodes. *Proc. SPIE* 3139, 135–140 (1997).
11 Parkyn, W.A., and Pelka, D.G. TIR lenses for fluorescent lamps. *Proc. SPIE* 2538, 93–103 (1995).
12 Wang, K., Luo, X., Liu, Z., Liu, S., Zhou, B., and Gan, Z. Optical analysis of an 80-W light-emitting-diode street lamp. Opt. Eng. 47, 013002–013002–13 (2008).
13 Piegl, L., and Tiller, W. The NURBS Book, 2nd ed. Berlin: Springer, 1996.
14 Wang, K., Chen, F., Liu, Z., Luo, X., and Liu, S. Design of compact freeform lens for application specific light-emitting diode packaging. *Opt Expr.* **18**, 413–425 (2010).
15 US National Transportation Safety Board (NTSB). Access the NTSB Aviation Accident Database. Washington, DC: NTSB, 2008.
16 Mills, A. Solid state lighting: a world of expanding opportunities at LED 2002. *III-Vs Rev.* **16**(1), 30–33 (2003).
17 International Civil Aviation Organization (ICAO). *Aerodrome Design and Operations APP 2-7, Annex 14,* vol. 1. Montreal: ICAO, 2004.
18 Ross, S.M. *A First Course in Probability,* 8th ed., 267–269. Upper Saddle River, NJ: Pearson (2004).
19 Chen, F., Wang, K., Qin, Z., Wu, D., Luo, X., and Liu, S. Design method of high-efficient LED headlamp lens. *Opt Expr.* **18**, 20926–20938 (2010).
20 Alvarez, J.L., Hernandez, M., Benítez, P., and Miñano, J.C. TIR-R Concentrator: a new compact high-gain SMS design. *Proc. SPIE* 4446 (2002).
21 Muñoz, F., Benítez, P., Dross, O., Miñano, J.C., and Parkyn, B. Simultaneous multiple surface design of compact air-gap collimators for light-emitting diodes. *Opt. Eng.* **43**, 1522–1530 (2004).

22 OSRAM OSTAR. SAT LE UW S2W-NZPZ-FRKV. Data sheet. http://www.osram-os.com

23 Sun, C.C., Lee, T.X., Ma, S.H., Lee, Y.L., and Huand, S.M. Precise optical modeling for LED lighting verified by cross correlation in the midfield region. *Opt. Lett.* **31**, 2193–2198 (2006).

24 Schruben, J.S. Formulation of a reflector-design problem for a lighting fixture. *J. Opt. Soc. Am.* **62**, 1498–1501 (1972).

10

Freeform Optics for LED Lighting with High Spatial Color Uniformity

10.1 Introduction

Light-emitting diodes (LEDs), with increasing luminous efficiency in recent years, have greater applications in daily life, such as road lighting, backlighting for LCD display, headlamps of automobiles, and interior and exterior lighting.[1–3] As people have more and more requirements for LED lighting quality, the spatial color uniformity (SCU) of LED lighting has become an important factor in assessing the quality of LED lighting, and along with the light efficiency and light shape controllability is one of the three key factors of high-quality LED lighting. However, among these three factors, the SCU is the most commonly overlooked one. With the development of LED lighting in the market of indoor lighting and high-end lighting, the LED packaging module gradually revealed the problem of poor color quality, especially for phosphor conversion white LEDs,[4,5] and this has become an obstacle to achieve high-quality LED lighting.

The widely adopted phosphor-coating process in terms of a freely dispersed coating is easy to operate and low cost, but it results in yellow rings in the radiation pattern,[5,6] which reduces SCU. This problem can be overcome by phosphor conformal coating, including electrophoresis,[7] slurry, settling,[8] evaporating solvent,[9] and wafer-level coating.[10] Besides conformal coating, phosphor layers with convex shapes[5] and transparent refractive-index-matched microparticle (TRIMM)-doped diffusers[11] also can enhance SCU significantly. Although these methods are effective to achieve high SCU, most of them are complicated and costly. Due to the introduction of new processes, most of these methods have the problems of high cost of technological innovation, complicated packaging processes, and high requirements for control precision at high cost – some methods even at the expense of luminous efficiency for the purpose of improving the SCU.

In this chapter, we will propose a new method by which a freeform lens is used to improve the SCU of the phosphor freely dispensing LED packaging module. We will analyze the mechanism to improve LED SCU and design a compact LED packaging module with a freeform lens. This method has the advantages of simple process, low cost, high light efficiency, and remarkable effect. This method can be widely applied to LED packaging modules or LED lighting designs to improve LED illumination quality.

Freeform Optics for LED Packages and Applications, First Edition. Kai Wang, Sheng Liu, Xiaobing Luo and Dan Wu.

10.2 Optical Design Concept

Figure 10.1a shows that the traditional LED mainly consists of a board, an LED chip, a phosphor layer, and a hemisphere silicone lens. The phosphor is freely dispersed on the chip. As shown in Figure 10.1a, light within the radiation angles of 0~90° will exit through the hemispherical surface of the lens and not overlap. Therefore, as shown in Figure 10.2, optical power of P_0 at the center of the far field consists of B_0 and Y_0, which represent the optical power of blue light (380–490 nm) and yellow light (490~780 nm) emitted from the chip and the phosphor, respectively, at the direction of 90°. Similarly, the optical power of P_1 at the edge of the far field consists of B_1 and Y_1. In this analysis, we introduce the yellow–blue ratio (YBR) to illustrate the variation of correlated color temperature (CCT).[5] SCU is the ratio of the minimum YBR to the maximum YBR in the range of 0~180° as follows:

$$SCU = \frac{YBR_{\min}}{YBR_{\max}} \tag{10.1}$$

However, in the traditional LED, YBR of P_1 (Y_1/B_1) is much larger than that of P_0 (Y_0/B_0) owing to phosphor freely dispersed coating,[5,6] which results in yellow rings.

In this chapter, we try to eliminate yellow rings by overlapping light with different radiation angles. Figure 10.1b shows a modified LED integrated with a freeform lens, which has two discontinuous surfaces, the side and the top surface. Light within the radiation angles of 0°~θ exits from the LED, goes through the side surface, and covers

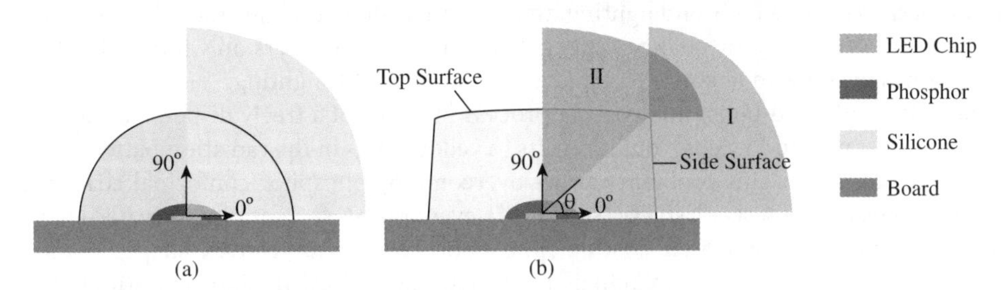

Figure 10.1 Schematic of light output of (a) the traditional LED and (b) the modified LED.

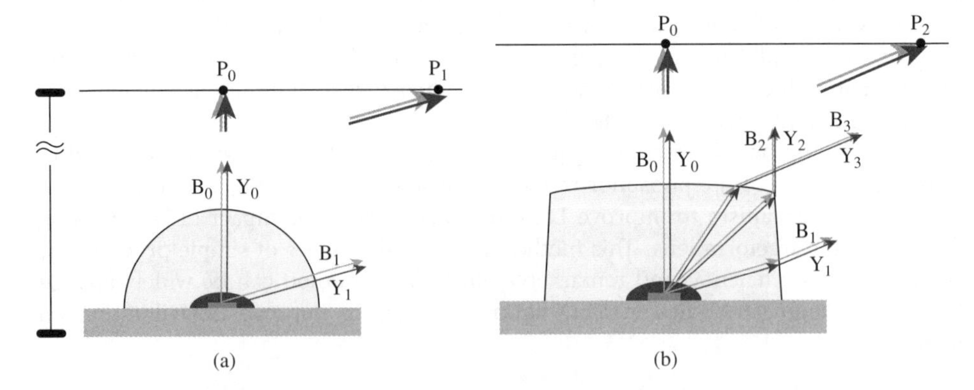

Figure 10.2 Schematic of optical power of center and edge points in the far field of (a) the traditional LED and (b) the modified LED.

the emergence angles of 0~90°. Also, light within 0~90° exits through the top surface and covers the same range of emergence angles of 0~90°. Therefore, as shown in Figure 10.1b, the far field of the modified LED consists of two parts, part I and part II. Light in these two parts overlaps each other. The optical power of P_0 in Figure 10.2b includes B_0, Y_0, B_2, and Y_2, while the optical power of P_2 includes B_1, Y_1, B_3, and Y_3. Because $Y_2/B_2 > Y_0/B_0$ and $Y_3/B_3 < Y_1/B_1$,[5,6] we can obtain the following formulas:

$$\frac{Y_0}{B_0} < \frac{Y_0 + Y_2}{B_0 + B_2} \tag{10.2}$$

$$\frac{Y_1 + Y_3}{B_1 + B_3} < \frac{Y_1}{B_1} \tag{10.3}$$

which means the difference of the YBR between the central point and the edge point of the far field becomes smaller after adopting the freeform lens, which enhances the SCU. Furthermore, it is theoretically possible to make the YBR at different points of the far-field equal, leading to perfect color distribution of the LED (see Eq. 10.2). The SCU could reach as high as 1.

$$\frac{Y_0}{B_0} < \frac{Y_0 + Y_2}{B_0 + B_2} = \frac{Y_1 + Y_3}{B_1 + B_3} < \frac{Y_1}{B_1} \tag{10.4}$$

10.3 Freeform Lens Integrated LED Module with a High SCU

10.3.1 Optical Design, Molding, and Simulation

First, two optical models of the traditional and the modified white LEDs are built as shown in Figure 10.3a and 10.3b. The size of the chip is 1 × 1 mm. The thicknesses of the layers are N-GaN 4 μm, MQW 100 nm, P-GaN 300 nm, and Si 100 μm. The absorption coefficients and refractive indexes for N-GaN, MQW, and PGaN are 5, 8, and 5 mm^{-1} and 2.42, 2.54, and 2.45, respectively.[13] The shape of the phosphor layer is a spherical cap with a height of 0.4 mm and a radius at the base of 1.5 mm, which is formed by a freely dispersed coating. The phosphor concentration is 0.35 g/cm^3. Based on the Mie scattering model, the absorption and reduced scattering coefficients of phosphor are 3.18 and 5.35 mm^{-1} for blue light and 0.06 and 7.44 mm^{-1} for yellow light.[14] The refractive index

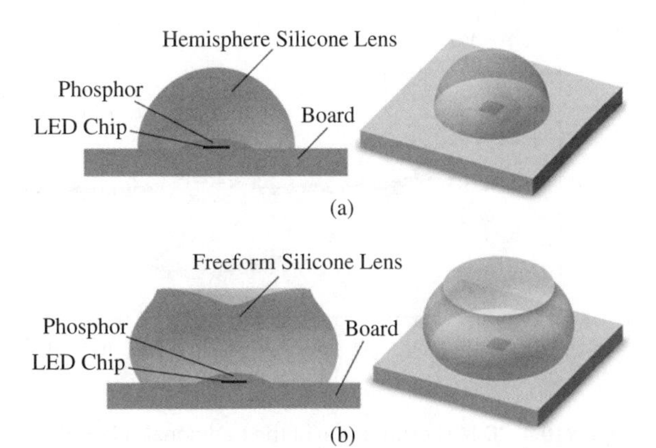

Figure 10.3 Optical models of (a) the traditional LED and (b) the modified LED.

of silicone is 1.50. The heights are 3 mm for both the hemisphere and freeform silicone lenses, which are easy to manufacture by a molding process.

Next, blue and yellow light are separately calculated by a Monte Carlo ray-tracing method. To simplify the calculation, specific wavelengths of 465 and 555 nm are used in the calculation to represent blue and yellow light, respectively, which have been verified as feasible during a simulation.[5] The phosphor layer first collects the absorbed blue light and then re-emits the yellow light from the top and bottom surfaces. Optical power from 0° to 180° in the space is collected to analyze the color distribution.

The side and top surfaces of the freeform lens are designed according to the freeform algorithm mentioned in Chapter 3. Light exiting from both the side and the top surfaces of the lens will uniformly irradiate on the target plane. We discuss the situation that the light source output beam angle $\theta = 40°$, and the freeform lens uniform light pattern output beam angle is 130°. Simulation results are shown in Figure 10.4. We can find that the YBR of the traditional LED is 3.71 at the center, while it rapidly reaches as high as 11.09 at the edge, resulting in a low SCU of only 0.334. After adopting the freeform lens, the YBR of the modified LED increases from 3.71 to 5.26 at the center and decreases from 11.09 to 5.12 at the edge compared with the traditional LED. There is a shift toward yellow (lower CCT) for the light emitted between 40° and 140° and a shift toward blue (higher CCT) out of this region. Therefore, the difference between the YBR of the center and the edge becomes quite small in the modified LED. The SCU increases significantly from 0.334 to 0.957, and the enhancement reaches as high as 186.5%. Moreover, the comparisons of YBR distributions in three-dimensional full space of these two LED modules are shown in Figure 10.5, from which we can find that the SCU of the modified LED module is much better than that of the traditional LED module in the full space.

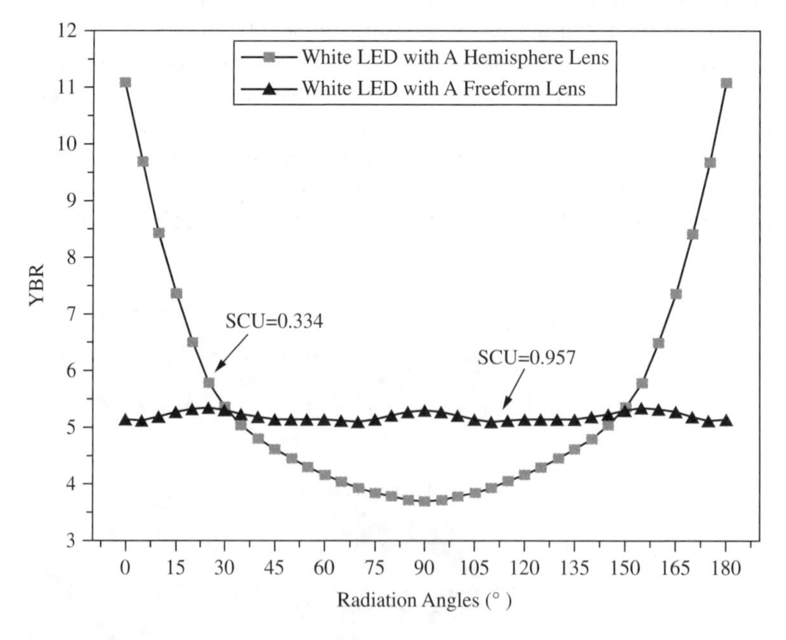

Figure 10.4 2D SCU distribution of the traditional LED and the modified LED.

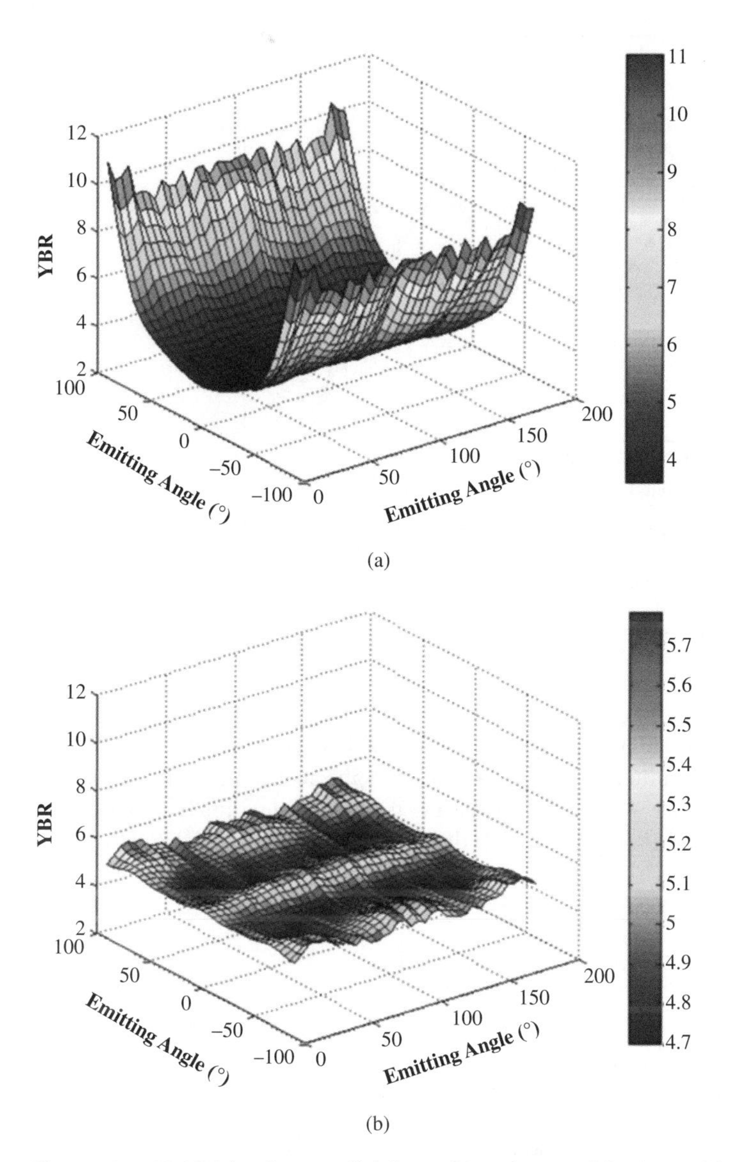

Figure 10.5 3D SCU distribution of (a) the traditional LED and (b) the modified LED.

Besides good SCU performance, the light extraction efficiency of the modified LED reaches as high as 99.5% of that of the traditional LED, whose light extraction efficiency is defined as a reference of 100%. This demonstrates that the freeform lens is barely lower than the output light efficiency of the LED packaging module. Moreover, light intensity distributions (LIDs) of LEDs are very important for researchers when designing optical systems of LED fixtures. As shown in Figure 10.6, the LID of the traditional LED is close to Lambertian, while the modified LED is of the batwing type, which is suitable for some applications that require large view angles, such as LED general lighting (e.g., bulb light or down light), LED backlighting, and LED road lighting.

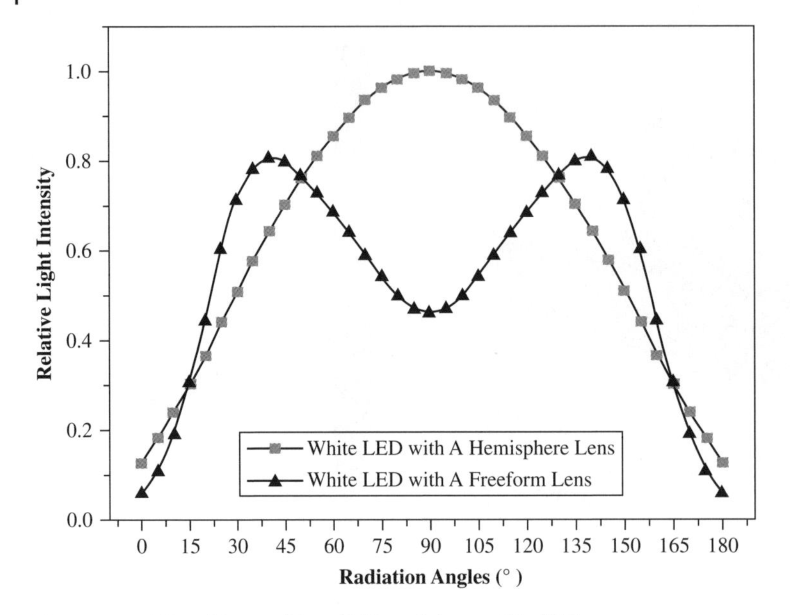

Figure 10.6 LIDs of the traditional LED and the modified LED.

10.3.2 Tolerance Analyses

The effects of various shapes of phosphor layers on the SCU were also studied, including changes of height (h) and radius (a) of the base of the spherical cap like the phosphor layer. Figure 10.7 depicts that when the height is 0.4 mm and the radius increases from 0.9 to 1.9 mm, the SCU of the traditional LED decreases from 0.633 to 0.272. However,

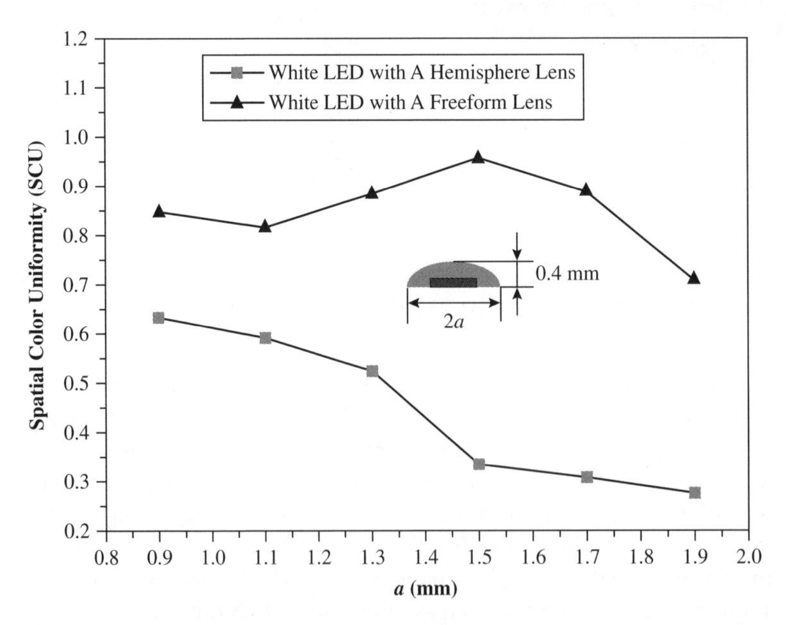

Figure 10.7 Effects of radii of phosphor layers on the SCU of LEDs.

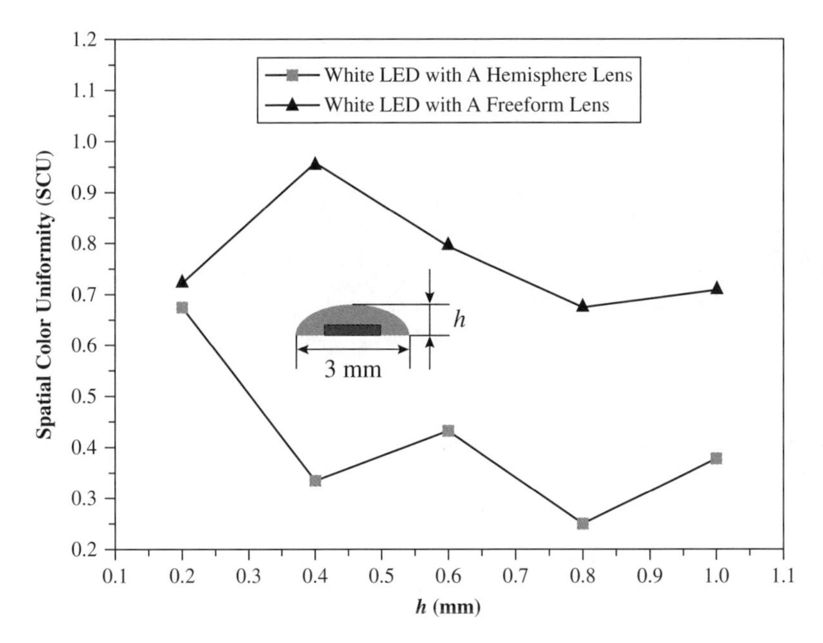

Figure 10.8 Effects of heights of phosphor layers on the SCU of LEDs.

in the same circumstances, the SCU of the modified LED is still at a high level of around 0.8. Effects of the heights of phosphor layers on SCU are shown in Figure 10.8. We can find that when the radius is 1.5 mm and the height increases from 0.2 to 1.0 mm, the SCU of the traditional LED decreases from 0.676 to 0.379, while most SCUs of the modified LED keep at around 0.7. Therefore, the SCU of the modified LED is not only at a high level but also stable when the shape of the phosphor layer changes.

10.3.3 Secondary Freeform Lens for a High SCU

From the analysis presented in this chapter, the major reason that leads to SCU having a lower value for a conventional LED packaging module is that the YBR value at a large angle is much higher than the YBR values in the middle region and even as high as thrice that of the middle region. From Figures 10.4 and 10.5, we can find that, for a conventional LED packaging module when the spatial angle is between 45° and 135°, the YBR value is around 4; the lowest value is 3.71, while the highest is 4.63; and the SCU corresponding to this region is 0.801 and relatively uniform compared with the whole LED packaging module. The SCU is acceptable within this region. Therefore, when we design the lens, we adopt the spherical lens surface within the region of 0~45° to let the light ray emit directly from the light source. The main design task lies in the control of light rays with a large emitting angle between 45° and 90°. The light rays within this region will be deflected to the central area by a freeform lens and overlapped by the light rays emitted from the central region, achieving high SCU.

Based on the above analysis, we design a freeform PMMA secondary lens that combines the freeform surface with the spherical surface. Its top surface (corresponding to the range 0~45°) is spherical with a radius of 100 mm, and its side surface is a freeform

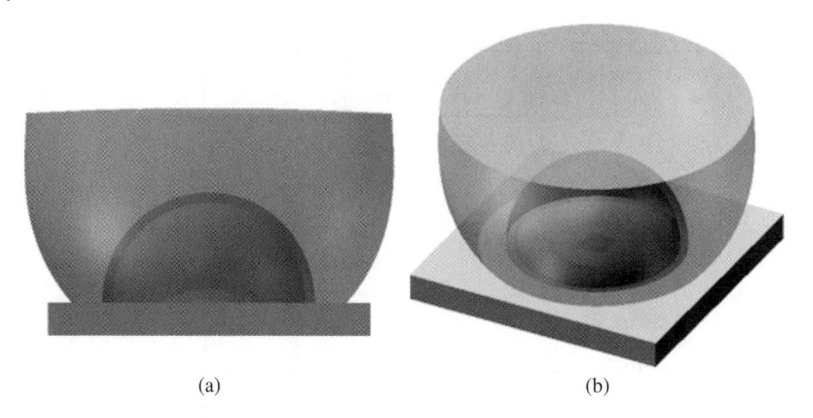

Figure 10.9 Optical model of the new LED module based on the secondary freeform lens with a combination of freeform and spherical surfaces.

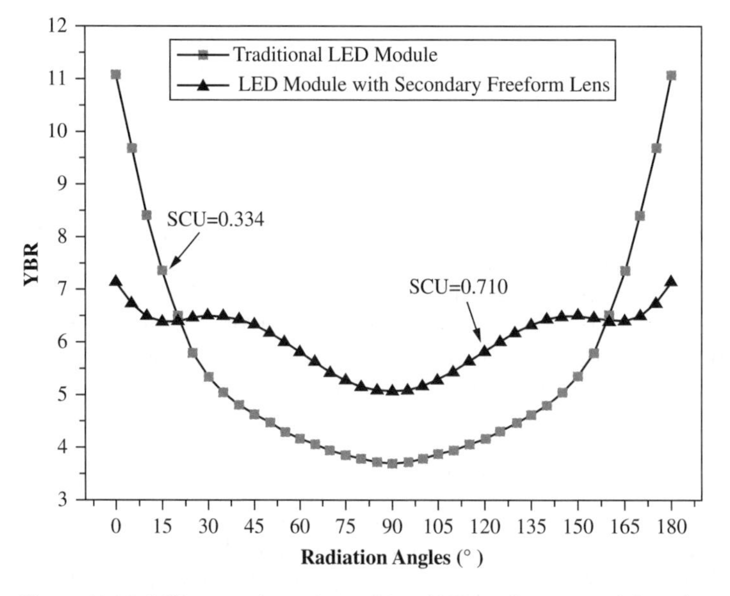

Figure 10.10 SCU comparison of a traditional LED packaging module with a new LED packaging module (freeform surface and spherical) (2D).

surface corresponding to an emitting angle of 90°. The lens bottom has a concave hemispherical cavity with a radius of 3.3 mm, to facilitate the installation of the LED packaging module shown in Figure 10.9. The new SCU lens simulation results are shown in Figures 10.10 and 10.11. SCU reaches a high level of 0.710, and the enhancement reaches 112.6% when compared to traditional LED modules.

10.3.4 Experimental Analyses

The three lenses discussed in this chapter all can greatly improve the SCU of the LED module. Due to the relatively small light-emitting angle of the freeform lens composed of one spherical surface and one freeform surface, it can be adopted for more lighting

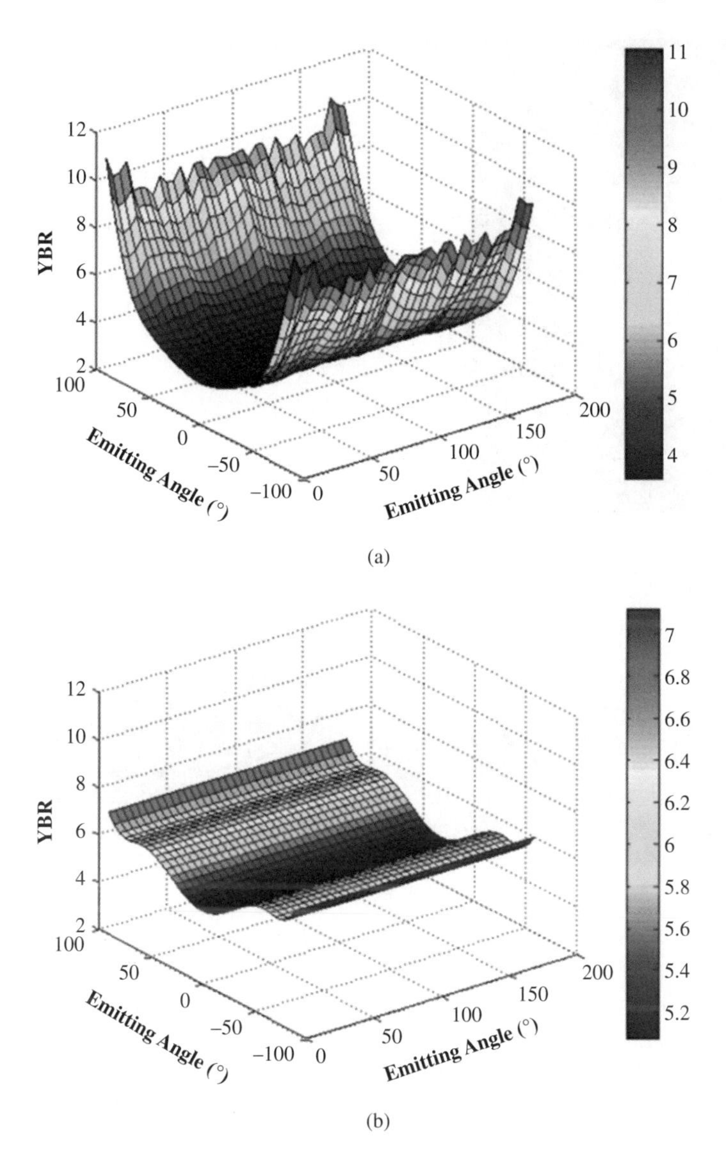

Figure 10.11 SCU comparison of (a) a traditional LED packaging module with (b) a new LED packaging module (freeform surface and spherical) (3D).

applications and therefore is selected for experimental analysis. A sample lens produced by mechanical processing methods is mounted on the LED packaging module, shown in Figure 10.12. Three LED packaging modules are adopted as light sources in the experiments. The phosphor layer of one LED module (sample 1) is coated by the freely dispensing coating method, whereas the other two LED modules (sample 2 and sample 3) adopt a conformal coated phosphor layer, as shown in Figure 10.13.

The entire experimental test is as follows. Firstly, LED packaging modules are fixed on a heat sink with fins by thermal paste. Heat generated by the LED will be emitted into the air through fins, and the LED junction temperature under working condition will

Figure 10.12 SCU freeform lens with a combination of freeform and spherical surfaces.

(a) (b) (c)

Figure 10.13 LED packaging modules of (a) sample 1, (b) sample 2, and (c) sample 3.

be below its threshold in order to guarantee LED optical performance stability during testing. Then the LED packaging module and the heat sink are fixed on a measuring rotation test platform with an accuracy of 1°, which can achieve a space of 0~180° rotation. The main focus of this study is to investigate LED light color uniformity in the far field. and therefore the rotating structure is set in 1 m distance from the test plane. The experiment uses a digital colorimeter for measurement of colorimetric and photometric parameters of light emitted from the LED. CCT, xy color coordinates, $u'v'$ color coordinates, and illuminance values of light can be measured simultaneously incident on the probe and stored in the computer. The colorimeter is fixed on the test plane, and its center position where the probe lies is above the LED light. Then, we start to test from 0° with 5° intervals to measure the light emitted from the LED, mainly focusing on brightness and color information until measured at 180°, receiving a total of 37 test point values to complete the test. The optical efficiency of the lens uses an EVERFINE LED photoelectric testing system with a 0.5 m integrating sphere with a HAAS high-accuracy and fast spectral radiometer for testing. Firstly, we measure the luminous flux of the LED without a lens. and then we test the luminous flux of the LED with a lens; the flux ratio of these two is defined as the light output efficiency (LOE) of lens.

When the entire spatial color distribution test is finished, we can calculate the root mean square (RMS) value $\Delta u'v'$ of the spatial color distribution on each test angle θ:

$$\Delta u'v'(\theta) = \sqrt{(u'(\theta) - u'_{weighted})^2 + (v'(\theta) - v'_{weighted})^2} \tag{10.5}$$

where $u'_{weighted}$ and $v'_{weighted}$ are:

$$u'_{weighted} = \frac{\sum_{\theta} E_N(\theta)u'(\theta)}{\sum_{\theta} E_N(\theta)} \tag{10.6}$$

$$v'_{weighted} = \frac{\sum_{\theta} E_N(\theta)v'(\theta)}{\sum_{\theta} E_N(\theta)} \tag{10.7}$$

where $E_N(\theta)$ represents the normalized light intensity or illuminance of the testing point corresponding to testing angle θ. In addition, according to the US Department of Energy's (DOE) requirement of solid-state lighting, the variation of chromaticity $\Delta u'v'$ must be within 0.004.

Sample 1's LED module is tested first, and the LOE of the freeform lens is 95.98%. From Figure 10.14, we can find that the LED module without a lens will generate a light pattern where there is bluish color in the central region and yellowish color in the edge region, resulting in an obvious yellow ring. After adding a freeform lens, the yellow ring disappears, and the color of the whole light pattern is consistent. Color uniformity is significantly higher than that without a lens and looks more comfortable. Figure 10.15 shows the experimentally measured spatial distribution of CCT comparison. We can see that when the lens is not added, the highest CCT is 4382 K and the lowest CCT is 2981 K, decreasing from the center area to the edge area, and the CCT changes significantly, especially in the large emitting angle. In contrast, after adding the freeform lens, the highest CCT is 4328 K and the lowest CCT is 3799 K, and the CCT distributed relatively uniformly overall, better than the case without a lens. Figure 10.16 shows the

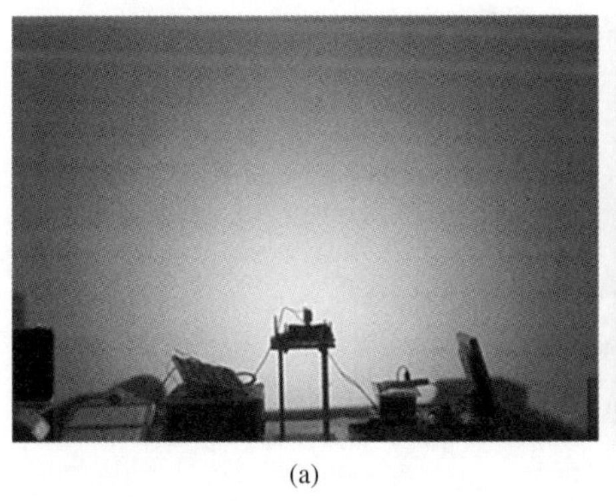

(a)

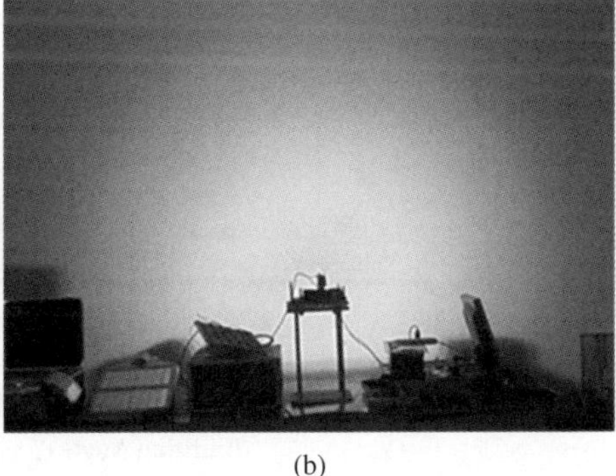

(b)

Figure 10.14 Light pattern for sample 1's LED at a distance of 1 m: (a) no lens is added, and (b) a freeform lens is added.

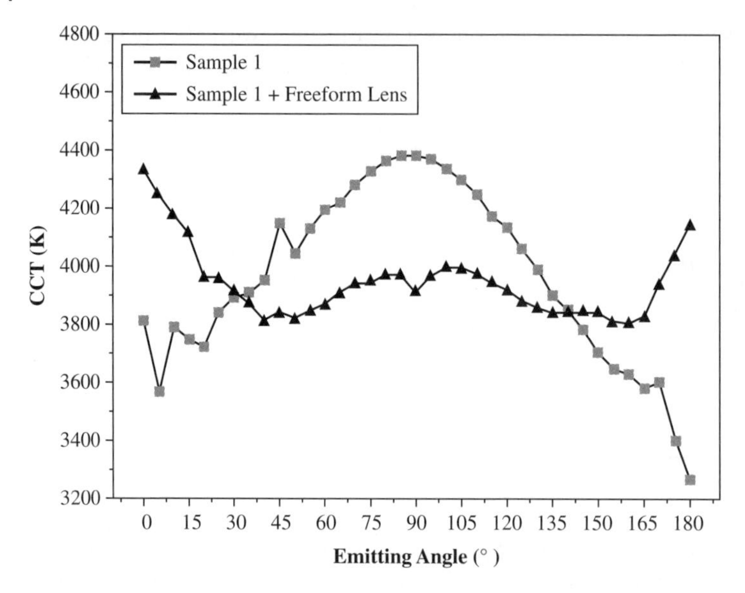

Figure 10.15 Spatial CCT distribution comparison of sample 1's LED.

$\Delta u'v'$ distribution calculated based on the experimental data obtained. When there is no lens, $\Delta u'v'$ basically is higher than 0.004. In 37 test points, only five test points near the center area meet the requirements. When the freeform lens is added later, most of the test points' $\Delta u'v'$ are less than 0.004, and the number of test points that meet the requirements increases from 5 to 24. SCU is improved significantly.

With the same method, we also test the sample 2 LED module, and the lens optical output efficiency is 95.07%. Based on the conformal phosphor coating process, sample

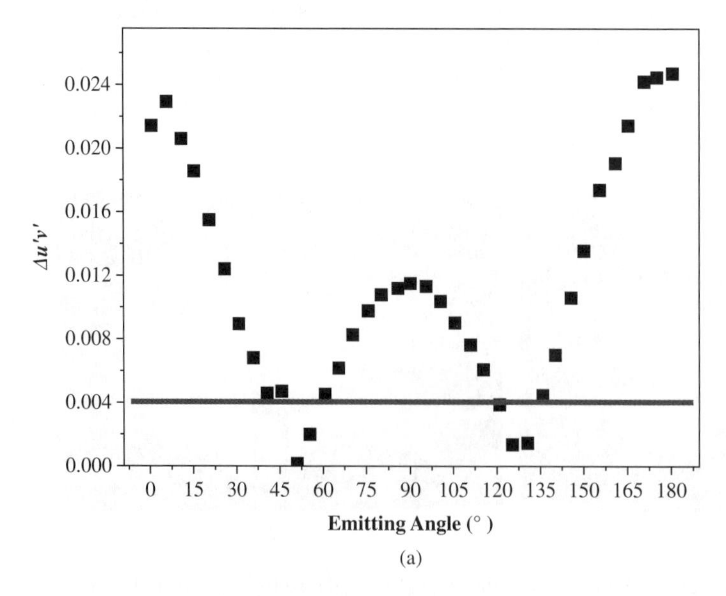

Figure 10.16 $\Delta u'v'$ distribution of the sample 1 LED (a) without a lens, and (b) with a freeform lens.

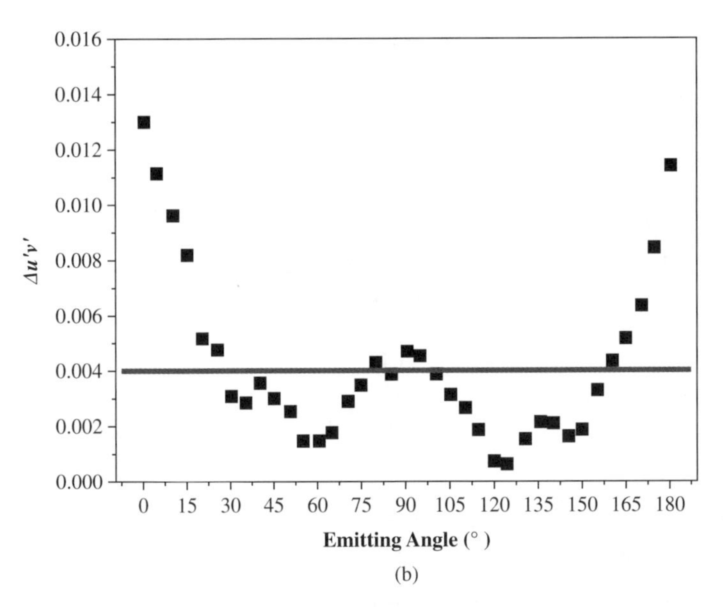

Figure 10.16 (*Continued*)

Figure 10.17 Light pattern for the sample 2 LED at a distance of 1 m: (a) no lens is added, and (b) a freeform lens is added.

2's LED is regarded as one of the LED packaging modules with the best color qualities. From the comparison of light patterns shown in Figure 10.17, we can find that since the conformal phosphor coating process is adopted, color uniformity of the light pattern of the sample 2 LED module is at a relatively high level, whether the freeform lens is added or not. It's hard to observe obvious difference just visually. Figure 10.18 shows the experimentally measured spatial CCT distribution. When no lens is added to the LED module, there is a little change in the center area but a clear downward trend at the edge of the light pattern. When the lens is added to the LED module, the color at the center area also changes a little, but at a large angle the CCT values increase obviously, which effectively avoids the problem of yellow ring caused by CCT decreases at a large angle. Figure 10.19 shows the $\Delta u'v'$ distribution. When the lens is not added, the SCU

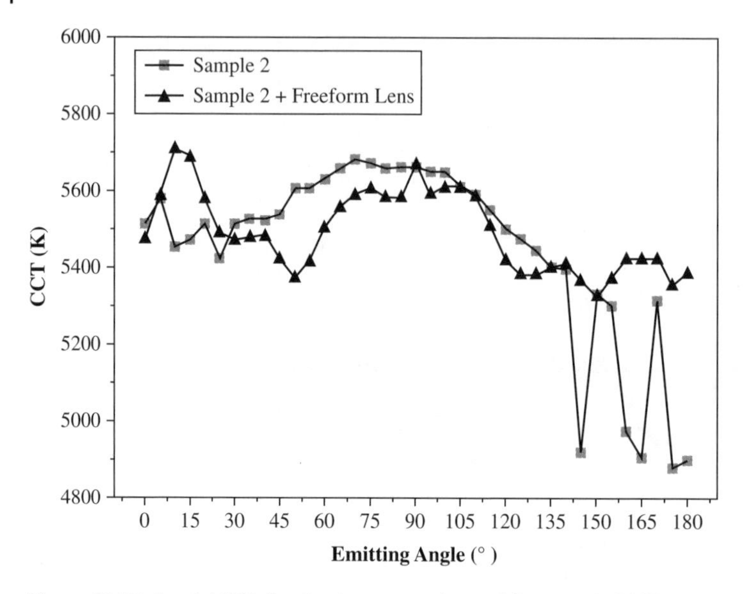

Figure 10.18 Spatial CCT distribution comparison of the sample 2 LED.

of sample 2's LED is obviously better than that of sample 1's LED, and the number of test points whose $\Delta u'v'$ values are less than 0.004 reaches 22. When the freeform lens is added, the SCU is further improved on the bases of conformal coating. The number of test points whose $\Delta u'v'$ are smaller than 0.004 increases to 32, and the color looks more uniform.

Similarly, a sample 3 LED module is also tested. It is found that when the lens is not added there is a very nonuniform CCT spatial distribution. As shown in Figures 10.20,

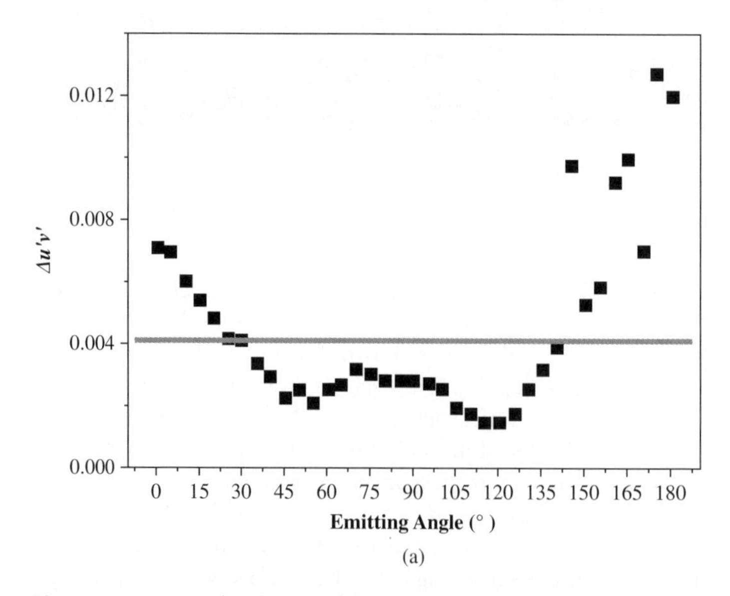

Figure 10.19 $\Delta u'v'$ distribution of the sample 2 LED (a) without a lens, and (b) with a freeform lens.

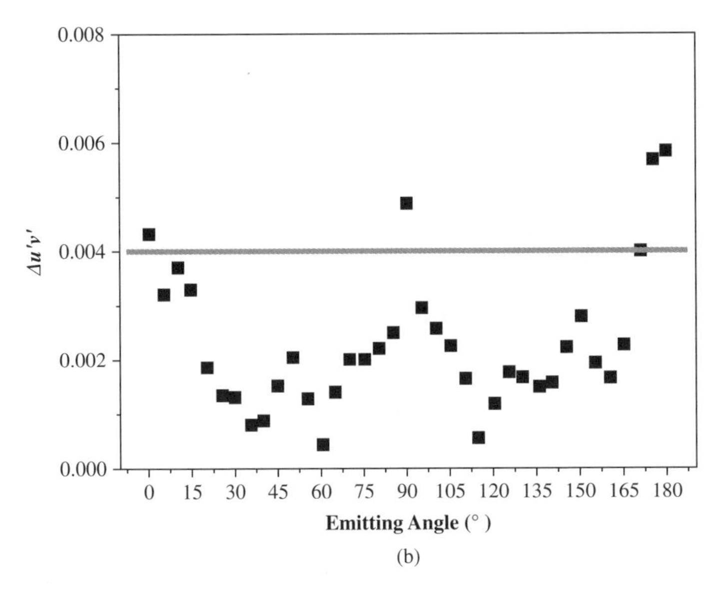

Figure 10.19 (*Continued*)

Figure 10.20 Light pattern for the sample 3 LED at a distance of 1 m: (a) no lens is added, and (b) a freeform lens is added.

10.21, and 10.22, CCT at the central area is at a high level but decreases dramatically at the large-angle area. The maximum CCT at the central area reaches as high as 6918 K, while the minimum CCT at the edge is only 4542 K. There is a difference of 2376 K, and an obvious yellow ring exists at the edge of the light pattern. In 37 test points, there are only 4 points whose $\Delta u'v'$ values are less than 0.004.

The reasons that cause the sample 3 LED module's uneven spatial color distribution are mainly related to its packaging structure. Although the phosphor conformal coating process is adopted, the blue light emitted by the LED chip and the yellow light emitted from the phosphor layer do not all directly exit from the packaging lens. At a large angle, a part of the light is reflected multiple times by the metal reflector cup in the LED module before exiting from the top surface of the packaging lens. Since there are some differences among the emitting positions of blue light and yellow light, after multiple

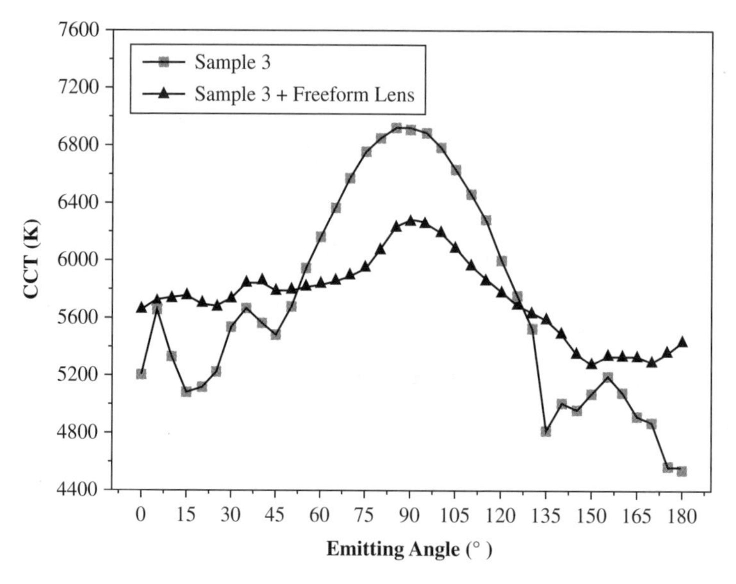

Figure 10.21 Spatial CCT distribution comparison of the sample 3 LED.

reflections through the side wall of the reflective cup, positions of equivalent emitting points of blue light and yellow light shift even larger, leading to the fact that at wide emitting angles, yellow light and blue light do not completely overlap, causing nonuniform spatial color distribution.

After adding the freeform lens, we can find that the CCT spatial distribution becomes more uniform. The maximum CCT at the central area drops from 6918 K to 6281 K, while the minimum CCT at the edge rises from 4542 K to 5269 K. The gap between the highest value and the lowest value is narrowed to 1012 K. The light pattern has better SCU. The yellow-ring phenomenon is eliminated with the freeform lens. Meanwhile, the number of test points that satisfy the requirements in order to achieve $\Delta u'v'$ less than 0.004 increases from 4 to 12, and most of the test points have values in the vicinity of 0.004, indicating a uniform light pattern.

From the above analyses, we can find that, for the phosphor freely dispensing coated LED packaging module, after adding the designed freeform lens, the number of test points meeting the requirements of SCU $\Delta u'v'$ increases significantly from 5 to 24, which is also larger than that of the sample 2 LED module after adopting the phosphor conformal coating process (with 22 points). Moreover, for the sample 2 LED module with high color quality, the points meeting the requirement of $\Delta u'v'$ also increase from 22 to 32, enhancing the color quality of the light pattern further. For the sample 3 LED module, although obvious enhancement has been achieved, the freeform lens can be optimized further to improve the SCU more. Therefore, using a specially designed freeform lens can significantly improve the SCU of the LED packaging module. Compared with existing methods (e.g., conformal coating, remote phosphor coating, and diffusion), this method has advantages such as simple process, low cost, high luminous efficiency, significant effect, and so on, and can be widely used in LED packaging modules or LED lighting design due to greatly improving the quality of LED lighting.

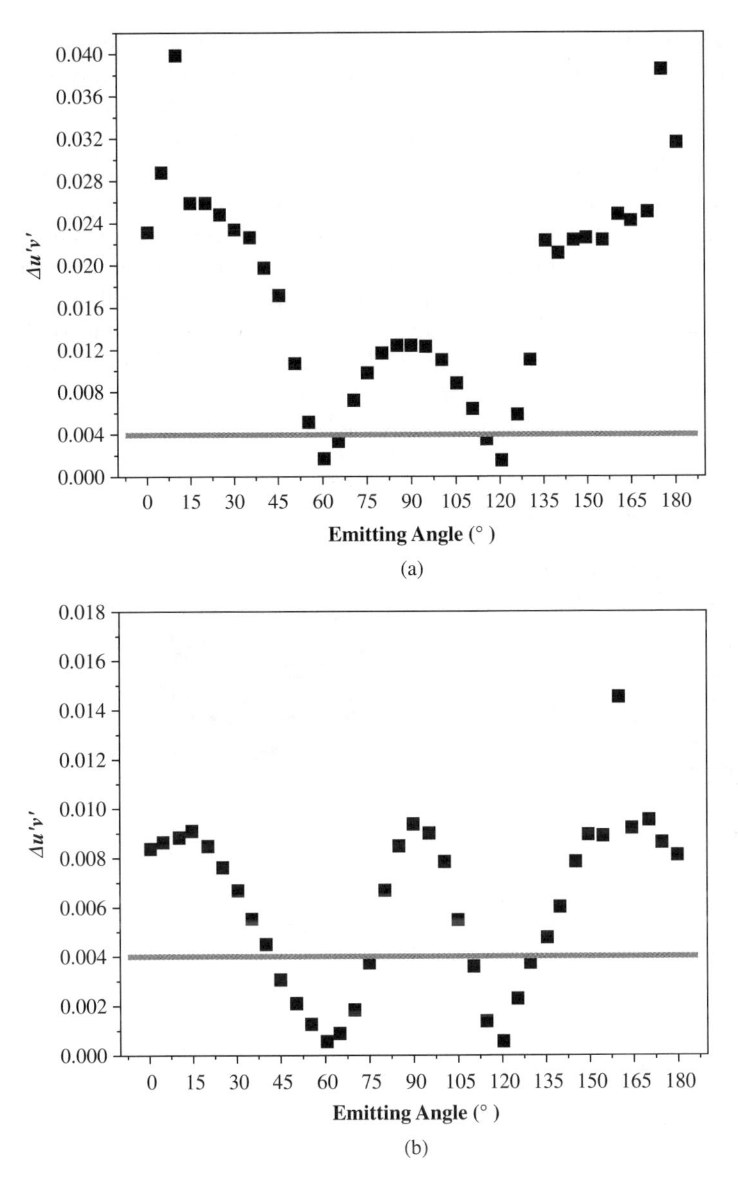

Figure 10.22 $\Delta u'v'$ distribution of the sample 3 LED (a) without a lens, and (b) with a freeform lens.

10.4 TIR-Freeform Lens Integrated LED Module with a High SCU

10.4.1 Introduction

Total internal reflection (TIR) components, which constitute a major part of optical devices distinct from reflectors and Fresnel lenses, have been applied successfully to numerous products with directional emergent light beams, such as collimators for small sources (LEDs and HID lamps), injectors for fiber-optics illumination systems, and solar concentrators.[15–17] The desired uniform illumination within a limited beam angle is easy to achieve by integrating LEDs with freeform TIR lenses.[17–22] However, most TIR designs do not pay attention to the color uniformity and have a considerable dimension, which limits their applications. Since phosphor-converted white LEDs have a color variation over an angle, it is vital to design a compact illumination optical system with a consideration of SCU. Figure 10.23 shows a classic TIR lens and its simulation analysis of the SCU. The classic TIR lens has a considerable dimension of 10.26 mm radius × 10.66 mm height, and its divergence half angle is limited to 42° due to collimated rays irradiating on the top surface. Figure 10.23b gives a model of a phosphor-converted white LED, which has been verified many times in the literature.[23,24] Simulation results of the white LED module integrated with the classic TIR lens are shown in Figure 10.23c. Here, the term *YBR* shown in the vertical axis means the intensity ratio of yellow light

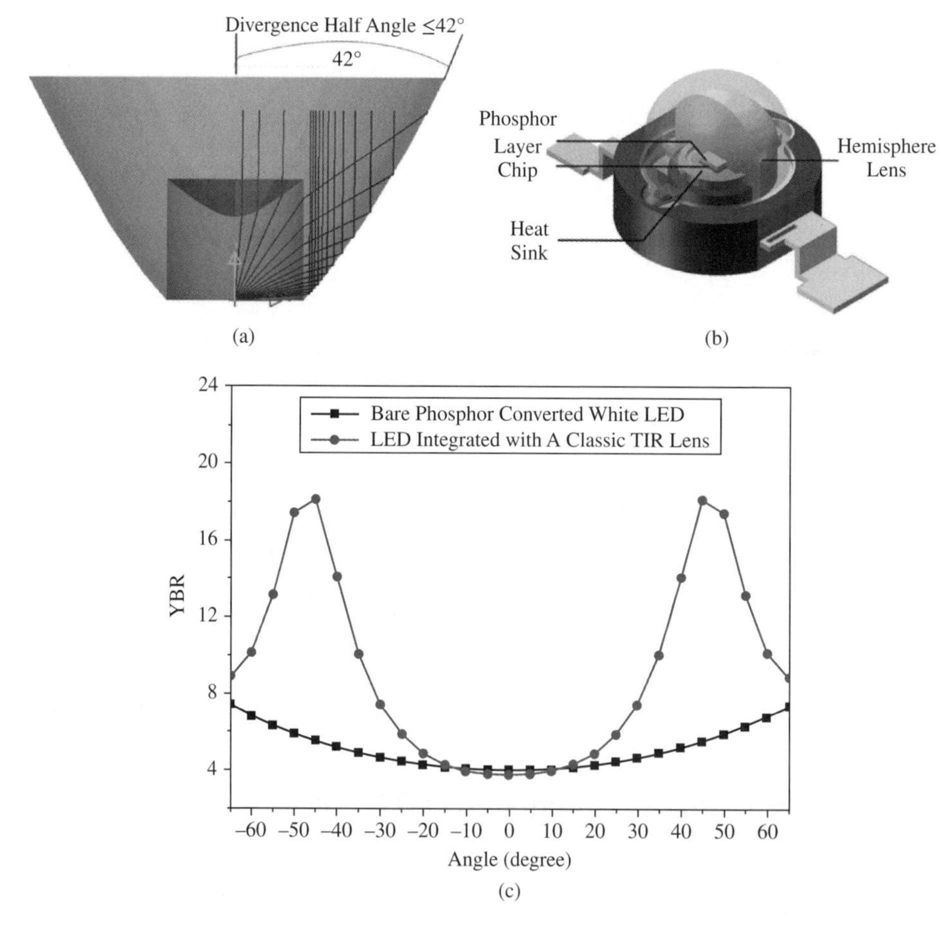

Figure 10.23 (a) Classic TIR lens; (b) traditional white LED packaging; and (c) YBR distributions of a traditional LED and the LED integrated with a classic TIR lens.

to blue light in the far field of a white LED, which indicates the color temperature of the LED. Flat YBR distribution within the viewing angle implies good SCU. It can be found that the LED module integrated with the classic TIR lens has a large color variation (circle dot curve), compared with the bare white LED (square dot curve), and leads to yellow rings in a high probability within the viewing angles −60°~60°.

In this section, according to the above viewpoint of color mixing, a phosphor-converted white LED module integrated with a modified TIR-freeform lens made of poly(methyl methacrylate) (PMMA) is presented and optimized to achieve compact size and high-quality white-light output. This white LED packaging module can be used to generate well-performing LED luminaires applying to general lighting with a large viewing angle.

10.4.2 Design Principle for a High SCU

White LED technology is at the point of surpassing traditional light technologies such as incandescent and fluorescent lamps in light output and lifetime. Phosphor-converted white LED technology is widely adopted, and such LEDs are created by coating a blue LED die with a layer of yellow phosphor and possibly an additional layer of red phosphor. This phosphor coating converts part of the blue light into yellow or red light, resulting in white light. The converted blue light depends on the distance that a light ray travels through the phosphor layer. Therefore, it is reasonable that the color of the output white light is angle-dependent. Such a property may cause the LEDs to be disqualified in general lighting. The worse condition is that the serious nonuniform color distribution could be perceived by people and is harmful to human eyes.

Figure 10.24 shows the design principle of our optical design. A verified and accurate phosphor-converted white LED model is applied in our study, as shown in Figure 10.23b. The white LED model is mainly composed of a chip, a phosphor layer, and a hemisphere lens. Simulation analysis indicates that a traditional white LED without any secondary optics has a nonuniform color pattern, low YBR in the center, and high YBR at the edge, as shown in Figure 10.24a. The light emitted normal to the surface is more bluish white, while the light emitted nearly parallel to the surface is more yellowish. Accordingly, we propose a designated color-mixing principle, as shown in Figure 10.24b. The TIR surface "mirrors" one part of rays (yellowish-white light) to "mix" with the other part of rays (bluish-white light), which enhances the SCU of the LED integrated with the modified TIR-freeform lens. Based on the designated light-mixing principle, a flatter YBR distribution can be obtained. Theoretically, this color-mixing way can perform better compared with the method mentioned in Section 10.4.1 and lead to a uniform YBR distribution more easily. The flat circle dot curve in Figure 10.24b is an ideal YBR distribution after the traditional LED integrated with the assumed designed TIR component.

10.4.3 Design Method of the Modified TIR-Freeform Lens

Since both the circular target and luminous intensity distribution of the white LED are rotational-symmetric, only the contour line of the modified TIR-freeform lens' cross

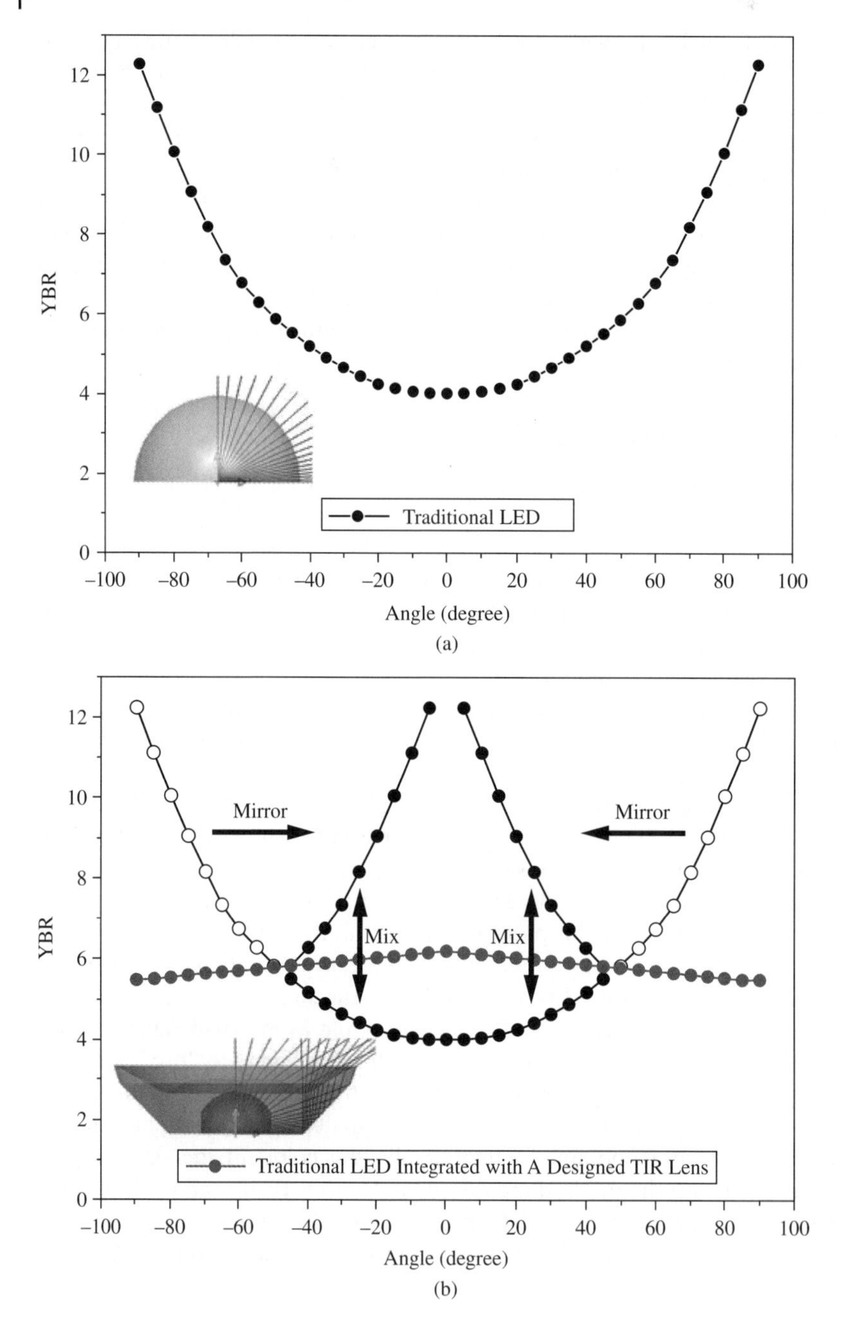

Figure 10.24 Design principle of the modified TIR-freeform lens. (a) Original YBR distribution; and (b) desired YBR distribution.

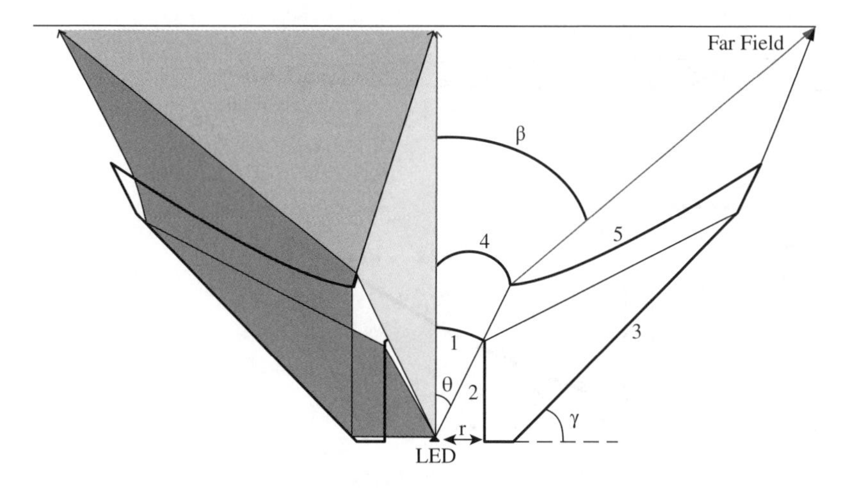

Figure 10.25 Sketch map of designed ray paths through the modified TIR-freeform lens.

section needs to be calculated. Then, rotating this contour line can generate the lens entity.

Figure 10.25 shows the schematic of the designed ray paths through the modified TIR-freeform lens. The emitted luminous flux of the initial light source is split into two parts by Surface 1 and Surface 2 with an angle θ, the sparse meshed and dense meshed parts. Both parts of optical power irradiate on the same far target plane or into the same divergence half angle (β) through the modified TIR-freeform lens. Surface 1 is a concave spherical surface, and Surface 2 is a cylindrical wall. The concave spherical surface will not change the transmission directions of light rays, so that we can focus on the design of Surface 4. The cylindrical wall is a considerate choice in our design, which makes the incident angles on the total-reflection surface (Surface 3) large enough to generate total reflections. Surface 3 is a circular truncated conical surface with an inclination angle γ, which could save the height of our designed component, compared with the freeform TIR surface shown in Figure 10.23b. As a consequence, the incident light on the top surface is not collimated, which results in a larger divergence half angle (the divergence half angle of classic lenses with collimated light irradiating on the top surface has a maximum value of 42°). Finally, Surface 4 and Surface 5 mix the two parts of optical power within a divergence half angle (β). Therefore, we propose a design method for modified TIR components with discontinuous refractive top surfaces to achieve compact size and high SCU.

Based on the aforementioned design principle and method, the lens design procedure includes three main steps.[27,28] Firstly, divide the light source and illumination target plane into equal numbered grids with equal luminous flux and area, respectively. Energy conservation is an assumption in the design. Secondly, establish the light energy mapping relationship by Snell's law and the edge ray principle, and figure out discrete points of the lens' contour line.[27,29] Finally, construct the lens by the lofting method. Since Surfaces 1, 2, and 3 are either conical or spherical surfaces, the calculation of Surface 4 and Surface 5 is relatively more significant. When the incident rays arrive at the top surface, Surface 4 and Surface 5 redistribute the rays onto the prescribed target plane to achieve uniform illuminance.[27] Figure 10.26a shows the generation method of the

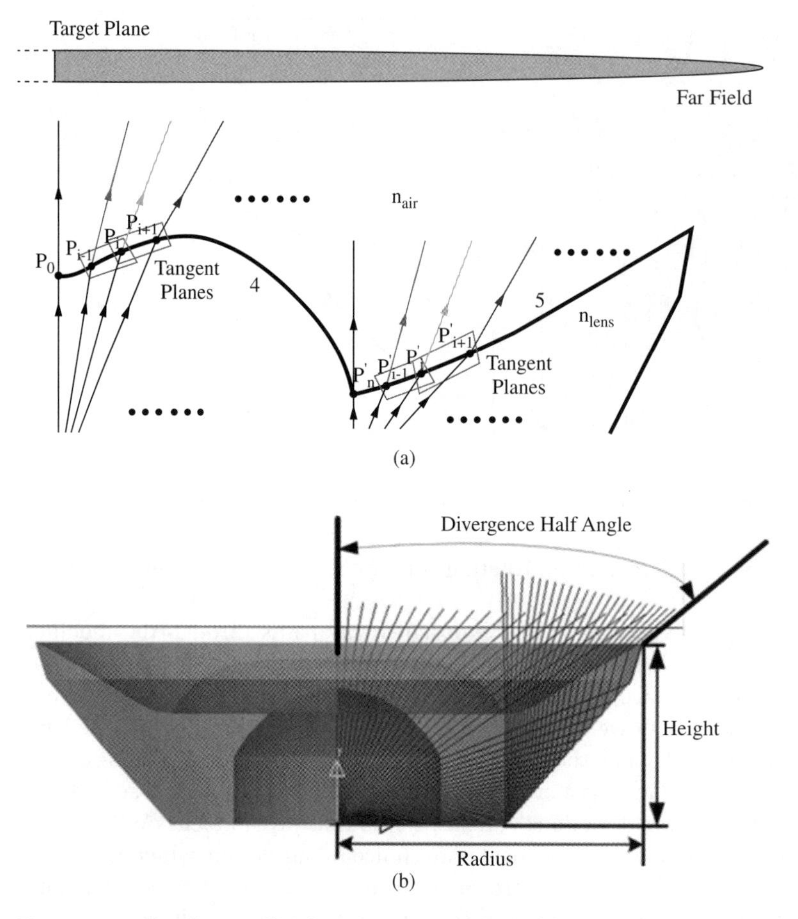

Figure 10.26 Sketch map of (a) the generation method of the discrete points of the top surface; and (b) a designed TIR lens with ($\theta = 60°$, $\gamma = 45°$, $\beta = 65°$). The modified TIR-freeform lens has a larger divergence half angle than the classic TIR lens.

discrete points (P_i). We can calculate these discrete points by the incident and exit rays using the inverse procedure of Snell's law.

As shown in Figure 10.26a, the emergent rays with the same colors own the same emergent directions and are set to mix with each other in the far field, so that the low YBR and high YBR will be mixed to generate a medium YBR. Figure 10.26b shows a designed modified TIR-freeform lens with ($\theta = 60°$, $\gamma = 45°$, and $\beta = 65°$), whose radius is 10.23 mm and height is 5.49 mm.

10.4.4 Optimization Results and Discussions

In order to find an optimum modified TIR-freeform lens, optimization by changing the parameters is essential. For the sake of simplicity, only the divergence half angle (β) is optimized to minimize the dimension and maximize the SCU. θ and γ are set as 60° and 45°, respectively. In our work, normalized standard deviation (NSD)[30] is used to evaluate the variation of YBR; it is defined as:

$$NSD = \frac{1}{E_{YBR}} \sigma(YBR) \tag{10.8}$$

where E_{YBR} is the average value of YBRs; and σ (YBR) means the standard deviation of the YBR distribution. A lower NSD represents a smaller YBR variation within the viewing angles and a higher SCU.

Analysis results of modified TIR-freeform components by changing the divergence half angle from 40° to 65° based on the Monte Carlo ray-tracing method (1 million rays) indicate that the height and radius distributions of modified LED packages along with β have a minimum radius at 50° and a minimum height at 55°, while the NSD distribution curve along with β is monotone, decreasing to a certain degree, as shown in Figure 10.27a. When we take SCU in the first place, the enhancement of SCU with an optimum appropriate divergence half angle (65°) reaches as high as 84% in terms of the NSD of the lens' YBR distribution, from 0.888 to 0.429; simultaneously, compared with the classic TIR lens (10.26 mm radius × 10.66 mm height), we can achieve a dimension of 10.23 mm radius × 5.49 mm height, a smaller size in vertical height (~0.52) within the same horizontal radius at $\beta = 65°$.

Figure 10.27b shows the detailed simulative YBR distributions of the traditional LED and modified TIR-freeform lenses with $\beta = 65°$. The YBR distribution of the modified TIR-freeform lens has increased values in the center and reduced values at the edge, which verifies our design idea. Since little optical power is transmitted to the edge angle, the YBR values within more than 75° are meaningless and are not presented in the figure. In addition, illuminance analysis of the LED package with the modified TIR component with $\beta = 65°$ is also carried out. As shown in Figure 10.28, compared with classic lenses, an enhanced performance of illuminance uniformity (enhancement ~20%) on the desired target plane is achieved with the modified LED package. Since the modified TIR-freeform lens owns a larger divergence angle, it's reasonable that its central illuminance is lower than that of the classic lens.

Although we adopt the extended white LED model, as the two parts of luminous flux through Surface 4 and Surface 5 are set to overlap and compensate for each other on the target plane, respectively, the relative uniform illuminance distribution can still be achieved easily even without any illumination optimization.

An effective color-mixing method is proposed to enhance the SCU of phosphor-converted white LEDs, and a modified TIR-freeform lens is designed to integrate with the white LED to verify the design idea. From the above simulation results, we conclude that the proposed TIR-freeform optics not only can be used for traditional directional lighting (small divergence angle) but also can be applied to general lighting (large divergence angle) to achieve high SCU. Analysis results of the modified TIR component based on the Monte Carlo ray-tracing method indicate that by comparing with the traditional LED integrated with the classic TIR lens. The modified LED package has the advantages of low profile, small volume, and high illuminance uniformity. More importantly, an enhanced SCU is achieved based on the modified TIR-freeform lens, and the optimum divergence half angle is 65°. In summary, the phosphor-converted white LED integrated with the compact modified TIR-freeform lens can provide high-quality lighting within a flexible divergence angle, which can apply to many illumination fields such as local lighting and general lighting.

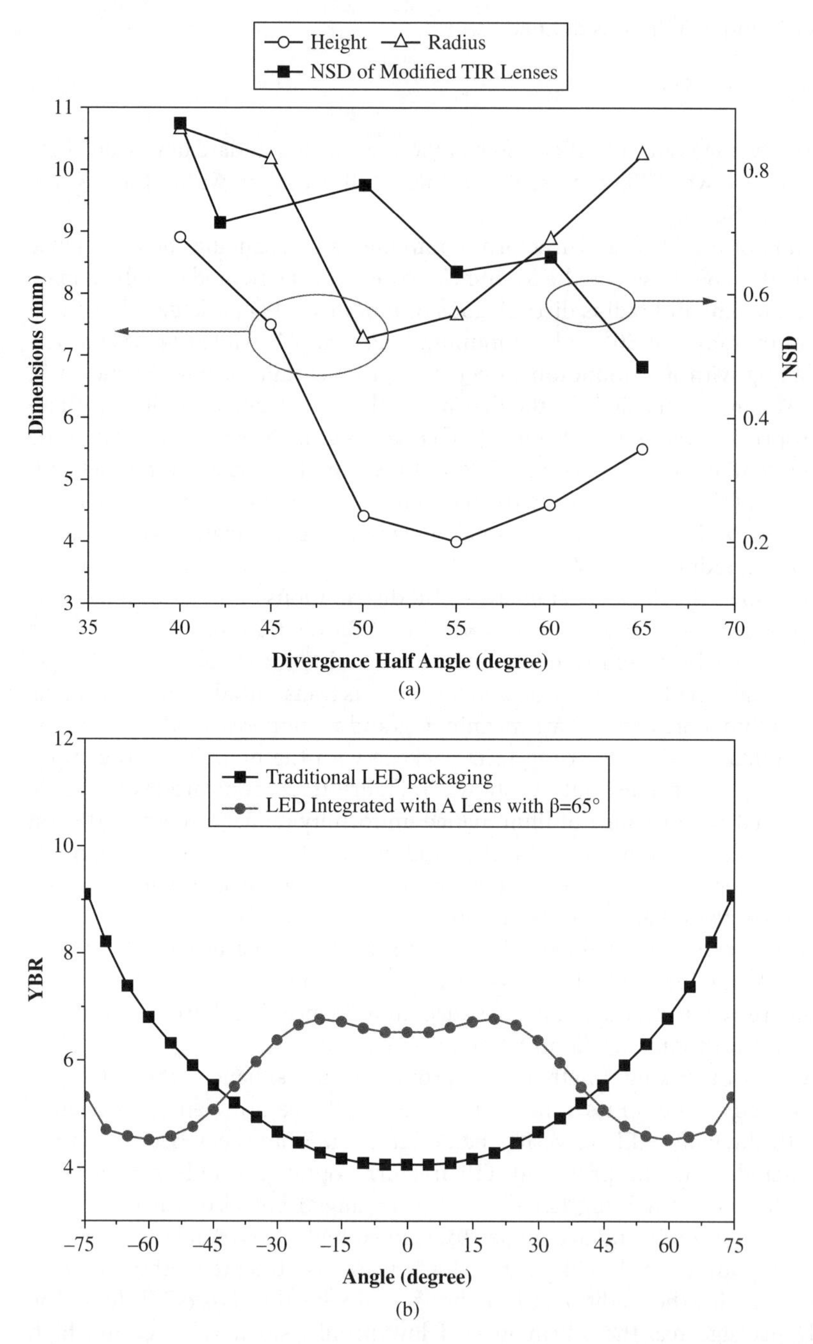

Figure 10.27 (a) Optimization processes of minimizing modified TIR-freeform lens dimensions as well as maximizing the SCU; and (b) YBR distributions of the traditional LED and modified TIR-freeform lens with $\beta = 65$.

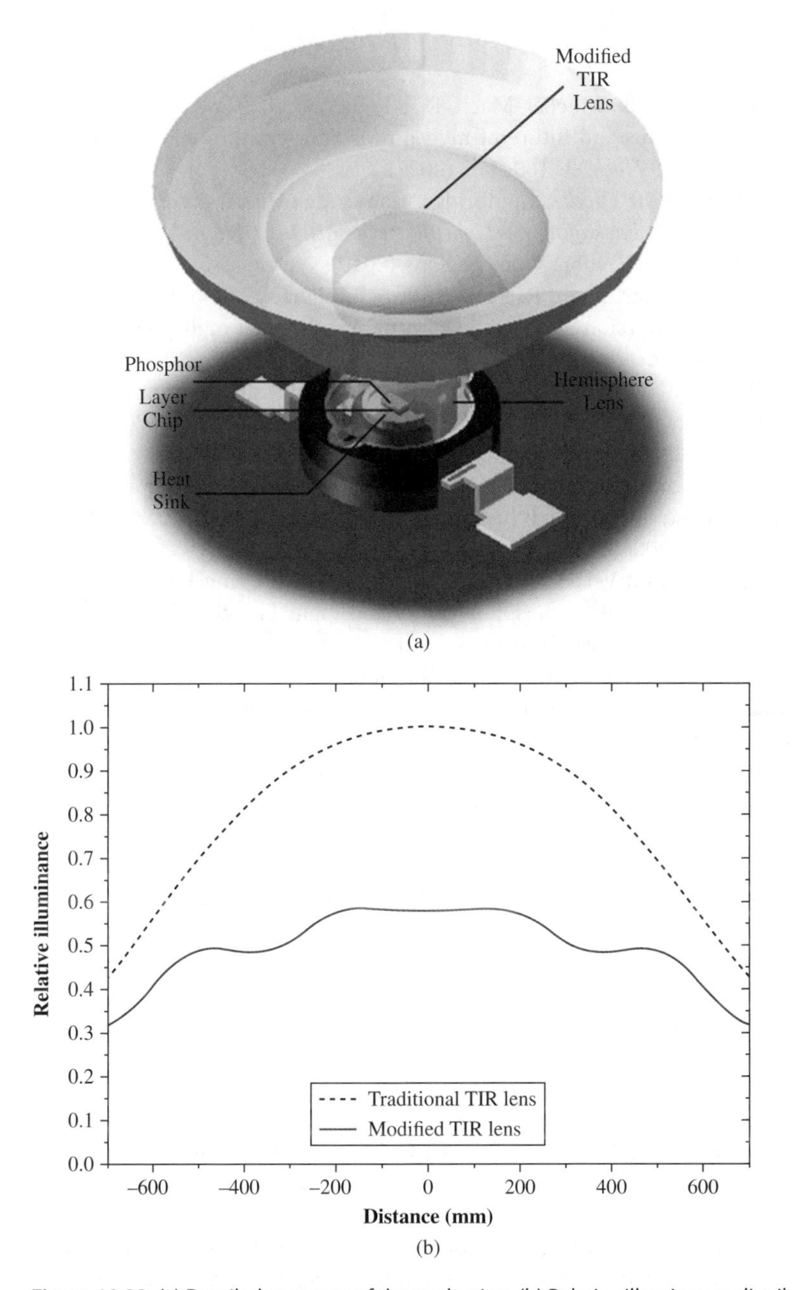

Figure 10.28 (a) Detailed structure of the packaging. (b) Relative illuminance distributions on a 700 × 700 mm target based on the extended light source 1 m away. Uniformity of illuminance is enhanced by 10.6% from 53.1% to 63.7%. The larger divergence angle of the modified TIR-freeform lens results in lower central illuminance compared with the classic TIR lens, while the marginal illuminance of the modified TIR-freeform lens on a large enough receiver is higher (not shown here).

References

1 Krames, M.R., Shchekin, O.B., Mueller-Mach, R., Mueller, G.O., Zhou, L., Harbers, G., and Craford, M.G. Status and future of high-power light-emitting diodes for solid-state lighting. *J. Disp. Tech.* **3**, 160–175 (2007).
2 Schubert, E.F. *Light-Emitting Diodes*. Cambridge: Cambridge University Press (2006).
3 Liu, S., and Luo, X.B. *LED Packaging for Lighting Applications: Design, Manufacturing and Testing*. Hoboken, NJ: John Wiley & Sons (2010).
4 Sommer, C., Wenzl, F.P., Hartmann, P., Pachler, P., Schweighart, M., and Leising, G. Tailoring of the color conversion elements in phosphor-converted high-power LEDs by optical simulations. *IEEE Photon. Tech. Lett.* **20**, 739–741 (2008).
5 Liu, Z., Liu, S., Wang, K., and Luo, X. Optical analysis of color distribution in white LEDs with various packaging methods. *IEEE Photon. Tech. Lett.* **20**, 2027–2029 (2008).
6 Jordan, R.C., Bauer, J., and Oppermann, H. Optimized heat transfer and homogeneous color converting for ultrahigh brightness LED package. *Proc. SPIE* **6198**, 61980B – 61980B – 12 (2006).
7 Collins, W.D., Krames, M.R., Verhoeckx, G.J., and Leth, N.J.M. Using Electrophoresis to Produce a Conformally Coated Phosphor-Converted Light Emitting Semiconductor. US Patent 6,576,488 (10 June 2003).
8 Yum, J.-H., Seo, S.-Y., Lee, S., and Sung, Y.-E. Comparison of Y3Al5O12:Ce0.05 phosphor coating methods for white-light-emitting diode on gallium nitride. Proc. *SPIE* **4445**, 60–69 (2001).
9 Negley, G.H., and Leung, M. Methods of Coating Semiconductor Light Emitting Elements by Evaporating Solvent from a Suspension. US Patent 7,217,583 (15 May 2007).
10 Braune, B., Petersen, K., Strauss, J., Kromotis, P., and Kaempf, M. A new wafer level coating technique to reduce the color distribution of LEDs. *Proc. SPIE* 64860X – 64860X – 11 (2007).
11 Wu, H., Narendran, N., Gu, Y., and Bierman, A. Improving the performance of mixed-color white LED systems by using scattered photon extraction technique. *Proc. SPIE 6669*, 666905–666905 – 12 (2007).
12 Chen, F., Liu, S., Wang, K., Liu, Z., and Luo, X. Free-form lenses for high illumination quality light-emitting diode MR16 lamps. *Opt. Eng.* **48**, 123002–123002 – 7 (2009).
13 Lee, T.-X., Gao, K.-F., Chien, W.-T., and Sun, C.-C. Light extraction analysis of GaN-based light-emitting diodes with surface texture and/or patterned substrate. *Opt. Expr.* **15**, 6670–6676 (2007).
14 Liu, Z.-Y., Liu, S., Wang, K., and Luo, X.-B. Studies on optical consistency of white LEDs affected by phosphor thickness and concentration using optical simulation. *IEEE Trans. Comp. Pack. Tech.* **33**, 680–687 (2010).
15 Wang, G., Wang, L., Li, F., and Zhang, G. Collimating lens for light-emitting-diode light source based on non-imaging optics. *Appl. Opt.* **51**, 1654–1659 (2012).
16 Chen, J.-J., Wang, T.-Y., Huang, K.-L., Liu, T.-S., Tsai, M.-D., and Lin, C.-T. Freeform lens design for LED collimating illumination. *Opt. Expr.* **20**, 10984–10995 (2012).
17 Parkyn, W.A., and Pelka, D.G. Compact nonimaging lens with totally internally reflecting facets. *Proc. SPIE* **1528**, 70–81 (1991).

18 Fournier, F.R., Cassarly, W.J., and Rolland, J.P. Fast freeform reflector generation using source-target maps. *Opt. Expr.* **18**, 5295–5304 (2010).
19 Fournier, F.R. A review of beam shaping strategies for LED lighting. *Proc. SPIE* 817007–817007 – 11 (2011).
20 Zhenrong, Z., Xiang, H., and Xu, L. Freeform surface lens for LED uniform illumination. *Appl. Opt.* **48**, 6627 (2009).
21 Zhao, S., Wang, K., Chen, F., Wu, D., and Liu, S. Lens design of LED searchlight of high brightness and distant spot. *J. Opt. Soc. Am. A* **28**, 815–820 (2011).
22 Sun, L., Jin, S., and Cen, S. Free-form microlens for illumination applications. *Appl. Opt.* **48**, 5520–5527 (2009).
23 Liu, Z., Liu, S., Wang, K., and Luo, X. Measurement and numerical studies of optical properties of YAG:Ce phosphor for white light-emitting diode packaging. *Appl. Opt.* **49**, 247–257 (2010).
24 Wang, K., Wu, D., Chen, F., Liu, Z., Luo, X., and Liu, S. Angular color uniformity enhancement of white light-emitting diodes integrated with freeform lenses. *Opt. Lett.* **35**, 1860–1862 (2010).
25 Zhu, L., Wang, X.H., Lai, P.T., and Choi, H.W. Angular uniform white light-emitting diodes with an internal reflector cup. *SID Symp. Dig. Tech. Papers* **41**, 1013–1015 (2010).
26 Wu, H., Narendran, N., Gu, Y., and Bierman, A. Improving the performance of mixed-color white LED systems by using scattered photon extraction technique. *Proc. SPIE* 6669, 666905–666905 – 12 (2007).
27 Wang, K., Chen, F., Liu, Z., Luo, X., and Liu, S. Design of compact freeform lens for application specific light-emitting diode packaging. *Opt. Expr.* **18**, 413–425 (2010).
28 Ding, Y., Liu, X., Zheng, Z., and Gu, P. Freeform LED lens for uniform illumination. *Opt. Expr.* **16**, 12958–12966 (2008).
29 Bortz, J., and Shatz, N. Generalized functional method of nonimaging optical design. *Proc. SPIE* 6338, 633805–633805 – 16 (2006).
30 Kim, B., Kim, H., and Kang, S. Reverse functional design of discontinuous refractive optics using an extended light source for flat illuminance distributions and high color uniformity. *Opt. Expr.* **19**, 1794–1807 (2011).

A

Codes of Basic Algorithms of Freeform Optics for LED Lighting

Based on MATLAB Software

1. Circularly Symmetric Freeform Lens for Large Emitting Angles

Point Source (Light Energy Distribution is Lambert Type)

```
%_____________Define original data and variable_____________
N_latitude=1000; % Number of subarea
halfdeg=pi/3; % Set half emitting angle

%_____________Find Km_____________
% Divide light energy distribution of source into N_latitude
   grids equally
% Km: Slopes of rays of source grids
Kma_radian=0;

for k=1:N_latitude
   Kma_radian(k+1)=abs(acos(cos(2*Kma_radian(k))
    -2/N_latitude)/2);
end

Kma_degree=Kma_radian*180/pi;
Km=cot(Kma_radian);


%_____________Find Q_____________
% Divide area of target plane into N_latitude grids equally
% Q: Positions of target grids
d=1000; % Height of target plane. Unit: mm
radius=d*tan(halfdeg);
S=pi*radius*radius;
Q=[1:1:N_latitude+1;1:1:N_latitude+1];
Q(2,:)=d;
```

Freeform Optics for LED Packages and Applications, First Edition. Kai Wang, Sheng Liu, Xiaobing Luo and Dan Wu.

```
Q=Q';
Q(1,1)=0;


for i=1:N_latitude
   Q(i+1,1)=sqrt(Q(i,1)*Q(i,1)+(S/(pi*N_latitude)));
end

%_____________Find P_____________
% P; Positions of points on the freeform curve
syms x0 y0 N1 N2 K

n1=1.49; % Refractive index - input rays
n2=1.00; % Refractive index - output rays
h=7; % Height of central point of the freeform lens.
      Unit: mm

% Define the original coordinate of point P
P=[0,h];
Pstore(1,1)=P(1);
Pstore(1,2)=P(2);

% Define light source position
S=[0,0];

% Calculate P
for j=1:1:N_latitude
   % Input rays
   I=P-S;
   % Output rays
   O=Q(j,:)-P;

   %Normal vector
   I=I/((I(1)^2+I(2)^2)^0.5);
   O=O/((O(1)^2+O(2)^2)^0.5);
   N=(n1*I-n2*O)/((1+(n1/n2)*(n1/n2)-2*(n1/n2)*dot(O,I))
     ^0.5);
   N=N/((N(1)^2+N(2)^2)^0.5);
   Nstore(j,1)=N(1);
   Nstore(j,2)=N(2);

   %Calculate the point Pj+1
   [x,y]=solve('(y-y0)=(-N1/N2)*(x-x0)','y=K*x','x','y');

Pstore(j+1,1)=subs(x,{x0,y0,N1,N2,K},{P(1),P(2),N(1),N(2),
 Km(j+1)}); check
```

```
Pstore(j+1,2)=subs(y,{x0,y0,N1,N2,K},{P(1),P(2),N(1),N(2),
 Km(j+1)});

   P(1)=Pstore(j+1,1);
   P(2)=Pstore(j+1,2);

end

xlswrite('P',Pstore); % Output positions of points of the
 freeform curve
```

2. Noncircularly Symmetric Discontinuous Freeform Lens

Point Source (Light Energy Distribution is Lambert Type)

```
%_____________Define original data and variable_____________
N_latitude=102; % Number of subarea along latitude
N_longitude=256;% Number of subarea along longitude 2^m,
 m>=2

a=30000; % Length of the target a=L. Unit: mm
b=10000; % Width of the target b=W. Unit: mm
H=8000; % Height of the target. Unit: mm
s=a*b; % Area of the target

%_____________Find Km_____________
% Divide light energy distribution of source into N_latitude
 grids equally
% Km: Slope of ray of each source grid

Kma_radian=0;

for k=1:N_latitude
   Kma_radian(k+1)=abs(acos(cos(2*Kma_radian(k))
    -2/N_latitude)/2);
end

Kma_degree=Kma_radian*180/pi;
Km=cot(Kma_radian);


%_____________Define P, Q, N and Dangle_____________

% P1store: Positions of points on the longitude freeform
   curve
```

```
P1store=[1:1:(N_latitude+1)*(N_longitude/4+1);1:1:
 (N_latitude+1)*(N_longitude/4+1);1:1:(N_latitude+1)*
 (N_longitude/4+1)];
P1store=P1store';


% N1store: Calculated normal vectors
N1store=[1:1:(N_latitude+1)*(N_longitude/4+1);1:1:
 (N_latitude+1)*(N_longitude/4+1);1:1:(N_latitude+1)*
 (N_longitude/4+1)];
N1store=N1store';

% Nrealstore: Real normal vectors
Nrealstore=[1:1:(N_latitude-2)*N_longitude/4;1:1:
 (N_latitude-2)*N_longitude/4;1:1:(N_latitude-2)*
  N_longitude/4];
Nrealstore=Nrealstore';

% Dangle: Deviation angle
Dangle=[1:1:(N_latitude-2)*N_longitude/4];
Dangle=Dangle';

DangleD=[1:1:(N_latitude-2)*N_longitude/4];
DangleD=DangleD';

% Q1store: Positions of target plane grids
Q1store=[1:1:N_latitude*(N_longitude/4+1);1:1:N_latitude*
 (N_longitude/4+1);1:1:N_latitude*(N_longitude/4+1)];
Q1store=Q1store';

%____________Calculate Q1____________
syms s1 m1 a1 b1 i1

for th=0:1:N_longitude/4 % th: Sequence number of light
 source/target plane from x axis

   % Calculate the length and the width of the trap
   for i=1:N_latitude-1
       [x,y]=solve('x*y=(s1/m1)*i1','x/y=a1/b1','x','y');
       c1=subs(x,{s1,m1,a1,b1,i1},{s,N_latitude-1,a,b,i});
       c2=subs(y,{s1,m1,a1,b1,i1},{s,N_latitude-1,a,b,i});
       AA(i)=abs(c1(1));
       BB(i)=abs(c2(1));
   end

   % Define the original position of Q
   Q=[1:1:N_latitude;1:1:N_latitude;1:1:N_latitude];
```

```
    Q(3,:,:)=H;
    Q=Q';
    Q(1,1)=0;
    Q(1,2)=0;


    % Calculate Q
    if th<=(N_longitude/8)
       Deg=atan((0.5*b/(N_longitude/8)*th)/(0.5*a));
        % Deg: Angle of target line from the x axis
       for i=2:N_latitude
          Q(i,1)=0.5*AA(i-1);
          Q(i,2)=(0.5*AA(i-1))*tan(Deg);
       end
    else

Deg=atan((0.5*b)/((N_longitude/4-th)*((0.5*a)/
 (N_longitude/8))));
        for i=2:N_latitude
           Q(i,1)=(0.5*BB(i-1))*cot(Deg);
           Q(i,2)=0.5*BB(i-1);
        end
     end

% Calculate Q1store

    for i=1:N_latitude
       Q1store(th*N_latitude+i,:)=Q(i,:);
    end

end

% j: The number of curves along longitude direction,
 j=0 for the first curve
j=0;
jump=0;


while jump==0
% _____________Calculate first curve of P1store using the
 first curve method_____________

syms A1 B1 C1 x01 y01 z01 Vx1 Vy1 Vz1

n1=1.49; % Refractive index - input rays
```

```
n2=1.00; % Refractive index - output rays
h=7; % Height of central point of the lens
DeviationAngle=6; % Deviation angle between real normal
 vector and calculated normal vector


th3=j*(N_latitude+1);

P1=[0,0,h];
P1store(1+th3,1)=P1(1);
P1store(1+th3,2)=P1(2);
P1store(1+th3,3)=P1(3);

S=[0,0,0];
for i=1:1:N_latitude
  % Input rays
  I1=P1store(i+th3,:)-S;

  % Output rays
  O1=Q1store(i+j*N_latitude,:)-P1store(i+th3,:);

  % Normal vectors
  I1=I1/((I1(1)^2+I1(2)^2+I1(3)^2)^0.5);
  O1=O1/((O1(1)^2+O1(2)^2+O1(3)^2)^0.5);
  N1=(n1*I1-n2*O1)/((1+(n1/n2)*(n1/n2)-2*(n1/n2)*dot(O1,I1))
   ^0.5);
  N1=N1/((N1(1)^2+N1(2)^2+N1(3)^2)^0.5);
  N1store(i+th3,1)=N1(1);
  N1store(i+th3,2)=N1(2);
  N1store(i+th3,3)=N1(3);

  % Calculate the point Pj+1
[x,y,z]=solve('A1*(x-x01)+B1*(y-y01)+C1*(z-z01)=0','x/Vx1=y/
 Vy1','y/Vy1=z/Vz1','x','y','z');

P1store(i+1+th3,1)=abs(subs(x,{A1,B1,C1,x01,y01,z01,Vx1,Vy1,
Vz1},{N1store(i+th3,1),N1store(i+th3,2),N1store(i+th3,3),
P1store(i+th3,1),P1store(i+th3,2),P1store(i+th3,3),
sin(Kma_radian(i+1))*cos(j*(pi/2)/(N_longitude/4)),
sin(Kma_radian(i+1))*sin(j*(pi/2)/(N_longitude/4)),
cos(Kma_radian(i+1))}));

P1store(i+1+th3,2)=abs(subs(y,{A1,B1,C1,x01,y01,z01,Vx1,Vy1,
Vz1},{N1store(i+th3,1),N1store(i+th3,2),N1store(i+th3,3),
P1store(i+th3,1),P1store(i+th3,2),P1store(i+th3,3),
sin(Kma_radian(i+1))*cos(j*(pi/2)/(N_longitude/4)),
```

```
sin(Kma_radian(i+1))*sin(j*(pi/2)/(N_longitude/4)),
cos(Kma_radian(i+1))}));

P1store(i+1+th3,3)=abs(subs(z,{A1,B1,C1,x01,y01,z01,Vx1,Vy1,
Vz1},{N1store(i+th3,1),N1store(i+th3,2),N1store(i+th3,3),
P1store(i+th3,1),P1store(i+th3,2),P1store(i+th3,3),
sin(Kma_radian(i+1))*cos(j*(pi/2)/(N_longitude/4)),
sin(Kma_radian(i+1))*sin(j*(pi/2)/(N_longitude/4)),
cos(Kma_radian(i+1))}));

end

   N1store(N_latitude+1+th3,1)=cos(j*(pi/2)/(N_longitude
    /4));
   N1store(N_latitude+1+th3,2)=sin(j*(pi/2)/(N_longitude
    /4));
   N1store(N_latitude+1+th3,3)=0;

% _____________Calculate the next curve of P1store using the
 second curve method_____________

syms A2 B2 C2 x02 y02 z02 Vx2 Vy2 Vz2

for th=j:1:N_longitude/4-1

    for i=1:1:N_latitude

[x,y,z]=solve('A2*(x-x02)+B2*(y-y02)+C2*(z-z02)=0','x/Vx2=y/
 Vy2','y/Vy2=z/Vz2','x','y','z');

P1store(i+(th+1)*(N_latitude+1),1)=abs(subs(x,{A2,B2,C2,
x02,y02,z02,Vx2,Vy2,Vz2},{N1store(i+th*(N_latitude+1),1),
N1store(i+th*(N_latitude+1),2),N1store(i+th*(N_latitude+1),
3),P1store(i+th*(N_latitude+1),1),P1store(i+th*(N_latitude
+1),2),P1store(i+th*(N_latitude+1),3),sin(Kma_radian(i))*
cos((th+1)*(pi/2)/(N_longitude/4)),sin(Kma_radian(i))*sin
((th+1)*(pi/2)/(N_longitude/4)),cos(Kma_radian(i))}));

P1store(i+(th+1)*(N_latitude+1),2)=abs(subs(y,{A2,B2,C2,x02,
y02,z02,Vx2,Vy2,Vz2},{N1store(i+th*(N_latitude+1),1),
N1store(i+th*(N_latitude+1),2),N1store(i+th*(N_latitude+1),
3),P1store(i+th*(N_latitude+1),1),P1store(i+th*(N_latitude
+1),2),P1store(i+th*(N_latitude+1),3),sin(Kma_radian(i))*
cos((th+1)*(pi/2)/(N_longitude/4)),sin(Kma_radian(i))*sin
((th+1)*(pi/2)/(N_longitude/4)),cos(Kma_radian(i))}));
```

```
P1store(i+(th+1)*(N_latitude+1),3)=abs(subs(z,{A2,B2,C2,x02,
y02,z02,Vx2,Vy2,Vz2},{N1store(i+th*(N_latitude+1),1),
N1store(i+th*(N_latitude+1),2),N1store(i+th*(N_latitude+1),
3),P1store(i+th*(N_latitude+1),1),P1store(i+th*(N_latitude
+1),2),P1store(i+th*(N_latitude+1),3),sin(Kma_radian(i))*
cos((th+1)*(pi/2)/(N_longitude/4)),sin(Kma_radian(i))*sin
((th+1)*(pi/2)/(N_longitude/4)),cos(Kma_radian(i))}));

       I1=P1store(i+(th+1)*(N_latitude+1),:)-S;

O1=Q1store(i+(th+1)*N_latitude,:)-P1store(i+(th+1)*
 (N_latitude+1),:);

       I1=I1/((I1(1)^2+I1(2)^2+I1(3)^2)^0.5);
       O1=O1/((O1(1)^2+O1(2)^2+O1(3)^2)^0.5);

N1=(n1*I1-n2*O1)/((1+(n1/n2)*(n1/n2)-2*(n1/n2)*dot(O1,I1))
 ^0.5);
       N1=N1/((N1(1)^2+N1(2)^2+N1(3)^2)^0.5);
       N1store(i+(th+1)*(N_latitude+1),1)=N1(1);
       N1store(i+(th+1)*(N_latitude+1),2)=N1(2);
       N1store(i+(th+1)*(N_latitude+1),3)=N1(3);
    end

     % Calculate the last point of the curve

       i=N_latitude+1;

[x,y,z]=solve('A2*(x-x02)+B2*(y-y02)+C2*(z-z02)=0','x/Vx2=y
 /Vy2','y/Vy2=z/Vz2','x','y','z');

P1store(i+(th+1)*(N_latitude+1),1)=abs(subs(x,{A2,B2,C2,x02,
y02,z02,Vx2,Vy2,Vz2},{N1store(i-1+(th+1)*(N_latitude+1),1),
N1store(i-1+(th+1)*(N_latitude+1),2),N1store(i-1+(th+1)*
(N_latitude+1),3),P1store(i-1+(th+1)*(N_latitude+1),1),
P1store(i-1+(th+1)*(N_latitude+1),2),P1store(i-1+(th+1)*
(N_latitude+1),3),sin(Kma_radian(i))*cos((th+1)*(pi/2)/
(N_longitude/4)),sin(Kma_radian(i))*sin((th+1)*(pi/2)/
(N_longitude/4)),cos(Kma_radian(i))}));

P1store(i+(th+1)*(N_latitude+1),2)=abs(subs(y,{A2,B2,C2,x02,
y02,z02,Vx2,Vy2,Vz2},{N1store(i-1+(th+1)*(N_latitude+1),1),
N1store(i-1+(th+1)*(N_latitude+1),2),N1store(i-1+(th+1)*
(N_latitude+1),3),P1store(i-1+(th+1)*(N_latitude+1),1),
P1store(i-1+(th+1)*(N_latitude+1),2),P1store(i-1+(th+1)*
(N_latitude+1),3),sin(Kma_radian(i))*cos((th+1)*(pi/2)/
```

```
(N_longitude/4)),sin(Kma_radian(i))*sin((th+1)*(pi/2)/
(N_longitude/4)),cos(Kma_radian(i))}));

P1store(i+(th+1)*(N_latitude+1),3)=abs(subs(z,{A2,B2,C2,x02,
y02,z02,Vx2,Vy2,Vz2},{N1store(i-1+(th+1)*(N_latitude+1),1),
N1store(i-1+(th+1)*(N_latitude+1),2),N1store(i-1+(th+1)*
(N_latitude+1),3),P1store(i-1+(th+1)*(N_latitude+1),1),
P1store(i-1+(th+1)*(N_latitude+1),2),P1store(i-1+(th+1)*
(N_latitude+1),3),sin(Kma_radian(i))*cos((th+1)*(pi/2)/
(N_longitude/4)),sin(Kma_radian(i))*sin((th+1)*(pi/2)/
(N_longitude/4)),cos(Kma_radian(i))}));


N1store(i+(th+1)*(N_latitude+1),1)=cos((th+1)*(pi/2)/
 (N_longitude/4));

N1store(i+(th+1)*(N_latitude+1),2)=sin((th+1)*(pi/2)/
 (N_longitude/4));

          N1store(i+(th+1)*(N_latitude+1),3)=0;

% _____________Calculate the deviation angle _____________
    for i=2:1:N_latitude-1


v1=P1store(i+1+th*(N_latitude+1),:)-P1store(i+th*(N_latitude
 +1),:);

v2=P1store(i+(th+1)*(N_latitude+1),:)-P1store
 (i+th*(N_latitude+1),:);

      Nreal=cross(v1,v2);
      Nreal=Nreal/((Nreal(1)^2+Nreal(2)^2+Nreal(3)^2)
       ^0.5);
      Nrealstore(i-1+th*(N_latitude-2),1)=Nreal(1);
      Nrealstore(i-1+th*(N_latitude-2),2)=Nreal(2);
      Nrealstore(i-1+th*(N_latitude-2),3)=Nreal(3);

      N11=N1store(i+th*(N_latitude+1),:);
      Dangle(i-1+th*(N_latitude-2))=acos(dot(N11,Nreal));

DangleD(i-1+th*(N_latitude-2))=abs(Dangle(i-1+th*(N_latitude
 -2))/pi*180);

       if DangleD(i-1+th*(N_latitude-2))>DeviationAngle

          jump=1;
```

```
          break;
        end
    end

    if jump==1
       break;
    else
       j=j+1;
    end
end
jump=jump-1;
%____________________________________________________________
%____________________________________________________________
end
```

3. Reversing the Design Method for the LED Array Uniform Illumination Algorithm of a Freeform Lens for the Required Light Intensity Distribution Curve (LIDC)

Point Source (Light Energy Distribution is Lambert Type)

```
%_____________Define original data and variable_____________
N_latitude=500; % Number of subarea along latitude
%______________Least square curve fitting________________
% Input the optimized LIDC data in the form of EXCEL table
xy=xlsread('Optimized-LIDC'); % Optimized-LIDC is the name
 of the EXCEL table
xy=xy';
N=46; % Number of points input
M=10; % Mth order polynomial

for i=0:M
```

```
  for j=1:N
     A(j,i+1)=xy(1,j)^i;
  end

end

Y=xy(2,:)';

% Coefficient of Mth order polynomial
a=inv(A'*A)*(A'*Y);

%______________Find Km________________
syms x0 xx x x1

y1=(a(1)+a(2)*x0+a(3)*x0^2+a(4)*x0^3+a(5)*x0^4+a(6)*x0^5
 +a(7)*x0^6+a(8)*x0^7+a(9)*x0^8+a(10)*x0^9+a(11)*x0^10)*
 sin(x0);

% Flux of each source subarea
Fi=double(int(y1,0,pi/2)/N_latitude);

x1=0;
Kma_radian_T=0;
a1=a(1);
a2=a(2);
a3=a(3);
a4=a(4);
a5=a(5);
a6=a(6);
a7=a(7);
a8=a(8);
a9=a(9);
a10=a(10);
a11=a(11);

for j=1:N_latitude/2

f=@(x)-a3*x^2*cos(x)+a1*cos(x1)-24*a5*sin(x)*x+5*a6*sin(x)*
x^4-120*a6*x*cos(x)-15120*a10*sin(x1)*x1^4-60480*a10*x1^3*
cos(x1)+181440*a10*sin(x1)*x1^2+362880*a10*x1*cos(x1)+a11*
x1^10*cos(x1)-10*a11*sin(x1)*x1^9+720*a11*sin(x1)*x1^7+
5040*a11*x1^6*cos(x1)-30240*a11*sin(x1)*x1^5-151200*a11*
x1^4*cos(x1)+604800*a11*sin(x1)*x1^3+1814400*a11*x1^2*cos
(x1)-3628800*a11*sin(x1)*x1-60*a6*sin(x)*x^2+6*a7*sin(x)*
x^5+30*a7*x^4*cos(x)-120*a7*sin(x)*x^3-360*a7*x^2*cos(x)
+720*a7*sin(x)*x+7*a8*sin(x)*x^6+42*a8*x^5*cos(x)-210*a8*
```

```
sin(x)*x^4+2520*a8*sin(x)*x^2+5040*a8*x*cos(x)-a9*x^8*cos
(x)+8*a9*sin(x)*x^7+56*a9*x^6*cos(x)-336*a9*sin(x)*x^5-
1680*a9*x^4*cos(x)+6720*a9*sin(x)*x^3+20160*a9*x^2*cos(x)
-40320*a9*sin(x)*x-a10*x^9*cos(x)+9*a10*sin(x)*x^8+72*a10*
x^7*cos(x)-504*a10*sin(x)*x^6-3024*a10*x^5*cos(x)+15120*
a10*sin(x)*x^4+60480*a10*x^3*cos(x)-181440*a10*sin(x)*x^2
-362880*a10*x*cos(x)-a11*x^10*cos(x)+10*a11*sin(x)*x^9+90*
a11*x^8*cos(x)-720*a11*sin(x)*x^7-5040*a11*x^6*cos(x)+
30240*a11*sin(x)*x^5+151200*a11*x^4*cos(x)-604800*a11*sin
(x)*x^3-1814400*a11*x^2*cos(x)+3628800*a11*sin(x)*x+60*a6*
sin(x1)*x1^2+120*a6*x1*cos(x1)+a7*x1^6*cos(x1)-a5*x^4*cos
(x)-6*a7*sin(x1)*x1^5-30*a7*x1^4*cos(x1)+120*a7*sin(x1)*
x1^3+360*a7*x1^2*cos(x1)-720*a7*sin(x1)*x1+a8*x1^7*cos(x1)
-90*a11*x1^8*cos(x1)-7*a8*sin(x1)*x1^6-42*a8*x1^5*cos(x1)
+210*a8*sin(x1)*x1^4+840*a8*x1^3*cos(x1)-2520*a8*sin(x1)*
x1^2+20*a6*x^3*cos(x)-5040*a8*x1*cos(x1)+a9*x1^8*cos(x1)
-8*a9*sin(x1)*x1^7-56*a9*x1^6*cos(x1)+336*a9*sin(x1)*x1^5+
1680*a9*x1^4*cos(x1)-6720*a9*sin(x1)*x1^3-20160*a9*x1^2*
cos(x1)+40320*a9*sin(x1)*x1+a10*x1^9*cos(x1)-9*a10*sin(x1)
*x1^8-72*a10*x1^7*cos(x1)+504*a10*sin(x1)*x1^6+3024*a10*
x1^5*cos(x1)-5*a6*sin(x1)*x1^4+a2*x1*cos(x1)-a6*x^5*cos(x)
+24*a5*sin(x1)*x1-12*a5*x1^2*cos(x1)-4*a5*sin(x1)*x1^3+a5*
x1^4*cos(x1)-6*a4*x1*cos(x1)-3*a4*sin(x1)*x1^2+a4*x1^3*cos
(x1)-2*a3*sin(x1)*x1+a3*x1^2*cos(x1)+12*a5*x^2*cos(x)+4*a5
*sin(x)*x^3+6*a4*x*cos(x)+3*a4*sin(x)*x^2-a4*x^3*cos(x)+
2*a3*sin(x)*x-a8*x^7*cos(x)+a6*x1^5*cos(x1)-20*a6*x1^3*
cos(x1)-a7*x^6*cos(x)-a2*x*cos(x)-840*a8*x^3*cos(x)-a2*sin
(x1)-24*a5*cos(x)-6*a4*sin(x)+2*a3*cos(x)+a2*sin(x)-362880*
a10*sin(x1)-3628800*a11*cos(x1)+40320*a9*cos(x1)+6*a4*sin
(x1)-720*a7*cos(x1)+5040*a8*sin(x1)+362880*a10*sin(x)+
3628800*a11*cos(x)-2*a3*cos(x1)-120*a6*sin(x1)+120*a6*sin
(x)+720*a7*cos(x)-5040*a8*sin(x)-40320*a9*cos(x)+24*a5*cos
(x1)-a1*cos(x)-Fi;
    x2=fzero(f,0.5);
    Kma_radian_T(j+1)=x2;
    x1=x2;
end

for j=N_latitude/2+1:N_latitude

f=@(x)-a3*x^2*cos(x)+a1*cos(x1)-24*a5*sin(x)*x+5*a6*sin(x)*
x^4-120*a6*x*cos(x)-15120*a10*sin(x1)*x1^4-60480*a10*x1^3*
cos(x1)+181440*a10*sin(x1)*x1^2+362880*a10*x1*cos(x1)+a11*
x1^10*cos(x1)-10*a11*sin(x1)*x1^9+720*a11*sin(x1)*x1^7+
5040*a11*x1^6*cos(x1)-30240*a11*sin(x1)*x1^5-151200*a11*
x1^4*cos(x1)+604800*a11*sin(x1)*x1^3+1814400*a11*x1^2*cos
(x1)-3628800*a11*sin(x1)*x1-60*a6*sin(x)*x^2+6*a7*sin(x)*
```

```
x^5+30*a7*x^4*cos(x)-120*a7*sin(x)*x^3-360*a7*x^2*cos(x)
+720*a7*sin(x)*x+7*a8*sin(x)*x^6+42*a8*x^5*cos(x)-210*a8*
sin(x)*x^4+2520*a8*sin(x)*x^2+5040*a8*x*cos(x)-a9*x^8*cos
(x)+8*a9*sin(x)*x^7+56*a9*x^6*cos(x)-336*a9*sin(x)*x^5-
1680*a9*x^4*cos(x)+6720*a9*sin(x)*x^3+20160*a9*x^2*cos(x)
-40320*a9*sin(x)*x-a10*x^9*cos(x)+9*a10*sin(x)*x^8+72*a10*
x^7*cos(x)-504*a10*sin(x)*x^6-3024*a10*x^5*cos(x)+15120*
a10*sin(x)*x^4+60480*a10*x^3*cos(x)-181440*a10*sin(x)*x^2
-362880*a10*x*cos(x)-a11*x^10*cos(x)+10*a11*sin(x)*x^9+90*
a11*x^8*cos(x)-720*a11*sin(x)*x^7-5040*a11*x^6*cos(x)+
30240*a11*sin(x)*x^5+151200*a11*x^4*cos(x)-604800*a11*sin
(x)*x^3-1814400*a11*x^2*cos(x)+3628800*a11*sin(x)*x+60*a6*
sin(x1)*x1^2+120*a6*x1*cos(x1)+a7*x1^6*cos(x1)-a5*x^4*cos
(x)-6*a7*sin(x1)*x1^5-30*a7*x1^4*cos(x1)+120*a7*sin(x1)*
x1^3+360*a7*x1^2*cos(x1)-720*a7*sin(x1)*x1+a8*x1^7*cos(x1)
-90*a11*x1^8*cos(x1)-7*a8*sin(x1)*x1^6-42*a8*x1^5*cos(x1)
+210*a8*sin(x1)*x1^4+840*a8*x1^3*cos(x1)-2520*a8*sin(x1)*
x1^2+20*a6*x^3*cos(x)-5040*a8*x1*cos(x1)+a9*x1^8*cos(x1)
-8*a9*sin(x1)*x1^7-56*a9*x1^6*cos(x1)+336*a9*sin(x1)*x1^5
+1680*a9*x1^4*cos(x1)-6720*a9*sin(x1)*x1^3-20160*a9*x1^2*
cos(x1)+40320*a9*sin(x1)*x1+a10*x1^9*cos(x1)-9*a10*sin(x1)
*x1^8-72*a10*x1^7*cos(x1)+504*a10*sin(x1)*x1^6+3024*a10*
x1^5*cos(x1)-5*a6*sin(x1)*x1^4+a2*x1*cos(x1)-a6*x^5*cos(x)
+24*a5*sin(x1)*x1-12*a5*x1^2*cos(x1)-4*a5*sin(x1)*x1^3+a5*
x1^4*cos(x1)-6*a4*x1*cos(x1)-3*a4*sin(x1)*x1^2+a4*x1^3*cos
(x1)-2*a3*sin(x1)*x1+a3*x1^2*cos(x1)+12*a5*x^2*cos(x)+4*a5
*sin(x)*x^3+6*a4*x*cos(x)+3*a4*sin(x)*x^2-a4*x^3*cos(x)
+2*a3*sin(x)*x-a8*x^7*cos(x)+a6*x1^5*cos(x1)-20*a6*x1^3*
cos(x1)-a7*x^6*cos(x)-a2*x*cos(x)-840*a8*x^3*cos(x)-a2*sin
(x1)-24*a5*cos(x)-6*a4*sin(x)+2*a3*cos(x)+a2*sin(x)-362880*
a10*sin(x1)-3628800*a11*cos(x1)+40320*a9*cos(x1)+6*a4*sin
(x1)-720*a7*cos(x1)+5040*a8*sin(x1)+362880*a10*sin(x)+
3628800*a11*cos(x)-2*a3*cos(x1)-120*a6*sin(x1)+120*a6*sin
(x)+720*a7*cos(x)-5040*a8*sin(x)-40320*a9*cos(x)+24*a5*cos
(x1)-a1*cos(x)-Fi;
    x2=fzero(f,1);
    Kma_radian_T(j+1)=x2;
    x1=x2;
end

Kma_radian_T(N_latitude+1)=pi/2;

Kma_radian_S=0;
for k=1:N_latitude

Kma_radian_S(k+1)=abs(acos(cos(2*Kma_radian_S(k))
 -2/N_latitude)/2);
```

```
end

Kma_degree_S=Kma_radian_S*180/pi;
% Slope of each light ray
Km_S=cot(Kma_radian_S);


%______________Find P________________
syms x0 y0 N1 N2 K

n1=1.54; % Refractive index - input rays
n2=1.00; % Refractive index - output rays
h=1.8; % Height of central point of the lens

% Define the original coordinate of point P
P=[0,h];
Pstore(1,1)=P(1);
Pstore(1,2)=P(2);

% Define light source position
S=[0,0];

% Calculate P
for j=1:1:N_latitude

    % Input rays
    I=P-S;

    % Output rays
    O(1)=sin(Kma_radian_T(j));
    O(2)=cos(Kma_radian_T(j));

    % Normal vector
    I=I/((I(1)^2+I(2)^2)^0.5);
    O=O/((O(1)^2+O(2)^2)^0.5);
    N=(n1*I-n2*O)/((1+(n1/n2)*(n1/n2)-2*(n1/n2)*dot(O,I))
     ^0.5);
    N=N/((N(1)^2+N(2)^2)^0.5);
    Nstore(j,1)=N(1);
    Nstore(j,2)=N(2);

    % Calculate the point Pj+1
    [x,y]=solve('(y-y0)=(-N1/N2)*(x-x0)','y=K*x','x','y');

Pstore(j+1,1)=subs(x,{x0,y0,N1,N2,K},{P(1),P(2),N(1),N(2),
 Km_S(j+1)});
```

```
Pstore(j+1,2)=subs(y,{x0,y0,N1,N2,K},{P(1),P(2),N(1),N(2),
 Km_S(j+1)});

    P(1)=Pstore(j+1,1);
    P(2)=Pstore(j+1,2);

end
```

Index

Freeform Optics for LED Packages and Applications, First Edition. Kai Wang, Sheng Liu, Xiaobing Luo and Dan Wu.

m

n

w

z